World Geomorphological Landscapes

Series Editor

Piotr Migoń, Institute of Geography and Regional Development, University of Wrocław, Wrocław, Poland

Geomorphology – 'the Science of Scenery' – is a part of Earth Sciences that focuses on the scientific study of landforms, their assemblages, and surface and subsurface processes that moulded them in the past and that change them today. Shapes of landforms and regularities of their spatial distribution, their origin, evolution, and ages are the subject of geomorphology. Geomorphology is also a science of considerable practical importance since many geomorphic processes occur so suddenly and unexpectedly, and with such a force, that they pose significant hazards to human populations. Landforms and landscapes vary enormously across the Earth, from high mountains to endless plains. At a smaller scale, Nature often surprises us creating shapes which look improbable. Many geomorphological landscapes are so immensely beautiful that they received the highest possible recognition – they hold the status of World Heritage properties. Apart from often being immensely scenic, landscapes tell stories which not uncommonly can be traced back in time for millions of years and include unique events. This international book series will be a scientific library of monographs that present and explain physical landscapes across the globe, focusing on both representative and uniquely spectacular examples. Each book contains details on geomorphology of a particular country (i.e. The Geomorphological Landscapes of France, The Geomorphological Landscapes of Italy, The Geomorphological Landscapes of India) or a geographically coherent region. The content is divided into two parts. Part one contains the necessary background about geology and tectonic framework, past and present climate, geographical regions, and long-term geomorphological history. The core of each book is however succinct presentation of key geomorphological localities (landscapes) and it is envisaged that the number of such studies will generally vary from 20 to 30. There is additional scope for discussing issues of geomorphological heritage and suggesting itineraries to visit the most important sites. The series provides a unique reference source not only for geomorphologists, but all Earth scientists, geographers, and conservationists. It complements the existing reference books in geomorphology which focus on specific themes rather than regions or localities and fills a growing gap between poorly accessible regional studies, often in national languages, and papers in international journals which put major emphasis on understanding processes rather than particular landscapes. The World Geomorphological Landscapes series is a peer-reviewed series which contains single and multi-authored books as well as edited volumes.

World Geomorphological Landscapes – now indexed in Scopus®!

Paolo Billi
Editor

Landscapes and Landforms of the Horn of Africa

Eritrea, Djibouti, Somalia

Springer

Editor
Paolo Billi
International Platform for Dryland Research
and Education
University of Tottori
Tottori, Japan

ISSN 2213-2090 ISSN 2213-2104 (electronic)
World Geomorphological Landscapes
ISBN 978-3-031-05486-0 ISBN 978-3-031-05487-7 (eBook)
https://doi.org/10.1007/978-3-031-05487-7

This Springer imprint is published by the registered company Springer Nature Switzerland AG
The registered company address is: Gewerbestrasse 11, 6330 Cham, Switzerland

Series Editor Preface

Landforms and landscapes vary enormously across the Earth, from high mountains to endless plains. At a smaller scale, nature often surprises us creating shapes which look improbable. Many physical landscapes are so immensely beautiful that they received the highest possible recognition—they hold the status of World Heritage Sites. Apart from often being immensely scenic, landscapes tell stories which not uncommonly can be traced back in time for tens of million years and include unique geological events such as meteorite impacts. In addition, many landscapes owe their appearance and harmony not solely to the natural forces. For centuries, and even millennia, they have been shaped by humans who have modified hillslopes, river courses and coastlines, and erected structures which often blend with the natural landforms to form inseparable entities.

These landscapes are studied by geomorphology—'the science of scenery'—a part of Earth Sciences that focuses on landforms, their assemblages, surface and subsurface processes that molded them in the past and that change them today. To show the importance of geomorphology in understanding the landscape and to present the beauty and diversity of the geomorphological sceneries across the world, we have launched a book series *World Geomorphological Landscapes*. It aims to be a scientific library of monographs that present and explain physical landscapes, focusing on both representative and uniquely spectacular examples. Each book will contain details on geomorphology of a particular country or a geographically coherent region. This volume covers Horn of Africa—an area, which is one of the least known and nowadays very difficult to access parts of the continent. And yet, it contains very diverse and fascinating geomorphology that includes the towering faulted escarpment of the Eritrean Highland, parts of the Afar Triangle with its active volcanoes and salt lakes, fascinating dryland rivers throughout the region and inselberg-dotted plains of Somalia. It is also home to some important archeological discoveries connected with the hominin history. But this book not only shows the geomorphological legacy of long-term landscape evolution and climate change, but also indicates relevance of geomorphology to successful environmental management of today. Several chapters examine issues of soil erosion, water resources and geomorphological hazards, all vital to the populations living in this unsettled part of the world.

The World Geomorphological Landscapes series is produced under the scientific patronage of the International Association of Geomorphologists (IAG)—a society that brings together geomorphologists from all around the world. IAG was established in 1989 and is an independent scientific association affiliated with the International Geographical Union (IGU) and the International Union of Geological Sciences (IUGS). Among its main aims are to promote geomorphology and to foster dissemination of geomorphological knowledge. I believe that this lavishly illustrated series, which keeps to the scientific rigor, is the most appropriate means to fulfill these aims and to serve the geoscientific community. To this end, my great thanks go to Prof. Paolo Billi, a person long involved in geomorphological research in the Horn of Africa, for agreeing to coordinate this unique volume in the series, which in a way complements the previous volume about Ethiopia. I am also very grateful to all individual authors who accepted invitations to contribute and shared their research experience from these remote areas.

In contrast to many other countries, the Horn of Africa is far less known regarding its geomorphology at the local scale and many of its regions are still *terra incognita*. Moreover, persistent political unrest makes the region very difficult to visit nowadays and large parts of Somalia, in particular, are essentially off-limits. Therefore, it was not feasible to strictly follow the format of previous volumes in the series and offer a wide range of site-specific stories. However, I am sure the readers will warmly welcome this unique presentation of the geomorphic environment of the Horn of Africa—perhaps the only comprehensive source of reference about this vast part of Africa.

Wrocław, Poland Piotr Migoń

Preface

The Horn of Africa shows a wide diversity of landscapes and landforms. Eritrea, Djibouti and Somalia owe their large-scale geomorphological features mainly to the complex crustal dynamics, characterized by large-scale uplifting, rifting and faulting, to the alternation of cycles of peneplanation, sea transgression and regression, marine and continental sedimentation. Quaternary faulting propelled intense volcanic activity and, in combination with erosion and hydrological processes, influenced alluvial, colluvial, lacustrine and coastal sedimentation. All these events affected Eritrea, Djibouti and Somalia in different ways and at different rates in space and time providing them with markedly contrasting, and often spectacular, physiographic and geomorphological features.

The Oligo-Miocene doming of the Arabian-Ethiopian region is, however, the most important geological event in the whole Horn of Africa. This remarkable uplifting was also associated with a three-lateral rifting and spreading converging toward the Afar depression and resulting in the Red Sea and Danakil troughs, the Gulf of Aden trough and the Ethiopian rift valley. The uplifting created high plateaus in Eritrea and northern Somalia. In Eritrea, the relief contrast between the plateau and the coastal lowlands is striking, with a jump of 2400 m in about 90 km from Asmara to Massawa on the Red Sea coast. The rift faults fragmented the Eritrean plateau and largely controlled the drainage network development and pattern.

Unlike Ethiopia, where the rifting is characterized by the emplacement of huge trap basalt amounts, basalts are present only in northern Somalia and in relatively small areas south of Asmara in Eritrea.

The geological evolution of the Horn of Africa also records long intervals of tectonic quiescence during which the whole landmass underwent vast denudation and the formation of peneplains. Outstanding examples of the Triassic peneplanation of the Precambrian crystalline basement are evident in Eritrea, whereas the Early Cretaceous to Oligocene marine regression made Somalia a very extensive table land that still today characterizes the largest part of this country. By contrast, Djibouti landscape is mainly the result of recent (Miocene to Quaternary) volcanic activity which has provided the country with most of its present landscape.

A complementary, and not secondary, factor in shaping the landforms we can observe today in Eritrea, Djibouti and Somalia is certainly the climate which varies from subhumid in southern Somalia to hyper-arid in Djibouti, the Danakil depression and along the Eritrean and northern Somalia coastal belt.

From this very short summary of the main geological events and climate diversity that characterize the eastern countries of the Horn of Africa, it is easy to prefigure the richness of their geological and geomorphological features.

Unfortunately, in the last two decades, critical conditions of social and political instability made impossible to carry out protracted and extensive field investigations in the region. Though also this book reflects such a difficult situation, ultimately, the efforts made by the authors resulted in a motivating collection of studies on different aspects of landscape and on the main factors influencing its development, including climate and man activity.

Moreover, most of the book chapters reports about remote areas that seldom have been visited by earth science scholars and whose peculiar landforms were before only succinctly described. The scientific and informative value of this book stands, therefore, not only in the variety of the research topics investigated, but also in the scarcity and dispersion of scientific publications on landscape and landforms of this part of the world.

Tottori, Japan Paolo Billi

Contents

About the Editor

Prof. Paolo Billi has been working in the University of Florence and the University of Ferrara in Italy and closed his academic career at the International Platform for Dryland Research and Education of the University of Tottori in Japan. His main research topics include fluvial geomorphology, sediment transport and sediment yield. He carried out his research activities on rivers in Italy and in the Horn of Africa. He is also editor of Springer volume on *Landscape and Landforms of Ethiopia*.

Climate Variability in the Horn of Africa Eastern Countries: Eritrea, Djibouti, Somalia

1

Paolo Billi

Abstract

Eritrea, Djibouti and Somalia have peculiar physiographic characteristics and, though their climate is mainly controlled by the north–south–north movement of the Intertropical Convergence Zone (ITCZ), the large difference in elevation (3000 m from sea level to the highest peak) and the variable distance from the ocean also play an important role, accounting for considerable spatial variability in climate characteristics. Gridded and ground instrumental data of mean monthly temperature and precipitation were the main platform to describe the climate of Eritrea, Djibouti and Somalia. The basic precipitation and temperature data were used to obtain other parameters such as rainfall erosivity and aridity indices. Data of other basic parameters such as wind or humidity are available only for a very small number of meteo-stations, and their time series are very short. The gridded data time series span a century-long interval (1901–2015) and were used to investigate mean annual temperature and precipitation trends. A complementary analysis of climate variations was carried out using old (early twentieth century) data reported by Fantoli (Elementi preliminari del clima dell'Etiopia. Sansoni, Firenze, 1940; Contributo alla climatologia della Somalia: riassunto dei risultati e tabelle meteorologiche e pluviometriche. Ministero degli Affari Esteri, Cooperazione Scientifica e Tecnica, Roma, 1965) and modern data. The gridded data time series show a marked increase in temperature in the three countries. A decreasing trend of precipitation is evident for Eritrea, whereas an increasing trend is observed for Somalia. Rainfall erosivity is relatively low, despite the presence of vast degraded areas. The aridity indices indicate that the region is already under desertification conditions and the situation is expected to get worse with a decrease in water resources and severe threats to agricultural productivity.

Keywords

Temperature • Rainfall • Rainfall erosivity • Aridity

1.1 Introduction

Eritrea, Djibouti and Somalia take up a wide belt around Ethiopia, making the transition from the ocean to the inner highlands. Altitudinal differences are pronounced and find their maximum expression in Eritrea, where the largest range of elevations is from −75 m at Lake Kulul, in the Eritrean portion of the Danakil depression, to 3018 m asl at Emba Soira, the highest mountain peak of Eritrea. Also in Djibouti and Somalia, there are large differences in elevation between sea level and the highest mountains such as Mt. Shimbiris (2460 m) and Mt. Mousa Ali (2063 m) in Somalia and Djibouti, respectively. These mountain tops are near the sea, 42, 33 and 82 km, respectively, which makes the local physiography and landscape even more impressive. Eritrea and Djibouti are mainly mountainous countries. In Eritrea, lowlands are found only in the western part of the country, crossed by the Barka River, and along the coastal belt. Djibouti has no significant flat land, whereas flat lowlands make up the largest proportion of Somalia, which has high mountain ranges only along the northern coast. Such a highly variable physiography, the influence of the ocean, the presence of fairly large mountain ranges and the effects of the global atmospheric circulation result in a wide variability of the average climate characteristics across the region. Eritrea, Djibouti and Somalia are also subjected to marked seasonality, interannual and longer-term variability of rainfall and other climatic parameters that stand as clear symptoms of ongoing climate change (Schreck and Semazzi 2004), with negative implications on natural systems and the

P. Billi (✉)
International Platform for Dryland Research and Education, Tottori University, Tottori, Japan
e-mail: bli@unife.it

P. Billi (ed.), *Landscapes and Landforms of the Horn of Africa*, World Geomorphological Landscapes,
https://doi.org/10.1007/978-3-031-05487-7_1

appearance of stronger constrains to sustainable development. The countries in concern include both drylands and flood-prone areas. According to international organisations (e.g. UNEP 1992; EU 2007; IPCC 2019) and several studies (e.g. Middleton and Thomas 1997; Huq et al. 2006; Sivakumar and Stefanski 2007), in developing countries climate change will lead to an increased susceptibility to soil erosion, flood hazard, droughts and shortage of cultivable land and will widen social inequalities, which in turn will increase human pressure on land and will speed up land degradation. Though in the countries considered here the available data series show large gaps, discontinuities and different lengths and even if the density of meteo-stations is low and not uniform, an attempt is made in this chapter to address the variability of the main climatic parameters and their long-term trends, with implications for soil erosion and land degradation.

1.2 Study Area, Data and Methods

Somalia is the largest country, with a surface of 637,657 km^2, Djibouti is the smallest with 23,200 km^2 (almost the size of Sardinia island in Italy) and the area of Eritrea is 125,700 km^2. With the exception of a very small portion in southern Somalia, all three countries are included within the equator and the Tropic of Cancer (Fig. 1.1).

Eritrea shows the largest physiographic variability, with lowlands in the west and along the coast, the central highlands and the Asmara plateau, at elevations ranging from 2000 to 2400 m asl, with the western slope and the steep eastern escarpment connecting the highlands to the western and eastern lowlands, respectively, and the Danakil depression, where the lowest elevation is −75 m below sea level (Fig. 1.2). Such variability of physiography and elevation over short distances is a crucial factor in determining different climatic regimes of this country.

Djibouti is mainly a mountainous country, though with the exception of a couple of peaked mountain ranges (Mt. Mousa Ali in the north and Goda mountains in the centre) with elevations close to 2000 m asl and other few mountains, especially in the portion north of the Tadjoura Gulf, with elevations around 1700 m asl, the majority of the country is below 1000 m asl. The overall landscape of the country consists of low mountains and hills with a few flat-bottomed depressions interposed (Fig. 1.1).

Somalia is mainly a flat land. Mountain ranges are present only in the northern portion facing the Gulf of Aden. Mt. Shimbiris is the highest peak (2460 m asl), located close to the sharp edge of the Gulf of Aden. An escarpment connects the uplifted land with the ocean across a horizontal distance of only 32 km. The central and southern portion of Somalia is a gently inclined plain, with the highest elevations <500 m asl, bounded eastward by the Indian Ocean (Fig. 1.1). The more uniform physiography of Somalia is reflected by less pronounced variation of the main climate characteristics.

The three countries are bound eastward by the sea (the length of coastline is 1083 km in Eritrea, 314 km in Djibouti and 3333 km in Somalia). The vicinity to the sea and coastline orientation also have substantial influence on the climate of coastal towns and settlements.

1.2.1 Data Type and Sources

Though the study area is vast (786,557 km^2 in total), the density of meteo-stations is rather low. Climatic data are very heterogeneous (Fessehaye et al. 2019). Only very few stations have long records spanning over three decades; otherwise, the data records are punctuated by many gaps of different size, the total length and the beginning and end of the time series are highly variable and asynchronous, with some stations having been moved to a different, though near, location. Attempts to fill the data gaps with standard correlation methods proved to be acceptable only in a few cases and were in general poorly successful. This results in a certain degree of approximation in data analysis. Notwithstanding such limitations, it is possible to depict spatial variability of the main climatic factors and of some derived parameters across the countries. The investigation of trends is affected by the same limitations and though gridded data at country scale level are available, the most reliable information is still provided by the few meteo-stations with longer data records.

The data used in this study were obtained from different sources. They include original data of the National Meteorology Agency of Ethiopia for some meteo-stations of Eritrea as recently as the year 2000, i.e. shortly after the separation of this country from the Ethiopian Federation. Since then, unfortunately, ground data of Eritrea are very scanty (Fessehaye et al. 2019) due to economic resource shortage. A similar, or even worse situation, is present in Somalia, where several years of civil war have hindered the collection of meteo-data. Since 2002, the SWALIM project of FAO re-established some meteo-stations and provided the conditions for their maintenance (Muchiri 2007). Other average climate data were obtained from Internet sources, e.g. climate-data.org. All data of this organisations derive from a climate model with a resolution of 30 arc seconds. The model uses weather data from thousands of weather stations from all over the world. The data refer to the 1982–2012 interval but are also refreshed from time to time. Other monthly data were obtained from the Global Historical Climatology Network (GHCN) (Menne et al. 2012). For long-term analysis of precipitation and temperature trends at country level, data covering the 1901–2015 interval were

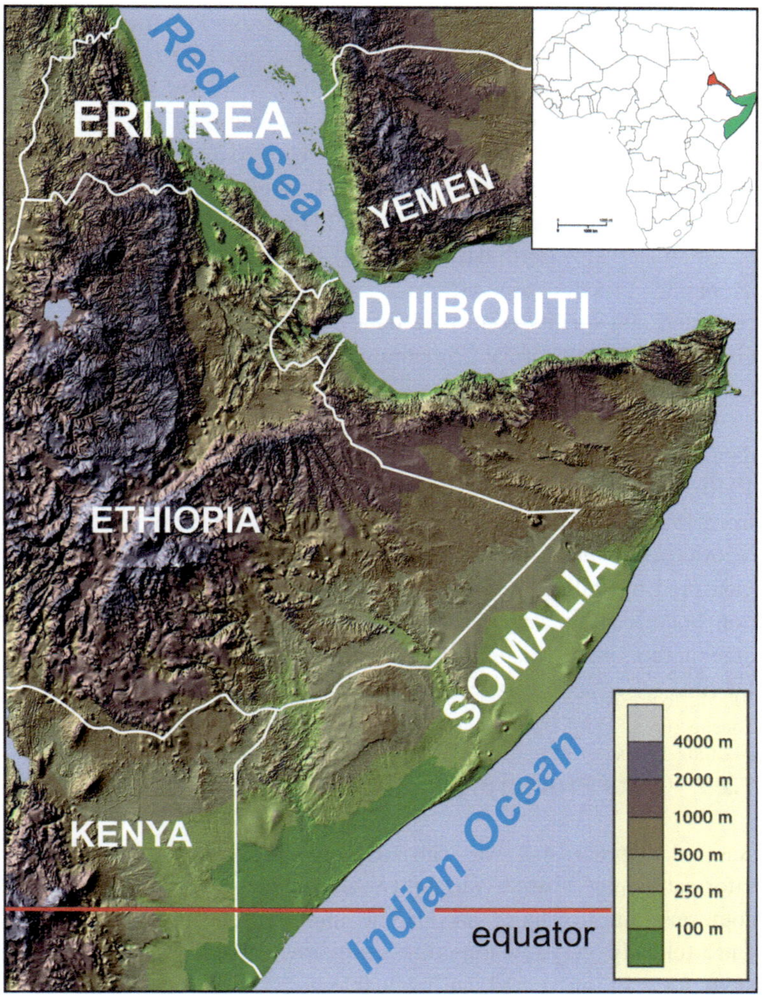

Fig. 1.1 Location map of the study countries: Eritrea, Djibouti, Somalia

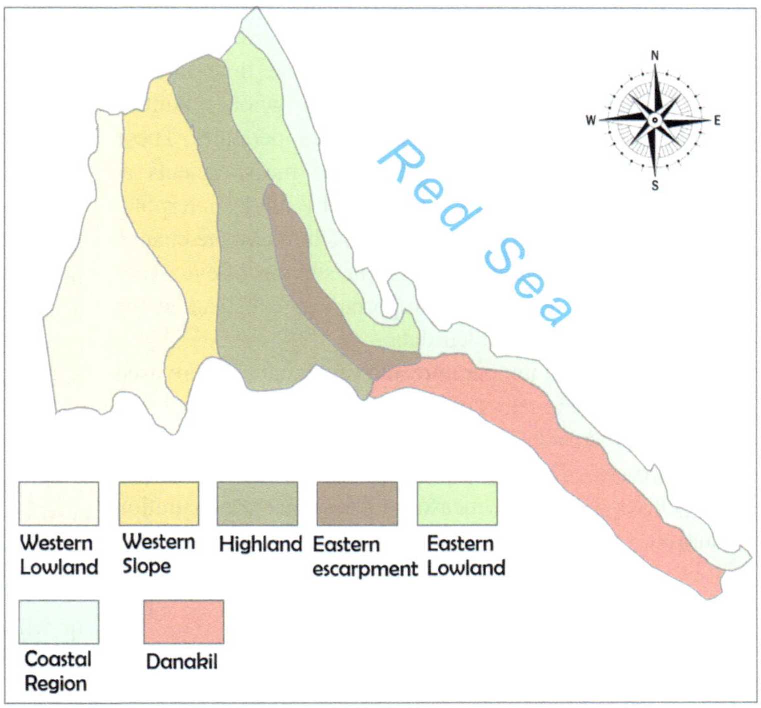

Fig. 1.2 Main physiographic areas of Eritrea

downloaded from the Climate Change Knowledge Portal of the World Bank (2019). These time series are based on gridded data interpolated into 0.5° latitude/longitude grid cells and derived from observational, quality-controlled data by the Climatic Research Unit (CRU) of University of East Anglia (UEA) in UK (Harris et al. 2014). Aggregated data were also obtained from Stat World (2019), and these are based on Berkeley Earth flat data (2019). The Berkeley Earth averaging process generates a variety of output, bias-corrected and quality-controlled data. Data sources include GHCN (Global Historical Climatology Network) and NCAR (National Centre for Atmospheric Research).

To investigate climate change, of particular interest are the data collected by Fantoli for Eritrea and Somalia (Fantoli 1940, 1965). This author collected several data sets ranging from 1912 to 1958 for some Somali stations and even older records for a few Eritrean stations. Most of the data were measured by Italian colonial authorities during the occupation, but this author also presents some older data, collected by other authors and British institutions, that he used to construct the time series and average data reported in his publications.

1.2.2 Data Processing Methods

Basic data processing was carried out to provide general information on spatial variability of climate across the countries and to have an insight into climatic trends. Unfortunately, very few long time series are available for the study area. In this regard, an invaluable help came from the data of Fantoli (1940, 1965) by which it was possible to depict the climate of a century ago on the basis of a few stations that provided instrumental data. The gridded data of the World Bank (2019) cover a very wide time interval (1901–2015), but they are reported as whole country-averaged values of precipitation and temperature. These data, though based on existing ground measurements at several meteo-stations, were generated by model interpolation and can be used in a general approach to climate change analysis in the three countries. Yet, the support of even very few instrumental data sets is important to substantiate the trends obtained from gridded data.

The basic data of temperature and precipitation were used to derive other parameters and to describe climate variability among the meteo-stations considered and their variation through time. The Precipitation Concentration Index (PCI) (Oliver 1980) is a measure of seasonal concentration of rainfall. It is defined as:

$$\mathrm{PCI} = 100\left\{\sum p_i^2 / \left(\sum p_i\right)^2\right\} \tag{1.1}$$

in which p_i is the mean monthly precipitation of each month.

When the rainfall amount is the same in every month, the PCI takes its minimum value of 8.3. By contrast, in the hypothetical situation of the whole annual precipitation falling in one month, the PCI assumes the value of 100. According to Oliver (1980), a PCI less than 10 indicates uniform distribution; a value from 10 to 15 denotes moderately seasonal distribution; a value from 15 to 20 indicates seasonal distribution, whereas an index above 20 reflects strong seasonal effects, with increasing values indicating increasing monthly rainfall concentration.

Very similar to the PCI in its structure, but defined for soil erosion investigation purposes, is the Modified Fournier Index (MFI) (Arnoldus 1980):

$$\mathrm{MFI} = \sum p_i^2 / P \tag{1.2}$$

in which p_i is the amount of rain in the ith month (mm) and P is the annual precipitation (mm).

Though these two indexes are very similar and related by a factor $1/P$, no significant correlation was found by Gabriels (2006). The MFI has been developed and used as a surrogate of rainfall erosivity index (e.g. CORINE 1992; Lee and Heo 2011). It is classified into five classes (Table 1.1). One limitation of the MFI is that it does not take into account the aridity of climate, in which short intense storms may be very effective in terms of soil erosion (Gabriels 2006). Therefore, this index should be used with caution and for areas with similar climatic conditions (Gabriels 2006).

Another rainfall erosivity index, developed for a simpler calculation of the R-factor in the USLE (and, for this reason, very often used in the literature), is the erosivity R-factor proposed by Renard and Freimund (1994):

$$R = 0.0483 P^{1.61} \tag{1.3}$$

in which P is annual precipitation (in mm) and R has the units of MJ mm ha^{-1} h^{-1} yr^{-1} for annual precipitation less than 850 mm, which is the case of the three countries.

Desertification is big threat for many sub-Saharan countries, and the study area is not an exception. A very commonly used parameter to measure the tendency to desertification is the aridity index expressed by the UNEP (1992) as P/PET, that is the ratio between annual precipitation and the potential evapotranspiration calculated with the Thornthwaite (1948) method. An area is considered under desertification conditions when $P/\mathrm{PET} < 0.6$.

In this study, the De Martonne (1925) Aridity Index was also used:

$$\mathrm{AI} = P/(T + 10) \tag{1.4}$$

in which P is annual precipitation in mm and T is mean annual air temperature in °C.

Table 1.1 Classification of MFI (from Gabriels 2006)

MFI	Description	Class
<60	Very low	1
60–90	Low	2
90–120	Moderate	3
120–160	High	4
>160	Very high	5

Table 1.2 Relationship between the De Martonne (1925) aridity index and irrigation requirement (Baltas 2008; ARPAV 2019)

AI	Condition	Irrigation
<5	Arid	Indispensable
5–10	Semi-arid	Indispensable
10–20	Dry sub-humid	Very useful
20–30	Sub-humid	Often useful
30–50	Humid	Not required

Though it is rather old, this aridity index was selected because its values can be associated with irrigation requirements (Baltas 2008; ARPAV 2019) (Table 1.2) and, therefore, it can be used to investigate current and future conditions of cultivation stress and water demand.

1.3 General Climate and Atmospheric Circulation

Climate is highly variable within the study area, but also within each of the countries. Such variability is mainly expressed by monthly distribution of rain, which is strongly controlled by the general atmospheric circulation, including factors such as the position of the Intertropical Convergence Zone (ITCZ) (Muchiri 2007), the Indian ocean temperature (Goddard and Graham 1999) or the strength of El Nino Southern Oscillation (ENSO) (Indeje et al. 2000). The vicinity to the ocean, the occurrence of highly elevated mountain ranges and plateaus, and marked relief contrasts contribute to the high variability of climate characteristics of Eritrea, Djibouti and Somalia.

The interannual movement of the ITCZ is the main engine of the study area climate, propelling distinct changes in wind direction throughout the year. In its oscillation across the equator, the ITCZ passes over the study area twice a year. This migration causes the onset of winds from the north-east, when the ITCZ is located in the south, and from the south-west, when it is in a northern position. This pattern results in two rainy seasons (typically there is a small and a main rainy season), that occur in March–June in southern portion of the study area, when the ICTZ moves northward, in July–September in the north and then in October–November in the south, as the ITCZ moves to the south. Some regional variations lead to different timing of rainy seasons or to the occurrence of only one rainy season (e.g. in the Eritrean highlands and western lowlands, or in the southernmost tip of Somalia–Kismayo). In spring (April–June), the ITCZ lies across the southern border between Ethiopia and Somalia. At this time, pressure is high in the subtropical latitudes of the Sahara and southern Africa, whereas a strong cyclonic cell develops over the equatorial region. Here, south-easterly and north-easterly winds converge causing the vertical upward motion of air, the formation of clouds and precipitation, especially in Somalia. The moist, south-easterly winds bring rain also to the central highlands of Ethiopia, producing the small rainy season (*belg* in the local language) (Fazzini et al. 2015).

In summer (June–September), the ITCZ reaches its northernmost position. Northern Ethiopia and Eritrea are under the influence of south-west equatorial westerlies and monsoon-type southern winds from the Indian Ocean. These winds are rich in moisture, and when they ascend over the highlands of Ethiopia and Eritrea, they produce heavy rains giving rise to the main rainy season (responsible for 50–80% of the annual precipitation) in this area. The Eritrean north-eastern escarpment and lowlands, being rain shadow areas, are dry, though in the Eritrean escarpment the condensation of humidity coming from the very near Red Sea, also in the dry season, results in the landscape greener than expected and in favourable condition for cultivation.

In the same period, in Somalia, the south-west monsoon and the southerly winds converge in a narrow belt along the coast producing some rain, whereas the south-westerly winds are pushed up by high-elevation areas of northern Somalia and inland Djibouti, leading to local precipitation (Muchiri 2007) and resulting in a bimodal distribution of rain similar to that of inland Ethiopia. In October, the ITCZ moves back southwards, producing over Somalia the same conditions as in the spring months. The transit of the ITCZ over Somalia gives way to two distinct rainy seasons, locally named as *Gu* (March–May), the main wet season, and *Deyr* (October–November), the secondary rainy season.

In winter, the ITCZ is to the south of the study area, which goes under the influence of continental air currents originated from high-pressure centres over north Africa and Arabia deserts and experiences dry and cloudless conditions and high temperatures. Exceptions to this general climatic situation are the Eritrean escarpment and the coastal plains of Eritrea and Djibouti, where the influence of the Red Sea and the development of local eddies produce moderate winter rains, though in some places such as Filfil, located about 30 km NNE of Asmara at the base of the escarpment, mean annual rainfall may exceeds 1000 mm (EMA 1988; Sati 2008).

Previous work demonstrated that other factors such as El Niño Southern Oscillation (ENSO) and warm than usual sea surface temperature (SST) in the western equatorial Indian Ocean may have an important effect in the interannual variations of precipitation amounts (Liebman et al. 2014). During El Niño Southern Oscillation (ENSO) conditions, a short rainy season with precipitation above the average is recorded in the eastern Horn of Africa (Liebman et al. 2014). Using data from 31 meteo-stations in Somalia, Hutchinson (1992) found that rainfalls are only moderately predicted by changes in the Southern Oscillation Index (SOI), i.e. the atmospheric component of ENSO, but rainfall was low in association with high values of the SOI, that is La Niña conditions. Mason and Goddard (2001), however, suggested caution in stressing climate response to strong ENSO events and pointed out that the areas affected by ENSO may change from season to season and that not necessarily less than average precipitation is to be expected during La Niña events and vice versa. This conclusion is also supported by Indeje et al. (2000), though they confirm that some relationships exist between seasonal rainfall in East Africa and phases of ENSO. Behera et al. (2005) highlighted an even more important role of the Indian Ocean sea surface temperature (SST) in affecting the amount of short rains in East Africa. The east–west dipole of SST anomalies, commonly indicated as the Indian Ocean Dipole (IOD), has been shown to be significantly correlated with rainfall variability in East Africa and Indonesia (Saji et al. 1999), with above average rainfall in the former during a positive event and the opposite in the latter. During El Niño years, the Indian Ocean becomes warmer and moist convergent currents flow over East Africa (Goddard and Graham 1999). In this regard, there is some obvious link between ENSO and IOD. Nevertheless, IOD and ENSO events, though sometimes may co-occur, are independently evolving phenomena and, according to the statistical model of Behera et al. (2005), 79% of extreme short rains occur during IOD years.

The complexity, overlapping and counteracting of such atmospheric forcings and the strength and frequency of their teleconnections over the eastern Horn of Africa are reflected by the variability of the average climatic conditions depicted essentially in terms of monthly patterns of temperature and precipitation. In 1997, the FAO produced an agro-ecological map of Eritrea that, though focussed on agro-ecological zones, in practice can be also considered a good climatic map (Fig. 1.3), as it will be shown in the next sections. A similar eco-climate map was produced by the FAO-SWALIM project in 2007 (Muchiri 2007) for Somalia (Fig. 1.4). These two maps show that in both Eritrea and Somalia arid and semi-arid conditions predominate, whereas humid and sub-humid climates are restricted to small and very small proportion of the territory of Eritrea and Somalia, respectively. In Djibouti, warm desert climate (BWh of Koeppen classification) prevails over warm semi-arid climate (BSh of Koeppen classification).

1.4 Temperature

As anticipated in Sect. 1.2.1, the temperature data used in this study are heterogeneous. They come from different sources and were measured by ground instrumentation or obtained from gridding estimates. In all three countries, long time series of climate data are very uncommon and unevenly distributed, so the combined use of different data was a necessary compromise, which Ledesma and Futter (2017) consider as acceptable.

Eritrea is the country in which temperature is spatially more variable (Fig. 1.5). This is due to its higher relief

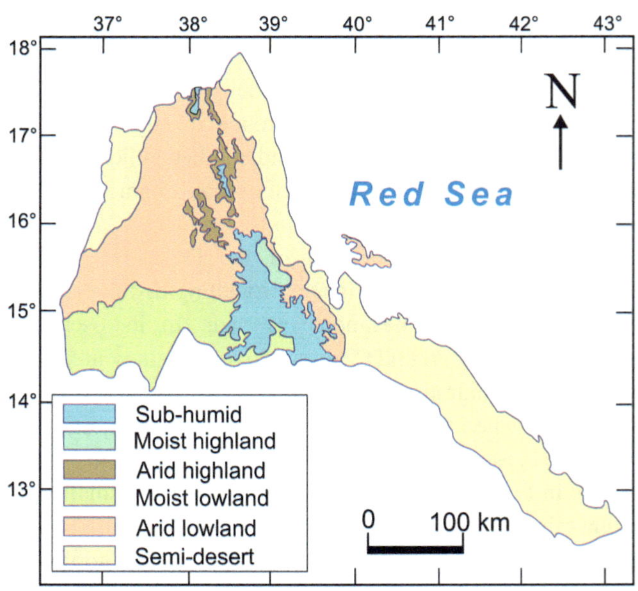

Fig. 1.3 Agro-ecological map of Eritrea

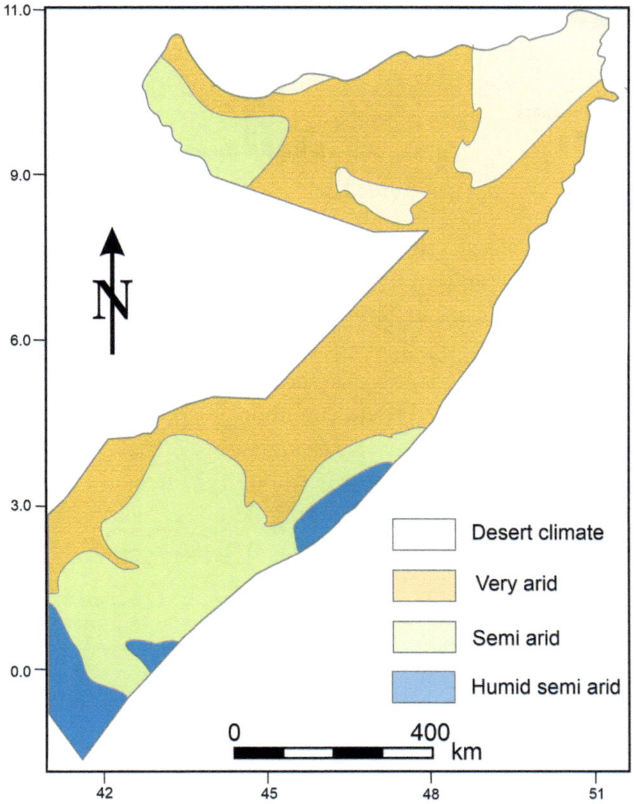

Fig. 1.4 Eco-climate map of Eritrea (modified from Muchiri 2007)

contrast. In Djibouti, the spatial viability of temperature is reduced (Fig. 1.6) because of the small size of this country and smaller relief contrast, whereas the rather uniform physiography of Somalia (with the exception of northern highlands) results in a rather uniform distribution of temperature (Fig. 1.7). Djibouti and Somalia are hottest countries. Their mean temperature, obtained by averaging the mean annual temperature of all the meteo-stations in a country, is 26.6 and 26.1 °C, respectively. Eritrea is the coolest with 22.9 °C. Actually, for the Eritrean Danakil, a very hot part of this country, climate data are not recorded on a regular basis and, for this reason, also the gridded data are not reliable. According to WMO data, reported by the Norwegian Meteorological Institute (NMI 2020) for the plain of Badda lake (about 100 m below sea level) in the centre of Eritrean Danakil depression, in the 1961–1990 interval the mean annual temperature was 29 °C, with the mean monthly maximum temperatures of 40–41 °C from June to August. In the Samoti plain (see Chap. 7 of this book for more information about this area), located about 18 km north-east of Badda, in November, the author of this paper commonly measured daytime temperatures of 42–44 °C and night-time temperatures of 28–30 °C. Travellers of the past and very sporadic earlier studies (Pedgley 1967) reported peak temperatures as high as 50 °C for the Dallol area in Ethiopia, only 16 km and 50 km south of the border with Eritrea and Badda lake, respectively. High temperatures, close to those in Danakil, are also reported for Djibouti where, in summer, the monthly mean maximum temperatures range from 37 to 41 °C, i.e. show values very similar to those recorded at Assab and Massawa (Figs. 1.5 and 1.6).

The hottest towns in the study area are Luuq and Berbera in Somalia (mean annual temperature Ta = 30.9 and 30.2 °C, respectively), followed by Beylul (30.6 °C) on the southern Red Sea coast of Eritrea, 45 km north-west of Assab, and Djibouti City (29.9 °C). The highest monthly temperatures are recorded at Luuq (42.2 °C in March) in Somalia, at Djibouti City (41.3 °C in July) and at Tennessee (40.9 °C in May) in Eritrea, close to the Sudan border.

The coolest town is Saladero (Ta = 15.6 °C; 2325 m asl) in the Eritrean highlands, about 16 km south-west of Asmara, whereas the coolest town in Somalia is Warmaddow (Ta = 16.8 °C; 2100 m asl), in the coastal range about 60 km east of Bosaso and Diyara (Ta = 20.9 °C; 1546 m asl) in the mountains of the Day Forest National Park of Djibouti. The lowest minimum monthly temperatures are recorded in January in Asmara (6.3 °C) (Fig. 1.5), Ceerigaabo (6.7 °C) (Fig. 1.7) and Day (10.9 °C) (Fig. 1.6) in Eritrea, Somalia and Djibouti, respectively.

The smallest value of the annual temperature range (Tr) is measured at Ceeldheer (Tr = 2.6 °C; Ta = 27.3 °C) on the Somali coast about 286 km north-east of Mogadishu. In Eritrea, the smallest value of Tr is recorded at Algena (Tr = 4.1 °C; Ta = 18.3) (Fig. 1.5), whereas it is 8.3 °C at Ali Sabieh (Ta = 25.6 °C) in Djibouti (Fig. 1.6). The largest annual temperature excursion is recorded at Xamass (Tr = 14.4 °C; Ta = 26.2 °C; 588 m asl) on the Gulf of Aden rift escarpment (Fig. 1.8), a few kilometres to the north-west of Burco/Burao; the second largest value of Tr is found in Djibouti City (Tr = 11.3; Ta = 29.9; 19 m asl) and the third one at Massawa on the Eritrean coast (Tr = 10.3 °C; Ta = 29.4 °C; 10 m asl).

In Eritrea, the annual temperature range tends to decrease with elevation ($R^2 = 0.61$) (Fig. 1.9) and to increase with mean annual temperature ($R^2 = 0.57$) (Fig. 1.10). Instead, no correlation was found for Somalia and Djibouti. This latter country has the largest proportion (44%) of stations with the mean annual temperatures Ta > 28 °C. In Eritrea, 27% of the stations record annual temperatures higher than 28 °C, but the proportion of stations with an annual temperature range less than 5 °C is 34% (Tr > 5 °C in all the meteo-stations of Djibouti). In Somalia, the mean annual temperatures are a bit milder (only in 20% of stations Ta > 28 °C) but hot temperatures are persistent as 36% of stations record a Tr < 5 °C and 69% of stations record a Ta > 26 °C. The World Climate Research Programme (WCRP-WMO 2011) reported daily peaks of 47–49 °C, recorded at Massawa and Assab, and extreme low values of −6.5 and −5.4 °C measured in December–January in

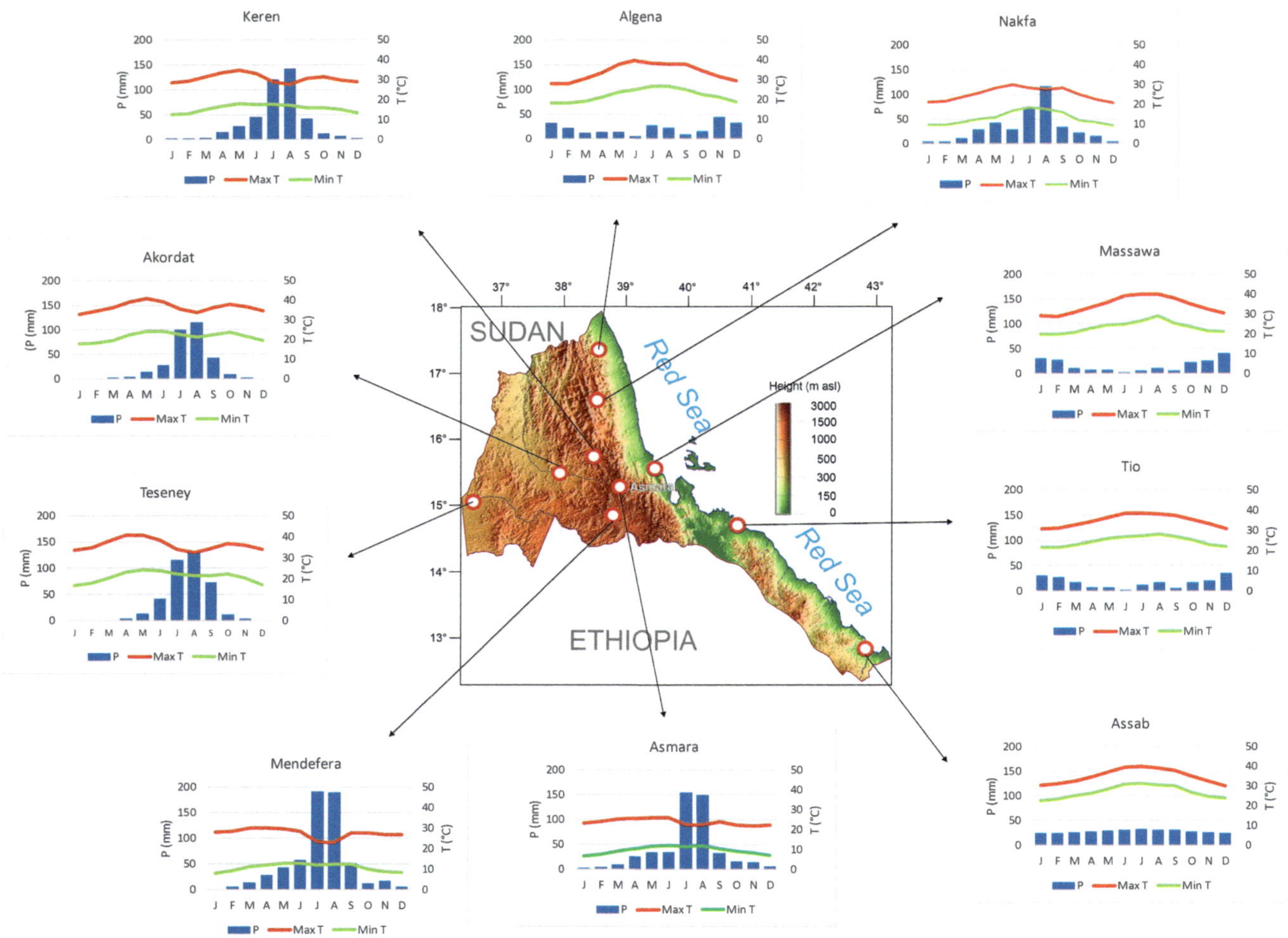

Fig. 1.5 Representative climograms of the main physiographic and climatic areas of Eritrea

Asmara. Northern Somalia experiences extreme daily temperatures, with peaks of 45 °C in July along the coastal plains of the Gulf of Aden, and December lows below zero in the highlands (Hadden 2007). The highest diurnal temperature of 52 °C, however, was measured near Lake Assal in Djibouti (Nicholson 2011).

The three countries are characterised by a significant ($0.86 < R^2 < 0.95$) temperature gradient. The decrease of mean annual temperature with elevation (commonly referred to as "lapse rate") is 5.6, 5.9 and 5.3 °C every 1000 m in Eritrea, Djibouti and Somalia, respectively (Fig. 1.11).

1.5 Precipitation

1.5.1 General Considerations

As schematically presented in Sect. 1.3, the influence of a few main climate drivers at planetary scale, combined with the local topography and the distance from the ocean, results in different monthly rainfall patterns across the eastern countries of the Horn of Africa.

The precipitation spatial variability of Eritrea is mainly controlled by elevation, the proximity to the Red Sea and the influence of the Saharan high on the western lowlands. The eastern escarpment is, therefore, dominated by winter rains, hot temperatures and greater cloud cover, whereas the highlands have a bimodal distribution of monthly precipitation with the main, larger rainy reason in summer (July and August) and a small rainy season in spring (March–May). The summer rains result from the migration of the Intertropical Convergence Zone (ITCZ) to the north of Eritrea, and the convergence associated with it releases generalised rains in the area. The big rains run out when the ITCZ regresses southward in September. The spring rains result from high, cold westerlies encountering warm moist air flowing from the south propelled by the Arabian high pressure, weakening as it moves southward to the Indian Ocean. In the western lowlands, the majority of rain falls from July to September, when the ITCZ is in a northern position. In the eastern, coastal belt,

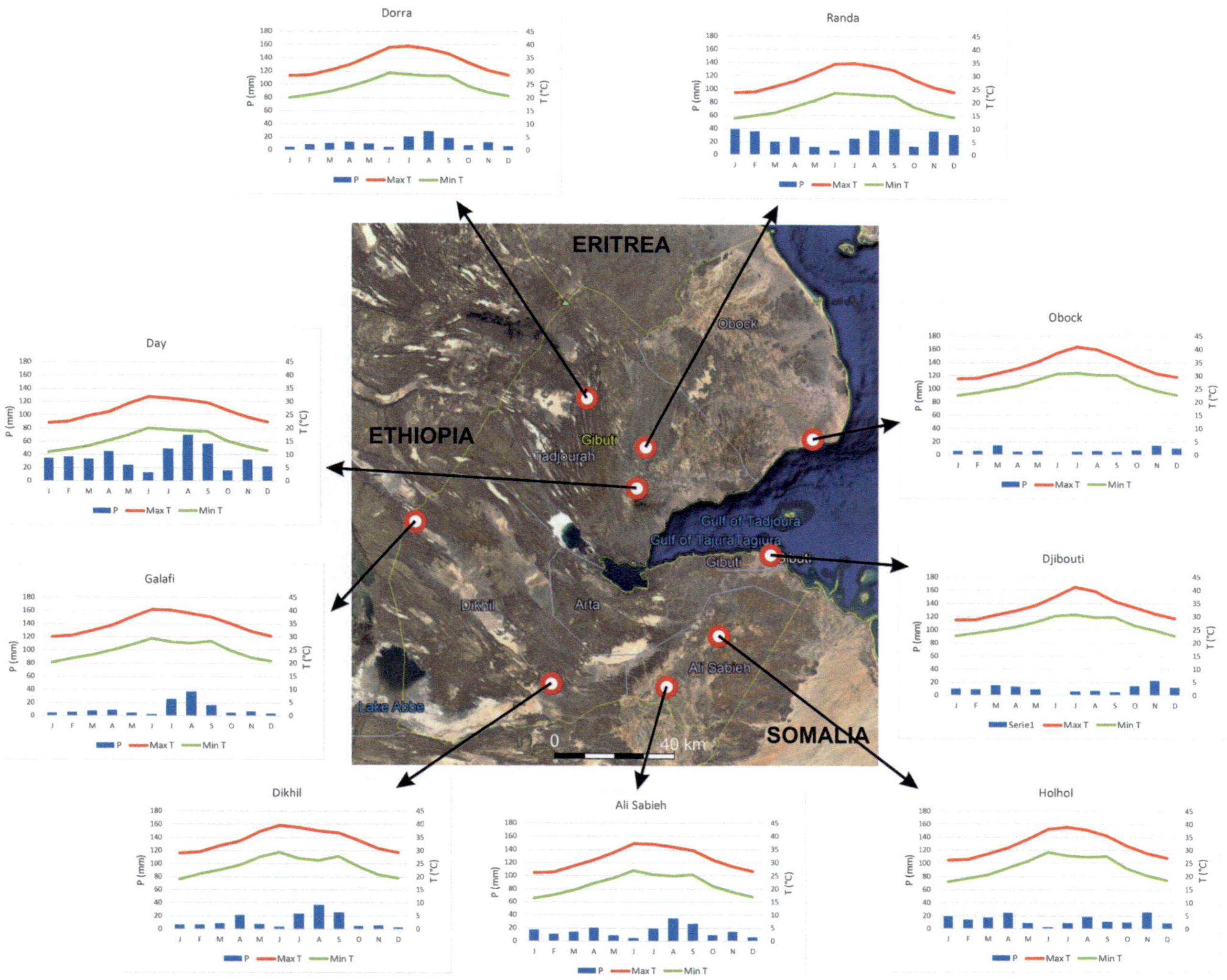

Fig. 1.6 Representative climograms of the main physiographic and climatic areas of Djibouti

precipitation is scarce and it is mainly concentrated in winter, when the Indian Ocean easterlies and the Atlantic westerlies converge (Ethiopian Mapping Authority 1988; Van Buskirk and Amare 1994) (Fig. 1.5).

In Somalia, rains are highly variable in space and time (Fig. 1.7). They are mainly controlled by the northward and southward movement of the ITCZ, giving way to two distinct rainfall seasons: the spring main rainy season (April–June) (*Gu* in the local language) and the small rains (October–November) (*Deyr* in the local language). The position of the ITCZ and the pressure characteristics of the air masses over Arabia and India influence also the prevailing winds over Somalia. In January, the north-easterly winds prevail. They originate over Arabia; hence, they are dry, hot and cloudless (Fig. 1.12). In April and October, the ITCZ is passing over Somalia in its northward and southward movement, respectively, and south-easterly and north-easterly winds meet in a convergence zone resulting in the upward movement of air which, on its turn, causes condensation, cloud formation and rainfall (Fig. 1.12). In July, moist winds flow from the Indian Ocean and, as they cross the equator, become the south-west monsoon. A narrow belt of convergence is then formed, especially along the southern Somali coast, which receives substantial amounts of rain (Fig. 1.12) (Muchiri 2007).

The precipitation pattern of Djibouti shares the climatic characteristics recorded by the coastal meteo-stations in southern Eritrea and northern Somalia. In the inland areas of Djibouti, the monthly rainfall pattern may be both unimodal or bimodal, but the main rainy season is always in July–August. In the coastal areas, rain is very small and falls mainly in winter (Fig. 1.6).

It is worth recalling here that the same considerations about the general quality of temperature data are valid for precipitation as regards the source of data (ground instrumental measurement and gridded data), length of observations and rain gauges distribution over the territory.

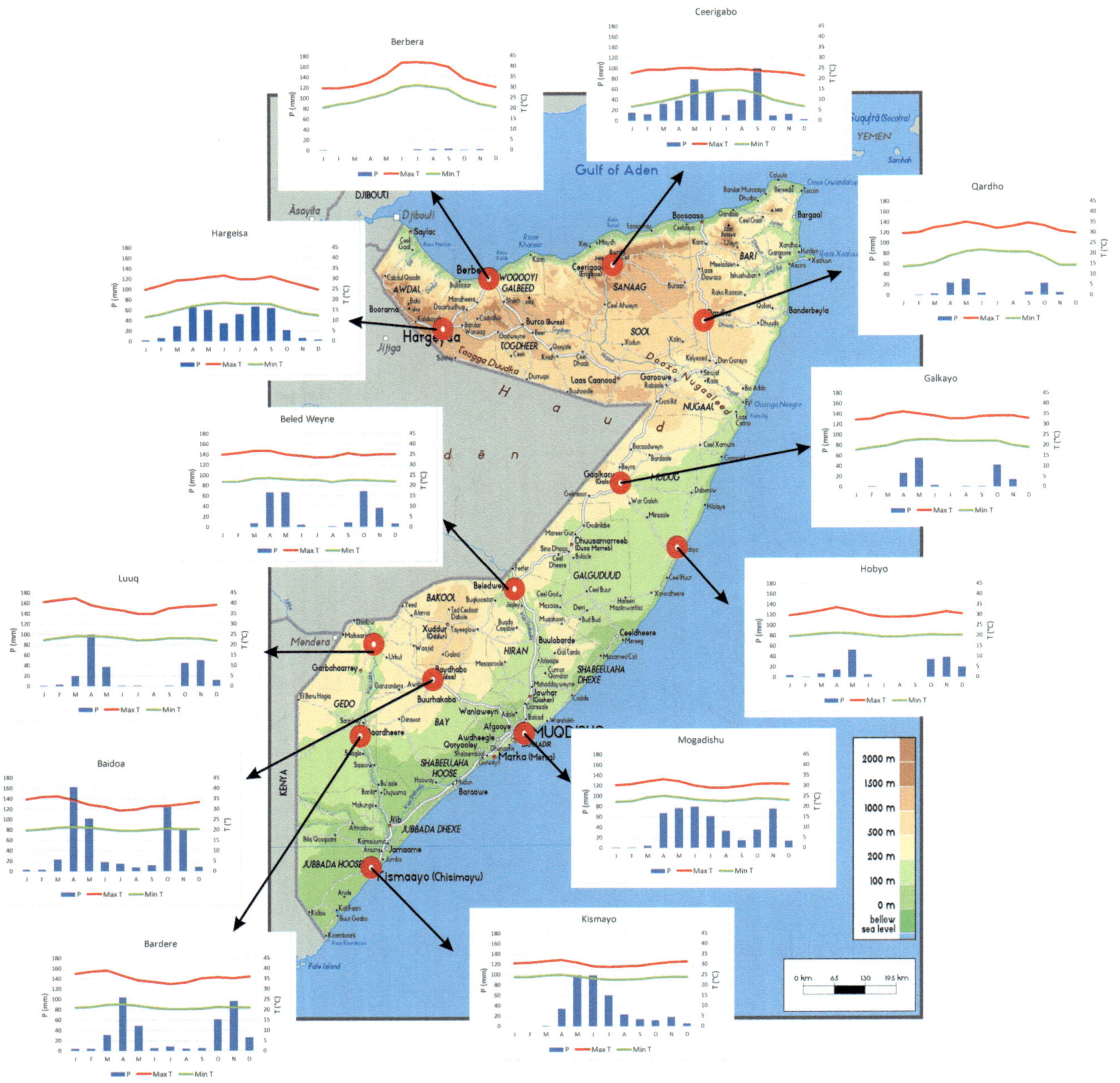

Fig. 1.7 Representative climograms of the main physiographic and climatic areas of Somalia

The Precipitation Concentration Index (PCI) (Oliver 1980) varies from 10.2 to 26 across Eritrea and indicates a wide range of rain seasonality (from moderate to strongly seasonal distribution). A slightly smaller PCI range (9.5–24.5) is calculated for Somalia, which implies a rain seasonality grade similar to that of Eritrea. In Djibouti, the PCI range is smaller (9.5–13.3) indicating a moderate seasonality of rain. These results reflect the complex interactions of factors controlling precipitation over the eastern Horn of Africa.

The first synthetic result, obtained from averaging the annual precipitation (P) of all the meteo-station used in this study, indicates Eritrea as the country that receives the largest amount of rain, 421.9 mm, followed by Somalia, 308.7 mm, and Djibouti, 240.3 mm. These differences are also confirmed by the spatial distribution of rain in the three countries. In Eritrea, in fact, 76% of the rain gauges considered record an annual precipitation higher than 250 mm, whereas in Somalia and Djibouti the percentages decrease to 57 and 37%, respectively. The higher percentage in Eritrea is probably due to the lack of data in the driest zones of this country, i.e. the Danakil and the western lowlands (Fig. 1.3), and to a larger number of meteo-stations located at higher elevations than in the other two countries (Fig. 1.11).

Fig. 1.8 Northern Somalia escarpment facing the Gulf of Aden near Sheikh

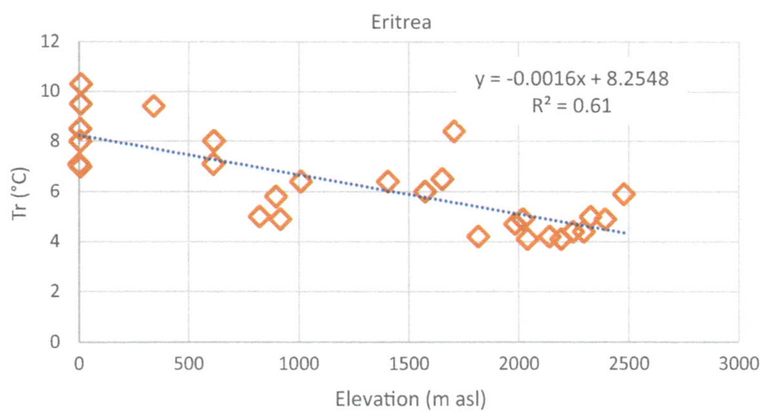

Fig. 1.9 Change of annual temperature range (Tr) with elevation in Eritrea

Among the meteo-stations considered, the wettest one is Nefasit (766 mm), in the main escarpment, about 14 km east of Asmara. Nefasit is not located at 1652 m asl, i.e. an intermediate elevation for Eritrea, but its higher than average precipitation is due to an additional rainfall contribution from the condensation of the moist winds coming from the Red Sea in winter. The highest annual precipitation of Djibouti is recorded at Diyara (439 mm) in the Day Forest National Park, whereas the wettest meteo-station of Somalia is Buur Gaabo (595.5 mm), on the southern coast of Somalia, about 56 km from the border with Kenya. The meteo-station with the lowest rainfall of Eritrea is Beylul (69 mm), located on the Red Sea coast, about 46 km north-west of Assab. As already reported, the lack of data from the Eritrean Danakil does not allow to make any consideration about rainfall in this area, but presumably there are sites which may receive less than 50 mm of rain in a year (Fig. 1.13). The same argument could be brought about for central Somalia, where we find the lowest precipitation for this country at Abudwak (35 mm), or the

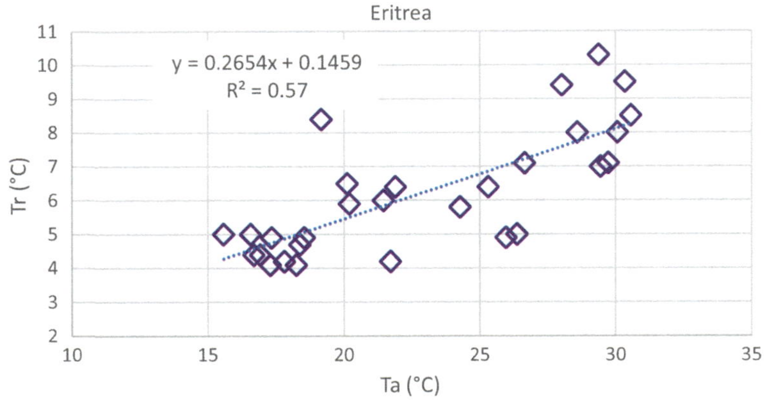

Fig. 1.10 Change of annual temperature range (Tr) with mean annual temperature (Ta) in Eritrea

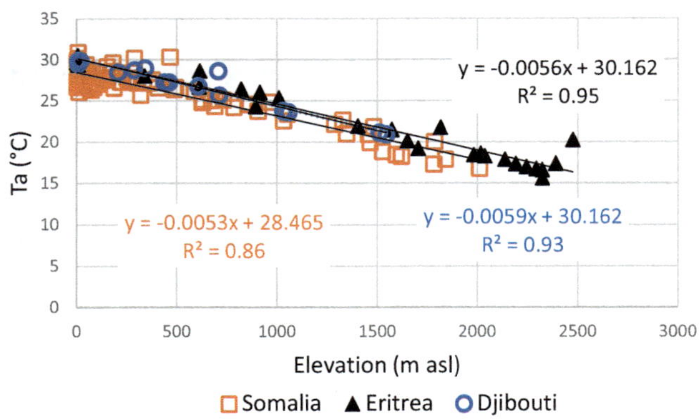

Fig. 1.11 Lapse rate in Eritrea, Djibouti and Somalia

northern coastal areas (Fig. 1.14). The lowest precipitation of Djibouti is recorded at Obock (91.1 mm) on the northern coast, opposite to Djibouti City.

In the areas of Eritrea with a unimodal monthly precipitation pattern, the summer (July–September) main rainy season accounts from 60 to 80% of the annual precipitation. In Somalia, the main rainy season (*Gu*, April–June) and the minor one (*Deyr*, October–November) on average amount to 46% (range 18–62%) and 25.5% (range 5–46%) of the annual precipitation, whereas in Djibouti no specific main rainy season can be clearly identified since precipitation is rather uniformly distributed in every month, as confirmed by the average PCI = 10.6.

Unfortunately, hourly rainfall intensity data are not available and also daily data are scarce. According to the WCRP-WMO (2011) report, extreme intensities of 143 and 124 mm in 3:30 and 3:50 h, respectively, have been recorded in Asmara. Extreme daily intensities were measured in Massawa (113–130 mm/24 h and 364 mm/48 h), Akordat (113 mm/24 h) and Teseney (109 mm/24 h) (WCRP-WMO 2011). For this study, daily rainfall data of Somalia were available only for Galkayo, Garowe, Hargeisa and Mandera (a small town in Kenya near the triple border junction with Somalia and Ethiopia, on the Dawa River, the largest tributary of the Juba River). The highest precipitation in 24 h (89.7 mm/24 h) was measured in Garowe (P = 121.2 mm yr^{-1}), a small town in the Nogal valley, about 155 km from the Indian Ocean coast. Other meteo-stations recorded slightly lower intensities, e.g. 82.9 mm/24 h in Hargeisa (P = 327 mm yr^{-1}), 64.3 mm/24 h in Mandera (P = 584 mm yr^{-1}) and 60.7 mm/24 h in Galkayo (P = 183 mm yr^{-1}), located about 200 km south of Garowe and about 190 km far from the Indian Ocean (Fig. 1.7). It is worth noticing that the highest daily intensity was measured in an arid area (Garowe), where the annual precipitation is the least. No daily rainfall data were instead available for Djibouti. The town of Gode is located in Ethiopia, but it is only 122 km from the border with Somalia and subjected to an arid climate (annual precipitation is only 237 mm), similar to that of the Somali territories beyond the border. In Gode, the maximum daily intensity ever recorded by the local meteo-station was 174 mm/24 h; one can therefore expect similar values for inland areas of central Somalia.

Drought is a recurrent event in the three countries, with hundreds of thousands of affected people (Fig. 1.15). According to Nicholson (2014), droughts in the Horn of Africa are a consequence of years with annual rainfall 30–75% below normal. Other authors (e.g. Marthews et al. 2015) attribute an important role in the failure of the long rains to human influence: higher temperatures caused by human activities have been resulting in an increased net incoming radiation at the surface. Lott et al. (2013) framed the reduced precipitation during the main rainy season in the Horn of Africa within the global warming rather than invoked the effects of a local anthropogenic change.

Djibouti shows a distinctive (R^2 = 0.88) control of elevation on annual precipitation gradient (Fig. 1.16), whereas the change of annual precipitation is poorly correlated with elevation in Eritrea (R^2 = 0.49). In Somalia, a weak, inverse precipitation gradient exists only if the data of the dry and hot coastal stations and those of the Bur Region are neglected and only meteo-stations at elevation <500 m asl are considered (Fig. 1.17). This very unusual result may

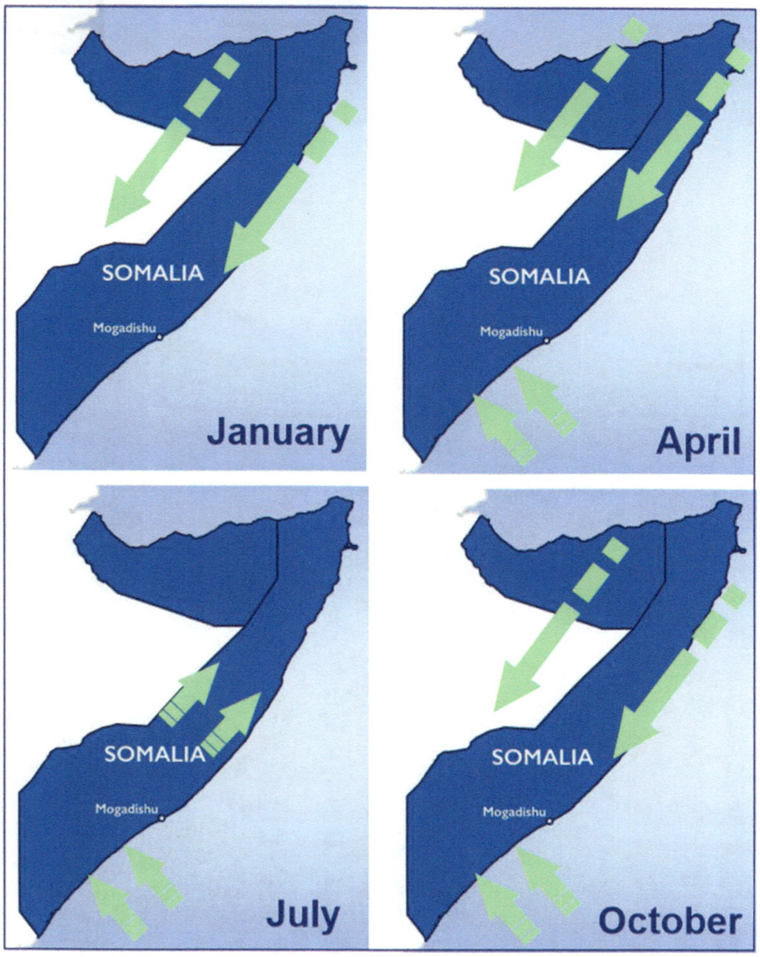

Fig. 1.12 Prevailing winds over Somalia in different seasons (modified from Muchiri 2007)

depend on the high spatial variability of precipitation across Somalia, associated with the distance from the ocean. A large portion of Somalia is a uniform tableland, gently inclined towards the ocean, and the elevation of inner meteo-stations is typically lower than 500 m asl. The influence of the moist winds coming from the ocean tends to progressively decrease as they move to the hotter inland and lose humidity all the way long.

1.5.2 Rainfall Erosivity

According to FAO (Nana-Sinkam 1995), many African countries have already lost a significant quantity of their soils, subject to various forms of degradation. Many areas in Africa are said to be losing over 50 tonnes of soil per hectare per year and the countries of Horn of Africa are not exempt from such a calamity (Figs. 1.18 and 1.19). Soil degradation caused by rainfall-induced erosion, desertification, deforestation, and poor agricultural practices is undermining the very resources on which people depend for their survival. Land degradation is typically expressed by the occurrence of deep gullies, the origin of crusts that water cannot infiltrate, and laterite that hand tools and plant roots cannot penetrate.

A complex combination of several factors determines the rate of soil erosion. The Universal Soil Loss Equation (USLE) (Wischmeier and Smith 1958) and its revised forms (e.g. RUSLE) (Renard et al. 1997) were the first successful attempt to quantify the role of the main factors involved and to suggest the way to measure them. Rainfall is definitely the triggering factor in hydrological soil erosion. Rainfall erosivity is the capacity of rain to detach soil particles and to generate overland flow which, in turn, entrains soil particles. Rainfall erosivity is a function of physical characteristics of precipitation and its energy (Nyssen et al. 2005). The procedure to calculate the USLE rainfall factor (*R*-factor) proposed by Wischmeier and Smith (1958) is rather complex and surrogate or alternative factors, easier to be calculated, have been proposed in the literature. Most of them are based on the annual precipitation (e.g. Renard and Freimund 1994; Eq. 1.3) or on the concentration of monthly rain as the Modified Fournier Index (MFI) proposed by Arnoldus (1980) (Eq. 1.2). Lee and Heo (2011) tested eight alternative methods (including the MFI and the Renard and Freimund's

Fig. 1.13 Drylands in the Eritrean Danakil about 150 km south of Massawa

R-factor) to calculate rainfall erosivity against more than 20 years of actual rainfall erosivity calculated by high resolution precipitation data in South Korea. They concluded that both annual precipitation and alternative parameters could be used for Korea, but their regression models had limitations when used to predict actual rainfall erosivity in other locations and the simplified methods for estimating rainfall erosivity should be used with caution.

In Horn of Africa countries, rainfall data sets are commonly incomplete and cover short periods. Hourly data are not available, and the only reliable precipitation data are those about monthly rainfall (and daily data, but only for few meteo-station). For these reasons, though the argument of Lee and Heo (2011) is embraceable, in order to explore the geomorphological effectiveness of rainfall in the study area, Arnoldus's (1980) MFI and the *R*-factor of Renard and Freimund (1994) were used. In Eritrea, the MFI ranges between 8 and 153, with 45% of the meteo-stations recording MFI values higher than 120, i.e. from moderate to high erosivity (Gabriels 2006) (Table 1.1). The *R*-factor ranges between 44 and 2126 MJ mm ha^{-1} h^{-1} yr^{-1} and values higher than 1150 MJ mm ha^{-1} h^{-1} yr^{-1}, indicating moderate to high erosivity (Panagos et al. 2017) (Table 1.3), are calculated for 31% of the meteo-stations. It is worth noticing that while the lowest values of both parameters are calculated for the area of Beylul, where also the lowest annual precipitation is recorded, the highest value of the MFI is calculated for the meteo-station of Bet Ansian (a small village located on the southern part of the central highlands, close to the Ethiopian border), where the second largest annual precipitation is recorded, whereas the highest value of the *R*-factor is measured at the meteo-station with the highest annual precipitation, which is not surprising since this parameter is directly proportional to the annual rain (Eq. 1.3).

Djibouti is characterised by very low ($9 < \text{MFI} < 44$) (Table 1.1) and very low to moderate ($69 < R\text{-factor} < 868$) (Table 1.3) rainfall erosivity. Both the lowest and highest

Fig. 1.14 Arid coastal belt a few kilometres south of Berbera in northern Somalia

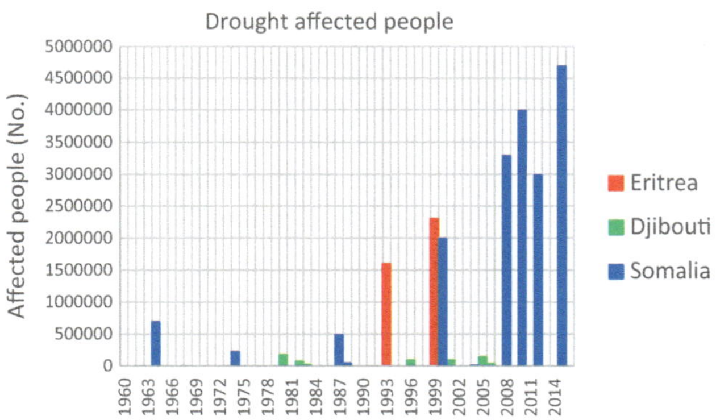

Fig. 1.15 Number of drought affected people in the study countries since 1960

erosivity of both parameters are calculated for the sites of Obock, on the northern coast, and Diyara, in the Day Forest National Park, respectively. The range extremes of both parameters coincide with the smallest and largest annual rainfall. The range of rainfall erosivity calculated for Somalia is larger than for Djibouti, but narrower than for Eritrea. The MFI, in fact, ranges from 5 to 109, i.e. from very low to moderate erosivity (Table 1.1), and the *R*-factor ranges from 15 to 1417 MJ mm ha^{-1} h^{-1} yr^{-1}, that is from moderate to high erosivity (Table 1.3). The MFI extremes do not coincide with the lowest and largest values of annual precipitation, but are very close to them. The former are recorded at Bosaso, on the northern coast and at Ischia (in the local language it means water spring) Baidoa in the Bur Region.

A rough comparison with the actual conditions of land degradation and soil erosion of the sites, where the extreme low or high rainfall erosivity was observed, seems to confirm the conclusion of Lee and Heo (2011) that these alternative methods to express rainfall erosivity are probably not very effective in addressing the real amount of energy transferred by the rain to the soil. In Horn of Africa countries, the wide variability in temperature, rainfall distribution, geographic position and elevation and, ultimately, human impact, play an important role as well. More studies in arid and semi-arid regions are however needed to investigate the effectiveness of the MFI and the *R*-factor in providing even broad reference values of rainfall erosivity.

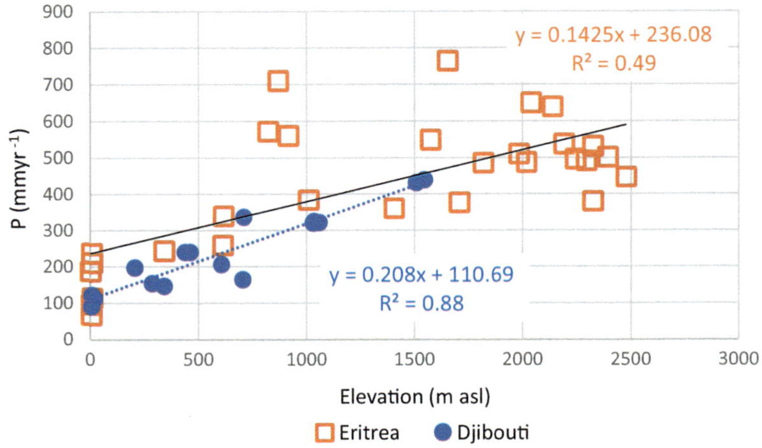

Fig. 1.16 Variation of mean annual precipitation with elevation in Eritrea and Djibouti

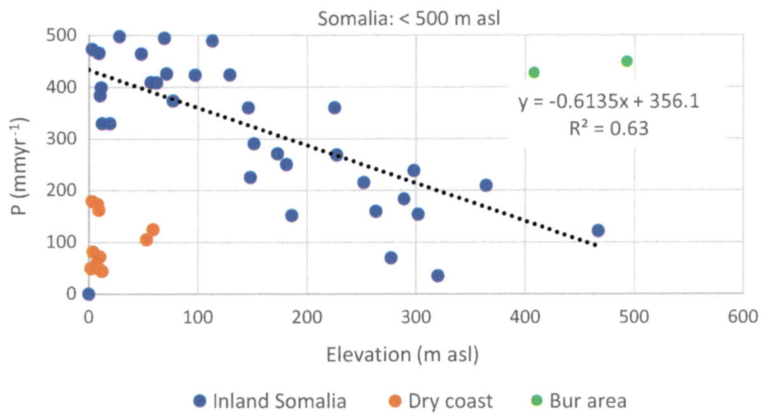

Fig. 1.17 Variation of mean annual precipitation with elevation in Somalia

1.6 Relative Humidity

Only very few data of relative humidity data are available for Eritrea and Djibouti. The higher values are recorded on the coast in February–April, with 74–77 and 69–71% at Massawa and Assab, respectively (Fig. 1.20). In Asmara, relative humidity shows a peak in August and September (73–78%), whereas in other months it ranges between 43 and 62% (Fig. 1.20). Lower values of relative humidity (29–61%) (Fig. 1.20) are recorded at Teseney, in the Eritrean western lowlands (Fig. 1.5) close to the Sudan border. In Djibouti, relative humidity is rather constant 70–75% every month, with the exception of summer months (June–September), during which relative humidity values range from 40% in July to 55% in September.

The data of relative humidity are more abundant for Somalia thanks to the data set compiled by FAO-SWALIM (Muchiri 2007). The Somali sites with higher relative humidity are found on the coast, whereas those with lower values are located inland or in areas at higher elevation, as illustrated by the analysis of selected meteo-stations reported in Fig. 1.21. Mogadishu, Hobyo and Kismayo experience high relative humidity conditions all year round ranging from 75 to 80%. Berbera and Bosaso have similar values but in summer (June–September) relative humidity falls down to 53–59% (Fig. 1.21a). By contrast, inland areas and more elevated sites in Somalia have more pleasant climate, characterised by lower values of relative humidity ranging from 47 to 64% (Fig. 1.21).

1.7 Wind

Wind data are also scarce and, in a few cases, they are not measured with standard instrumentation and procedures (Rosen et al. 1999; Muchiri 2007). According to the Government of Eritrea (2007) report on Wind Energy Applications in Eritrea, a total of 20 wind recording stations were operating in 2007. Unfortunately, the data published in that report refer to a small number of stations and cover a very short interval of 3–4 years (2000–2004). More data are available for Somalia thanks, once again, to the FAO-SWALIM project (Muchiri 2007). All the wind data available for this study refer to average monthly wind velocities.

Fig. 1.18 Degraded land south of Mendefera in the southern highlands of Eritrea close to the border with Ethiopia

On the basis of data available for this study, higher wind velocities of Eritrea are recorded at Massawa and Asmara (Fig. 1.22), with 5.6 m s^{-1} average velocity measured at Massawa from April to July. At Asmara, wind velocity is slightly lower than in Massawa, but more constant in every month, with values around 4.6 m s^{-1}. The inland meteo-stations of Akordat and Teseney record lower but rather invariable wind velocities around 2.0 and 1.8 m s^{-1}, respectively, throughout the whole year. Intermediate velocities (around 2.8 m s^{-1}) are observed at Edd, a small village on the Red Sea coast, about 150 km north of Assab. Wind directions are not available.

Faster winds are measured in Djibouti where the meteo-stations of Obock and Djibouti City report a range of 3.1–5.1 m s^{-1} (higher velocities occurring in July and September) and 3.3–5.28 m s^{-1} (higher velocities recorded in August and September), respectively (Fig. 1.23).

In Somalia, wind velocity and direction are substantially influenced by the movement of the ITCZ and the local physiography (especially in the north) (Muchiri 2007). The strongest winds of Somalia (8.4–11.5 m s^{-1}) are recorded in summer (June–August), during the south-westerly monsoon, primarily in the northern part of the country, but also in Mogadishu, though with slower winds (5.6–7.8 m s^{-1}) (Fig. 1.24). Mogadishu experiences the highest wind velocity in winter (8.9 m s^{-1}, in February), whereas Luuq and Jowhar record similar, lower velocities (2.2–4.2 m s^{-1}). At Hobyo, higher values, close to those in Mogadishu, with a peak of 6.7 m s^{-1} are recorded in July–August and a low value of 4.2 m s^{-1} typifies November and December (Fig. 1.24).

General information about the prevailing wind direction can be obtained from the orientation of aeolian landforms. In the southern coastal strip of Eritrea, barchanoid dunes (Fig. 1.25) indicate a prevailing, strong wind from south-east. A similar orientation of sand dunes can be observed in the Samoti plain in the northern part of the Eritrean Danakil (see Chap. 7). The wind responsible for the morphology and movement of these dunes is locally known as "*khamsin*". It is a dry, hot and rather intense wind (gusts may reach a velocity of 22 m s^{-1}) coming from south-east and blowing for a few days without interruption. Sometimes, it is also able to generate dust storms. The *khamsin* is active from late winter to late spring, and it is generated by high-pressure cells over the northern Arabian Peninsula and Mediterranean deep depressions moving eastward. In the

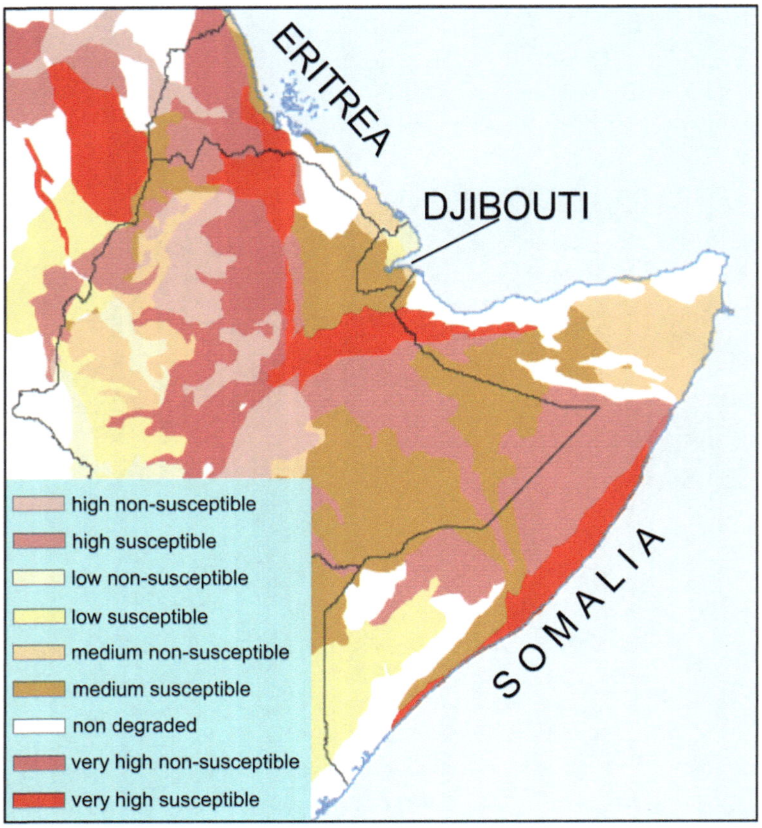

Fig. 1.19 Map of the susceptibility to erosion in the study countries (modified from UNEP 2010)

Table 1.3 Classification of erosivity *R*-factor based on Panagos et al. (2017)

R-factor	Description
<200	Very low
200–400	Low
400–1150	Moderate
1150–3100	High
>3100	Very high

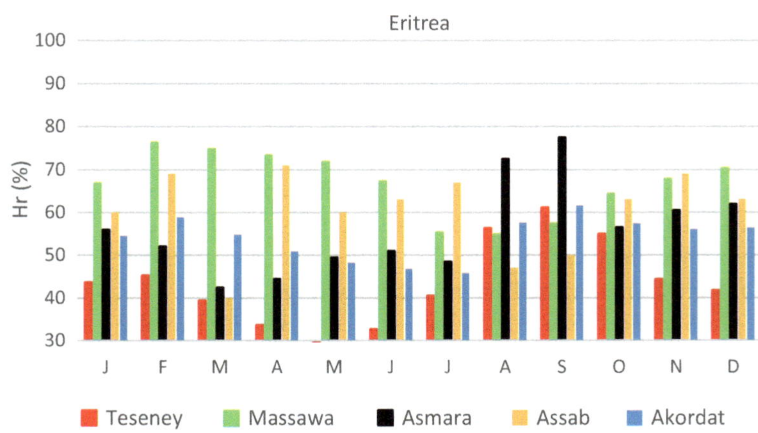

Fig. 1.20 Mean monthly relative humidity in selected meteo-stations of Eritrea

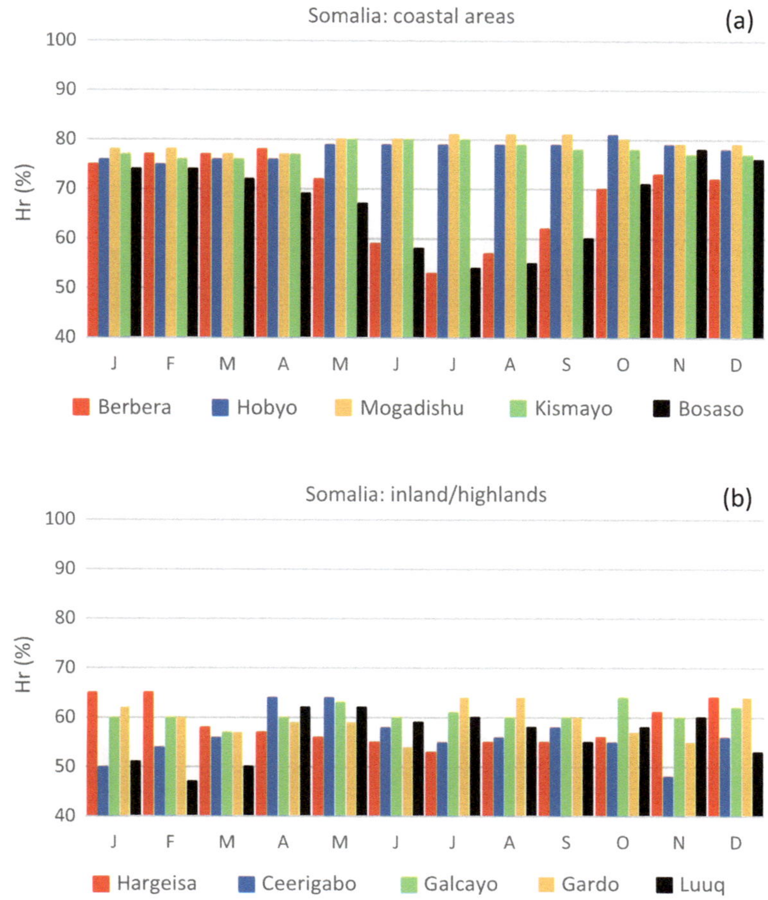

Fig. 1.21 Mean monthly relative humidity of selected meteo-stations in the coastal areas (**a**) and in the inland and highlands (**b**) of Somalia

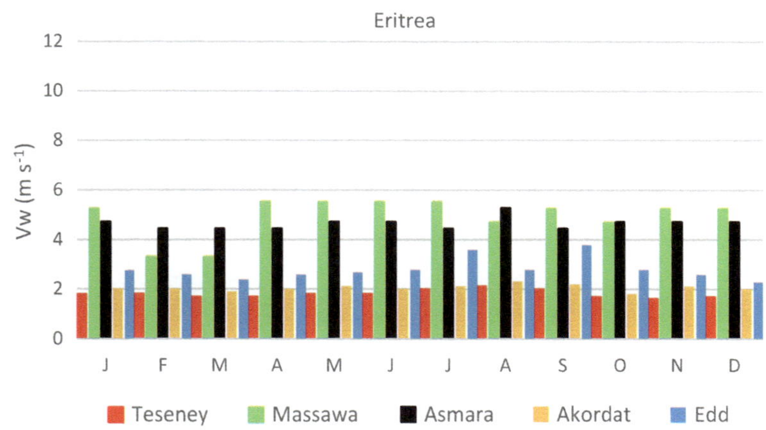

Fig. 1.22 Mean monthly wind velocity in selected meteo-stations of Eritrea

central coastal belt, instead, linear dunes (Fig. 1.26) indicate winds with a north–south direction. According to the observations of Fantoli (1940), at Massawa and on the coast, the dominant winds are from the south-east. In winter, they alternate with a mainly northerly airflow produced by the North African anticyclone and directed towards the Red Sea low pressure trough. As a result, the northern part of the Red Sea is exposed to predominantly north-west to north winds that can also reach the northern coast of Eritrea (Fantoli 1940; Edwards 1987).

This alternation may be responsible for the formation of linear dunes. According to a few authors (e.g. Tsoar 1989; Lancaster 2011), field observations indicate that linear dunes are formed under conditions of a bidirectional wind regime, with the two modes separated by 90° or more, and are parallel to the resultant wind. Recent studies, however,

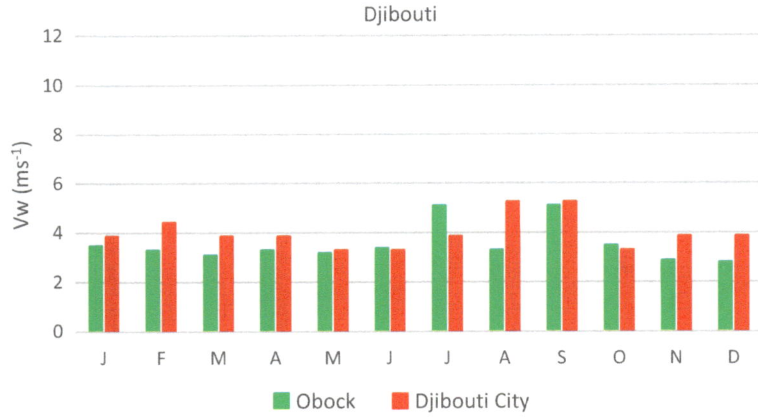

Fig. 1.23 Mean monthly wind velocity of selected meteo-stations in Djibouti

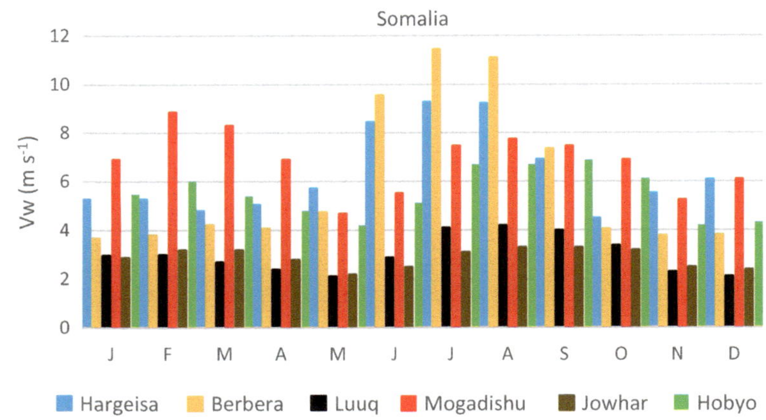

Fig. 1.24 Mean monthly wind velocity of selected meteo-stations in Somalia

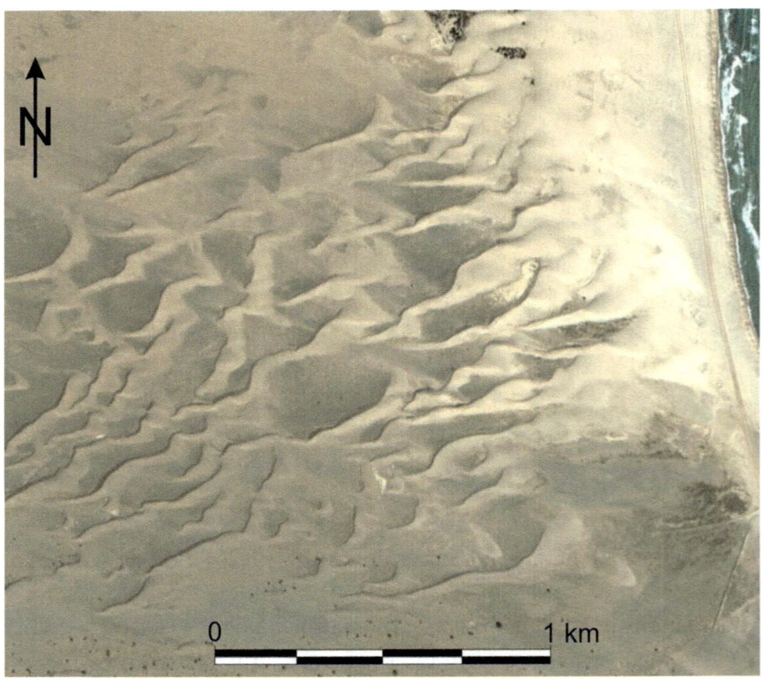

Fig. 1.25 Barchanoid, wind dunes in the southern coastal belt of Eritrea

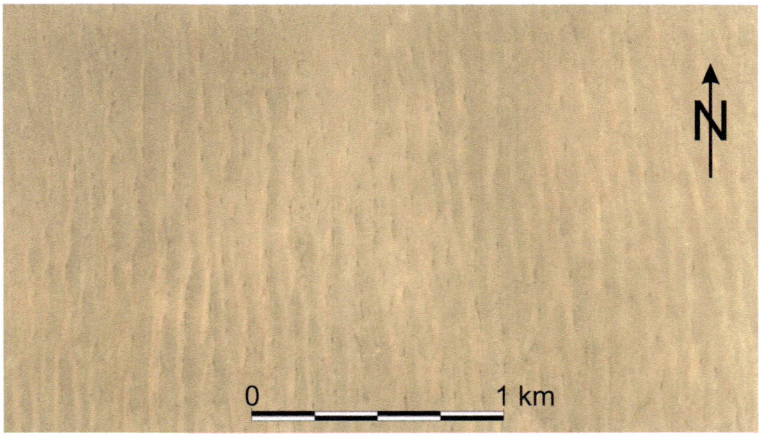

Fig. 1.26 Linear dunes in the central part of the Eritrean coastal belt

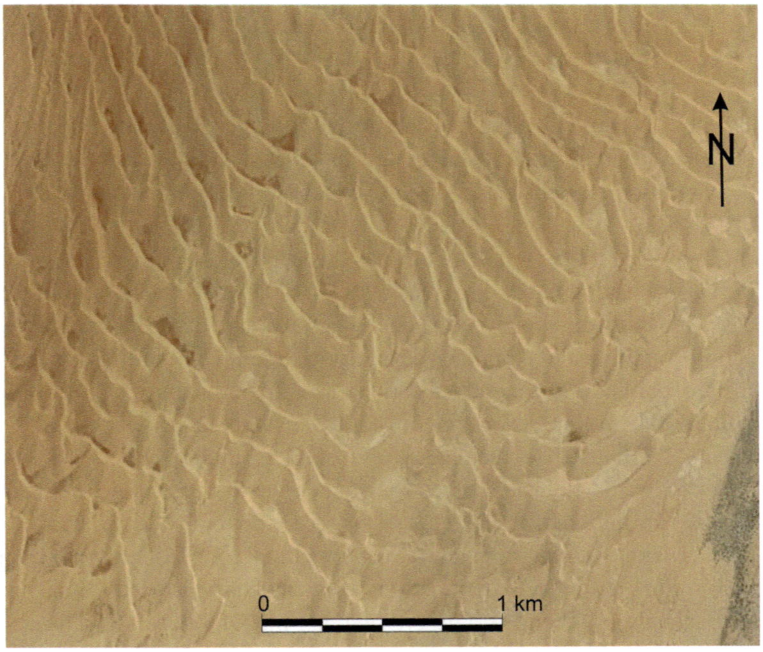

Fig. 1.27 Barchanoid dunes field in the southern coastal belt of Somalia

suggest that partial induration of sediment may produce linear dunes parallel to the dominant wind (Rubin and Hesp 2009).

In central and southern Somalia, several barchanoid dune fields along the coastal belt indicate strong wind from the south-west (Fig. 1.27), whereas along the northern coast facing the Gulf of Aden windblown dunes are no so common (and smaller than in the south) and indicate winds from the north-east (Fig. 1.28), i.e. the prevailing directions of the summer and winter monsoon, respectively.

1.8 Aridity and Desertification

The countries of Horn of Africa are subjected to different climates, but they have one characteristic in common, which is aridity. Aridity is a general concept and there are several definitions of aridity, all equally scientifically correct. The choice of which definition or best index to use depends on the purpose of the investigation to be carried out (Nicholson 2011). In this study, the choice of aridity indices was driven primarily by the kind, quality and quantity of data available and, subordinately, the purpose of describing the aridity component of climate of the three countries and the reflections in terms of soil erosion, water resources, landforms and geomorphological processes. The indices used in this study are the desertification index, DI = P/PET (UNEP 1992) in which P is mean annual precipitation and PET is annual potential evapotranspiration calculated with the method of Thornthwaite (1948) and the De Martonne (1925) aridity index (AI, Eq. 1.4). The UNEP desertification index was selected because it is commonly used in many studies on climate change and is considered a good indicator of the risk of desertification onset. The De Martonne aridity index was

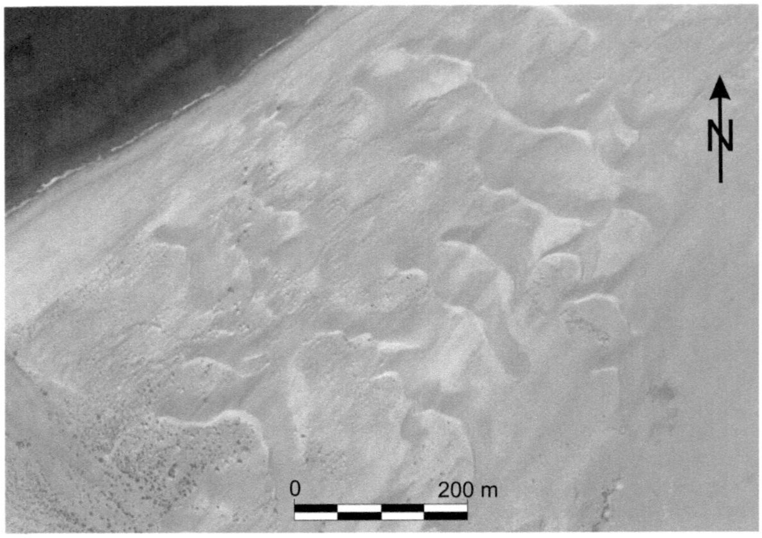

Fig. 1.28 Barchanoid dunes in the central part of the northern Somalia coastal belt facing the Gulf of Aden

selected because it requires easily accessible data and because different aridity classes are associated with water demand for various crops and irrigation requirements (Table 1.2).

The UNEP (1992) indicates 0.6 as the threshold value of the desertification index, i.e. areas with DI > 0.6 are prone to desertification or already affected by aridity and desertification conditions, making them unsuitable for cultivation. In Eritrea, one third of the meteo-station has a DI less than 0.6, in four of them (one seventh) DI is 0.61–0.63 and the highest value of DI is 0.81 calculated for Nefasit. Among these meteo-stations, Nefasit records the highest annual precipitation (766 mm) and a PET of 948 mm, i.e. about 40% less than average annual potential evaporation of Eritrea (1563 mm—calculated from data of all meteo-stations used in this study). Djibouti is a desert land and the DI values of all meteo-stations of this country are below 0.6, with the maximum value of 0.47. PET is also rather high, ranging from 1019 to 3353 mm (mean value of 2083 mm), with the highest value recorded in Djibouti city. The desertification situation of Somalia is similar to that of Djibouti. In fact, no meteo-station records a DI value higher than 0.60 and the highest value is 0.55, recorded at Boorama, located in the north-western highlands. In Somalia, the country average PET (1792 mm yr^{-1}, using the country-averaged gridded data and 1840 mm yr^{-1}, obtained from averaging the mean PET of selected meteo-station reported by Muchiri 2007) is lower than in Djibouti, but higher than in Eritrea (1697 mm yr^{-1}, using the country-averaged gridded data and 1563 mm yr^{-1}, obtained from averaging the mean PET of selected meteo-stations). The highest PET of Somalia is calculated for Lughaye (3961 mm yr^{-1}, on the coast about 100 km from the border with Djibouti), and the lowest is measured at Warmaddow (783 mm yr^{-1}), in the mountains (2011 m asl), 60 km east of Bosaso. Both meteo-stations receive very little rain in a year (174 and 211 mm, respectively), but Lughaye is on the coast (8 m asl) and has a mean annual temperature of 30.9 °C, whereas that of Warmaddow is much cooler, 16.8 °C. Yet, it is worth recalling the lack of climatic data for the Eritrean Danakil, whose PET values are expected to be very high, likely higher than those calculated for the meteo-station of Djibouti inland.

In the three countries, the aridity index of De Martonne (1925) assumes the highest and lowest values in the same meteo-stations, for which the extreme values of the desertification index are calculated. Though AI and DI are calculated in a completely different way and have only the annual precipitation at the numerator in common, the two indices are highly correlated (Figs. 1.29 and 1.30). AI and DI are strongly correlated (R^2 = 0.98) for Eritrea and Djibouti, and an individual interpolating function can be used for both countries to derive DI from AI (Fig. 1.29):

$$\mathrm{DI} = 0.0076\mathrm{AI}^{1.4833} \quad (1.5)$$

In Somalia, the correlation between AI and DI is also high, but the data plot suggests different interpolation functions for the meteo-stations located in the northern part of the country (latitude higher than 7° N) and those located in the central and southern part. The two interpolating functions are:

$$\mathrm{DI} = 0.0342\mathrm{AI} - 0.0329 \quad \left(R^2 = 0.97\right) \quad (1.6)$$

and

$$\mathrm{DI} = 0.0218\mathrm{AI} - 0.0209 \quad \left(R^2 = 0.95\right) \quad (1.7)$$

Equation 1.6 can be used for the northern locations and Eq. 1.7 for the central to southern areas of Somalia.

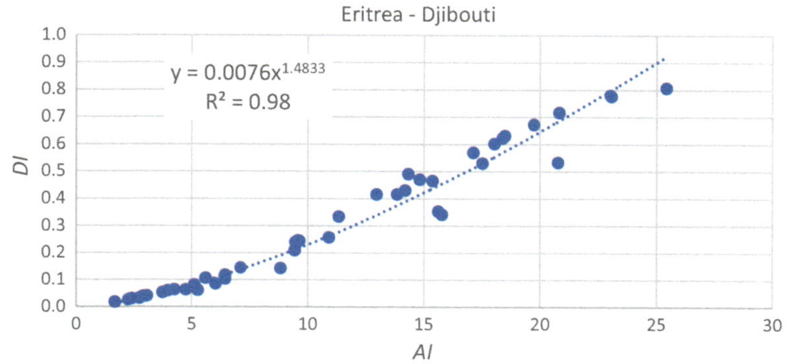

Fig. 1.29 Correlation between the De Martonne (1926) aridity index (AI) and the UNEP (1992) desertification index (DI) for Eritrea and Djibouti

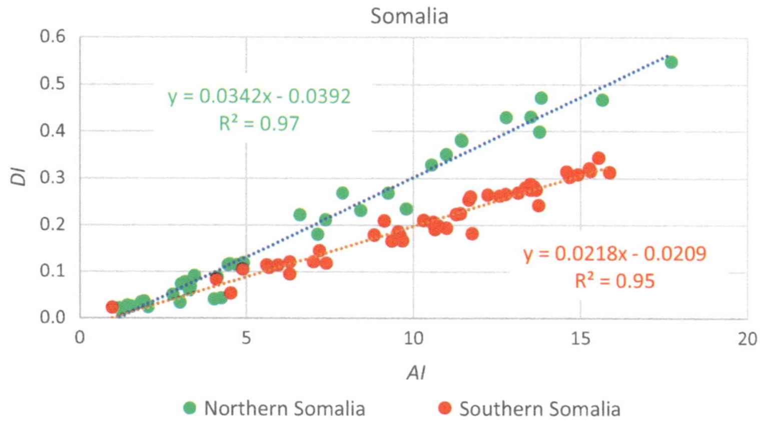

Fig. 1.30 Correlation between the De Martonne (1926) aridity index (AI) and the UNEP (1992) desertification index (DI) for northern and southern Somalia

These results clearly indicate that many parts of the countries are under conditions of severe aridity and desertification.

1.9 Climate Change

Accurate and reliable investigations on climate change imply the availability of time series of validated climatic data as longest as possible. Unfortunately, this is not the case for the countries of Horn of Africa. The World Bank (2019) Climate Change Knowledge Portal gives access to very long (1901–2015) gridded data time series for mean monthly temperature and precipitation. Though the length of the time series is suitable for climate change analysis, their representativeness is questionable, as it will be shown in the next section. These data, in fact, though based on gridded data produced by the Climatic Research Unit (CRU) of University of East Anglia (UEA), UK, are released as country averages and it is not clear by which instrumental data are they validated. Nevertheless, these data can be useful to depict a trend, though very general, of the main climatic parameters. In addition to this simple analysis, a comparison between instrumental old data recorded earlier than 1960 and collected by Fantoli (1940, 1965) and recent data averaged for three decades is carried out in an attempt to depict a more realistic change.

1.9.1 Temperature Change

The gridded data indicate a clear increasing trend of mean annual temperature for Horn of Africa. The increase is more marked in Eritrea, where temperature started to rise since the early years of the twentieth century (Fig. 1.31). The time series plot, however, also shows that the series consists of two "independent" parts, both showing an increasing trend (Fig. 1.31b). The decrease in mean temperature recorded in the early 1940s is too marked to be natural and is probably due to a change in the position of some meteo-stations or to the replacement of old thermometers with more modern ones. For instance, the coordinates of a few meteo-stations reported by Fantoli (1940) are different from the current ones. Though the time series is split in two patterns, their increasing trends are unequivocal. The increase in temperature, however, is more marked in the later pattern, with a net increase of about 2 °C since the 1940s (Fig. 1.31b).

Also, the data for Djibouti are better interpolated by two distinct trends (Fig. 1.32), though in this case there is no information about a change of the measuring devices or their location. The first pattern covers the period from 1901 to about 1963, and it is characterised by a slightly decreasing trend. From 1964 on, the temperature returns to increase, reaching a mean annual temperature that is about 1 °C higher than in the 1940s (Fig. 1.32). Very similar is the

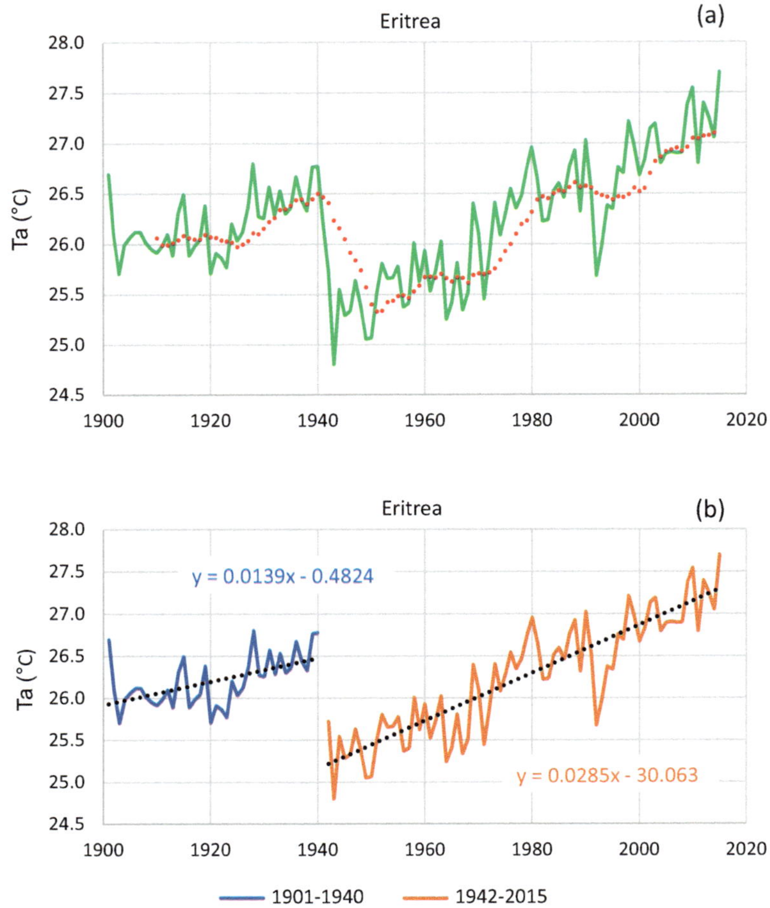

Fig. 1.31 Century long variation of mean annual temperature gridded data in Eritrea: **a** original, uninterrupted time series (the red dotted line is the 10 years moving average); **b** the original time series split into two main portions to highlight the big, probably unnatural, drop reported a for the early 1940s. Notice the more marked temperature increase of the latest portion of the time series

situation in Somalia, in which the time series of mean annual temperature follows two distinct patterns (Fig. 1.33). The first one, from 1901 to 1963, does not show any evident trend, whereas the second one, from 1964 to 2015, is characterised by an increasing trend, which is less marked than in Eritrea and Djibouti. The increase in temperature since 1964–2015 is, in fact, only of 0.6 °C (Fig. 1.33).

It is worth noticing the similarity between the mean annual temperature trends observed in the three countries, notwithstanding the differences in size, physiography, geographical position and ocean influence. Moreover, it is striking that the time series pattern for each country consists of two distinctive parts, and those of Djibouti and Somalia encompass exactly the same interval (Figs. 1.32 and 1.33).

Unfortunately, no century-long time series of ground measured temperature data are available for the Horn of Africa countries. Nevertheless, the data reported by Fantoli (1940, 1965) can be used to make a comparison with modern averages, typically referred to the last three decades. Fantoli's data cover different intervals within the first 60 years of the twentieth century. Unfortunately, old data from only two stations are available for Eritrea (Asmara and Keren), eight occur for Somalia and none for Djibouti (Table 1.4). Table 1.4 shows that the difference in the mean annual temperature (Tma) between old and recent data is very small. In seven meteo-stations out of ten, Tma was higher in the past. The largest differences are observed at Bardera (today's Tma = −0.75 °C) and Keren (today's Tma = −0.34 °C). The three meteo-station with a Tma of today higher than in the past are Luuq, Baidoa and Hobyo (the largest difference is +0.26 °C). However, if we consider average maximum and minimum annual temperatures, we find that seven meteo-stations out of ten have higher maximum temperatures today, whereas eight meteo-stations record lower minimum temperatures today than in the past. These results also indicate that the temperature range is wider today than in the past at eight out of ten meteo-stations.

The temperature differences for the individual months also are very small, commonly in the 0.1–0.2 °C range. The largest differences are the maximum temperatures in March at Bardera (+2.5 °C in the past) (Fig. 1.34) and at Asmara in September (+1.2 °C today) (Fig. 1.35). Though this comparison between old and recent data is based on a very few meteo-stations, the different conclusions that can be drawn with respect to the gridded data are substantial. These latter data indicate a marked temperature increase with a clear hike

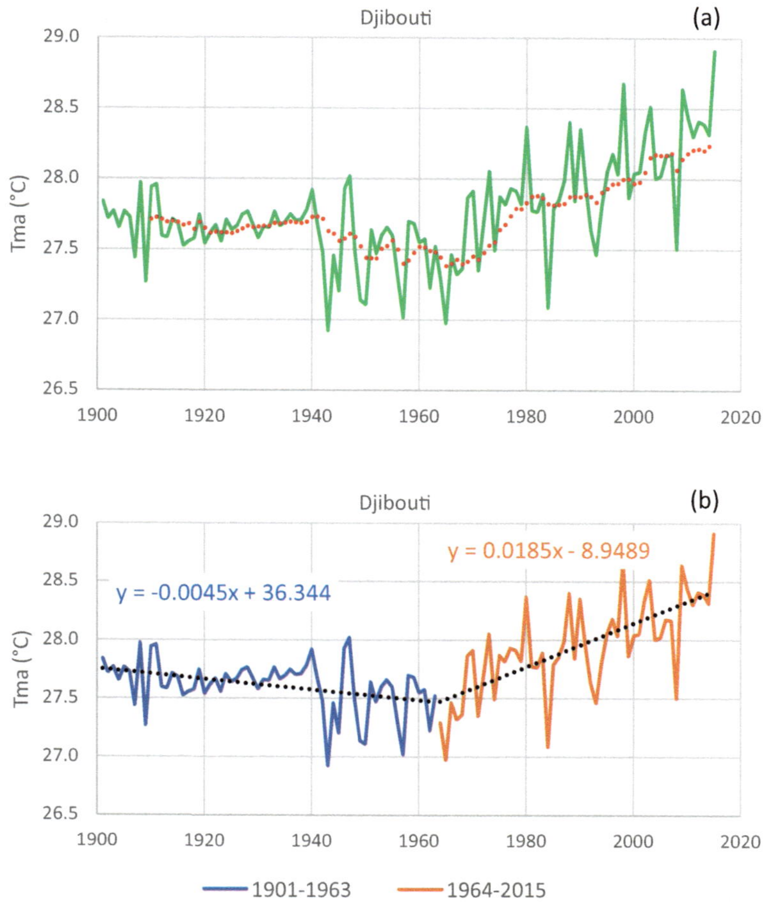

Fig. 1.32 Century long variation of mean annual temperature gridded data in Djibouti: **a** original, time series (the red dotted line is the 10 years moving average); **b** the original time series split into two main portions to highlight the marked temperature increase following the early 1960s

after the early 1960s, whereas the comparison with the meteo-stations raw data seems to indicate no substantial temperature increase. For the meteo-station of Asmara, its position change has already been addressed, but for the other sites no explanation is at hand. The gridded data are anchored on ground instrumental measurements, but in the studied countries, marked differences in local conditions may make it difficult to model data that respond to so many factors. Further studies are, however, necessary to strengthen the validity of temperature gridded data in Eritrea, Djibouti and Somalia.

1.9.2 Precipitation Change

The gridded data indicate a moderate decrease in annual precipitation for Eritrea of about 37 mm during the 1901–2015 interval (Fig. 1.36), which corresponds to a decrease of about 13% with respect to the long-term average. In Djibouti, no substantial change is observed (Fig. 1.37), whereas in Somalia an increasing trend is recorded with a rise, during the last century, of about 13 mm, i.e. about +5% in respect to the long-term average (Fig. 1.38).

In Eritrea, the main rainy season occurs in summer (July–September) and contributes 50–80% of the annual precipitation. Thus, the change in annual precipitation depends strictly on the change in the summer rain (Fig. 1.39) and any negative trend of the latter has negative repercussions for agriculture and water resources. The summer rain data indicate a negative trend with a decrease of about 38 mm that is a rainfall loss of 19% with respect to the long-term average during the 1901–2015 interval (Fig. 1.40). In Djibouti, the summer (July–September) rain shows a marked decreasing trend too, but it is counterbalanced by an increasing trend of precipitation from October to June (Fig. 1.41). In Somalia, the *Gu* (March–May) and the *Deyr* (October–November) rainy seasons together make up about 90% of annual precipitation and both show an increasing trend, though that of the *Deyr* rain is characterised by a higher increasing rate (Fig. 1.42).

As for the temperatures, in order to verify the precipitation trends based on the gridded data, rain gauge measurements used to build old (Fantoli 1940, 1965) and recent (commonly the last three decades) rainfall data are also compared. Only the old data of three meteo-stations (Asmara, Massawa and Akordat) are available and in two

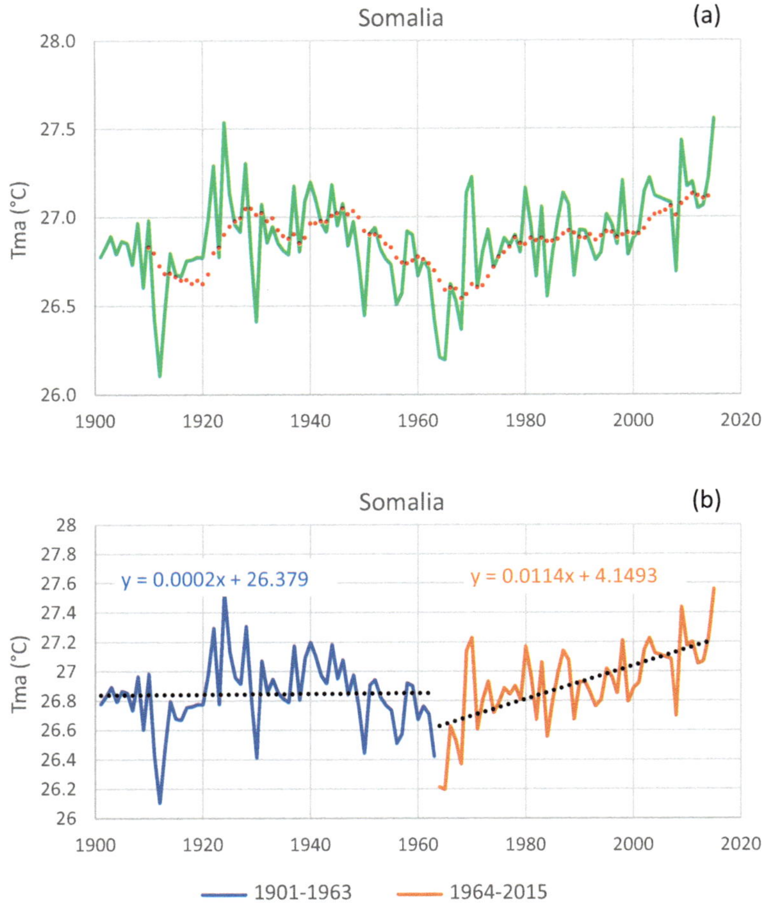

Fig. 1.33 Century long variation of mean annual temperature gridded data in Somalia: **a** original, time series (the red dotted line is the 10 years moving average); **b** the original time series split into two main portions to highlight the marked temperature increase following the early 1960s

of them a significant decrease of rainfall with respect to the past is recorded, particularly at Akordat, where the contemporary annual precipitation is 105.6 mm less than in the past (−22.7 mm in Asmara), whereas in Massawa a small increase of 22 mm in the current precipitation is observed (Table 1.5). Djibouti records the worst decrease of rainfall, −126 mm, whereas Somalia shows a balanced situation, with four meteo-stations out of nine (Mogadishu, Bardera, Beledweyne and Oddur) recording higher annual rainfall today than in the past, four (Hobyo, Baidoa, Luuq, Kismayo) recording a decrease and one (Galkayo) showing no change. The average of the increased amounts is around 55 mm (highest increase 71.5 mm at Oddur), whereas the average decreased amount is about 22 mm (highest decrease −31.3 mm at Luuq).

The largest monthly difference with past rainfall is observed at Bardera and Mogadishu in November, with about 40 mm more today than in the past (Figs. 1.43 and 1.44), and −53 mm in August at Akordat (Fig. 1.45) and −40 mm in May at Kismayo (Fig. 1.46).

Unlike the temperature, the comparison of old and recent instrumental data confirms the trends emerged from the analysis of gridded data. The intervals covered by the old temperature and precipitation data are almost the same and, notwithstanding the non-uniformity of the old data, the better agreement with the precipitation trends seems to indicate that the models used to produce the country-averaged precipitation data are better calibrated for the study area than those which generated the temperature data.

1.9.3 Rainfall Erosivity Change

To investigate the change of rainfall erosivity, only the variation of the MFI calculated from gridded data was used. This choice is dictated by the fact that the *R*-factor (Renard and Freimund 1994) is a direct derivation from annual precipitation and, of course, it reflects exactly the pattern and trends observed for rainfall. The MFI, too, is influenced by the annual rainfall, but this index is more independent since it also considers the monthly distribution of rain.

In Eritrea, the country-averaged annual values of the MFI show a decreasing trend and the index, with the exception of a few peaks, is permanently less than 90, i.e. the value which identifies the threshold between moderate to low rainfall erosivity (Table 1.1) (the blue dashed line in Fig. 1.47). In Djibouti, by contrast, a decreasing trend of MFI is very marked, leading to conditions of very low rainfall erosivity

Table 1.4 Comparison of old temperature data (Fantoli 1940, 1965) with recent data, typically referred to the last three decades. Temperatures in °C

Station	Time range	Average max *T*	Average min *T*	Mean Tma
Asmara	1890–1930	23.2	10.3	16.8
	Recent	23.7	9.5	16.6
Keren	1919–1937	30.9	16.1	23.5
	Recent	30.6	15.7	23.2
Mogadishu	1911–1958	30.2	23.9	27.1
	Recent	30.6	23.5	27.0
Hobyo	1925–1958	30.2	23.1	26.6
	Recent	30.5	22.9	26.7
Bardera	1915–1958	36.3	21.8	29.0
	Recent	35.4	21.2	28.3
Baidoa	1922–1958	32.2	20.2	26.2
	Recent	32.6	20.3	26.4
Luuq	1923–1958	37.1	24.0	30.5
	Recent	38.3	23.0	30.7
BeledWeyne	1927–1958	34.9	22.6	28.7
	Recent	34.8	22.5	28.6
Galkayo	1933–1958	33.9	20.9	27.4
	Recent	33.7	21.0	27.4
Kismayo	1933–1958	29.7	24.2	27.0
	Recent	30.2	23.7	27.0

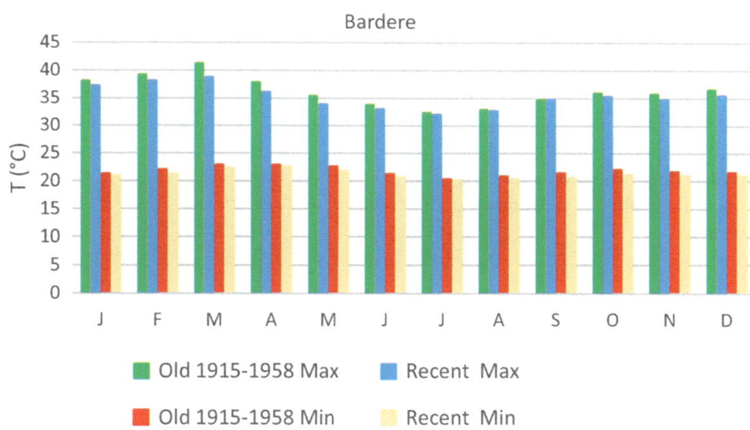

Fig. 1.34 Comparison between old (Fantoli 1940, 1965) and recent (the last three decades) data of mean monthly maximum and minimum temperature measured at Bardera in Somalia

(Fig. 1.48). Somalia is characterised by an increasing trend of the MFI, though its values are constantly less than 60, which is the threshold between low and very low erosivity (Fig. 1.49).

As expected, these results confirm that the MFI is tightly linked with rainfall and its trends cannot but confirm those of annual precipitation. However, a more interesting result is the unexpected, insufficient capability of the MFI in addressing rainfall erosivity in the studied countries, where severe hydro-meteoric erosion is common and is the most important erosion process which causes extensive land degradation (Fig. 1.18). The country-averaged MFI values are low and likely underrepresent the actual situation of rainfall erosivity. The analysis of the *R*-factor data lead to the same qualitative conclusion and further studies are necessary to evaluate the capability of these two parameters in addressing rainfall erosivity, at least, in the eastern Horn of Africa.

1.9.4 Desertification Trends

The UNEP (1992) adopted the ratio between annual precipitation (*P*) and potential evapotranspiration (PET) as a basic parameter to define the aridity of an area. The UNEP

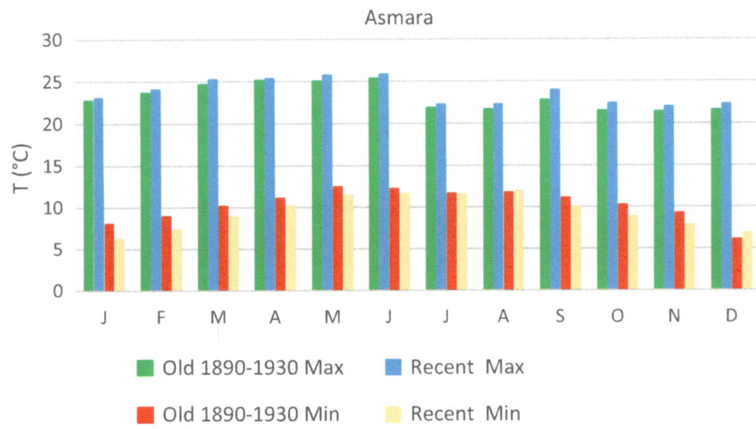

Fig. 1.35 Comparison between old (Fantoli 1940, 1965) and recent (the last three decades) data of mean monthly maximum and minimum temperature measured at Asmara in Eritrea

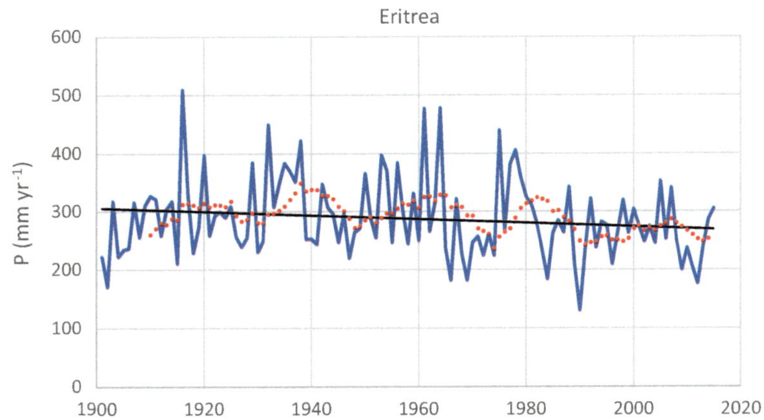

Fig. 1.36 Century long time series of annual precipitation gridded data of Eritrea (the red dotted line is the 10 years moving average)

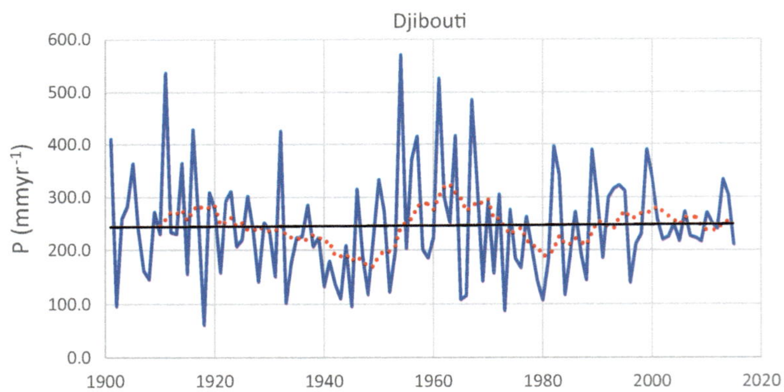

Fig. 1.37 Century long time series of annual precipitation gridded data of Djibouti (the red dotted line is the 10 years moving average)

indicates the value of 0.6 of the desertification index (DI = *P*/PET) as the threshold below which an area is under arid or very arid conditions or at risk of desertification. The DI time series based on the gridded data (World Bank 2019) of the three countries show that they were and still are under arid and semi-arid conditions leading to desertification, since their DI values are constantly very low and less than 0.6. Mean DI is, in fact, 0.17, 0.12 and 0.15 in Eritrea, Djibouti and Somalia, respectively.

Eritrea shows a clear negative trend (Fig. 1.50), which implies a general worsening of water deficit and severe limitations to farming. This result is the combination of the negative trend in annual precipitation with an increasing trend of the mean annual temperature. Djibouti is in a similar situation and, though the negative trend is gentler than in Eritrea, its starting conditions are that of an arid country, in which desertification is already a matter of fact (Fig. 1.51). Somalia has been also facing aridity and desertification conditions throughout the last century, but in this case the DI does not show any trend (Fig. 1.52).

The analysis of the De Martonne (1925) Aridity Index (AI) time series calculated on the basis of gridded data (World

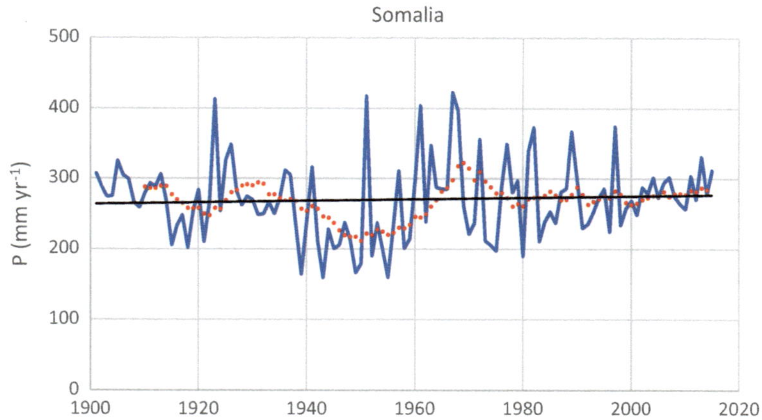

Fig. 1.38 Century long time series of annual precipitation gridded data of Somalia (the red dotted line is the 10 years moving average)

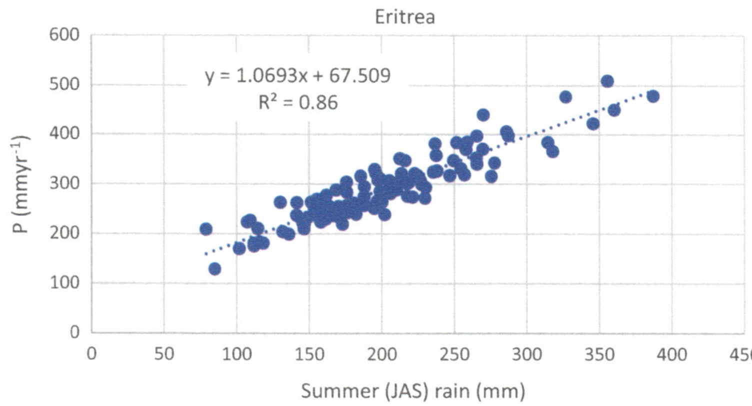

Fig. 1.39 In Eritrea, the summer rains (July, August and September) contribute for about 60–70% of the annual precipitation and explain 86% of its variability

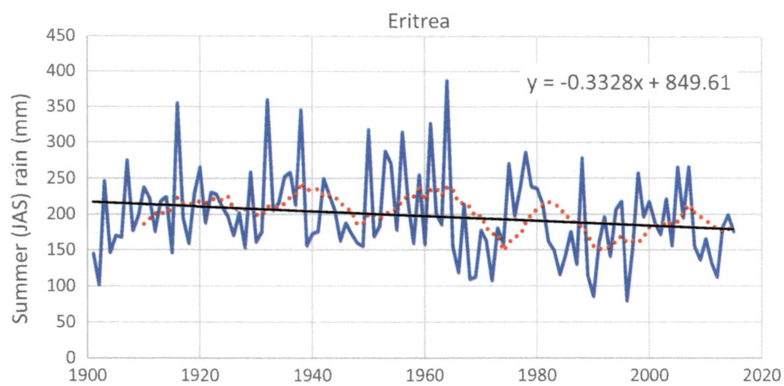

Fig. 1.40 Long-term variability of the summer rains (July, August and September) gridded data in Eritrea. A decreasing trend is evident, and it is confirmed by the 10 years moving average (the red dotted line)

Bank 2019) confirms the arid and semi-arid conditions of the region and the unavoidable necessity of irrigation for any farming development. The AI values, with only sporadic peaks, are commonly comprised within the 5–10 interval, which implies aridity conditions in which irrigation is indispensable. In Eritrea, the trend is negative (Fig. 1.53) and foreshadows worsening of the already difficult conditions of water deficit and scarce yields without an appropriate development of irrigation schemes. The AI time series of Djibouti does not show an evident trend, but the AI values, with the exception of a few peaks higher than 10, but less than 20 (Table 1.2), are almost all comprised within the 5–10 range, confirming arid and semi-arid conditions and indispensable irrigation (Fig. 1.54). By contrast, the AI data of Somalia indicate an increasing trend (Fig. 1.55), which is a positive situation for the agriculture of this country, but at least for a few decades, the AI values will be still below the threshold value of 10 and irrigation will continue to be necessary.

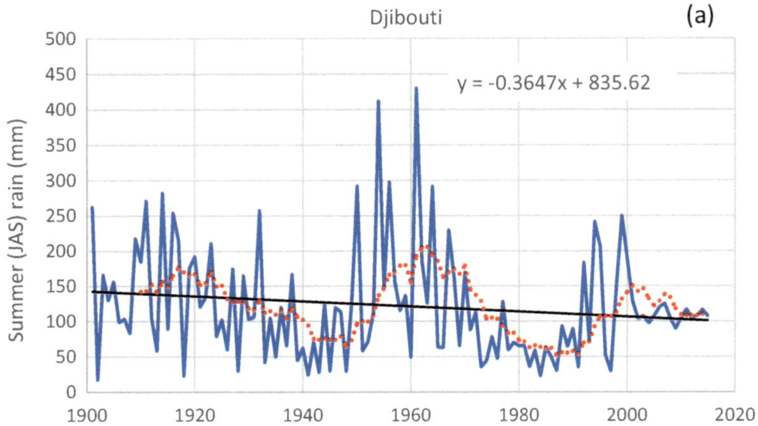

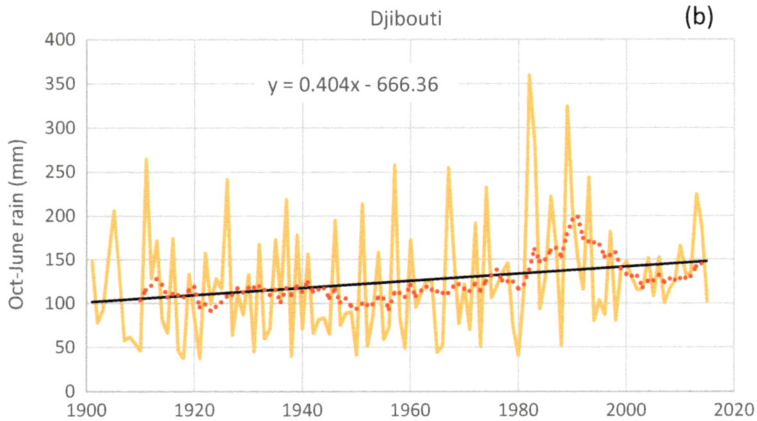

Fig. 1.41 Long-term variability of seasonal rains gridded data in Djibouti. The summer rains (July, August and September) time series (**a**) is characterised by a modest decrease, which is counterbalanced by the marked increase in the other months (October–June) rains (**b**). The red dotted lines are the 10 years moving averages

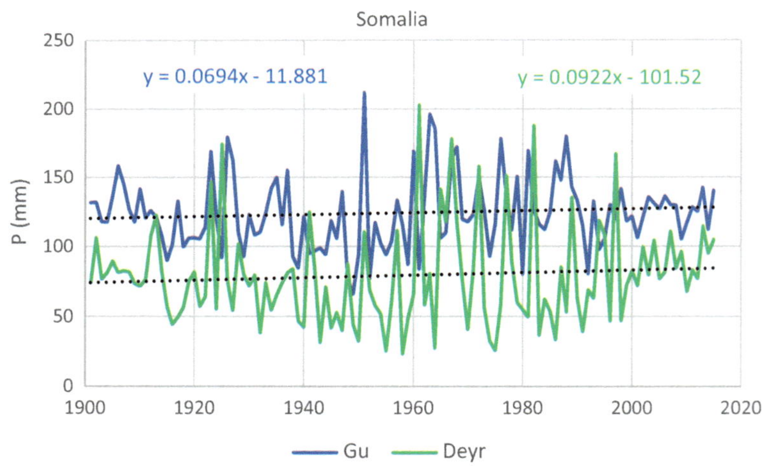

Fig. 1.42 Long-term variability of the two main rainy seasons in Somalia (gridded data). Both the *Gu* (March–May) and *Deyr* (October–November) rains show an increasing trend

1.10 Discussion

The climatic profile of the countries of Horn of Africa is that typical of arid environments, though sub-humid and hyper-arid areas are present as well. The annual temperature range is small if compared with continental and high-latitude climates. This result is not unexpected given the position within the equatorial realm and the influence of the ocean that is commonly less than 300 km far from the most inland meteo-stations. In Eritrea, the annual temperature range tends to decrease with elevation (R^2 = 0.61) (Fig. 1.8) and to increase with the mean annual temperature (R^2 = 0.57) (Fig. 1.9). The first result is opposite to the findings of Bandopadhyay (2016) for west Bengal. This author used the data of 20 meteo-stations, but the interval analysed was only 15 years (2000–2014). In Eritrea, the largest precipitation occurs in summer (July and August) on the highlands and

Table 1.5 Comparison of old, annual precipitation data (Fantoli 1940, 1965) with recent data, typically referred to the last three decades

Station	Time range	Annual *P* (mm)	Recent − Old (mm)
Asmara	1890–1930	491.7	−22.7
	Recent	469.0	
Akordat	1922–1935	416.6	−105.6
	Recent	411.0	
Massawa	1885–1915	163.0	+23.0
	Recent	186.0	
Djibouti	1900–1920	126.0	−126.0
	Recent	110.2	
Mogadishu	1910–1959	399.5	+66.5
	Recent	466.0	
Hobyo	1924–1958	199.8	−19.8
	Recent	180.0	
Bardera	1915–1960	384.3	+17.7
	Recent	402.0	
Baidoa	1922–1960	585.0	−19.0
	Recent	566.0	
Luuq	1922–1960	303.3	−31.3
	Recent	272.0	
Beledweyne	1927–1958	204.0	+66.0
	Recent	270.0	
Galkayo	1934–1958	148.0	−1.0
	Recent	147.0	
Kismayo	1896–1913	3836	−17.6
	Recent	366.0	
Oddur	1924–1937	377.2	+71.5
	Recent	448.7	

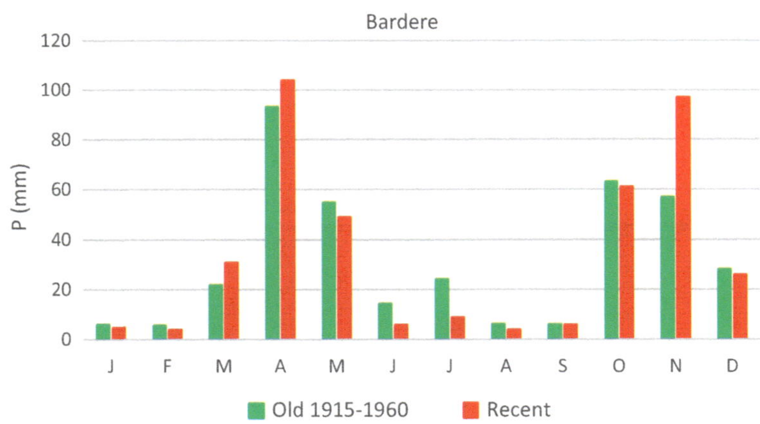

Fig. 1.43 Comparison between old (Fantoli 1940, 1965) and recent (the last three decades) data of mean monthly precipitation measured at Bardera in Somalia

the longer persistence of cloudy conditions contribute to reduce the daily temperatures, though in summer the insolation is expected to be higher (the hottest months are May and June). In the hottest areas, rainfall is scarce, but it occurs mainly in winter (December–February). The temperature reduction associated with precipitation is small but it occurs in the coolest months; hence, a larger than average range of the annual temperature is recorded in the hottest areas, where vegetation is scarce or absent, desert conditions prevail, and the albedo is very high in the other months.

Fig. 1.44 Comparison between old (Fantoli 1940, 1965) and recent (the last three decades) data of mean monthly precipitation measured at Mogadishu in Somalia

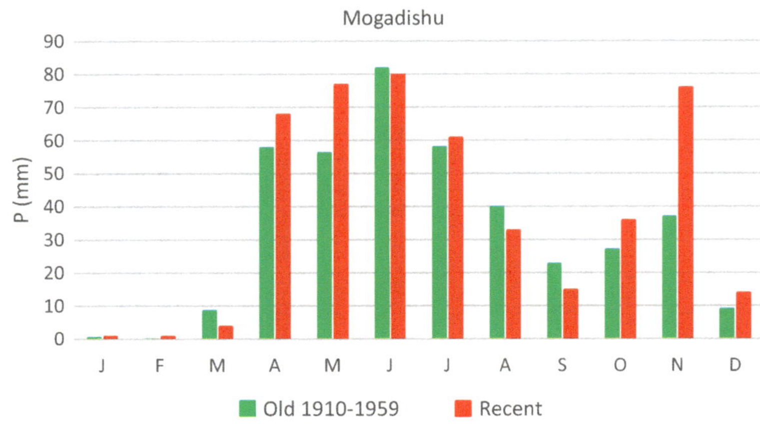

Fig. 1.45 Comparison between old (Fantoli 1940, 1965) and recent (the last three decades) data of mean monthly precipitation measured at Akordat in Eritrea

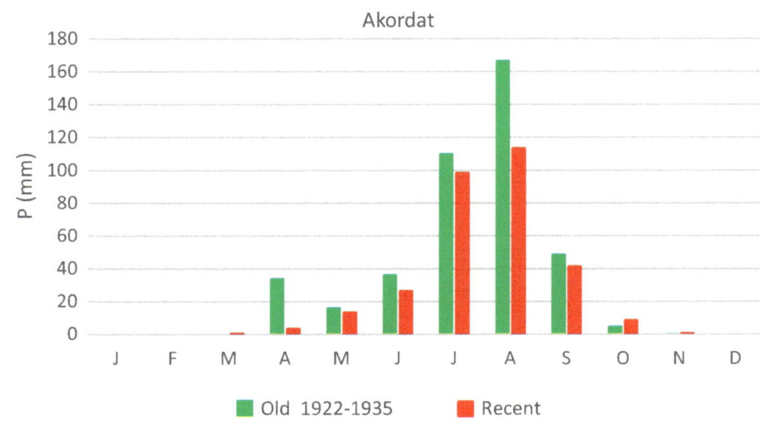

Fig. 1.46 Comparison between old (Fantoli 1940, 1965) and recent (the last three decades) data of mean monthly precipitation measured at Kismayo in Somalia

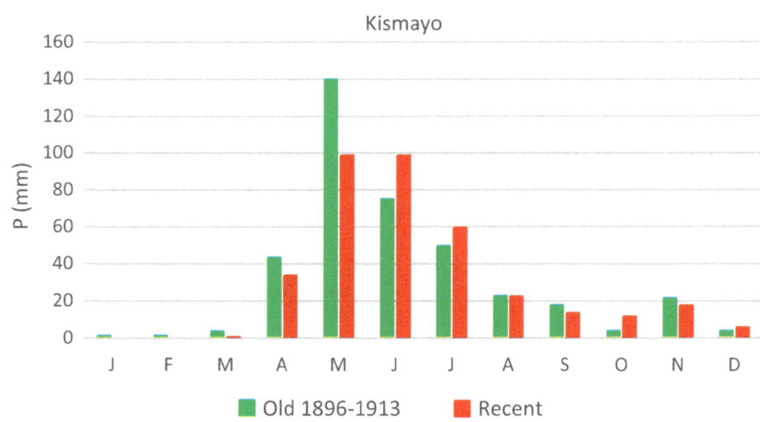

No correlation was instead found for Somalia and Djibouti. In Somalia, with the exception of the northern escarpment, the relief contrast is much smaller compared to Eritrea, whereas the mitigation effect of the Indian Ocean, combined with the cooling effect of the oceanic winds, is significant in a relatively flat land.

Djibouti shows a distinctive ($R^2 = 0.88$) annual precipitation gradient (Fig. 1.18), whereas elevation has a poor control on annual precipitation in Eritrea ($R^2 = 0.49$). The physiography of Djibouti is similar to that of the Eritrean escarpment, but here the cool moist winter winds results in larger amounts of annual rains than those associated with the monsoon in the highlands. In Eritrea, the meteo-stations with the largest annual precipitation, in fact, are located on the escarpment at intermediate elevations (1000–1500 m asl) compared to the highlands (around 2500 m asl). In Somalia, no clear rainfall gradient was found. However, if only the data from meteo-stations at elevation <500 m asl are considered and those of dry and hot coastal areas and those of the Bur region are excluded, a weak, inverse precipitation

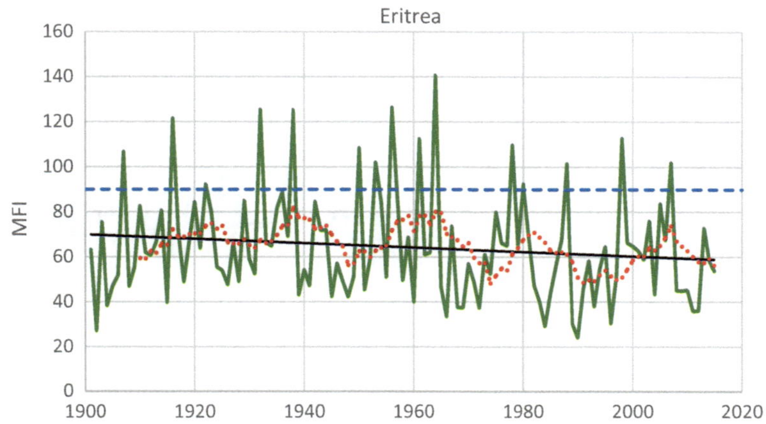

Fig. 1.47 Long-term variability of rainfall erosivity calculated by the Modified Fournier Index (MFI) (Arnoldus 1980) for Eritrea on the base of precipitation gridded data. The blue dashed line identifies the threshold between moderate to low rainfall erosivity (Table 1.1). The red dotted line is the 10 years moving average

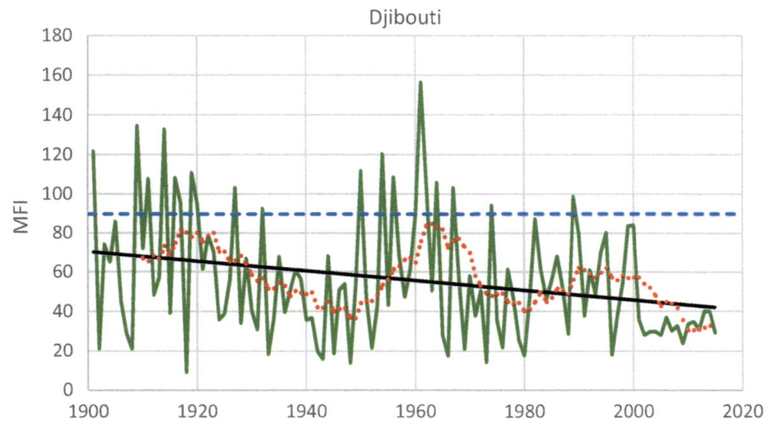

Fig. 1.48 Long-term variability of rainfall erosivity calculated by the Modified Fournier Index (MFI) (Arnoldus 1980) for Djibouti on the base of precipitation gridded data. The blue dashed line identifies the threshold between moderate to low rainfall erosivity (Table 1.1). The red dotted line is the 10 years moving average

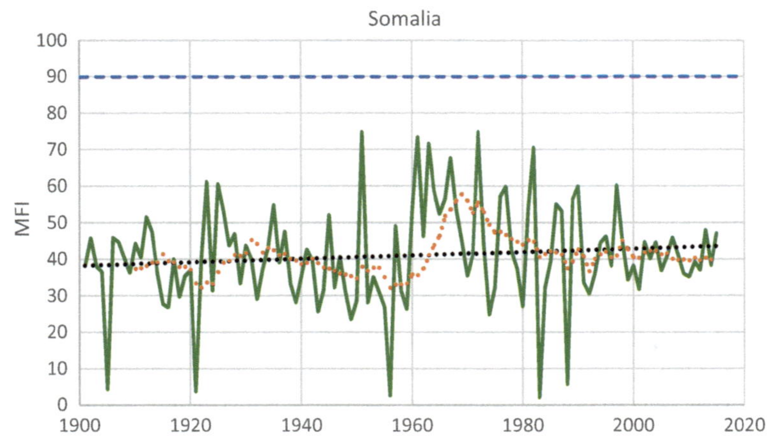

Fig. 1.49 Long-term variability of rainfall erosivity calculated by the Modified Fournier Index (MFI) (Arnoldus 1980) for Somalia on the base of precipitation gridded data. The blue dashed line identifies the threshold between moderate to low rainfall erosivity (Table 1.1). The red dotted line is the 10 years moving average

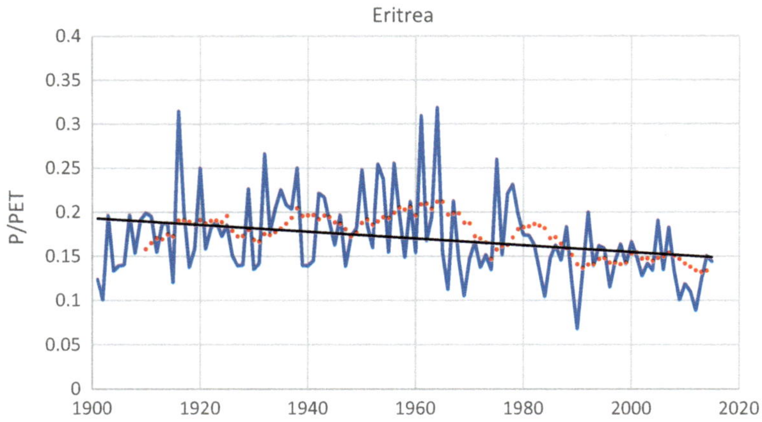

Fig. 1.50 Long-term variability of the UNEP's (1992) desertification index (DI = *P*/PET in which *P* is annual precipitation and PET is annual potential evapotranspiration) calculated for Eritrea on the base of gridded data. The red dotted line is the 10 years moving average

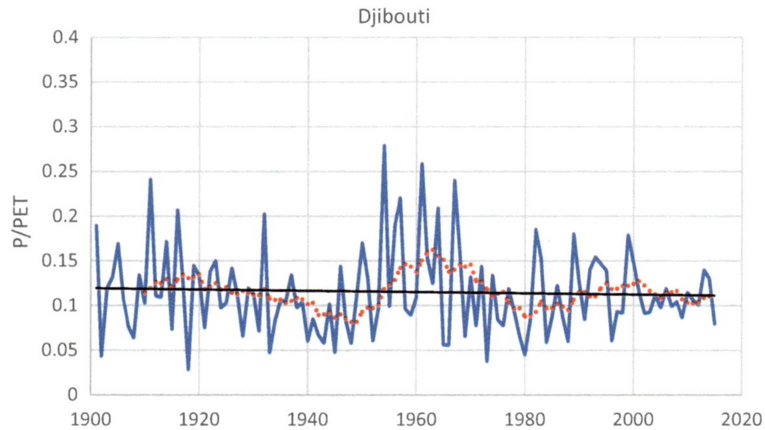

Fig. 1.51 Long-term variability of the UNEP's (1992) desertification index (DI = *P*/PET in which *P* is annual precipitation and PET is annual potential evapotranspiration) calculated for Djibouti on the base of gridded data. The red dotted line is the 10 years moving average

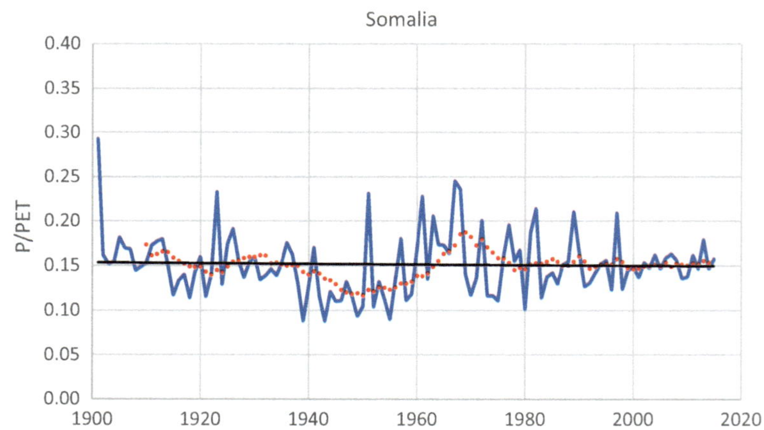

Fig. 1.52 Long-term variability of the UNEP's (1992) desertification index (DI = *P*/PET in which *P* is annual precipitation and PET is annual potential evapotranspiration) calculated for Somalia on the base of gridded data. The red dotted line is the 10 years moving average

gradient is obtained (Fig. 1.19). Such unusual result was also reported by Basist and Bell (1994) for British Columbia. These authors, however, provide the generic explanation that elevation is a comparatively weak overall predictor variable. This conclusion may be right for British Columbia, and the climate of which is very different from that of Somalia. This very unusual result obtained for Somalia may depend on the high spatial variability of precipitation across the country, which is also influenced by the distance from the ocean. A large portion of Somalia is a uniform tableland, gently inclined towards the ocean and the elevation of inner meteo-stations is typically lower than 500 m asl. The influence of moist winds coming from ocean tends to progressively decrease as they move to the hotter inland and lose humidity all the way long. A similar pattern was found by Heyward and Clarke (1996) in Sierra Leone. According to these authors, rainfall in the monsoon months tended to decrease with distance from the sea and the ocean facing

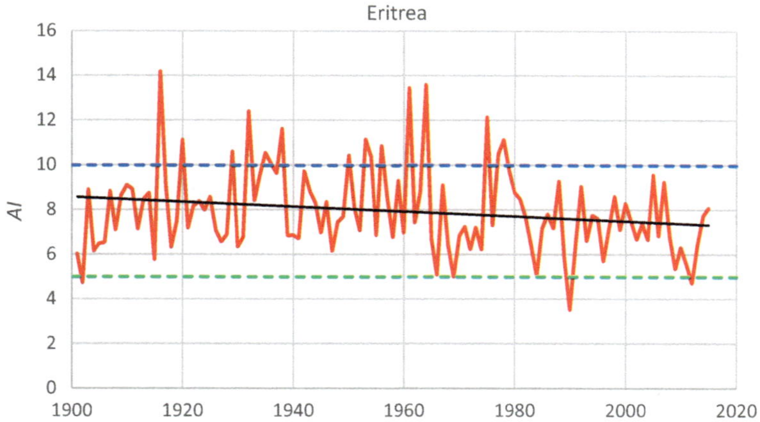

Fig. 1.53 Long-term variability of the De Martonne's (1926) aridity index calculated for Eritrea on the base of gridded data. The blue and the green dashed lines indicate the thresholds below which irrigation is very useful (AI = 10) and indispensable (AI = 5), respectively (Table 1.2). Irrigation is becoming more and more necessary

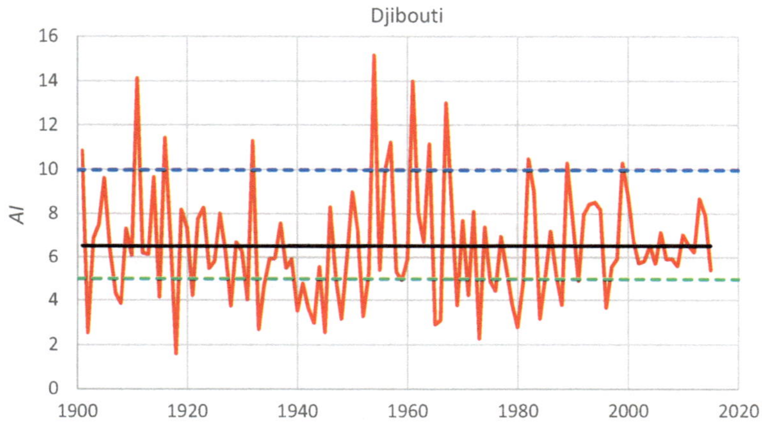

Fig. 1.54 Long-term variability of the De Martonne's (1926) aridity index calculated for Djibouti on the base of gridded data. The blue and the green dashed lines indicate the thresholds below which irrigation is very useful (AI = 10) and indispensable (AI = 5), respectively (Table 1.2)

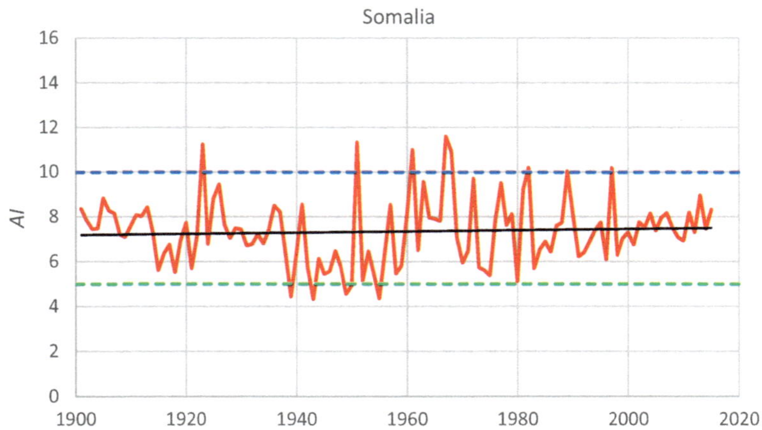

Fig. 1.55 Long-term variability of the De Martonne's (1926) aridity index calculated for Somalia on the base of gridded data. The blue and the green dashed lines indicate the thresholds below which irrigation is very useful (AI = 10) and indispensable (AI = 5), respectively (Table 1.2)

gauges were found to have a substantial control on the regression fit.

The average rainfall erosivity calculated for Eritrea, Djibouti and Somalia is 890, 361 and 559 MJ mm ha^{-1} h^{-1} yr^{-1}, respectively. These values are comparable with those calculated for central and eastern Europe (Panagos et al. 2015). A comparison of the erosional morphology of central Europe with that of the Horn of Africa countries, however, suggest a higher degree of land degradation and the occurrence of more intense and extensive erosion processes in Eritrea, Djibouti and Somalia, though other important factors such as sparse vegetation and the occurrence of poorly structured soils, especially in the drier areas, are to be taken into consideration as well. Nevertheless, the moderate to low rainfall erosivity values calculated with the alternative methods described in Sect. 1.2.2 (Arnoldus's 1980, MFI and the *R*-Factor calculated with the method of Renard and Freimund 1994) seem to support the conclusion of Lee and Heo (2011) that these alternative methods are probably not very effective in addressing the real amount of energy transferred by the rain to the soil. In many parts of the region, human impact also seems to play an important role in accelerating land degradation. In arid and semi-arid regions, vegetation is sparse and rainfall is scarce, erratic, but may be very intense. Under such bioclimatic conditions, the effectiveness of the MFI and the *R*-factor in addressing rainfall erosivity likely needs to be calibrated as the basic USLE or RUSLE methods should be used primarily to provide reference values of rainfall erosivity. Further studies on this topic are however needed.

The climate variations throughout the past century were analysed using both gridded data and a small number of representative meteo-stations data. The two data sets returned different results. The gridded data show a very clear and marked increase in temperature, whereas the ground-based instrumental data indicate that in the recent decades mean temperatures are imperceptibly higher than in the past (about a century ago). The gridded data of temperature show clearly that in the early 1940s something has caused a marked drop in Eritrea's mean temperature and the time series is broken into two distinct portions (Fig. 1.31). A similar situation, though not so clearly marked, occurred in the mean annual temperature for Djibouti and Somalia around the early 1960s (Figs. 1.32 and 1.33). In Eritrea, both portions of the time series show an increasing trend, whereas in Djibouti and Somalia a marked increasing trend is evident only after the early 1960s. The reason for a drop of 2 °C in mean annual temperature of Eritrea is not known. The coordinates of Asmara and Massawa meteo-stations reported by Fantoli (1940) are different from those of today. We can expect that other meteo-stations, on which the gridded data are based, were moved. Such a repositioning could also explain the slightly higher mean annual temperatures recorded in Eritrea before the 1940s (mean Ta = 26.2 °C) compared to those of the 1943–1985 interval (mean Ta = 25.9 °C). The temperatures of the last three decades, however, are higher (mean Ta = 26.8 °C) and show an increasing trend very similar to that observed for the data predating the 1940s (Fig. 1.31). The mean annual temperature time series of Djibouti and Somalia show a marked increasing trend after the early 1960s, but in the case of these countries no information about meteo-station repositioning is available. Given the vicinity of Djibouti to Eritrea, the similarity in the general climate conditions, and the similarity between the temperature patterns in Somalia and Djibouti it is difficult to attribute the drop in the mean temperature gridded data of Eritrea to natural causes. Probably other human-induced changes such as the replacement of old thermometers with more modern temperature recorders should be taken into consideration. This hypothesis seems to be confirmed by the gridded data time series of precipitation in which, unlike temperature time series, marked break points are not present. Further studies are, however, necessary to strengthen the validity of temperature gridded data.

Gridded and ground rain gauge data of annual precipitation show similar trends: a moderate decrease in Eritrea, no substantial change in Djibouti and an increasing trend in Somalia. Though in the last two decades Somalia has been affected by several devastating floods, the result of an increasing trend in annual precipitation is not expected and it is in contrast with the findings of many authors reporting worrying decreasing trends of rainfall in sub-Saharan countries (e.g. Agnew and Chappell 2000; Hulme 2003; Niang et al. 2014). According to Souverijns et al. (2016), the increase in precipitation in southern Somalia is due to the increase in temperature which results in enrichment of water vapour content and intensification of moisture transport in the Hadley circulation (Niang et al. 2014). This hypothesis is partly supported by Liebman et al. (2014), especially as regards the observed increase in the October–December precipitation (*Deyr*, small rains, in Somalia), which is mainly influenced by warming of the western Indian Ocean. Climate models predict that the sea surface temperature of the Indian Ocean will continue to increase (Meehl et al. 2007), but the link, if any, between the expected decrease of rainfall in the Horn of Africa, and especially in Ethiopia (Williams et al. 2012) and the increase in precipitation in southern Somalia is not clear yet.

1.11 Conclusions

Eritrea, Djibouti and Somalia are mainly subjected to a hot and semi-arid climate, though in some small areas sub-humid and hyper-arid conditions are present as well. The three countries have hot climate as more than half of the meteo-stations record a mean annual temperature higher than 26 °C. Monthly temperatures higher than 40 °C were measured at Luuq (42.2 °C in March) in Somalia, at Djibouti City (41.3 °C in July) and at Tennessee (40.9 °C in May) in Eritrea. Daily peaks of 47–49 °C were recorded at Massawa and Assab (also reaching 50 °C and probably even more in the Danakil), whereas lows below zero (as much as −6 °C) are not uncommon in Asmara and in the Eritrean eastern–southeastern highlands. The decrease of mean annual temperature with elevation is well established, with a decrease of 5.6, 5.9 and 5.3 °C every 1000 m in Eritrea, Djibouti and

Somalia, respectively. The analysis of century-long time series of gridded data showed that the distribution of mean annual temperature consists of two main patterns, the most recent of which is characterised by a marked increase in temperature (2 °C from the early 1940s to 2015 in Eritrea, 1 °C and 0.6 °C from early 1960 to 215 in Djibouti and Somalia, respectively). By contrast, the comparison of old data reported by Fantoli (1940, 1965) with the instrumental data for the last three decades indicates an imperceptible increase of temperature. Such discrepancy may depend on the use of more accurate recording instrumentation replacing old thermometers in the last decades, but further investigation on this problem is necessary.

Rain gauge data indicate that on average Eritrea receives the largest amount of rain, 421.9 mm yr^{-1}, followed by Somalia, 308.7 mm yr^{-1}, and Djibouti, 240.3 mm yr^{-1}. Large differences in annual precipitation are, however, observed between areas located in particular physiographic conditions or subjected to hot and dry climate as the Danakil and the coastal belt of Eritrea, Djibouti and northern Somalia (from the 766 mm yr^{-1} at Nefasit, located on the main escarpment, about 14 km east of Asmara to less than 50 mm yr^{-1} in the Danakil depression or the Daban basin south of Berbera in northern Somalia). Hourly rainfall intensity data are not available for the study area, but daily rainfall values ranging between 80 and 120 mm were recorded in the highlands of Eritrea and northern Somalia. Actually, field experience suggests that these daily rain amounts may have fallen in a much shorter time (1–6 h), resulting in very aggressive rainfalls. Rainfall erosivity was calculated using the Modified Fournier Index (MFI) (Arnoldus 1980) and the *R*-Factor of Renard and Freimund (1994). The results correspond to moderate erosivity, though large parts of the three countries are affected by severe land degradation and the landscape is deeply sculptured and shaped by active hydro-meteoric erosion. Though other factors, not last the human impact, should be considered as well, the lower than expected values of rainfall erosivity suggest to consider the MFI and the Renard and Freimund *R*-Factor methods as a preliminary approach to obtain just an approximate information about rainfall erosivity in arid and semi-arid areas. Further studies in different drylands are, however, necessary to calibrate the effectiveness of these methods, alternative to the detailed rain gauge recordings, to catch the actual rainfall erosivity in regions with little annual precipitation and a monsoon rainfall regime.

The analysis of 1901–2015 gridded data time series indicates a moderate decrease in annual precipitation for Eritrea, no substantial change in Djibouti and an increase in Somalia. A similar pattern is confirmed by comparing the old data of Fantoli (1940, 1965) with the rain gauge data for the last three decades. Both the decrease and the increase in rain in Eritrea and Somalia, respectively, are attributed by a few authors to the higher than average warming of the Indian Ocean.

Both the UNEP (1992) and the De Martonne (1925) aridity indices confirm that the studied countries are presently subjected to desertification conditions and that irrigation is indispensable to rise the agriculture productivity beyond the sustenance level. Eritrea shows a clear negative trend of the Desertification Index (DI) (UNEP 1992) which, combined with the projected decrease in annual precipitation and an increase in mean annual temperature, will lead to general worsening of the water deficit and to severe farming limitations. Similar is the situation of Djibouti and, though the negative trend of DI is gentler than in Eritrea, this result is not comforting since Djibouti is an arid country in which desertification is already a matter of fact. Somalia is also facing aridity and desertification conditions throughout the last century, but in this country the DI does not show any trend.

Acknowledgements The authors is grateful to Paolo Paron for providing useful information about climate data sources and the FAO-SWALIM report series.

References

Agnew CT, Chappell A (2000) Desiccation in the Sahel. In: McLaren SJ, Kniveton DR (eds) Linking climate change to land surface change. Kluwer Academic Publishers, New York, pp 27–48

Arnoldus HMJ (1980) An approximation of the rainfall factor in the universal soil loss equation. In: De Boodt M, Gabriels D (eds) Assessment of erosion. Wiley, Chichester, pp 127–132

ARPAV (2019) http://www.arpa.veneto.it/temi-ambientali/agrometeo/dati/cartografia/de-martonne

Baltas EA (2008) Climatic conditions and availability of water resources in Greece. Int J Water Resour Dev 24(4):635–649

Bandopadhyay S (2016) Does elevation impact local level climate change? An analysis based on fifteen years of daily diurnal data and time series forecasts. Pac Sci Rev A Nat Sci Eng 18(3):241–253

Basist A, Bell GD (1994) Statistical relationships between topography and precipitation patterns. J Clim 7:1305–1315

Behera SK, Luo JJ, Masson S, Delecluse P, Gualdi S, Navarra A, Yamagata T (2005) Paramount impact of the Indian Ocean dipole on the East African short rains: ACGCM study. J Clim 18:4514–4530

Berkeley Earth (2019) http://berkeleyearth.org/source-files/

CORINE (1992) Soil erosion risk and important land resources in the southern regions of the European Community. EUR 13233 EN. Office for Official Publications of the European Community, Luxembourg

De Martonne E (1925) Traité de Geographie Physique. A. Colin, Paris

De Martonne E (1926) Une nouvelle function climatologique: L'indice d'aridité. La Meteorologie 2:449–458

Edwards FJ (1987) Climate and oceanography. In: Edwards FJ, Head SM (eds) Red Sea. Pergamon Press, Oxford.

Ethiopian Mapping Authority (EMA) (1988) The national atlas of Ethiopia. EMA, Addis Ababa

EU (2007) Climate change impacts on developing countries—EU accountability. European Parliament's Committee on the Environment, Public Health and Food Safety, Brussels

Fantoli A (1940) Elementi preliminari del clima dell'Etiopia. Sansoni, Firenze

Fantoli A (1965) Contributo alla climatologia della Somalia: riassunto dei risultati e tabelle meteorologiche e pluviometriche. Ministero degli Affari Esteri, Cooperazione Scientifica e Tecnica, Roma

Fazzini M, Bisci C, Billi P (2015) The climate of Ethiopia. In: Billi P (ed) Landscapes and landforms of Ethiopia. Springer, Heidelberg, pp 65–88

Fessehaye M, Brugnara Y, Savage MJ, Brönnimann S (2019) A note on air temperature and precipitation variability and extremes over Asmara: 1914–2015. Int J Climatol 39:5215–5227

Gabriels S (2006) Assessing the modified Fournier index and the precipitation concentration index for some European countries. In: Boardman J, Poesen J (eds) Soil erosion in Europe. Wiley, Chichester, pp 675–684

Goddard L, Graham NE (1999) Importance of the Indian Ocean for simulating precipitation anomalies over eastern and southern Africa. J Geophys Res 104:19099–19116

Government of Eritrea (2007) Wind energy applications in Eritrea: mid-term review. Asmara

Hadden RL (2007) The geology of Somalia: a selected bibliography of Somalian geology, geography and earth science. Topographic Engineering Center, US Army Corps of Engineers, Alexandria, VA

Harris I, Jones PD, Osborn TJ, Lister DH (2014) Updated high-resolution grids of monthly climatic observations—the CRU TS3.10 dataset. Int J Climatol 34(3):623–642

Heyward D, Clarke RT (1996) Relationship between rainfall, altitude and distance from the sea in the Freetown Peninsula, Sierra Leone. Hydrol Sci J 41(3):377–384

Hulme M (2003) Climate change within the period of meteorological records. In: Adams WM, Goudie AS, Orme AR (eds) The physical geography of Africa. Oxford University Press, Oxford, pp 88–102

Huq S, Reid H, Murray LA (2006) Climate change and development links. Gatekeeper 123, London

Hutchinson P (1992) The southern oscillation and prediction of "Der" season rainfall in Somalia. J Clim 5:525–531

Indeje M, Semazzi FHM, Ogallo LJ (2000) ENSO signals in East African rainfall seasons. Int J Climatol 20:19–46

IPCC (2019) Climate change and land. Intergovernmental Panel on Climate Change, Geneva

Lancaster N (2011) Desert dune processes and dynamics. In: Thomas DSG (ed) Arid zone geomorphology: process, form and change in drylands. Wiley, Chichester

Ledesma JLJ, Futter M (2017) Gridded climate data products are an alternative to instrumental measurements as inputs to rainfall–runoff models. Hydrol Process 31:3283–3293. https://doi.org/10.1002/hyp.11269

Lee JH, Heo JH (2011) Evaluation of estimation methods for rainfall erosivity based on annual precipitation in Korea. J Hydrol 409:30–48

Liebmann B, Hoerling MP, Funk C, Bladé I, Dole RM, Allured D, Quan X, Pegion P, Eischeid JK (2014) Understanding recent eastern Horn of Africa rainfall variability and change. J Clim 27:8630–8644

Lott FC, Christidis N, Stott PA (2013) Can the 2011 East African drought be attributed to human-induced climate change? Geophys Res Lett 40:1177–1181

Marthews TR, Otto FEL, Mitchell D, Dadson SJ, Jones RG (2015) The 2014 drought in the Horn of Africa: attribution of meteorological drivers. In: Herring SC, Hoerling MP, Kossin JP, Peterson TC, Stott PA (eds) Explaining extreme events of 2014 from a climate perspective. Spec Suppl Bull Am Meteorol Soc 96(12):83–88

Mason SJ, Goddard L (2001) Probabilistic precipitation anomalies associated with ENSO. Bull Am Meteor Soc 82:619–638

Meehl GA, Stocker TF, Collins WD, Friedlingstein AT, Gaye AT, Gregory JM, Kitoh A, Knutti R, Murphy JM, Noda A, Raper SCB, Watterson IG, Weaver AJ, Zhao ZC (2007) Global climate projections. In: Solomon S, Qin D, Manning M, Chen Z, Marquis M, Averyt KB, Tignor M, Miller HL (eds) Climate change 2007: the physical science basis. Contribution of working group I to the fourth assessment report of the Intergovernmental Panel on Climate Change. Cambridge University Press, Cambridge, pp 749–845

Menne MJ, Durre I, Vose RS, Gleason BE, Houston TG (2012) An overview of the global historical climatology network-daily database. J Atmos Ocean Technol 29:897–910

Middleton N, Thomas D (1997) World atlas of desertification. Arnold, London

Muchiri PW (2007) Climate of Somalia. Technical report No W-01. FAO-SWALIM, Nairobi

Nana-Sinkam SC (1995) Land and environmental degradation and desertification in Africa: issues and options for sustainable economic development with transformation. Joint ECA/FAO agriculture division monograph no. 10. Addis Ababa

Niang I, Ruppel OC, Abdrabo MA, Essel A, Lennard C, Padgham J, Urquhart P (2014) Africa. In: Barros VR, Field CB, Dokken DJ, Mastrandrea MD, Mach KJ, Bilir TE, Chatterjee M, Ebi KL, Estrada YO, Genova RC, Girma B, Kissel ES, Levy AN, MacCracken S, Mastrandrea PR, White LL (eds) Climate change: impacts, adaptation, and vulnerability. Part B: regional aspects. contribution of working group II to the fifth assessment report of the Intergovernmental Panel on Climate Change. Cambridge University Press, Cambridge, pp 1199–1265

Nicholson SE (2011) Dryland climatology. Cambridge University Press, Cambridge

Nicholson SE (2014) A detailed look at the recent drought situation in the Greater Horn of Africa. J Arid Environ 103:71–79

NMI (2020) Weather statistics for Badda (Eritrea). www.yr.no/place/Eritrea/Other/Badda/statistics.html

Nyssen J, Vandenreyken H, Poesen J, Moeyersons J, Deckers J, Haile M, Salles C, Govers G (2005) Rainfall erosivity and variability in the northern Ethiopian highlands. J Hydrol 311:172–187

Oliver JE (1980) Monthly precipitation distribution: a comparative index. Prof Geogr 32(3):300–309

Panagos P, Ballabio C, Borrelli P, Meusburger K, Klik A, Rousseva S, Tadic MP, Michaelides S, Hrabalíková M, Olsen P, Aalto J, Lakatos M, Rymszewicz A, Dumitrescu A, Begueria S, Alewell C (2015) Rainfall erosivity in Europe. Sci Total Environ 511:801–814

Panagos P, Borrelli P, Meusburger K, Yu B, Klik A, Li KJ, Yang JE, Ni J, Miao C, Chattopadhyay N, Sadeghi SH, Hazbavi Z, Zabihi M, Larionov GA, Krasnov SF, Gorobets AV, Levi Y, Erpul G, Birkel C, Hoyos N, Naipal V, Oliveira PTS, Bonilla CA, Meddi M, Nel W, Al Dashti H, Boni M, Diodato N, Van Oost K, Nearing M, Ballabio C (2017) Global rainfall erosivity assessment based on high-temporal resolution rainfall records. Sci Rep 7:4175

Pedgley DE (1967) Air temperature at Dallol, Ethiopia. Meteorol Mag 96:265–271

Renard KG, Freimund JR (1994) Using monthly precipitation data to estimate the R factor in the revised USLE. J Hydrol 157:287–306

Renard KG, Foster GR, Weesies GA, McCool DK, Yoder DC (1997) Predicting soil erosion by water: a guide to conservation planning with the Revised Universal Soil Loss Equation (RUSLE). Agriculture handbook 703. United States Department of Agriculture, Washington

Rosen K, Van Buskirk R, Garbesi K (1999) Wind energy potential of coastal Eritrea: an analysis of sparse wind data. Sol Energy 66 (3):201–213

Rubin DM, Hesp PA (2009) Multiple origins of linear dunes on Earth and Titan. Nat Geosci 2(9):653–658

Saji NH, Goswami BN, Vinayachandran PN, Yamagata T (1999) A dipole mode in the tropical Indian Ocean. Nature 401:360–363
Sati VP (2008) Farming systems and strategies for sustainable livelihood in Eritrea. AJAFAND 8(2):2019–2237
Schreck CJ, Semazzi FHM (2004) Variability of the recent climate of eastern Africa. Int J Climatol 24:681–701
Sivakumar MVK, Stefanski R (2007) Climate and land degradation—an overview. In Sivakumar MVK, Ndiang'ui N (eds) Climate and land degradation. Springer, Berlin, pp 105–135
Souverijns N, Thiery W, Demuzere M, van Lipzig NPM (2016) Drivers of future changes in East African precipitation. Environ Res Lett 11:114011
Stat World (2019) Global temperatures. https://stat.world/biportal/?solution=Climate+Statistics&gclid=EAIaIQobChMI8f2dyerx6AIVRe7tCh1oWQLHEAMYASAAEgJsbPD_BwE
Thornthwaite CW (1948) An approach toward a rational classification of climate. Geogr Rev 38(1):55–94
Tsoar H (1989) Linear dunes—forms and formation. Prog Phys Geogr 13(4):507–528
UNEP (1992) World atlas of desertification. United Nations Environment Programme, Nairobi
UNEP (2010) Africa water atlas. Division of Early Warning and Assessment (DEWA), United Nations Environment Programme, Nairobi
Van Buskirk R, Amare A (1994) Rainfall statistics of Eritrea. University of Asmara, Asmara
WCRP-WMO (2011) Climate observations and regional modeling in support of climate risk management and sustainable development. WCRP informal/series report no. 20/2011. Geneva
Williams AP, Funk C, Michaelsen J, Rauscher SA, Robertson I, Wils THG, Koprowski M, Eshetu Z, Loader NJ (2012) Recent summer precipitation trends in the Greater Horn of Africa and the emerging role of Indian Ocean sea surface temperature. Clim Dyn 39:2307–2328
Wischmeier W, Smith D (1958) Rainfall energy and its relationship to soil loss. Trans Am Geophys Union 39:285–291
World Bank (2019) Climate change knowledge portal. https://climateknowledgeportal.worldbank.org/

Geology and Geomorphological Landscapes of Eritrea

2

Ernesto Abbate and Paolo Billi

Abstract

The landscape of Eritrea is highly variable and reflects the complex geological history of the area, which is only partially shared with the other regions of the Horn of Africa. The structural geomorphology of Eritrea was investigated through field surveys, literature reviews and a few topographic profiles oriented east–west across the country. The information collected led to the production of a new schematic geological map. The older crustal deformations controlled the orientation of the main fluvial systems, whereas later tectonic events affected their upstream drainage networks. The geological history of Eritrea is very long as it started in the Neoproterozoic, though it was punctuated by a few, more or less long intervals of quiescence. The modern landforms derive from the combined effects of the powerful uplift, to which the whole Horn of Africa was subjected throughout the Cenozoic, and the present arid climate. Fluvial erosion resulted in the accumulation of clastic deposits, a few thousands of meters thick, which gave rise to the coastal belt. In spite of an average denudation rate of 30 mm/ka (a value similar to those inferred for the upper Blue Nile and other areas with a similar structural setting), hard rocks such as the laterites, formed on top of old peneplain surfaces, are still preserved and their present-day elevation along east–west profiles witnesses an impressive upwards dislocation of about 2500 m associated with the impingement of the Afar Plume. In Eritrea, the emplacement of Trap basalts is spatially rather limited, especially if compared with the impressive expansion all across the neighboring Ethiopia. Most of volcanic activity of Eritrea is recent (Quaternary) and associated with the very last phases of the Danakil depression formation. Presently, arid conditions and a volcanic morphology provide the Eritrean Danakil with a unique and fascinating landscape.

Keywords

Crustal deformation • Structural morphology • Peneplain • Uplifting • Horn of Africa

2.1 Introduction

The coastal portion of the Horn of Africa is a wide belt (narrower in Eritrea and wider in Somalia) that includes the eastern side of the vast Ethiopian highlands, forming a horn-shaped peninsula protruding into the Indian Ocean. The many, large-scale geologic events that occurred since the Proterozoic to present have shaped this area into its major landforms. The doming that uplifted the Ethiopian highlands, the rifting and sea floor spreading of the Red Sea and the Gulf of Aden and the formation of the Afar triple junction depression are certainly the main tectonic events that produced a distinctive and variegated physiography on the area, characterized by marked relief contrasts (e.g., in the nearest southern highlands of Eritrea, from the Gulf of Zula to the highest peak of 2250 m asl, the horizontal distance is only 29 km), vast and gently inclined areas, narrow and elongated structural basins and new crust formation with the emplacement of a huge amount of volcanic products. At the same time, the endless action of erosion and deposition processes modified the landscape gifting us with spectacular deep valleys, high escarpments and weathering landforms, basin filling, alluvial fans and deltas, desert and coastal wind-blown dunes, sebkhas and a number of volcanoes that are too recent to be substantially affected by erosion.

E. Abbate (✉)
Department of Earth Sciences, University of Florence, Florence, Italy
e-mail: ernesto.abbate@unifi.it

P. Billi
International Platform for Dryland Research and Education, Tottori University, Tottori, Japan
e-mail: bli@unife.it

P. Billi (ed.), *Landscapes and Landforms of the Horn of Africa*, World Geomorphological Landscapes,
https://doi.org/10.1007/978-3-031-05487-7_2

While in Eritrea and Djibouti the major landforms are associated with the uplift of Proterozoic and Paleozoic metamorphic basement, emplacement of Trap basalt series, normal faulting and recent volcanism, Somalia is mainly a tableland, inclined to south-east (going down to the sea level from an altitude of 500–600 m asl) and underlain by sedimentary rocks, resultant from a long-lasting (Mesozoic to Early Cenozoic) transgression and regression cycle. Uplifted terranes and escarpments are present only in the northern part of Somalia and coincide with the border of the Gulf of Aden trough. In Somalia, volcanism is limited to the northern part, whereas the basement crops out only in the area between the lower reaches of the Wabe Shebelle and Juba rivers, called the *Bur* region, and a much smaller area near Hargeisa and west of Bosaso.

Such a complex and intense geodynamic activity was interrupted by phases of tectonic stability that resulted in periods of extensive peneplanations, the most important of which affected the crystalline basement at the end of the Paleozoic and the sedimentary rocks during and after the Lower Cretaceous to Oligocene regression.

The main landscapes and landforms of Eritrea, Djibouti and Somalia are the results of very many complex, long-lasting, overlapping and interacting geologic events and processes. This chapter will focus mainly on Eritrea, and an overview of the major landforms and their relation with its geological structure is given.

2.2 General Physiography: From Eritrea to Somalia

The triangular backbone of the Horn of Africa, formed by the Eritrean–Ethiopian–Somali plateau, is bordered on two sides for over 4000 km by coastal belts of different widths (Fig. 2.1). Those of northern Eritrea and northern Somalia are only a few tens of kilometers wide and are located at the foot of the plateau. Those, equally narrow, of southern Eritrea and Djibouti are characterized by more or less (Djibouti) defined mountain chains and alignments secluded from the plateau structure.

Large coastal plains of central and southern Somalia separate the extreme southern slopes of the Somali plateau from the Indian Ocean. They are some hundred kilometers wide and often covered by bush vegetation. Within the variegated morphology of the Horn of Africa, each of these coastal regions is associated with a characteristic hinterland. In northern Eritrea and northern Somalia, a steep slope links the narrow coastal plain to the plateau with an altitude difference of more than 2000 m. In central Eritrea, the elevation difference of 2500 m between Asmara on the plateau rim and the Massawa coastal plain is gained in about 50 km of horizontal distance The same rim, which is NNW/SSE oriented, i.e., parallel to the adjacent Red Sea trend, attains the highest elevation in Eritrea with the Amba Soira (3018 m asl). From this rim, the plateau, characterized by a rugged morphology, goes down toward the Sudanese lowlands with a low-elevation (500–600 m asl), c. 170 km wide, belt punctuated by scattered isolated hills and groups of small mountains (around 1000 m asl). Near Asmara, the plateau becomes narrow, down to 70 km, bulges up northward in the Nakfa area and shrinks close to the northern boundary with Sudan (Fig. 2.2).

South of the Gulf of Zula, a c. 70 km wide, NW–SE-oriented belt extends for about 300 km parallel to the southernmost Red Sea, from Zula to the Ali Sabieh Mts. (south of Djibouti). From a general flat morphology close to sea level, three mountain regions rise from north to south: the Danakil block in Eritrea with Mt. Aibaba (1350 m asl), the Randa-Obock region and the Ali Sabieh Mts., both in Djibouti. In northern Djibouti, Eritrea, Ethiopia and Djibouti political boundaries meet as in a triple point close to the prominent Mussa Ali volcano (2028 m asl).

In Somalia, a wide triangular region with the corners in Zeila, Cape Guardafui and Garoe/Eil has a northern belt overlooking the Gulf of Aden and extending for 800 km in a ENE direction (Fig. 2.1). The seaward portion is bordered by a narrow coastal plain (the Guban region from Seylac to Berbera), which is connected to an imposing ridge further inland by a steep slope. The Shiick High, the Ahl Madow, Al Mescat and Al Bahari mountains follow each other in a row from the Gulf of Berbera to Cape Guardafui at the Gulf of Aden/Indian Ocean junction. This alignment, constantly above 2000 m in elevation, is the rim of the Somali plateau which gently decreases southward (the elevation of 500 m is reached 500 km from the rim) toward the plains of central Somalia. The plateau is dissected by two transverse ESE trending valleys (Darror and Nogal) with homonymous ephemeral rivers reaching the Indian Ocean.

The rather harsh landscapes of northern Somalia turn into the wide plains of central and southern Somalia where the rivers from the arcuate structure of the Somali plateau in the Harar region flow toward the Indian Ocean over a length of about one thousand km. West of Mogadishu, in the Bur region, spectacular inselbergs breach the monotonous flat morphology of southern Somalia.

2.3 Geological Outline of Eritrea

Though with a variable areal distribution, the Proterozoic basement makes up the largest part of relief of Eritrea and forms an equilateral triangle with a north-pointing vertex across the Sudan border. Basement rocks are the backbone of northern and central Eritrea covering more than the 70% of the region north of the Danakil depression (Fig. 2.3).

Fig. 2.1 Physiographic map of the Horn of Africa

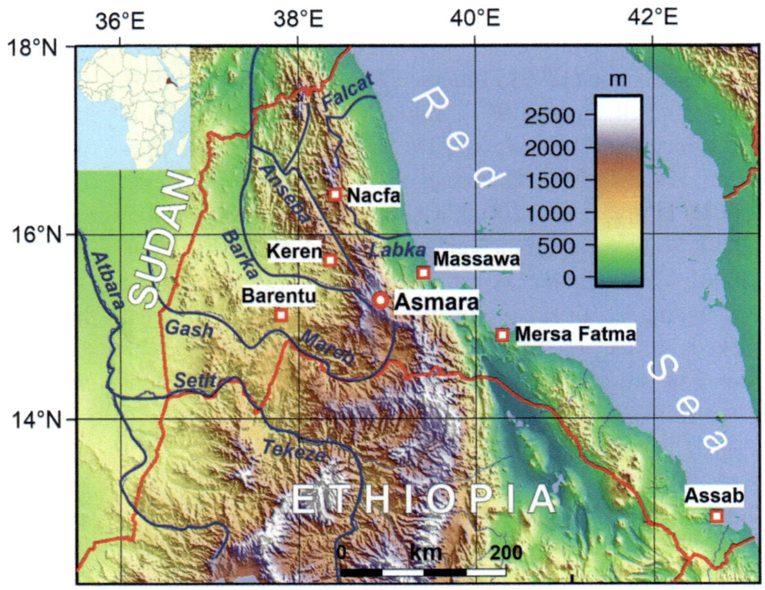

Fig. 2.2 Digital elevation map of Eritrea with the main physiographic elements

Investigations on the basement played an important role in the advancement of knowledge about the Eritrean territory, both for its undoubted geodynamic significance in the framework of the geology of East Africa and for its possible mining value. Starting from the pioneering geological observations by Baldacci (1891) and Bibolini (1920) in northern Eritrea, the petrographic studies on the basement rocks by Manasse (1909), Aloisi (1931) and Andreatta (1941) and the identification of the Barca suture zone by Francaviglia (1938), the attempts to recognize lithological succession and establish the chronology of the deformational and metamorphic events were discussed extensively in the first fundamental work on Eritrea's geology by Dainelli and Marinelli (1912) and in the second volume of the Geology of Eastern Africa by Dainelli (1943). In the 1:2,000,000 geologic map of the Horn attached to the volume, Dainelli (1943) made a first distinction between areas with different metamorphic grades. Subsequently, Mohr (1962), in Fig. 2.3 of his Geology of Ethiopia, proposed a schematic anticipatory map which takes into account the distinctions of metamorphic degree and the main foliations of the basement throughout East Africa and with particular detail in Eritrea. Later, authors of both papers (Gherardi 1951a, b) and regional maps (Kazmin 1973; Merla et al. 1979, as well as modern authors), maintained this distinction (Ghebreab et al. 2005; discussion in Andersson et al. 2006) (see later).

After 30 years of warfare, research on the basement geology resumed intensively in the nineties through the activity of the University of Asmara and the Eritrean Ministry of Mines with the collaboration of foreign universities, particularly from Sweden and the UK. Numerous publications span two decades of joint international studies, and a short list includes, among others, Berhe (1990), Drury and Berhe (1993), Drury et al. (1994), De Souza Filho (1995), Woldehaimanot (1995), Tadesse (1996), Ghebreab (1996), Talbot and Ghebreab (1997), Teklay (1997), De Souza Filho and Drury (1998), Drury and De Souza Filho (1998), Ghebreab (1998, 1999), Ghebreab and Talbot (2000), Beyth (2001), Drury et al. (2001), Ghebreab et al. (2002, 2005, 2009), Solomon and Ghebreab (2006) and Andersson et al. (2006). These studies confirmed that in the Eastern African Proterozoic continental framework, the Eritrean basement rocks were part of the southern Arabian-Nubian Shield, an accretion orogen (East African Orogen, Stern 1994) elongated from Sinai to Kenya (Fritz et al. 2013).

2.3.1 The Neoproterozoic Basement

Within the Neoproterozoic supercontinent Rodinia, Eritrea was in the central part of the Arabian-Nubian Shield (ANS). About one billion years ago, Rodinia began to rift apart, the Mozambique ocean started opening, and new margins (East and West Gondwana) accreted through ocean spreading, crustal growth, island arc- and plume-related magmatic activities (Teklay 2006). Various phases of convergence, which eventually gave rise to the East African Orogen (Pan-African Orogeny Kröner and Stern 2004; Johnson and Woldehaimanot 2003), followed from 850 to 550 Ma ago affecting the newly formed island arc, oceanic plateaus and ridges, old and new granitoid intrusions and volcano-sedimentary successions.

Although these lithostructural elements were intensively deformed and metamorphosed in various degrees, it has been possible to distinguish single tectonostratigraphic rock assemblages (terranes) with characteristic lithology, metamorphic grade, bounding tectonic lineaments and geodynamic histories. The terranes cover the whole northern and half of central Eritrea and are almost absent in the eastern/southern part of the country. They align NNE–SSW with a predominant west steeply dipping tectonic fabric and are separated by c. north–south trending transpressional suture belts. The width of the terranes is up to 100 km, and their length is up to 500 km. Some terranes and sutures continue along strike in north Ethiopia, Sudan and Saudi Arabia.

From the Sudan border to the east, the more prominent terranes are Barka, Hagar, Adhoba Abi, Nakfa, Arag and Ghedem (Fig. 2.4) (Drury and Berhe 1993; Drury and De Souza Filho 1998; Ghebreab et al. 2009). In particular, the Barka terrane is an intensely folded assemblage of amphibolite to granulite facies, with mafic gneisses, quartzites, marbles and a major olistostrome (De Souza Filho and Drury 1998). Initially, due to its location close to the stable Africa Sudanese crust, the Barka terrane was considered as a possible relic of an old continental crust, but more recent research proved that it was, like all the other northern Eritrean terranes, a product of the Pan-African event.

To the east of the Barka terrane and beyond the Barka river is the Hagar terrane dominantly composed by calc-alkaline metavolcanic and metavolcanoclastic rocks in greenschist facies, discontinuous marbles, cherts, oceanic basalts and pelagic sediments, basic–ultra-basic rocks and a major olistostrome. All these features point at accretionary wedge and fore-arc settings (Drury et al. 1994).

Further to the east, the Adobha Abi terrane exposes in a highly dismembered shear zone of calc-alkaline greenschist facies rocks and dispersed basic and ultrabasic ophiolites with oceanic island arc character (Woldehaimanot 2000).

Next is the Nakfa terrane which underlies most of the northern highlands of Eritrea with its c. fifty thousand square kilometers. It consists of greenschist facies calc-alkaline volcanic and volcanoclastic rocks mainly prevailing on tholeiitic volcanites, phyllites, slates, sandstones and conglomerates. It is interpreted as a relict island arc assemblage.

Peculiar in the Nakfa terrane is the great amount of syn- and post-tectonic granite and granitoid intrusions, particularly in its western half, commonly occurring as inselbergs

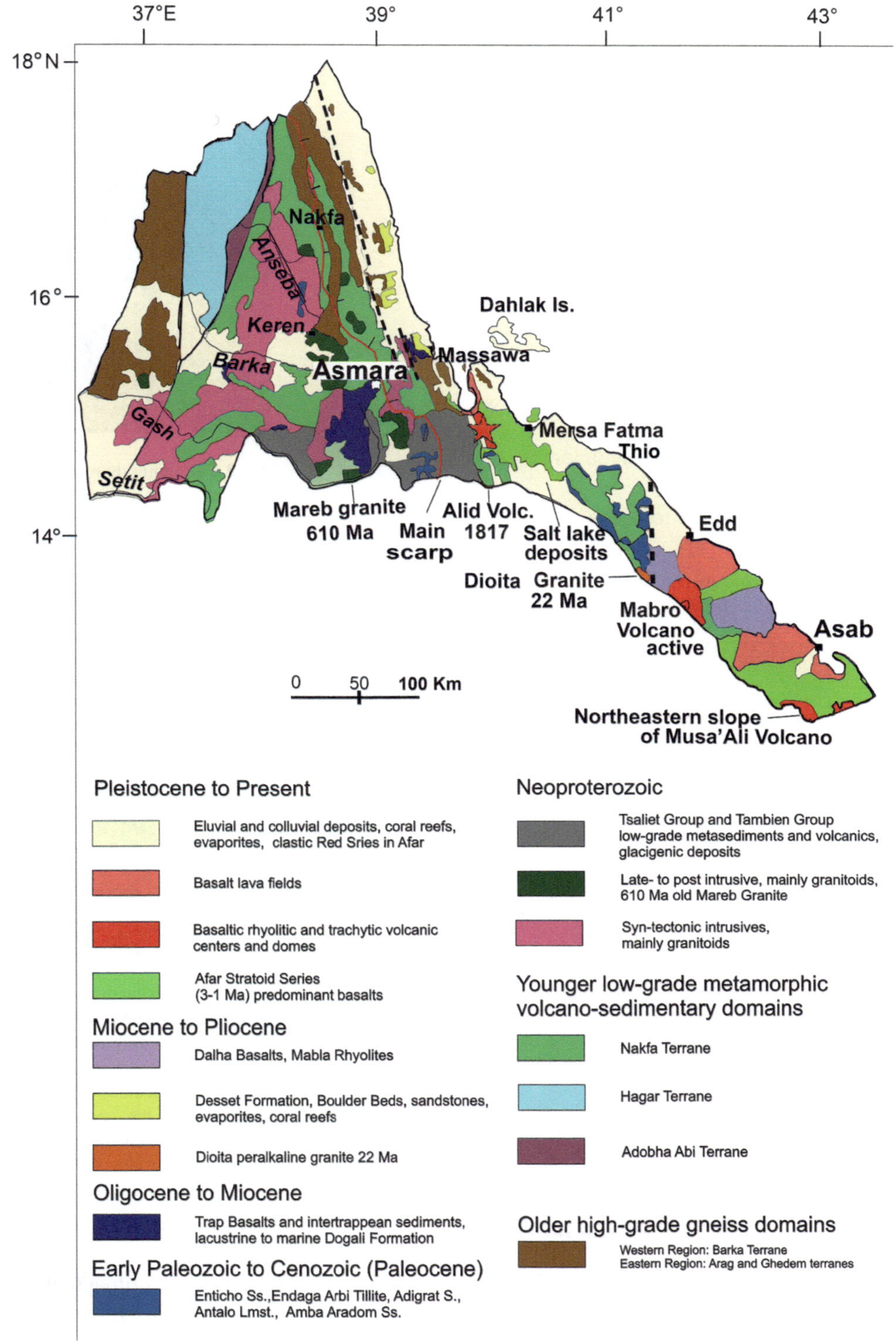

Fig. 2.3 Schematic geological map of Eritrea. The red solid line is the margin of the main escarpment; the black solid lines are the main faults, and the dash-dotted lines are the major lowland Tertiary faults. Red star: Buia *Homo erectus/ergaster* site and Alid volcano location. Sources of compilation: Brinkmann and Kürsten (1970), Kazmin (1973), CNR-CNRS (1971, 1975), Merla et al. (1979), Drury and Berhe (1993), Drury et al. (1994), Teklay (1997), Drury and De Souza Filho (1998), Sagri et al. (1998), Ghebreab and Talbot (2000), Beyth et al. (2003), Ghebreab et al. (2005, 2009), Kumpulainen et al. (2006, 2007), Avigad et al. (2007), Eritrea Department of Mines (2009), Bussert (2010, 2014)

(Fig. 2.5). Moreover, a N–S trending, tens of kilometers wide belt of "early" granitoids, has been reported by Teklay (1997, map in Drury and De Souza Filho 1998) from Nakfa to Barentu. They were connected with the first stages of the accretion processes dated at 811–846 Ma. Additionally, huge masses of "late granites" (650–630 Ma old—Kröner et al. 1991; Teklay 1997) mark the eastern part of the Nakfa terrane. One of these post-tectonic granites is Mt. Bizen (2435 m asl), famous for its precipitous slopes and the monastery on its top. It is located near Nefasit, on the Asmara-Massawa road (Fig. 2.6).

The area between Nakfa and Keren offers panoramic views of another remarkable landscape in the Rora range, where the weathered and dissected rocks of the basement

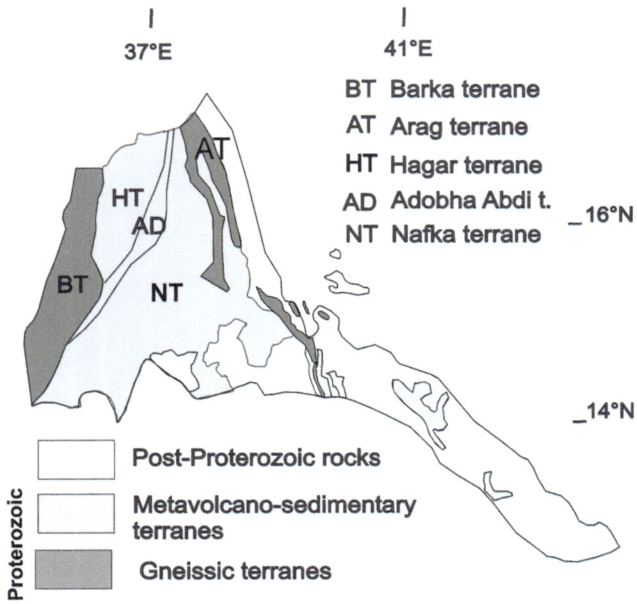

Fig. 2.4 Schematic map of major Neoproterozoic terranes of Eritrea (modified after Ghebreab et al. 2005; Avigad et al. 2007)

Fig. 2.5 Syn-tectonic granite near Keren. Notice the rounded morphology of the peak due to exfoliation weathering processes

(mainly metavolcanics) are a stand for remnants of a flat-lying sedimentary sequence (Fig. 2.7) (see more detail in the next sections).

The Nakfa terrane has an extension in the Danakil block, which is a continental fragment detached and rotated since the Miocene from the eastern margin of northern Ethiopia (Burek 1970) (Fig. 2.8). In the core of the Danakil block, a greenschist facies basement of limited extent with chlorite and sericite schists and granitoids is surrounded by Paleozoic

Fig. 2.6 Granitoids rocks (Ghebreab 1999) of the Nakfa terrane and the Bizen Monastry on the homonymous mountain (about 30 km East of Asmara)

Fig. 2.7 Roras landscape 15 km SW of Nacfa. An unmetamorphosed horizontal sequence mostly composed of reddish Cretaceous laterites, but likely including older Phanerozoic successions, rests on a deeply eroded Neoproterozoic Nakfa terrane, mainly composed of greenschists facies metavolcanics. The flat tops of the Roras, now intensely cultivated, in the past were used as fortification places and host historical and archeological sites

to Mesozoic sediments (Vinassa de Regny 1921–1924; Bannert et al. 1970; Kazmin 1973; Bunter et al. 1998; Sagri et al. 1998).

In central Eritrea to the east of the Nakfa terrane, the Arag and Ghedem terranes sporadically expose their high-grade gneisses, metavolcanics and pelitic metasediments along a belt from the Takora region, at the northernmost Eritrea/Sudan border, to the Ghedem promontory 20 km south of Massawa. In the map of Fig. 2.3, it is assumed that along the coastal belt, rocks of the Arag/Ghedem terranes are overlain, from the northernmost outcrops, by the synrift Miocene Desset Formation.

The afore described multi-terrane classification for the Eritrean Neoproterozoic has been reconsidered in the last

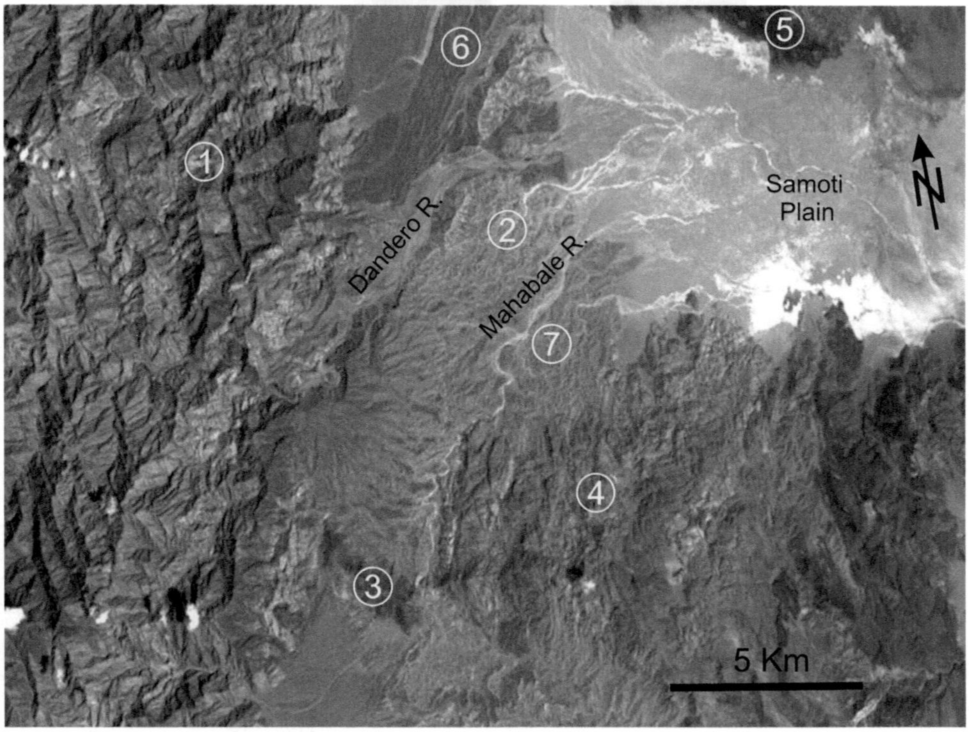

Fig. 2.8 Nakfa terrane along the western side of the Gulf of Zula/Alid volcano/Samoti plain corridor heading to the northern Afar depression. Starting from west: 1—on the left of the Dandero river Nakfa terrane with NNE-trending folded metamorphic limestone, sandstones, shales; 2—fluvial deposits of the Dandero and Mahabale rivers; 3—Adigrat sandstones brown outcrop across the valley; 4—on the high right reach of the Mahabale large greenish blue outcrop of the Nakfa terrane with metavolcanics and phyllites covered eastward by Neogene brownish rhyolitic domes and tuffs; 5—to the north of the whitish Samoti alluvial plain, basaltic lavas of the active Alid volcano limited to the west by a fault black; 6—in the northernmost central portion brownish boulder bed of the Dandero terraces; 7—Pleistocene sediments of the Dandero basin (Google Earth coordinates: 14°45′18″ N–39°57′29″ E)

twenty years and replaced by an alternative proposal of assembling the basement units within two comprehensive groups ("domains"): a high-grade gneiss domain, including the Barka, Arag and Ghedem terranes structurally overlain by a low-grade metavolcano-sedimentary domain, including the Hagar, Adoba Abi and Nakfa terranes, each domain comprising shear zone-bounded subdomains (Ghebreab 1996, 1999; Ghebreab et al. 2005). The low-grade domain largely widens in the central portion of northern Eritrea, and it is east–west longitudinally bounded by units of the high-grade domain (Fig. 2.3).

In the context of west/east Gondwana convergence, both domains are considered representatives of the same juvenile crust of accreted island arcs and granitoids (Fig. 2.9) (Ghebreab et al. 2005; Andersson et al. 2006; Avigad et al. 2007) with magmatism beginning at 800–850 Ma (Teklay 1997). Deformation patterns, metamorphic grades and crustal shortenings were predominantly acquired through suturing and escape tectonics, whereas large-scale extensional orogenic collapse brought high-grade lower crust slivers exposed to the surface (Fig. 2.10) (Ghebreab et al. 2005; Andersson et al. 2006). For instance, geochemical 40Ar/39Ar and field data indicate that some mylonitized Ghedem rocks were exhumed from a depth of c. 45 km to the upper crust (c. 15 km) between 593 and 567 Ma (Fig. 2.11) (Ghebreab et al. 2005).

In summary, from the early stages of rifting at ca. 870–840 Ma ago, the whole cycle of the East African Orogen amalgamation with the birth of the Gondwana (Stern and Johnson 2010) was achieved through distinct phases of exhumation at c. 550 Ma.

2.3.2 The Neoproterozoic and Paleozoic Glacial Episodes and Long Sediment Vacancies

The Neoproterozoic to Late Paleozoic history of the Eritrean crust records intervals during which wide peneplained areas were covered by continental and/or shallow marine sediments with traces of glacial activity. The discovery of these glacial deposits ignited a long debate, which started in the beginning of the last century when Italian prospectors and geologists active in northern Eritrea were reporting, within the Proterozoic metamorphic basement, conglomerates with coarse, predominantly quartzitic pebbles, some striated, in a hard, purple argillaceous-siliceous metamorphic matrix. Verri

Fig. 2.9 Tsada Amba carved in the granitoids of the Nakfa terrane with its monastery on top of the peak near Agaz SW of Keren

(1909) was the first to assume glacial origin for these coarse clastics. Later on, these conglomerates, named Rora Bagla conglomerates (Bibolini 1920, 1921) after the Bagla table-topped mountain (Rora means "mountain" in Tigre language) 25 km WSW from Nakfa, were commonly found associated with stromatolithic carbonates and interleaved within the basement schists in wide areas of the Hababland between Nakfa and the Haggar Mts., close to the Eritrea/Sudan border (Cecioni 1981), and more to the south in the Bizen Domain of the Heddas river 60 km SE of Asmara (Beyth et al. 2003), along the Mareb river near Adi Qwala, 80 km south of Asmara along the Eritrea/Ethiopia border, and in western Eritrea along the Setit river south of Barentu. Correlations were established with similar lithological successions belonging to the Tambien Group (Beyth 1972a, b) extensively outcropping in the Adwa-Adigrat area in northern Ethiopia. The Tambien Group is a 2–3 km thick, greenschists-grade metavolcanic and metasedimentary sequence deposited unconformably on the Nakfa terrane in an intra-oceanic arc platform setting during the last phases of the east/west Gondwana convergence (Avigad et al. 2007) (Fig. 2.12).

Unconformably above the Tsaliet metavolcanics indicating an igneous activity between 850 and 750 Ma (Tadesse 1996) lies the Tambien Group (Beyth 1972a, b). It comprises continental greywackes, slates- and shales-supported polymictic conglomerates, sometimes with striated pebbles, sporadic marine stromatolithic carbonates, and slates progressively becoming diamictitic with striated cobbles as large as 150 mm (Miller et al. 2003). These lithologies, sedimentary structures, paleogeographic reconstructions and chemostratigraphies ($\delta^{13}C$ and $^{87}Sr/^{86}Sr$ signatures of the carbonate rocks and $\delta^{13}C$ signature of the organic matter removed from the carbonates) are diagnostic of one or, possibly, two glacial events (Beyth 2001; Beyth et al. 2003; Miller et al. 2003, 2011; Alene et al. 1999, 2006; Avigad et al. 2007). Chronological constraints from syn-tectonic and post-tectonic granitoid intrusions, detrital zircon ages and correlations with global isotope stratigraphy provide for these glaciations, invading even the marine domain, a certain time interval bracketed between 720 and 660 Ma during the Cryogenian (Stern and Miller 2019). This interval recorded in the Eritrean Neoproterozoic coincides with that of many Neoproterozoic "Snow Ball Earth" deposits from other continents (Hoffman and Scharag 2002; Hoffman 2009) and is referred to as the Sturtian glacial episode (Stern and Miller 2019; Park et al. 2020). In addition, paleomagnetic

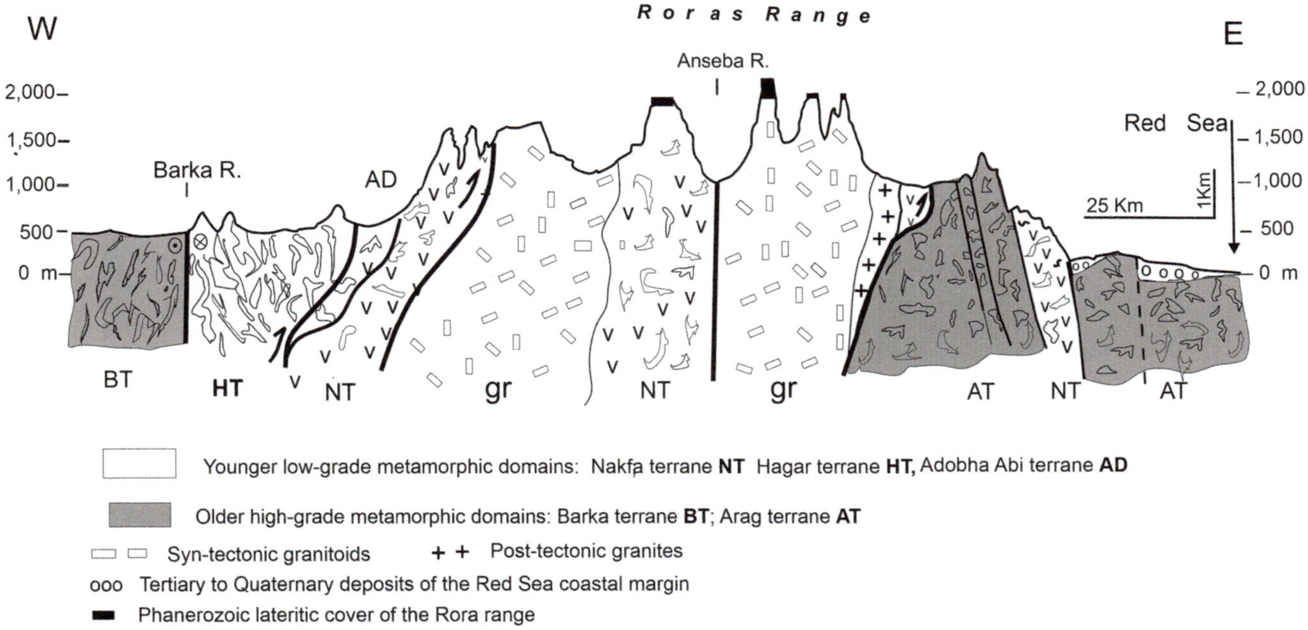

Fig. 2.10 General west to east, along the 16°9′ N parallel, geological section across the major Neoproterozoic structural units of the northern Eritrean plateau and adjoining Red Sea margin. For trace, see Fig. 2.20. *Legend* Older high-grade metamorphic domains: Barka terrane (BT) and Arag terrane (AT); younger low-grade metavolcanic and sedimentary domains: Nakfa terrane (NT) and Hagar terrane (HT). Main sources: Drury and Berhe (1993) and Ghebreab et al. (2005). Additional sources: Drury et al. (1994); De Souza Filho and Drury (1998)

Fig. 2.11 Folded gneissic rocks of the high-grade Ghedem Domain affected by Tertiary faulting along the western side of the Gulf of Zula-Samoti

reconstructions provide evidence that the East African portion of the Arabian-Nubian Shield was then drifting at low latitudes in a pan-glacial Earth (Hoffman et al. 1998; Stern and Miller 2019).

The Paleozoic deposits are exposed in sparse outcrops in central Eritrea from Mendefera southward to the Eritrean/Ethiopian border (Kumpulainen 2007), in the northern rim of the Danakil block (Kumpulainen 2007) and

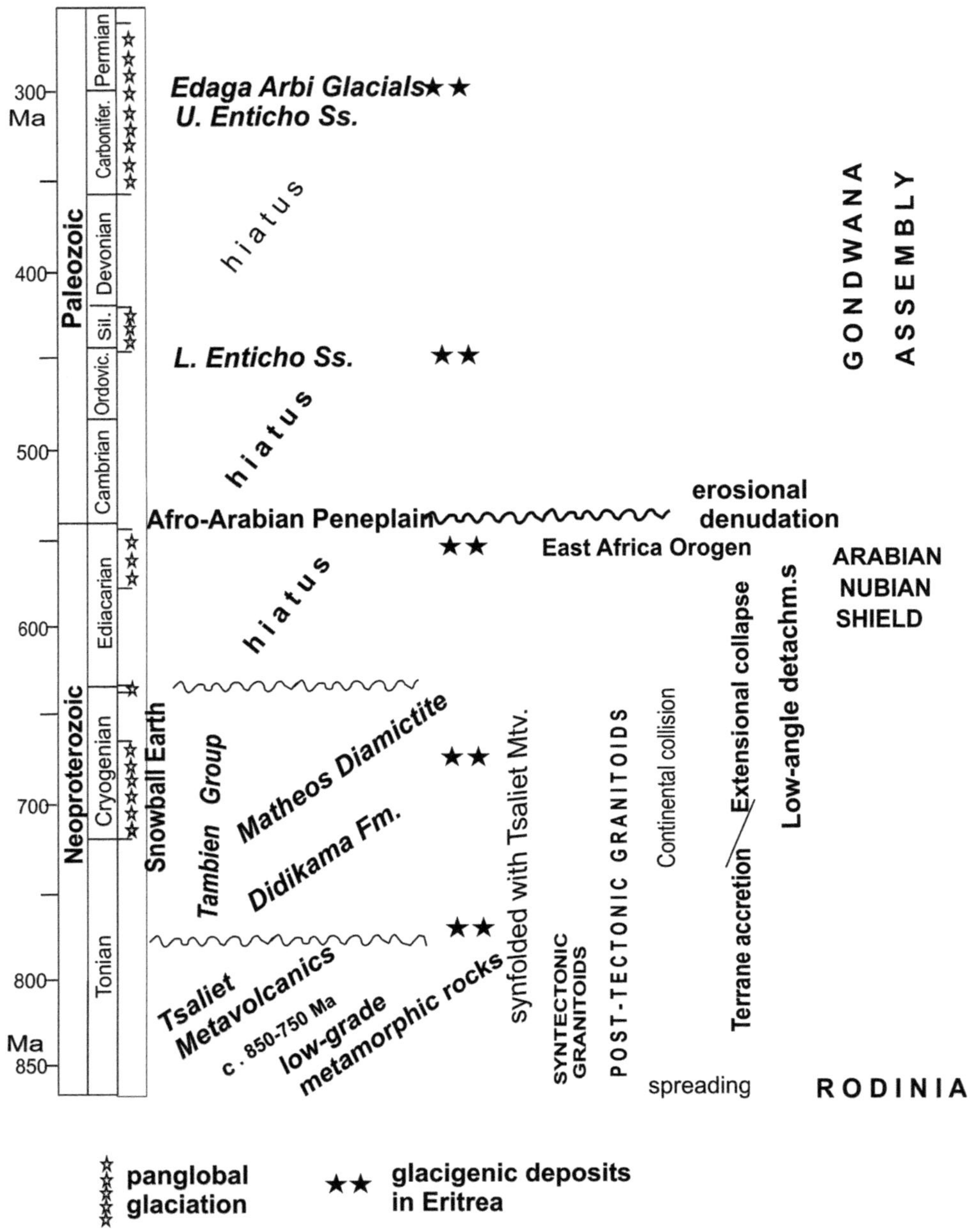

Fig. 2.12 Schematic stratigraphic chart for the Neoproterozoic and Paleozoic of Eritrea with pan-global glaciations and glacigenic deposits in Eritrea. Structural events of the East Africa Orogen in Eritrea

in a northernmost outcrop 20 km SSE of Asmara (Ghebreab and Talbot 2000). Those in central Eritrea are contiguous to the type area of Paleozoic deposits in the Adwa-Adigrat region of northern Ethiopia where Dow et al. (1971) and Beyth (1972a, b), drawing inspiration from Blanford (1869, 1870) and Merla and Minucci (1938), described the Enticho Sandstone. This is a c. 200-m-thick succession with a basal tillite including clasts up to boulders, a lower glacigenic sandstone unit and an upper shallow marine sandstone unit (Bussert and Dawit 2009) (Figs. 2.13 and 2.14). The Enticho Sandstone makes lateral and, in places, vertical upward transition to the Endaga Arbi Tillite, a glacial/periglacial deposit with shales and tillites with striated clasts and, at places, an upper shallow marine sandstone.

The age of both Paleozoic glacigenic formations is still uncertain due to complex local stratigraphy characterized by several facies changes, lack of detailed mapping and difficulty of finding diagnostic fossils. Based on correlation with glacial deposits in northern Africa and Arabia, using palynomorphs, hydrozoa and trace fossils determinations, some authors (Dow et al. 1971; Merla et al. 1979; Saxena and Getaneh 1983; Kumpulainen 2007) have assumed an Ordovician-Silurian age for the glacigenic event of the Endaga Arbi Tillite. Alternatively, a latest Carboniferous–Early Permian age of the Endaga Arbi Tillite has been documented through palynomorphs and trace fossils by Bussert and Schrank (2007). The same authors have tentatively suggested a younger Enticho Sandstone, heteropic of the

Fig. 2.13 Glacial grooves on the Early Paleozoic Enticho Sandstone near Eubucao pass (Amba Soira)

Fig. 2.14 Early Paleozoic Enticho Sandstone unconformably above the slates of the Neoproterozoic Nakfa terrane in the Dahan Dahan range (2800 m elev.), 12 km ENE of the Amba Soira

Carboniferous-Permian Endaga Arbi Tillite, and an older Enticho Sandstone of Ordovician-Silurian age. This would only in part agree with the Late Carboniferous–Early Permian paleomagnetic age estimation by Kidane (2013, 2014) for the Enticho Sandstone and for the Endaga Arbi glacials. Sacchi et al. (2007) also assumed that the occurrence of clastic volcanic material in some tillites could be better explained if referred to the diffuse Permian volcanic activity in Africa.

Although these uncertain litostratigraphic allocations are under investigation in Eritrea and Ethiopia, the two sets of dating (Ordovician-Silurian and Late Carboniferous–Early Permian) correspond chronologically to Paleozoic global glacial episodes (Bussert and Schrank 2007; Bussert and Dawit 2009; Bussert 2014) (Fig. 2.12). The petrography and geochemistry of the arenaceous glacial and periglacial material (Lewin et al. 2018) indicate a big difference in size between the large ice sheet covering Gondwana during the Ordovician-Silurian and the scattered local glaciers during the Carboniferous-Permian.

On a regional scale, the Paleozoic deposits, recently investigated in detail by Bussert and Schrank (2007), Hambrey (2007), Bussert and Dawit (2009), Bussert (2010),

Dawit (2010) and Lewin et al. (2018), rest unconformably on a striated Proterozoic basement (Arag terrane and Bizen Domain). This unconformity is the Afro-Arabian Peneplain (Miller et al. 2003), dated c. 550 Ma and traceable from Morocco to Oman on the beveled Arabian and Nubian Shields that already merged into the Pan-African crust at the end of the East African orogeny.

Only a few hundred meters of sediments, resting on the Afro-Arabian Peneplain, attest a c. 330 million years interval (from c. 550 Ma—terminal phases of the East Africa Orogen to c. 220 Ma—the base of the Adigrat Sandstone) with very low sedimentation rates. The sediments were ice-proximal assemblages deposited on continental platform, in some places bordered by the sea (Kumpulainen 2007). It is likely that regional uplift events caused only episodic denudation in a low relief landscape, with no thick molasse deposition.

For its relevance, the Afro-Arabian Peneplain can be qualified as a *paleogeological hiatus surface* (Friedrich 2019; Friedrich et al. 2018), i.e., a continental-scale unconformable contact.

2.3.3 A Rifted Gondwana and the Jurassic Sea

The prolonged tectonic quiescence of the Eastern Africa was interrupted at c. 250 Ma ago by the Permo-Triassic rifting of Gondwana which, starting from South Africa (Karroo), produced the separation of Tanzania, Kenya and Madagascar through a dense network of branching rifts.

The unconformity, at the base of this event more or less at the transition Paleozoic/Phanerozoic, was called in Ethiopia and in Eritrea "First erosional cycle" by Merla and Minucci (1938) and PS2 planation surface by Coltorti et al. (2007) (Fig. 2.15).

The clastic supply, provided by the diffuse rift-shoulder uplift (Purcell 2018) and designated as "Adigrat Sandstone" (Blanford 1870), was dispersed over very large areas by braided or low-sinuosity rivers. They were crossing a vast ramp gently sloping southeast toward the rim of the Nubian plate and the Indian Ocean. The Adigrat Sandstone, a medium-grained cross-bedded fluvial arenite, thick up to 700 m, with interbeds of conglomerates, and its Late Triassic to Middle Jurassic correlative arenites, are widespread in the whole Horn of Africa and Arabia, including the Mazera Sandstone of Kenya, the Negrenegre Sandstone of Tanzania, the Minjur Sandstone of Saudi Arabia and the Kohan Series of Yemen (Mohr 1962; Kamen-Kaye 1978; Bosellini 1989; Bosellini et al. 1997). In Eritrea, the Adigrat Sandstone is present in the Danakil block (Vinassa de Regny 1930; Bannert et al. 1970; Sagri et al. 1998).

The Adigrat Sandstone, which in northern Ethiopia unconformably covers the Enticho Sandstone and the Endaga Arbi Tillite, has an uncertain lithostratigraphic characterization in Ethiopia and in Eritrea as well. Recent papers (Merla et al. 1979; Garland 1980; Kumpulainen 2007; Bussert and Schrank 2007), in fact, assigned its classical type section, located close to Adigrat in the northernmost Ethiopia, to the Enticho Sandstones rather than to the Adigrat Sandstone. A thorough discussion on this matter, mainly due to the commonly irrelevant fossiliferous content of the Adigrat Sandstone and meager distinctive features, is given by Merla et al. (1979) and Kumpulainen et al. (2006). Yet, Kumpulainen et al. (2006) addressed some distinctive features between the Adigrat Sandstone and the Enticho Sandstone, the former first being an immature fluvial type of sandstone, and the second latter a cross-bedded friable, predominantly fluvial sandstone, but probably, also partly, also of probable marine origin.

Uncertainties also involve the age of the Adigrat Sandstones regarded as Triassic–Liassic (Kazmin 1973) or Trias to Dogger (Merla et al. 1979), Triassic–Middle Jurassic (Tefera et al. 1997) and Early Jurassic (Hankel 2013). Recently, through pollen assemblages, Dawit (2010) obtained for the classical sections near Mekele and at the Blue Nile gorge a range in age from Late Triassic to Middle Jurassic. Through accurate facies analysis, the same author has documented some shallow marine character (lagoon and shoreface transition) of the Adigrat Sandstone, thus confirming what assumed by previous researchers (Merla and Minucci 1938; Dainelli 1943; Dow et al. 1971; Kumpulainen et al. 2006). Likely, the marine inflows were coming from the Neotethys across Arabia.

The general subaerial conditions of the Adigrat Sandstones deposition (Getaneh 2002) ended during the Middle to Late Jurassic when the vast platform on the East Africa craton was progressively inundated from SE due to further rifting events connected with the Gondwana dismembering and to the Mesozoic eustatic sea-level rise (Haq et al. 1988).

In Eritrea, evidence of this ingression is preserved in the northern Danakil block (regional maps by Brinkmann and Kürsten 1970) and further north in the Kezan area c. 100 km NW of Massawa (Bunter et al. 1998). At Kezan Anadarco and Eritrean Ministry of Mines, geologists identified 100 m of Antalo Limestones with dark and pure limestones, dolomitized and cherty limestones. At Kazen, the Antalo Limestone represents the northernmost evidence of the time-transgressive flooding of the Jurassic sea onto Eastern Africa covering a distance of more than 1500 km from the present-day Indian ocean during 30 Ma (from Toarcian to Callovian).

After the pioneer approaches by Vinassa de Regny (1930) who stressed the intense deformation of the block, the outcrops in the northern Danakil block were recently studied by Sagri et al. (1998). Unlike the classical succession, in which a full-fledged marine deposition is preceded by some hundreds of meters of a peritidal and lagoonal unit with

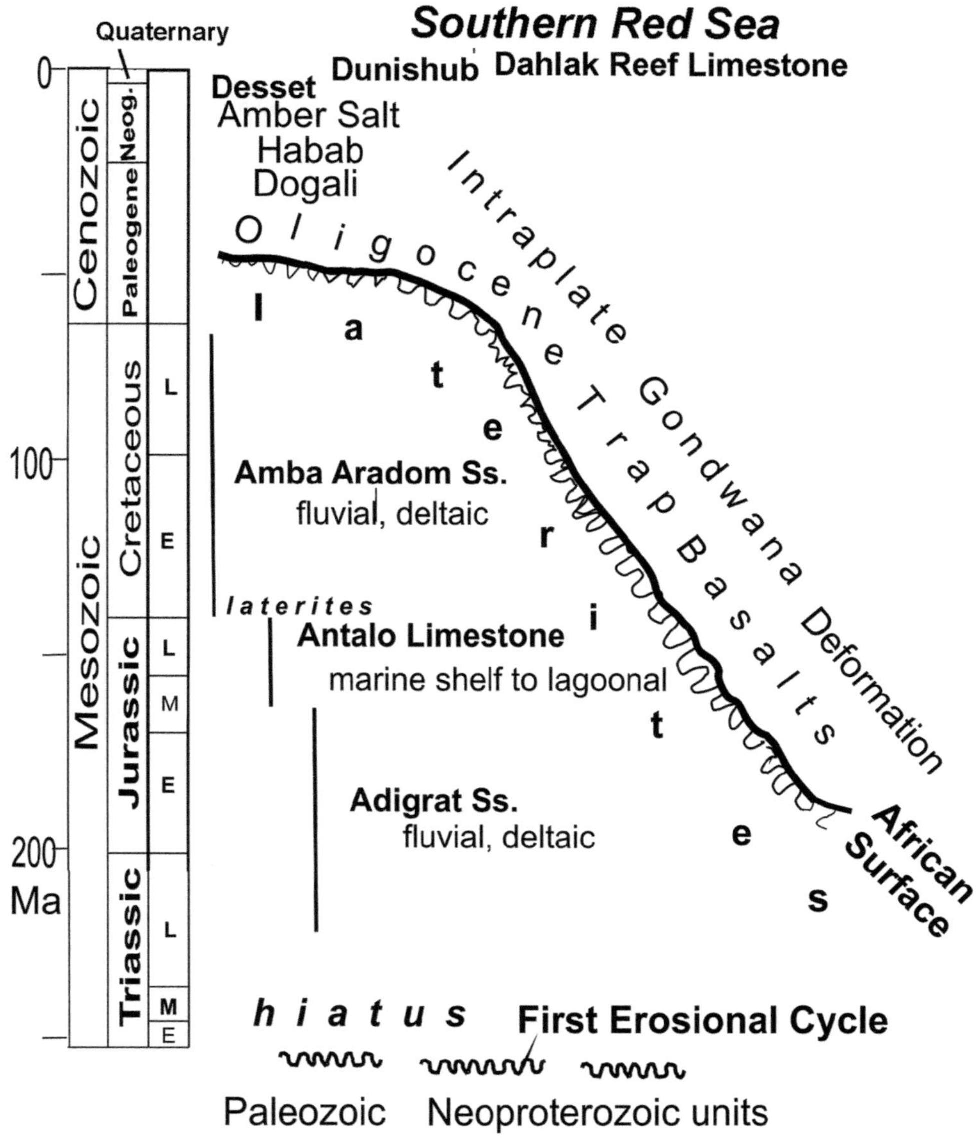

Fig. 2.15 Schematic stratigraphic chart for the Mesozoic and Cenozoic successions in Eritrea mainly relative to central Eritrea

characteristic shales and gypsum (e.g., Goatzion Formation—Assefa 1981; Russo et al. 1994), the succession near Adeilo includes just a few tens of meters of tectonized gypsum beds between the 700 m of Adigrat Sandstone s.l. (also including Paleozoic sediments) and the overlying marine Antalo Limestone (unit designated in Tigray by Blanford 1870). It comprises gray limestones, dark fetid marly limestones, marlstones, biocalcarenites, tempestite lumachellas and terrigenous calcareous sandy beds.

The Antalo Limestone is highly fossiliferous with nannoplancton, radiolarians, benthic foraminifers, brachiopods, pelicypods and ammonites ranging in age, in the Adeilo succession, from late Callovian to early Kimmeridgian (162–155 Ma) (Fig. 2.16) (see detailed biostratigraphy in Sagri et al. 1998).

The depositional environment of the Antalo Limestone varies repeatedly within shelf and open marine, in particular from inner to mid and outer ramp and, more rarely, to basin. Seven large-scale cycles from several tens to some hundreds of meters are framed between the major surfaces and make up seven depositional sequence due to the global fluctuations of the sea level over a period of time of about ten million years (Haq et al. 1988, Sagri et al. 1998).

In the Tigray type area, Bosellini et al. (1997) have defined as Antalo Supersequence a second-order depositional sequence including the Antalo Limestone and overlying Agula Shales with a basal gradual transition to the Adigrat Sandstone and an upper marked unconformity with the Aptian-Albian Amba Aradom Sandstone.

The Antalo Limestone in the Danakil Alps is at least one thousand meters thick (Hutchinson and Engels 1970; Sagri et al. 1998). This thickness, never recorded for this unit in the whole Horn of Africa, has prompted the idea of an ancestral Red Sea subsiding basin sub-parallel with and

Fig. 2.16 Antalo Limestones in the NW rim of the Danakil block, near Adeilo, 30 km SSW of Tiho. In this middle portion of the formation, thick beds of dark gray, sulfide-rich calcilutites and dark gray marly interbeds are partially replaced upward by yellowish calcilutites, marlstones and calcarenites. This succession, deposited in a subtidal environment, is particularly rich in ammonites and other molluscs, echinoderms, brachiopods, macro- and micro-foramifers and calcareous nannoplancton of Oxfordian-Kimmeridgian age (further details in Sagri et al. 1998). General dipping toward N

underlying the present-day Red Sea (Merla et al. 1979; Bosellini et al. 1995). A similar outstanding thickness (1700 m—Hutchinson and Engels 1970) for the Adigrat Sandstone at Ras Andabba has been disputed by Merla et al. (1979, p. 27).

In the southeastern part of the Danakil block, marly levels (Fig. 2.17) (possible correlative of the Agula Shales of the Mekele type section) mark the unconformable transition from the Antalo Limestone to the overlying short sections of the Amba Aradom Sandstone (Assefa 1981, formerly Upper Sandstone in Merla et al. 1979) composed of an alternation of fine to coarse-grained sandstones, siltstones and shales of braided fluvial environment. The succession Adigrat-Antalo-Amba Aradom was interpreted by Dainelli (1943) as an ideal complete transgressive-regressive cycle "with a chronostratigraphic and lithologic symmetry of ingression and regression deposits" (Merla et al. 1979) on a continental scale. The recognition of a distinct regional angular unconformity, marked by laterites, between an eroded Antalo Limestone and the Amba Aradom Sandstone in the type area of Mekele (Bosellini et al. 1997) has introduced a prominent element of discontinuity in the sedimentary cycle.

Toward the end of the Jurassic, there was a regional basin inversion with a progressive shallowing of the Antalo sea up to emersion, possibly due to a deepening of coastal basins along the embryonic Indian Ocean accompanied by basement highs rising. Because of its widespread occurrences, the Amba Aradom formation has many putative correlatives in the Horn of Africa, ranging in age from post-Kimmeridgian to pre-Middle Eocene. The most commonly accepted age for the Amba Aradom sandstone is Early Cretaceous. In the Harar region, Aptian to Albian marine macroforaminifer orbitolinids testify a short Early Aptian sea transgression during the Amba Aradom fluviatile deposition (Gortani 1973; Bosellini et al. 1999a; Bosellini et al. 2001). The similarity between Adigrat and Amba Aradom sandstones in the field and the difficulty of biostratigraphic or petrographic discriminations make uneasy to distinguish them when the interposed Antalo Limestone has been eroded or not deposited. In this case, the whole arenaceous body has been given different names, such as Tekeze Sandstone, Yesomma Sandstone (Figs. 2.18 and 2.19), Nubian Sandstone or other local names.

2.3.4 A Long Waiting Before the Volcanic Outburst: The Laterites

A layer of lateritic weathering typically covers truncated structures of the Eritrean and Ethiopian Proterozoic basement (among others, Mohr 1962; Davidson 1983). In the Ethiopian northern Tigray region close to central Eritrea, a laterite layer rests on mildly folded and eroded Mesozoic sedimentary sequences (Abul-Haggag 1961; Bosellini et al. 1997). From the Mareb river to the northern border, laterites are found in Eritrea conformably beneath the Trap basalts (Dainelli 1943). Moreover, a very informative map by Drury et al. (1994) shows that the laterites are diffusely present in northern Eritrea north of the line Massawa-Akordet representing the only pre-Oligocene sedimentary deposit.

The pre-Trappean laterite pedogenesis, firstly pointed out by Blanford (1870) and accurately described and discussed by Dainelli and Marinelli (1912) and Dainelli (1943), is commonly a few meters thick, but it can be over 100 m in

Fig. 2.17 Cycles of predominant yellowish marls and darker calcilutites make up the highest levels (Oxfordian-Kimmeridgian) of the Antalo Limestone plunging beneath the Edd-Thio coastal plain (Danakil block). In the background on the right the lower units of the Jurassic succession above the basement

places (Fig. 2.20). It can be red (more limonitic clay) above the crystalline schists or, more rarely, light colored (more kaolinitic clay) above a felsic substratum (Dainelli 1943). Laterite can appear as a massive clay, or as a densely brecciated clay with diffusely slickensided angular fragments (Figs. 2.21 and 2.22). Nice outcrops are in Asmara near the airport, on the hills downtown and near Kudus Geourgis Church 4 Km away on the road to Massawa (South of Asmara the contact with the overlaying Trap basalts is exposed along the Mareb river and near Shiket on the way to Adi Qwala).

In northern Eritrea, the pre-Trappean laterites compose the outstanding flat top of small, steep-sided plateaus forming the Roras Mts., a spectacular segmented range of Nakfa terrane granitoids and metavolcanics, elongated NNW for c. 100 km west of Nakfa with a constant elevation around 2500 m asl (Figs. 2.23 and 2.24).

Laterites are also present in Yemen and southern Sudan between the Early Oligocene basalts and the basement or Mesozoic strata (Davison et al. 1994; Al-Subbary et al. 1998; Kenea et al. 2001).

Laterite occurrences are particularly significant since they are considered indicative of the peneplanation of an area with extremely low relief and low elevation. In Eritrea and Ethiopia, the erosional event connected with the laterites has been variously labeled as "2° ciclo erosivo" (second erosive cycle) ("erosione prevulcanica", Merla and Minucci 1938); "pre-Trappean peneplanation" (Mohr 1962) and "PS4" (the last out of a sequence of PSs the oldest being pre-Paleozoic, Coltorti et al. 2007) and recognized on a continental scale as the African Erosion Surface (Burke and Gunnell 2008).

Andrews Deller (2006) has mapped the different mineralogical facies of the laterites in a test area SW of Asmara and has determined a minimum Ar/Ar laterization age of c. 40 Ma for a laterite beneath the Trap basalts of Asmara (Andrews Deller 2003). Also Perelló et al. (2020) obtained a K/Ar Paleocene, Thanetian (59 Ma) age from a supergene alunite in a laterite close to Asmara.

Typical of regions with hot and humid tropical climate, the laterites suggest a long tectonic and morphological stability of a territory with modest clastic inflows. Moreover, their altimetric significance (they developed close to sea

Fig. 2.18 Transition from the Kimmeridgian portion of the Antalo Limestones to the Yesomma Sandstone in the SE rim Danakil block, Belo Buye/Dahan region, 90 km SE of Tiho. The Yesomma Sandstone at the top of the mountain overlies a c. 20 m thick Tertiary basalt sill (brownish) with interposed varicolored shales and marlstones. Downsection, dark fetid limestones and yellowish marlstones are intruded by further basalt sills (light brown). General dipping toward NE

level) allows to assess the amount of uplift to which these laterite layers (and the basement if present) have been subjected. This possibility was put forward by Dainelli (1943) and applied to a vast area of Eritrea from the Mareb river to Nakfa by Drury et al. (1994). In the Asmara-Massawa transect, the elevation difference between laterite occurrences on the plateau and their correlative close to Massawa is c. 2.2 km driven by faulting in the Red Sea escarpment (Drury et al. 1994). Along the same sector, a similar magnitude of displacement is confirmed by the occurrence of large fossil mammals on the plateau and on the lowlands (see Sect. 2.4.1).

2.4 Syn-Rift Volcanic and Sedimentary Rocks on the Plateau and on the Lowland

2.4.1 Syn-Rift Successions on the Plateau

The Traps: flows and dikes

Since the Middle Eocene (about 40 Ma ago) in Kenya and southern Ethiopia and with a peak of activity in the Early Oligocene (about 30 Ma ago) in the remainder of Ethiopia, in Eritrea, Sudan and Yemen, a vast volcanism with predominant flood basalts resulted in the Ethiopian Large Igneous Province (ELIP). This volcanic assemblage is traditionally referred to as the Traps from Blanford's (1870) Trap Series. Following the classical single-stage Southern Red Sea evolution reconstruction with rifting since 30 Ma, more or less in connection with the arrival of the Afar plume,

Fig. 2.19 Details of the Antalo Limestones/Yesomma Sandstone transition in a Google Earth image (site to the north of the outcrop in Fig. 2.18). Pinkish, yellowish and whitish marlstones and shales with rare black-gray fetid calcilutites are followed by a 20-m-thick Tertiary basalt sill (see also Fig. 2.18). General dipping toward NE. (Google Earth coordinates: 13°59′12″ N–49°18′46″ E)

Fig. 2.20 Poorly indurated and thin ferricrete horizon (a) separates the Neoproterozoic basement (b) from the Asmara Traps (c) (courtesy Sara Passerini)

and spreading since 5 Ma (e.g., Bosworth and Burke 2005), the Traps and associated sediments are considered syn-rift. A different, more complex tectonic history could be derived if the double-spreading model, resumed through new data by Almalki et al. (2014) for the conjugate Arabian margin, could be documented and adopted also for the Eritrea margin (see Sect. 2.4.3).

The Trap volcanic cover is widespread in Ethiopia with sections some 2000 m thick (Abbate et al. 2015), whereas there are only few isolated outliers of some hundreds of meters thick in Eritrea. They are located south of the Barentu/Massawa line and north of the Mareb river. One of the largest and better exposed outcrop is the Asmara outlier (the Asmara plateau) which is 100 km long in a NS direction and 40 km across (Fig. 2.25). With the constant interposition of laterites, the Asmara outlier rests on the basement in its northern sector and on the Adigrat Sandstone in the southern sector.

In the Asmara outlier, Zanettin et al. (1999, 2006a, b) not only recognized one formation of the typical Ethiopian Trap succession (the transitional tholeiitic Aiba/Alaji formation with a K/Ar age of c. 29 Ma), but also established two new formations: the tholeiitic Asmara Basalt, coeval with the Aiba/Alaji Basalt and youngest in the succession, and the mildly alkaline Adi Ugri Basalt with an Ar/Ar age of c. 22 Ma in the upper portion. Four basalts above the laterite or immediately over gave Ar/Ar ages from c. 33 Ma to c. 28 Ma (Drury et al. 1994). Ignimbritic events intervene in the basaltic successions, such as the 24 Ma old Serae Rhyolite (Zanettin et al. 2006a) intercalated in the lower Adi Ugri Basalt and correlative of the alkaline trachyte-rhyolite lavas and domes of the Senafe and of the Adwa-Axum (Ethiopia) areas and of the somewhat older Alaji Rhyolites of the typical northern Ethiopia Traps. At Senafe, a spectacular dome impends over the town, and its related lavas rest on both the Enticho Sandstones and the basement with interposed laterites (Fig. 2.26).

With reference to the distinction of the Ethiopian basalts based on the TiO_2 content (Pik et al. 1998), the NW Ethiopian basalts and the Asmara and Sudanese outliers are included in the low-Ti province (Rooney 2017). In Eritrea, Teklay et al. (2005) found a temporal transition from low to high TiO_2 tholeiitic basalts at the Amba Tekera, 40 km NW of Asmara. The upper basalt levels of the Amba Tekera gave an Early Miocene whole-rock Ar/Ar age of ca. 20 Ma (Teklay et al. 2003).

Connected to the Trap volcanism is an impressive, 7-km-wide and at least 50-km-long, sub-vertical basalt dike swarm that forms small parallel ridges, a few km long, in the landscape between Debaroa and Asmara (Mohr 1999). They impressed Dainelli and Marinelli since “They look like gigantic walls with horizontal basalt prism” (Dainelli and Marinelli 1912). The dike swarm cuts across the Asmara Basalt in a NNE direction. By contrast, many basalt dike swarms, scattered along the escarpment between Asmara and the Gulf of Zula and considered as Trap feeders, are aligned with the Red Sea spreading axis. They are subvertical with an average NNW strike, often pervasively sheared with slickensides mainly along the chilled margins. The slickensides are often dike-parallel with mainly sinistral polarity. Dikes with dip-slip (subvertical) shearing are also observed, and the best evidence for strike slip is on the road Asmara-Massawa between 25 and 28 km downhill Nefasit

Fig. 2.21 Typical laterite-facies profile above the Neoproterozoic basement and beneath the Trap basalts of the Asmara plateau. In the foreground, the grayish laterite envelopes relics of weathered basement. At distance, it becomes mottled with an irregular mixture of bright whitish kaolinitic and red iron-mineral lumps. In the higher slopes, a brownish ferricrete horizon develops beneath the basalt flows. Mereb river valley 30 km south of Asmara (courtesy Sara Passerini)

(1993 campaign with P. Passerini). Strike slip movements parallel to, or at small angles with, the Afar rifts have been considered significant in stretching the continental crust (Abbate et al. 1995) (Fig. 2.27).

The Mendefera intertrappean bed

In the volcanic story of the Asmara plateau, the deposition of a continuous sedimentary level at the transition between the Asmara Basalt and the Adi Ugri Basalt denotes a long period of volcanic and tectonic quiescence (Abbate et al. 2012, 2014). This level, quite evident in the landscape (e.g., along the Mendefera-Barentu road), is a few tens of meters thick and consists of fluviolacustrine silts and mudstones, commonly red and subordinately green and grayish, with lignite seams and tree trunks (Fig. 2.28). Basalt flows and dikes and rhyolitic layers are also present. In a section near Mendefera, the time constrains of this intertrappean intercalation are an Ar/Ar age of c. 29.0 Ma for the top of the Asmara Basalt beneath the sediments and an Ar/Ar age of c. 23.6 Ma for the base of the overlying Adi Ugri basalts (Abbate et al. 2014). The same site has yielded the fossil remains of two Late Oligocene/Early Miocene proboscideans (teeth and maxillary bone of a deinothere and a gomphoteriid) (Abbate et al. 2012), confirming a previous finding by Vialli (1966). The presence of *Gomphotherium sp.* signals the oldest occurrence of this proboscidean before its dispersal toward Asia and Europe.

Further proboscideans remains were found in the lowland intertrappean beds of Dogali near Massawa (Sagri et al.

Fig. 2.22 Extremely fine-grained pinkish kaolinitic laterite horizon wasted by an enigmatic blueish fluid

1998; Shoshani et al. 2006), i.e., two thousand meters below the Mendefera fossils.

2.4.2 Syn-Rift Successions on the Lowland

The Dogali section

The variety of hard rocks and sediments found on the plateau also crop out, with very good exposures, in the classical section near Dogali (20 km east of Massawa) (Porro 1936; Kazmin 1975; Drury et al. 1994; Antonielli et al. 2009) (Fig. 2.29). From bottom, orthogneiss and metasediments of the Ghedem Domain and a laterite layer (Ghebreab and Talbot 2000) with some basement pebbles heralding a precursory plateau rifting (Ghebreab 1998) are overlain by c. 200 m of Trap basalts with the lowermost flow Ar/Ar dated 28.0 Ma (Drury et al. 1994). They are followed by the c. 500-m-thick Dogali Formation (Kazmin 1973) consisting, in the lower half, of lacustrine yellow, brown laminated

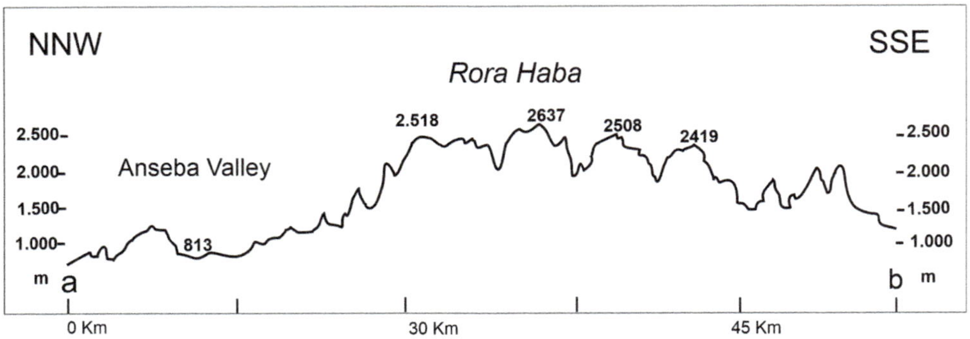

Fig. 2.23 NNW–SSE topographic section along the Roras Range in the Rora Haba area. Remnants of an original wide, faulted and weathered plateau with an average elevation of 2500 m are clearly visible in the central part of the profile. The location of the a–b section is reported in Fig. 2.24. Google Earth coordinates: a: 17°11′35.43″ N, 38°10′25.26″ E; b: 16°07′24.37″ N, 39°30′28.58″ E

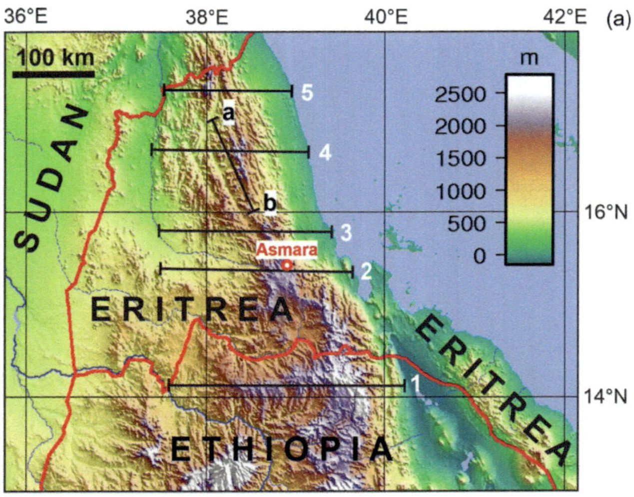

Fig. 2.24 Location of representative topographic profiles; profile a–b Fig. 2.23; profiles 1–5 Fig. 2.37

diatomite clay and silts alternating with basalt flows and tuffs. A channelized conglomerate up to 10 m thick with basalt pebbles and silicified tree trunks yielded Late Oligocene/Early Miocene primitive proboscidean remains (Sagri et al. 1998; Shoshani et al. 2006) and a basalt flow, 15 m above this fossiliferous level, has an Ar/Ar age of 26.8 Ma. In the upper portion of the formation, after 200 m of basalt, the succession includes an alluvial facies with sands and boulders up to 40 cm derived from basement rocks, basalts and felsic volcanites. The occurrence of basement rocks in the clastics indicates that faulting on the plateau and on the plateau shoulders was ongoing. The Dogali Formation terminates with a transition to cyclic arrangements of fluvial to coastal plain associations and episodic marine occurrences with coralline patch reefs, gypsum layers and mollusk shells (*Pecten*, *Ostrea*). Corals and mollusks give a Middle to Late (?) Miocene age (about 10 Ma) for the top of the formation (Zuffardi Comerci 1936; Montanaro Gallitelli 1939, 1973; Selli 1973).

Late Oligocene/Early Miocene paleoenvironmental reconstruction of central and northern Eritrea

The syn-rift Mendefera and Dogali sections with their plateau and lowland development can provide good hints for a Late Oligocene/Early Miocene paleoenvironmental reconstruction of central and northern Eritrea. Initially, the lowland and plateau regions of today were a unique, low-elevated wide area characterized by a relatively thin, volcanic cover (400–500 m as indicated by the Dogali section and the Asmara plateau region), underlain by laterites resting on basement rocks planed by older denudation cycles and dislocated by minor, local rifting. It is difficult to figure out the areal extension of the volcanic cover in Eritrea since the Trap volcanites are missing in northern Eritrea north of the Asmara plateau (Teklay et al. 2005) and resurface with a few tenths of meters thick outcrop only in the Red Sea Hills in Sudan (Kenea et al. 2001).

The basalt thicknesses on the plateau and on the lowland suggest that the volcanic outpourings were rather meager if compared with those of the nearby northern Ethiopian plateau. Moreover, they were intermittent (see intertrappean episodes), distributed in a rather long time interval between 29 and 20 Ma and possibly related to "narrow tongues" (Ebinger and Sleep 1998) of Afar plume material. As regards the environmental conditions, on the basis of similar lithostratigraphical and faunistical content, we can assume that the whole region was covered by a vegetation of moist tropical forest favorable to host proboscideans (see the Ethiopian Chilga lacustrine deposits, Tana Lake, Yemane et al. 1987).

The biophysical environment inferred for the northern and central Eritrea hinterland can also be assumed for the Sahamar-Sahel coastal region, the SSE/NNW elongated, narrow (250 by 25 km) coastal belt at the foot of the escarpment from Massawa to the Sudan border. For this poorly studied region, the available literature reports numerous patches of Neoproterozoic terranes, probably Arag, sporadic outcrops of laterites and Oligocene Dogali

Fig. 2.25 View of the Mendefera–Adi Quala plateau (50 km south of Asmara). The evident break in slope is due to the Trap basalts overlying the basement through a discontinuous interposed unit that in different near localities can be the Oligocene rhyolite body or Paleozoic sediments. An upper less clear break can be appreciated within the Traps and corresponds to an intertrappean sedimentary level

Fig. 2.26 View from the Mendefera-Adi Quala plateau toward SE across the Mareb valley with Miocene trachyrhyolitic volcanic plugs interspersed on the plain. In the near background, well evident is the triangular shape of the Metera plug incumbent on the Metera village and on the town of Senafe. Metera was an important Axumitic town (first centuries AD). In the extreme background, the ridges of the Eritrean plateau rim include the Amba Soira—the highest mountain peak (3250 m) in Eritrea

Fig. 2.27 Tertiary basalt swarm near Nefasit 40 km East of Asmara (see text)

Fig. 2.28 Red silts, mudstones and conglomerates of the intertrappean episode in the Mendefera Trap section. This intercalation yielded Oligocene big mammal faunas (Vialli 1966; Abbate et al. 2014). Despite their usual limited thickness, these intercalations cover a 1–2 Ma of volcanic quiescence

deposits, widespread fossiliferous Neogene to Pleistocene shallow marine and alluvial sediments (Porro 1936; Kazmin 1973; Merla et al. 1979; Drury et al. 1994). As shown in ENI (1971) and Kazmin's (1973) maps, high-grade basement chunks and Paleozoic metasediments can be found aligned in a 15-km-wide NNW-striking belt close to the foot of the escarpment.

Early to Late Miocene syn-rift deposition and volcanism

The syn-rift deposition continued during the Early to Middle Miocene with the deltaic to marine sandy and shaly Habab Formation. It can be seen in sparse exposures among the coast and widespread and very thick (up to 2500 m) covers in offshore drillings (Savoyat et al. 1989).

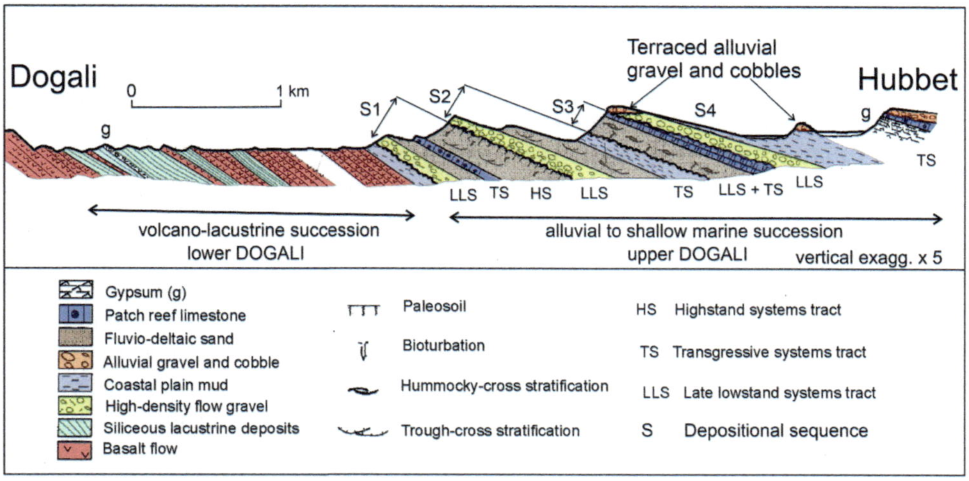

Fig. 2.29 Dogali succession restored in a SSW/NNE section between Dogali and Hibbet 15 km WNW of Asmara. In the upper—Dogali four depositional sequences (S1–S4). At the base of section—Paleocene to Eocene laterites with basalt clastics (Sagri et al. 1998)

The Late Miocene syn-rift rock units document an event that is well present in the Southern Red Sea drillings and have a notable effect on the whole Red Sea sedimentation: the restriction, up to stalling, of water circulation between the Red Sea and the Gulf of Aden. This was due to re-arrangements at the border of the Arabian/Nubian plates in the Bab El Mandeb and in northern Egypt (Bosworth and Burke 2005). This event prompted the deposition of the Amber Salt, three thousand meters of salt that become five thousand when involved in diapirism, as found in offshore drillings and geophysical prospecting.

A phase of intense rifting affected the plateau and the continental platform providing coarse clastic supply, also from the basement, to adjacent offshore and onshore basins (Desset Formation). In the Dogali area and along the whole Sahamar-Sahel, this formation is a prevailing sandy shallow marine unit with many pebbly inflows in the lower portion, sporadic coral reefs and evaporites and a few basalt flows (Bosworth et al. 2012). In the upper portion, it becomes deltaic and alluvial. According to Porro (1936), the Desset boulder beds have built a continuous hilly chain from the Sudanese border to the Gulf of Zula. The Desset Fm., more than 2000 m thick, is Late Miocene in age with a possible limited extension to the Early Pliocene (c. 5 Ma). In this case, the top of the Desset at the end of the syn-rift rock units was likely coeval with the first stages of the ocean spreading in the Southern Red Sea at 17° N lat. (Röser 1975).

The syn-rift Trap volcanism in central Eritrea can be matched in southern Eritrea with stages of the Afar plume activity represented by the Mabla Rhyolites and Dalha Basalts of Miocene to Pliocene age (CNR-CNRS 1971, 1975) widely outcropping in Djibouti but also present in the southernmost Eritrea near Edd. In the same area, there are outcrops of the Djioita granite (Varet, this volume) with an Ar/Ar age of 22 Ma. This body is correlative of similar granites of sialic crustal origin, such as the Affara-Dara, predominantly outcropping on the border of the old rifted Nubian continental crust and adjacent to new Afar units (Barberi et al. 1972).

Structural significance of the syn-rift successions during the Oligo-Miocene

The magmatic and sedimentary rock units and the structural setting of the syn-rift successions provide meaningful data for tracing and interpreting the paleogeographic events that marked the development of central and northern Eritrea during the Oligo-Miocene. As already outlined, between 29 and 20 Ma, this area, from the present-day continental platform to Sudan, was intermittently intruded and covered by the northernmost sprouts of the Afar plume, namely the Trap effusions (Ebinger and Casey 2001). In the open debate on what was the pre-Trap topography (e.g., Pik et al. 2003; Abbate et al. 2014; Coltorti et al. 2015; Sembroni et al. 2016), central and northern Eritrea is presumed to have been consisting of a wide low-elevation laterized peneplain covered by Trap basalts. After an early modest rifting event with basement coarse clastics emplaced before the Trap effusion, further sparse basalt basement coarse clastics supply is found along the Dogali succession. These clastics are high density flow gravels and alternate with much more common fine-grained lacustrine, fluvio-deltaic sediments. Two main coarse events are known: one toward the top of the Dogali Formation (c. 25 Ma) and a younger one, at the base of the Late Miocene Desset Formation (c. 10 Ma), deposited after fifteen million years without or with scarce coarse clastic sedimentation. We assume that the source area of the scarce clastic materials was the faulted eastern shoulder of a rising NNW/SSE elongated swell produced by the Afar plume impinging under the Eritrean crust (see later). Rifting, concomitant with the progressive uplift of the swell, produced a tectonic and erosional unroofing of the new highlands with sporadic clastic releases. The impingement of the Afar plume, the plate boundary forces producing the overall separation of western Arabia from Nubian Africa and, to a lesser degree, the crustal lateral extension induced by dike injections triggered and maintained these intermittent rifting processes with coarse sediments discharge.

Some datings are available from fission tracks analyses (Abbate et al. 2002; Ghebreab et al. 2002), which report a crustal cooling event with accelerated phase of denudation in the Early/Middle Miocene. Erosion resumed at the expenses of the Trap volcanites, which were covering the plateau. They were almost completely eroded and patches of laterites, probably lateritic ironstones or duricrusts such as those on the flat tops of the Roras (Dainelli 1943; Drury et al. 1994) were left above the basement (Fig. 2.30) (see also additional details in Sect. 2.3.4).

2.4.3 Post-rift Successions and Events

The Dahlak islands and a Danakil Homo

At the beginning of the Pliocene, when the Southern Red Sea began its active spreading offshore Eritrea, the Desset Fm. was overlain by the shallow marine sandy/shaly Dhunishub Fm. filling up as much as 2–3 km deep basins, adjacent to the Red Sea axial trough. Its littoral depositional environment promoted the growth of numerous carbonate platforms. The widest are the Eritrean Dahlak islands and the Saudi Farasan Islands (MacFadyen et al. 1930; Angelucci et al. 1981; Khalil 2012).

The Dahlak archipelago, with more than 200 islands (the largest island, Dahlak Khebir, is 760 km^2) and important trade center since the first centuries AD, sits on the Dahlak platform, a salient morphological modification in the offshore basins of central Eritrea (Fig. 2.31). Starting from 150 km north of Massawa (at c. 17° N), the continental platform widens abruptly up to 250 km and the conjugate shelf of the Saudi Farasan (the Roman garrison *Portus ferresanus*, II century AD) islands becomes similarly large. The interposed 5 Ma to present oceanic crust is c. 100 km wide.

The notable width of the two conjugate platforms was associated with a discontinuous spreading of the Southern Red Sea. According to Sultan et al. (1992) and Almalki et al. (2014), during the whole Miocene, there was a stalling in the ridge activity and the two platforms prograded and enlarged on the older oceanic crust until the spreading resumed during the Plio-Pleistocene. This still debated hypothesis is supported by magnetics and gravity data from oil exploration in the Farasan platform.

The vast shelf of the Dahlaks is less than 200 m deep and, in general, the islands rise 10–20 m asl. Only Dahlak Kebir attains a length exceeding 50 km. The islands are circular or semicircular and those with irregular edges also exhibit rounded bays and arcuate promontories and tongues. A peculiar feature is the domal structures ranging from a few meters to tens of meters, in places with the occurrence of diapiric salt. In the late thirties of the last century, deep oil drillings on the islands by AGIP encountered two thousand meters of Miocene salt after 100 m of Dahlak Reef Limestone and 200 m of clay, marl and sand (Carbone et al. 1998). The salt is correlated to the Amber Salt and the overlying unit to the Desset Formation. Diapirs rising from this impressive saline body generated the all-size domal structures in the Dahlak Reef Limestones (Fig. 2.32).

The substratum on which the present-day reefs grow is the Dahlak Reef Limestone of Pleistocene age (Montanaro Gallitelli 1973). In the nice map of Angelucci et al. (1981a, b), the Dahlak Reef Limestone is block faulted and gently folded. Yet, the strongest evidence of an ongoing active

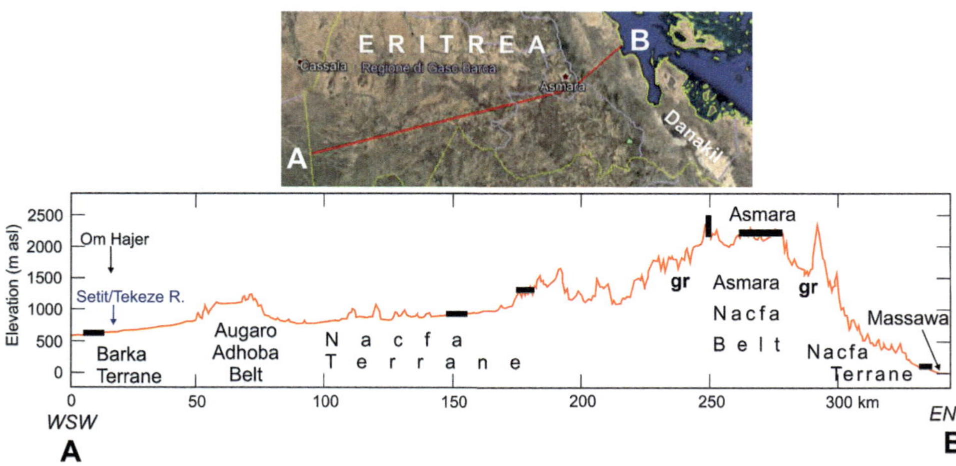

Fig. 2.30 This plateau cross section depicts the laterites associated with the pre-trappean surface recognized by Drury et al. (1994) in central Eritrea. From their original position, not far from the sea level, the laterites have been subjected to differential vertical movements described in text (Dainelli 1943; Drury et al. 1994). Terranes attribution based on Ghebreab and Talbot (2000) and Ghebreab et al. (2005)

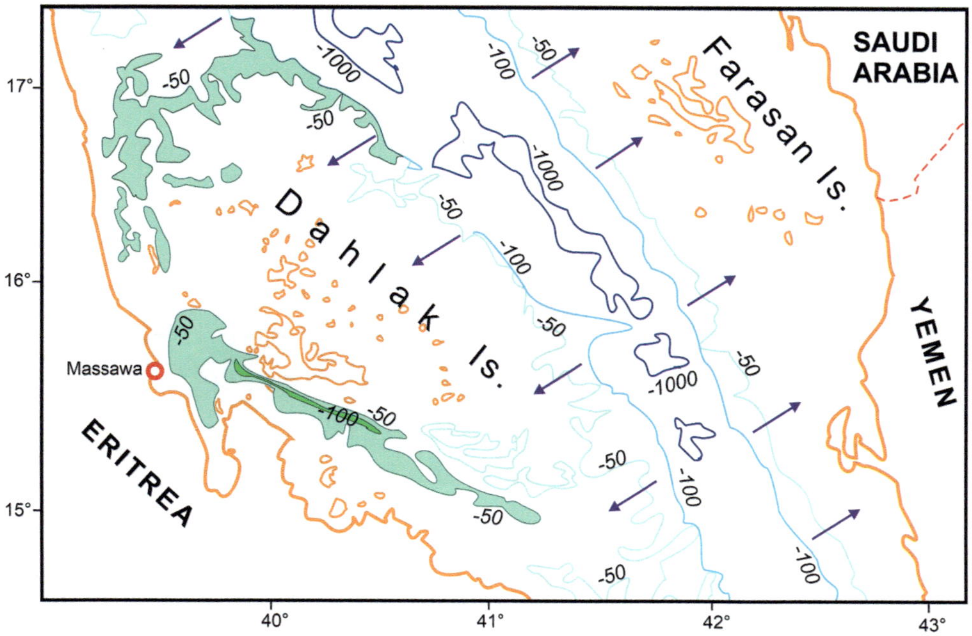

Fig. 2.31 The Dalhak and the Farasan Islands facing each other on conjugate margins separated by an active spreading axis since the Pliocene. The larger number of islets in both archipelagos is accounted for by a thinner (less than 100 m) present-day carbonate platform that can be easily pierced or, more generally, deformed by the upward rising salt domes originated from the extremely thick (more than 2000 m) evaporite level beneath the platform. The 50–100 deep narrow trough Massawa Channel (green in the map) separates the Dhalak islands from the coast (Massawa and Buri Peninsula). It does not find a similar feature in the Farasan Islands (modified after GPS Nautical Maps I-Boating Free Marine Navigation, 2022). The light blue lines are the −50 m isobaths; the blue lines are the −100 m isobaths; the dark blue lines are the −1000 m isobaths. The purple arrows indicate the Red Sea floor spreading

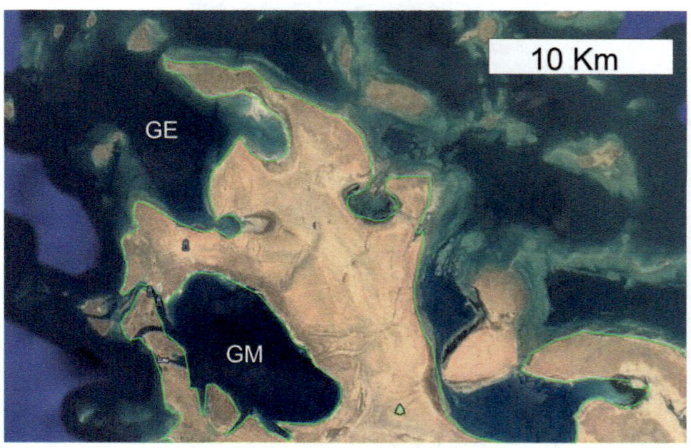

Fig. 2.32 Ghubbet Entatu (GE) and Ghubbet Mus Mefit (GM) are two bights in the NW portion of Dahlak Khebir, the largest island of the Dahlak archipelago. They are hollows as deep as ca. 50 m and with the longest axis of about 20 km. Ghubbet Mus Mefit is connected to the open sea through a tide channel. The ovoidal (as in Ghubbet Mus Mefit) or circular shapes of the hollows are common in many islands and islets of the archipelago. They are deep holes left by the leaching of the salt domes. See also Fig. 2.26

tectonics is the Massawa Channel, a NW/SE elongated rift which is 250 km long, 20 km wide, no more than 200 m deep. The channel starts from Massawa and, parallel to coast, reaches Tiho at a more or less constant distance of some tens of kilometers from the coast (Frazier 1970). Its occurrence can be related to the stretching of the Nubian plate margin in the process of separation of Arabia from Africa, and it can be seen as the first stage of a marginal basin (compare with eastern Ethiopian plateau margin adjacent to Afar, Zwaan et al. 2020). A rifting activity since the Late Oligocene is reported in the Southern Red Sea from oil prospecting by Hughes et al. (1991).

Close to Massawa, the regional NNW structural trend, marked by the alignment of the coastal belt, splits into two directions. One continues with almost the same NNW strike through the Gulf of Zula corridor and penetrates the hinterland toward the Afar axial range. The other points to SE, keeps parallel to the anticlockwise rotated Danakil block and goes on southward as far as Assab.

South of the Gulf of Zula, the NNW trend runs at the foot of the escarpment through small, clastic and post-rift basins. One of these is the Buia basin (Abbate et al. 2004; Ghinassi et al. 2015) (see also Chaps. 6 and 7 of this book for details) at the foot of the Alid volcano along the corridor Zula/Samoti/Badda that was a possible Pleistocene seaway connection between the Red Sea and the Danakil depression (Fig. 2.33). The Buia basin hosts sediments of alluvial plain (Fig. 2.34), ponds, playas and lacustrine delta environment. In a deltaic system, a *Homo erectus/ergaster* skull (Abbate et al. 1998; Macchiarelli et al. 1998), dated one million years

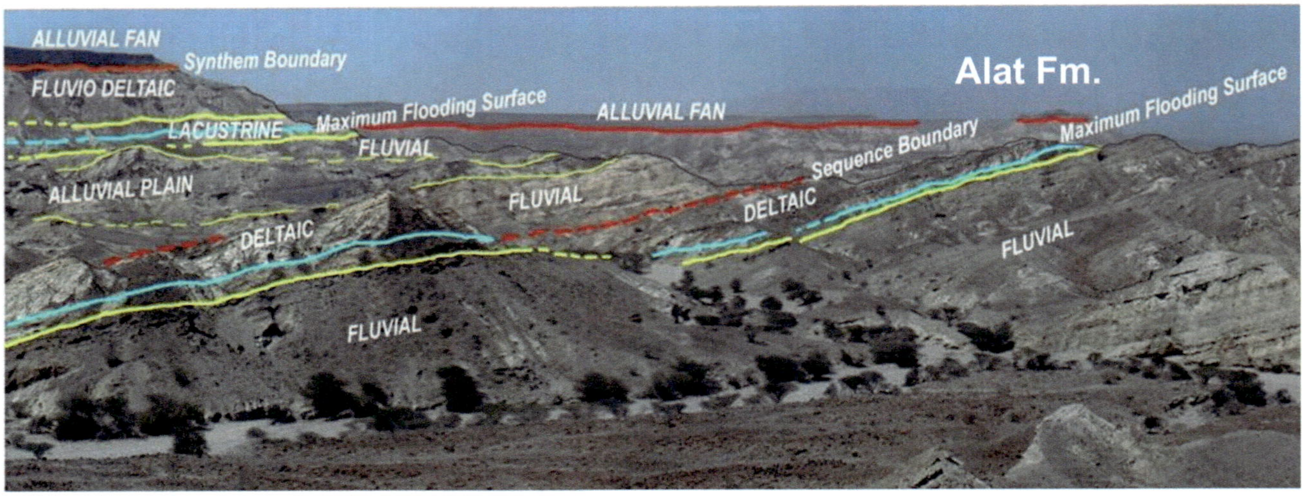

Fig. 2.33 Post-rift Neogene sediments at the apex of the Afar depression 20 km south of the Gulf of Zula. In the lowlands, at the foot of the highest rim of the Eritrean plateau between Senafe and Adi Keih, heavy sediment fluxes promote the birth of many alluvial basins, such as that of the Dandero river (in the foreground). The Dandero Group, a 500-m-thick sedimentary pile at the foot of the Mt. Aro, includes fluvial, alluvial plain, fluviodeltaic, deltaic and lacustrine deposits, irregularly repeated and intersected through time (Abbate et al. 1998). In a lacustrine to deltaic, one-million-year-old, unit (the Alat Formation, see notation on the right of the figure) an *erectus/ergaster Homo* cranium has been found (Macchiarelli et al. 1998)

Fig. 2.34 Middle Pleistocene deposits of the Dandero river filling a valley cut in the Proterozoic Nakfa green schists (20 km south of the Alid volcano)

through paleomagnetic (Albianelli and Napoleone 2004) and fission track analyses (Bigazzi et al. 2004), was found. The Buia *Homo* fills an important gap between early African *Homo erectus* (2.2–1.4 Ma) and the later African *Homo heidelbergensis* (0.65 Ma) (Fig. 2.35).

Fossils of big mammals and reptiles (Martínez-Navarro et al. 2004) and a rich dough of Acheulean stone tools (Martini et al. 2004) were also found together with the human remains (Fig. 2.35) South of Zula there are some northernmost sparse outcrops of the Red Series, a typical

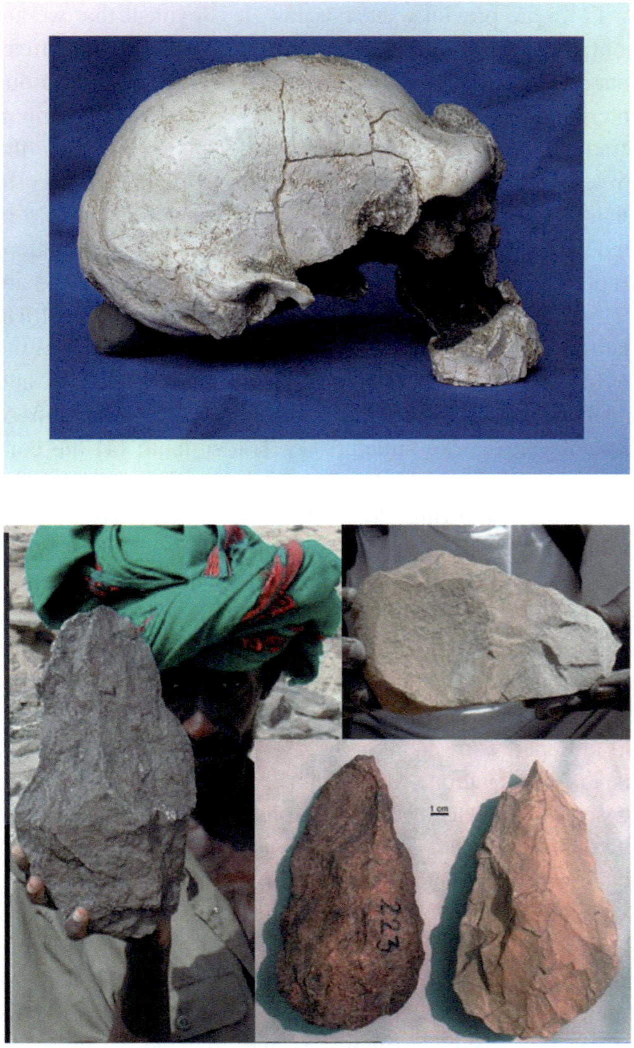

Fig. 2.35 *Erectus/ergaster Homo* cranium found in the Alat Fm. (Dandero Group) together with abundant Acheulean stone tools and a rich large mammal fauna

marker of the Afar realm (Barberi et al. 1970) with red conglomerates, sand and mudstones, sometimes gypsiferous and rare freshwater lumachella limestones, basalt flows and acidic tuff. With an estimated thickness of one thousand meters, the Red Series was deposited by alluvial fans, high-energy streams and in small ephemeral lakes. The radiometric dating of the basalts spans from the Early Miocene to the Pliocene (Bannert et al. 1970) sharing both syn- and post-rift tectonic regimes.

The second trend (NW/SE) is apparent on the coastal belt and its adjacent hinterland stretching from the Buri peninsula, through the elongated Danakil block, to the volcanic bodies of the Assab region. The post-rift volcanic activity is witnessed by stratoid basaltic lavas and ignimbritic sheets. They are representatives of the post-rift Afar Stratoid Series (Plio-Pleistocene, Barberi and Santacroce 1980) and by recent volcanic centers. The latter include, from north to south, the Jalua (Mesfin and Yohannes 2014), Alid (rhyolite dome with basaltic lava fields, active 1817; Marini 1938, 1998; Clynne et al. 1996), Dubbi, Nabro (eruption June 2011; Hamiel and Baer 2016) and Mallahle stratovolcanoes.

According to Ghebreab and Talbot (2000), a c. 60-km-long and E–W trending lineament would connect the Alid volcano with the volcanic plugs of the Senafe region.

Moreover, a spectacular range of numerous basaltic cinder cones and lava flows makes up the Assab volcanic field. It covers an area of c. 5000 km^2 and its lava flows reach the Red Sea.

The Neogene syn- and post-structural events: research history and a new proposal

The times of drift events had a significant role in the Neogene structural evolution of the Eritrean continental margin. The stretching during the separation of Arabia from Nubian caused progressive rifting in the hinterland and new structural discontinuities followed the c. NNW trend of the Precambrian basement, mainly determined by terrane boundaries. It is worth noticing that this direction parallels the Red Sea ridge axis and many dikes swarms on the escarpment.

Many suggestions have been put forward on the main processes originating the scarp and causing its dismantling or modification since the Miocene.

Abul-Haggag (1961) recognized that the eastern margin of the Eritrean plateau does not represent "an original fault-face" but just a retreating scarp and that the scarp still preserves a profile with successive benches due to step faulting.

According to Drury et al. (1994), a possible main lineament is the major normal fault or extensional system which from south of Asmara heads northward retracing Precambrian lineations. From the western border fault of the Damas graben, it proceeds to NNW at the foot of the northern plateau in the Samar-Sahel region. An eastward dipping detachment from the Damas western border is supposed to have activated a series of ramps and horst across the continental margin.

Talbot and Ghebreab (1997) and Ghebreab and Talbot (2000) assume a low-angle detachment across the continental margin with associated top-to-basin domino faults starting from the western border of the Damas graben. The lithosphere was extended and thinned, and core complexes of high-grade crustal rocks were exposed to surface. The structure evolved into a new deeper detachment producing an east-directed ramp system and a faulted monoclinal flexure considered as the Red Sea escarpment in Eritrea. Although some authors (e.g., Beyth et al. 2003) claim that the low-angle detachments and associated structures are the products of Neoproterozoic collapse events and not Tertiary deformations related to the opening of the Red Sea, Talbot and Ghebreab (1997) and Ghebreab and Talbot (2000) reconstructions are the most commonly accepted.

The discontinuous row of marginal basins from the Damas graben in central Eritrea to the Robit graben in central Ethiopia (Abbate et al. 2014; Zwaan et al. 2020) has been framed in the Southern Red Sea rifting by Tesfaye and Ghebreab (2013). They consider the western margin of these marginal basins as the breakaway zone of a ramp-flat detachment in which the Danakil horst was the flat.

The denudation, whether tectonical or erosional, persists to be a key question in the interpretation of the Eritrean high-elevation passive margin. In addition to classical regional geology investigations, this issue has been approached with apatite fission track (AFT) dating and (U/Th)He thermochronometry.

Abbate et al. (2002) and Balestrieri et al. (2005) sampled the Eritrean margin along four transects coast-perpendicular and latitudinally distributed from Keren to Buia (south of Massawa) for AFT and (U/Th)He analyses. These authors found a positive correlation between FT ages and sample elevation indicative of an erosive denudation of the margin starting from a main border lineament. They also assumed that during the Miocene, a remarkable uplift affected the margin, so that by Middle Miocene the material removed by erosion was c. a couple of kilometers thick. Moreover, the (U/Th)He ages of the transects typify a denudation of the escarpment predominantly evolved by downwearing with a main phase of post-break up erosion starting at about 15 Ma from a regional NNW-trending border lineament parallel to the coast and stretched from the Labka river (50 km NE of Keren) through Damas and Ghedem to Buia in the northernmost Danakil depression.

Ghebreab et al. (2002) reported that the ages of four samples collected in the Asmara-Massawa area were not positively correlated with sample elevation. This result would suggest a process of tectonic denudation through low-angle detachments rather than erosion (Ghebreab et al. 2002). Three AFT profiles carried out across the Ghedem Mt. a Proterozoic feature 20 km south of Massawa lead to a similar conclusion.

Drury et al. (2005, Comment) objected that Balestrieri et al. (2005) and Abbate et al. (2002) assumptions in favor of an erosional denudation of the Eritrean margin were founded on a correlation between FT ages and sample elevation erroneously considered positive. From this incorrect attribution derive further disputable aspects connected with the margin degradation, such as the position of the border faults constrained by the occurrence of the youngest FT ages, the FT age and interpretation of some inliers (Ghedem) and the time of the onset of post-break up erosion.

From the previous short review, it is evident that we are confronted with different conceptual models for the Eritrean margin, and we think that a crucial area subject to various interpretations is from Asmara to the sea. For this margin, a stratigraphic and structural detailed mapping of the Proterozoic units of the escarpment and of the Tertiary to Pleistocene units of the lowland would consistently help in understanding the Eritrean margin formation and evolution. Important issues to be carried out or further investigated are as follows: (1) a denser re-sampling for FT and (U/Th)He dating after discordant interpretations; (2) new datings of the mylonites related to low-angle detachments (Talbot and Ghebreab 1997; Ghebreab et al. 2005); (3) detection of possible breakaway structures or border fault; (4) the conceivable occurrence of post-Proterozoic cover shreds on the escarpment mirroring Drury et al. (1994) Oligocene datum on the western slope of the plateau.

Trying to clarify this uncertain picture, we propose a section (Fig. 2.36) from Asmara to the sea drawing particular attention to the relationships between the post-Proterozoic cover on the plateau (Mendefera section) and on the lowland (Dogali section). Although presently 50 km apart and separated by a difference of two thousand meters in elevation, we assume that during the Oligocene these two low-elevated successions were contiguous above the basement units. Their correlability is based on common features, such as the same stratigraphic succession with coeval Trap basalts and intertrappean sediments, the same fossiliferous content (*Deinotherium* and *Gomphotherium*) and common Late Oligocene radiometric dating of the volcanites. The local occurrence of some basement clastics below the basalt distinguishes the lowland section and indicates the inception of some rifting action. Traces of this event are missing in all the numerous exposures of the plateau outcrops.

In our reconstruction, the plume-generated uplifted region in its progressive rising got faulted along a main, steeply east-dipping extensional feature running NNW/SSE, located halfway between Asmara and Massawa, a few kilometers east of the Damas region. The probably upflexured footwall constituted the edge of the Eritrean plateau with its more than two thousand meters. On the hanging wall to the east of the fault, the lowland units were only scarcely affected by the upheaval or by the main fault, while the diffusely flexure and faulted footwall rocks, crowned by a retrogressing rim, were undergoing uplift and slope erosion becoming a morphologically rough escarpment.

As shown in Fig. 2.36, we assume that the new fault was located and developed along the Neoproterozoic "transition zone", a lithologically heterogeneous, intensely sheared, up to 6–7 km wide, NNW/SSE trending tectonic zone marking

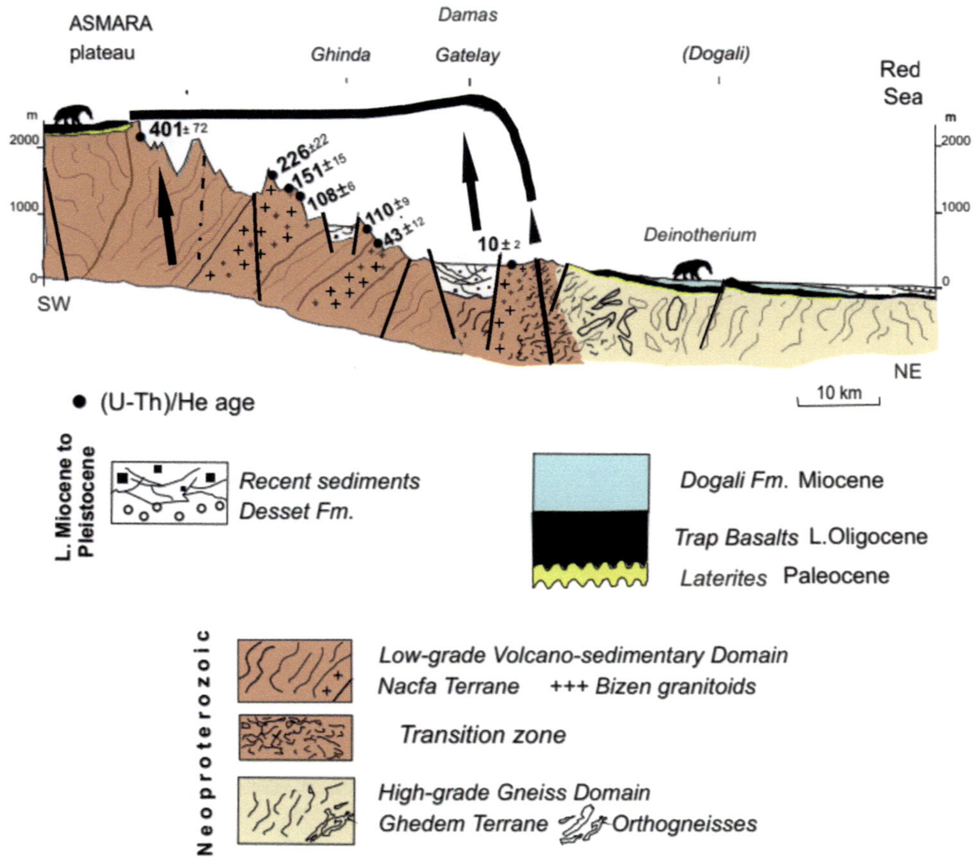

Fig. 2.36 General geological section across the Eritrean margin between Asmara and the Red Sea (20 km north of Massawa). Black dots: apatite fission track ages as in Abbate et al. (2002). The Afar plume impinging on the low-elevation Central Eritrea continental margin of the Nubian plate resulted in a major fault that disjointed by almost two thousand meters' difference in height the Asmara region from the Red Sea lowland. The footwall becomes the dissected margin of the Asmara plateau. Modified after Drury et al. (1994), Ghebreab (1998), Ghebreab and Talbot (2000), Ghebreab et al. (2005, 2007), Andersson et al. (2006)

the Neoproterozoic juxtaposition of the younger Nakfa terrane against the older Ghedem terrane (Ghebreab and Talbot 2000; Andersson et al. 2006; Ghebreab et al. 2007). This fault can be traced for more than one hundred kilometers from the Damas region to south of the Gulf of Zula. On its Neoproterozoic prominent low-angle deformations, younger steep normal faults were superimposed during the Tertiary extension related to the Red Sea rifting (Drury et al. 1994; Talbot and Ghebreab 1997).

The onset of the footwall exhumation, possibly eased by an isostatic response to extension, can be roughly time constrained by the age of the youngest common sedimentary, biostratigraphic or volcanic record on the plateau and on the lowland and/or by the thermal history of the footwall rocks. Based on the first approaches, the onset of the exhumation should postdate the Late Oligocene/Early Miocene age of the volcanites and of a common low-land environment suggested by stratigraphic considerations (see before). Moreover, in the Bizen granitoids footwall, close to the transition zone and at 300 m asl, the thermal history records a FT age of 10 ± 2 Ma and He ages of 5.5 and 9.2. A geothermal gradient of 25 °C/km and mean annual surface temperature of 25 °C requires 1.5–2.0 km of erosion removal since rifting began 15 Ma ago (Balestrieri et al. 2005), and this could approximate the difference in elevation between the plateau rim and the lowland units.

2.5 The Physiographic Evolution of the Eritrean Plateau and Its Denudation Rate

Physiographic variations through five profiles across central and northern Eritrea and northern Ethiopia

As anticipated in Sect. 2.4.2, at about the Late Oligocene the impinging Afar plume started to affect central and northern Eritrea that were a vast, low-elevated area since at least 50 Ma. Concurrently, the Eritrea continental margin was progressively rifted and pulled apart toward south Arabia.

The interaction between the plume-driven uplift and denudation starting from and almost even surface poses interesting questions about their effects on landscape morphology and dynamics.

Some hints are provided by a morphological appraisal of the present-day features of the Eritrean plateau through five WE topographical profiles with vertical exaggeration of c. 1:12 distributed along 300 km from the northernmost Ethiopia close to the Eritrea/Ethiopia border to the northern Eritrea close to the Eritrea/Sudan border (Fig. 2.37).

Section 1 runs all in the northernmost Ethiopia starting where the Setit (Tekeze) river enters Eritrea and terminates near the western border of the northern Afar depression. We

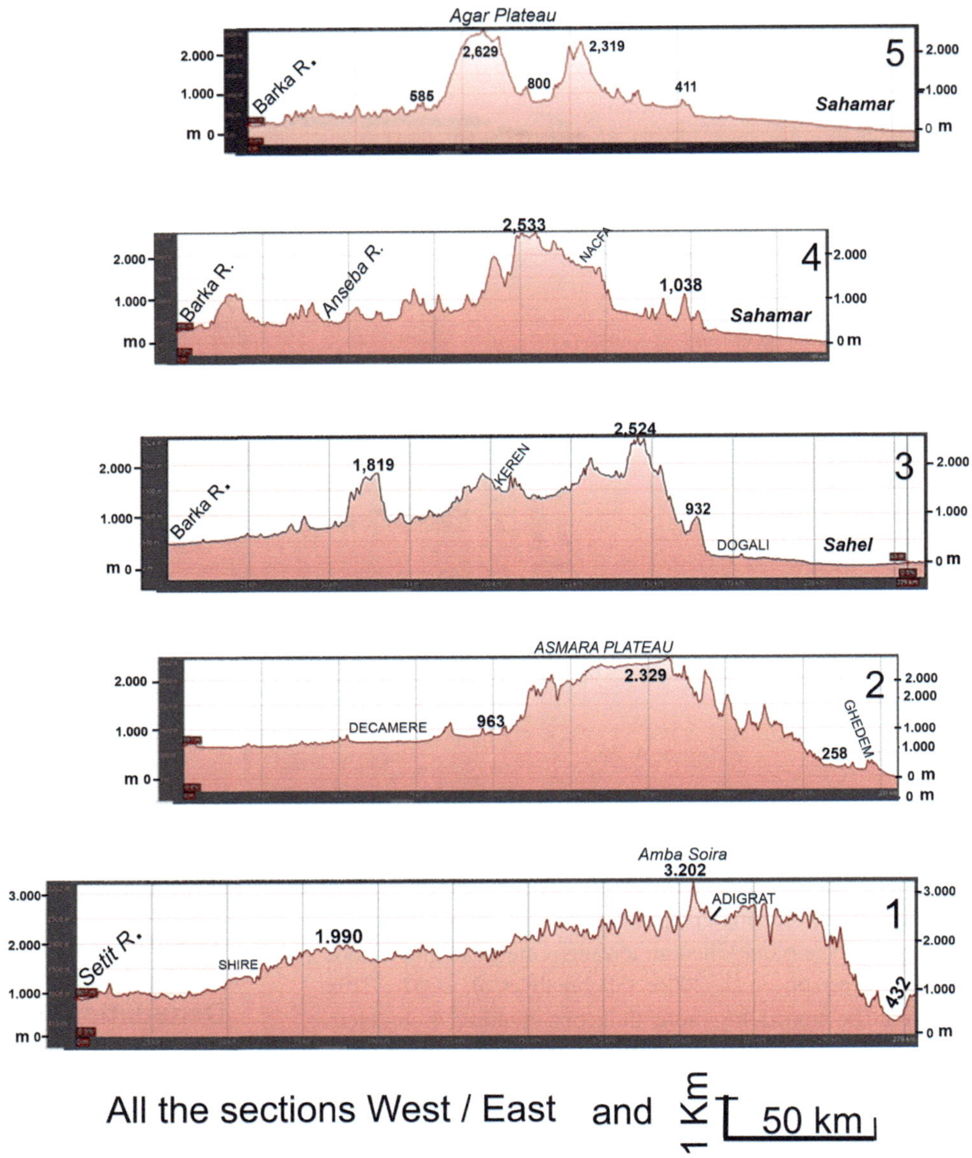

Fig. 2.37 Starting from Section 1 close to the Eritrea/Ethiopia border and heading north to Sudan the five sections record different physiographic features at growing latitudes along the Eritrean plateau showing similarities and differences of the crossed areas due to distinct responses to tectonic and magmatic events and erosive action

assume it keeps in some way the original physiographic conditions that appear more or less modified in Sections 2–5 across the Eritrean plateau. From the Adigrat-Axum region to the Shire region Section 1 shows a gentle westward slope across a WSW/ENE ridge transversal to the margin of the Eritrean/Ethiopian plateau with basement rocks covered by sub-horizontal Paleozoic sediments, large Paleogene Trap outcrops above laterites and, locally, more recent trachytes (see Garland 1980, section A–B). Large areas in the eastern half of the section have heights exceeding 2500 m, and north of the section the Amba Soira with its 3018 m is the highest elevation in Eritrea. It should be taken into account that without the large vertical scale exaggeration, the real inclination of the entire western flank of the swell (i.e., from Amba Soira to Shire) turns out less than one degree, and it is likely to envisage similar topographical conditions also for pre-erosion times with the diffuse, synchronous actions of plume-generated uplift and erosive processes.

As regards the altitude of this peneplain, we allow for a laterite level covered by a few hundred meters of Traps (see the Mendefera area) and obtain an original flat surface with a west/east extension of 250–300 km raising progressively for at least 2000 m while gradually eroded. The age of this plume-promoted upheaval is poorly constrained. A Late Oligocene/Early Miocene initiation was indicated in the previous paragraphs and the characteristics of the modern drainage system, in which several examples of ongoing river captures are present (see Chap. 4) would imply a persisting upheaval.

At contrast with most of the section, an abrupt descent is found at the extreme east of Section 1. It is marked by the Neogene Garsat graben, one of the northernmost marginal basins at the foot of the Eritrean/Ethiopian plateau (Zwaan et al. 2020).

Sections 2–5 are intended to show possible stages of the geomorphological evolution starting from a profile similar to that of Section 1.

The eastern half of Section 2 dominated by more or less eroded Trap basalts and post-tectonic granitoids reminds the corresponding portion in Section 1. In the western portion near Barentu, the essentially flat-lying surface with Trap basalts and laterites (see Fig. 2.37) can be correlated with similar occurrences at the top of the Asmara plateau.

Sections 2 and 3 provide a glimpse of some peculiarities relative to the coastal belt. Between these sections, almost in correspondence of an ideal Massawa-Keren line, there is a marked change in the coastal belt physiography. North of Massawa the Red Sea coast opens into the 30–40-km-wide and 300-km-long Sahamar-Sahel coastal plain. This plain maintains a constant width and a rectilinear seaward profile proceeding northward as far as the Tokar delta in Sudan. It is predominantly covered by Quaternary deposits, but outcrops of older units, those dated for the Proterozoic included, are also present. At the foot of the highlands, the plain is bound by an extended NS trending normal fault system connected to the rifting of the continental margin (Drury et al. 1994).

South of the Massawa-Keren line, for about 200 km toward SE as far as the Anfile Bay, the coastal lowland becomes narrow and indented with bays and cliffs. This change in the coastal physiography is matched inland by the reduction of width of the plateau, which rapidly decreases from the ca. 150–200 km south of Asmara to less than 50 km northward. Concomitantly, the N–S trending main escarpment makes an eastward shift (see Fig. 2.2). To the west, a distinctive morphological feature, trending approximately W–E, is the upper course of the Barka river which, after a long N–S path from the Eritrean border, turns eastward toward Asmara. Its active headwater erosion and the occurrence in the same region of WNW–ESE trending Tertiary dilatation fractures (Drury et al. 1994) resulted in the opening of a wide alluvial valley floor at the expense of the adjacent highland, which also reaches the foot of the Asmara plateau.

The occurrence of these west–east tectonic lines cutting across the central Eritrean plateau and the northern Ethiopian plateau can be matched with similar volcano-tectonic lineaments more to the south, such as the Tullu Wellel (Nekempti)-Yerer (Addis Ababa) and Bonga-Goba (Abbate and Sagri 1980; Mohr 1987; Wolde 1989; Abebe et al. 1998) interpreted to result from Cenozoic reactivation of pre-existing Neoproterozoic structures.

Sections 3, 4 and 5 cut across an area traditionally known as the Eritrean Plateau, but rightly named "highland" by Dainelli (1943) because of its rugged topography with reduced plateau remnants. Unlike Sections 4 and 5, Section 3 still shows a rough declining westward slope.

Section 4 intersects the valleys of the Anseba and Barka rivers, two main rivers of Eritrea flowing to the north, i.e., perpendicularly to the cross section. Their erosive action at the expenses of the western slope of the Rora range is evident. A faulted step, hosting the town of Nakfa, separates the highest (2523 m asl) central peak from the seaward basal hills.

Sections 3 and 4 have rather similar profiles with their eastern flank markedly down faulted.

Section 5 exhibits a topographic profile rather different from the previous ones. The uplifted portion has been severely eroded, and only two prominent, 2500 m high, mountains made of Neoproterozoic massive rhyolites of the Agar terrane survived substantial erosion.

Moreover, the highest elevations in the five sections are found with repeated values around 2500–3000 m generally close to the topographic step toward the lowlands. This suggests that the eastern border of a general, unique surface has been raised with the same intensity along a distance of at least 500 km from central to northern Eritrea. Moreover, if the size of the swell in the profiles 1–5 is compared, it is evident that, on the whole, it gets less prominent northward. Was the plume losing its capacity of building highlands in this direction? Yet, Zeyen et al. (1997) traced the path of the Afar plume across Sudan and as far as southern Egypt through heat flow analyses and pointed out that, in Eritrea, the volcanic activity connected in some way to the plume had already substantially diminished before entering Sudan. This suggests that northern Eritrea was less and less affected by plume activity without neglecting a northward, more effective denudation.

Denudational rates of elevated regions

It is possible to infer long-term denudational rates of elevated regions such as the Eritrean plateau. If one assumes the existence of a post-Proterozoic cover no more than 500 m thick (now completely eroded) on the plateau and erosion processes starting 15 Ma ago (Balestrieri et al. 2005), a denudation rate (DR) of 33 mm/ka is obtained, i.e., one-third of what was found by apatite/He data (Balestrieri et al. 2005) in the margin south of Asmara. Even lower DR (around 15 mm) is reported by McDougall et al. (1975) for the Blue Nile and Tekeze river basins since the Miocene. Pik et al. (2003) used the same thermochronometry for the Tekeze and upper Blue Nile basins and obtained DR values of 30–35 and 29–34 mm/ka, respectively. The average of all these data is 32 mm/ka, which is almost coincident with DR of 33 mm/ka calculated in this study for Eritrea. This slow denudation rate in a tectonic active region as the Eritrean plateau was perceived by Dainelli (1943) and Abul-Haggag (1961) and the latter author referred to it as "the slow tempo of erosion". The $^{40}Ar/^{39}Ar$ dating of low-Ti tholeiitic and high-Ti alkaline flood basalts in Eritrea of Teklay et al. (2003) suggests that erosion started about 20 Ma, leading to a lower value for the DR. On the base of these data, the DR of the Eritrean plateau can be approximate at 30 mm/ka. Further studies and dating are however necessary to fix this value with greater accuracy.

In the literature, denudation rate (DR) varies widely, commonly across two–three orders of magnitude. Such a variability depends on a number of factors including the method used to calculate the DR (Bishop 1985); the age and the kind of rocks considered; the tectonic history; the prevailing dislocation mechanism and climate. In spite of the many studies, there are still considerable uncertainties about the dominant factors controlling the DR.

Notwithstanding the impressive uplifting, the DR calculated for Eritrea is relatively low if compared with the mean estimate of 30 mm/ka obtained by fission tracks during the Cenozoic in regions of subdued relief, but it is not far from the estimated present-day global mean of 43 mm/ka (Summerfield 1991). Moreover, the DR of the Eritrean is an average value as it was likely lower at the beginning of the uplifting and higher when it reached its maximum elevation.

Ruxton and McDougall (1967) and Ahnert (1970) have shown that DR tends to increase with elevation though their values were obtained from modern erosion data and are one–two orders of magnitude higher than in Eritrea. Local relief (i.e., the difference between the maximum and minimum elevation in an area) is considered to be even more important than elevation in propelling the DR as found by Ahnert (1970) for 20 middle latitude basins. The current physiography of the Eritrea highlands is characterized by a large relief ratio (Fig. 2.30), and the DR of the recent past is expected to be much higher than the average of 33 mm/ka obtained across a time span of 15 Ma, especially in the river systems draining the escarpment as suggested by their steeper channels.

For river basin located in tectonically active areas, Summerfield (1991) reported present-day DRs higher than long-term average of in Eritrea, e.g., Indus 124 mm/ka, Ganges 271 mm/ka, Brahmaputra 677 mm/ka, Colorado 84 mm/ka and Nile 15 mm/ka. This latter value is very low because the sediment contribution of the White Nile (which is the largest upstream basin) is close to zero, whereas the most of the sediment supply to the Main Nile comes from the Blue Nile and the Atbara, whose catchments are much smaller (they account for about 18% of the upstream Nile basin) and whose sediment comes from the Ethiopian highlands (Sutcliffe and Parks 1999; Billi and Badri 2010). On the base of these considerations, we can therefore calculate the current DR of the northern Ethiopia highlands as 82 mm/ka, i.e., a DR value close to that measured for the Colorado or the Indus. Unfortunately, there is no sediment transport data for the Eritrean rivers, but it is reasonable to assume for the Eritrean highlands a current DR similar to that of the Ethiopian highlands. For the passive margin of south-east Australia, Persano et al. (2002) calculated an average DR of 45 mm/ka across the last 90–100 Ma. The south-eastern Australian margin is a typical high elevated passive margins characterized by a steep escarpment that separates the coastal plain from a low relief inland plateau. Such a physiographic setting and the structural events that generated it are similar to those of eastern Eritrea (Figs. 2.30 and 2.36), so it is not surprising that the active average DR of these areas is very similar.

Average DRs of the same order of magnitude of those calculated for Eritrea are reported by Zhang et al. (2003) for the crystalline rock region of the Huangling (DR = 17 mm/ka). This brief review of DR data reported in the literature for comparable settings indicates that the average DR calculated for Eritrea is not particularly low though phases with different intensity of erosion may have occurred in the past 15 Ma in relation to the variability of uplift rate and climate conditions. Further studies are, however, necessary to verify the control of denudation models on denudation rates.

In a tectonic active area, the river network development plays an important role in landmass denudation and, more specifically, in the export of the sediment delivered by erosion processes on slopes. Bankfull flow is a significant geomorphological factor since it maximizes the sediment transport of a river and, in the long term, transports the greatest cumulative volume of sediment (Wolman and Miller 1960). Bankfull flow is, therefore, the geomorphologically most active discharge contributing to landmass denudation and to the river network development.

The stress necessary to move the bed particles during bankfull floods is approximated by the bankfull Shields stress ($\tau*_{bf}$), defined by Pfeiffer et al. (2017) as follows:

$$\tau*_{\rm bf} = \rho R_{\rm bf} S/(\rho_{\rm s} - \rho) D_{50} \qquad (2.1)$$

in which ρ is the density of water, ρ_s is the density of sediment, R_{bf} is the hydraulic radius at bankfull stage, S is the channel slope, and D_{50} is the median size of bed material.

According to Sklar and Dietrich (2008), in channels of tectonically active areas, the stress needed to transport bed material is typically much higher than the stress to incise the rock. This is because in tectonically active areas, the sediment supply is generally high and rivers tend to maintain bankfull stresses much higher than the critical flow (i.e., the flow stress required to initiate the entrainment of the bed particles) (Pfeiffer et al. 2017). Under these conditions, the attainment of an equilibrium profile is delayed as the most of the river flow energy is dissipated to move the large quantities of sediment entering the channel and produced by active erosion processes.

Even a rapid glance at the Eritrean rivers reveals that those draining the escarpment are smaller, steeper in their headwaters, choked with sediment and probably have not yet reached an equilibrium profile (see also Chap. 4), whereas those draining the gently westward inclined plateau are much larger, have lower gradients (especially downstream of their headwaters) and their channels are not chocked with sediment as much as their eastern counterparts. Equation (2.1) indicates that channel slope and sediment size are important factors in controlling the bankfull stress. In the uplifting of Eritrea, associated with the Afro-Arabian doming, on the westward sloping monocline of the plateau, gentler channel gradients are expected than in the escarpment channels. On the plateau, the river systems developed on low gradients and the reduced sediment supply may have allowed them to develop an equilibrium profile. This geomorphological setting may provide a complementary explanation of the low denudation rates obtained with different methods reported in this chapter. On the escarpment, characterized by a faulted stepped morphology, the river headwater slopes were steeper since the beginning of the dome faulting and the escarpment formation. Here, the steep embryo channels started to develop their catchments and, given their steeper gradient, used a large proportion of their flow stress to remove the sediment supplied from slopes. The continuous downstream conveyance of the supplied material favored the backward migration of the headwaters and contributed to the shaping of the very steep escarpment as we know it today. This interpretation is confirmed by Pfeiffer et al. (2017) who found a decreasing trend of sediment transport capacity and erosion rates moving from tectonically active areas of North America west coast to the eastern, more stable inland.

2.6 Conclusions

The present, large-scale landscape of Eritrea is the result of a long and complex sequence of tectonic events interrupted by long periods of quiescence during which erosion prevailed with the formation of peneplains covered by lateritic duricrusts. These events are recorded in the uplifted basement which makes the backbone of northern and central Eritrea. The following Jurassic sea transgression that covered the largest part of the Horn of Africa and gave way to thick continental and marine sedimentary sequences is recorded only in small areas of southern Eritrea which mark the northernmost inland ingression of the Jurassic sea. After a long time of relative quiescence, the sea regression progressively affected the whole Horn of Africa and led to the onset of the Afro-Arabian doming and rifting, which are the most important tectonic events that caused the remarkable uplifting of the Eritrean plateau and the formation of the Red Sea. These large-scale tectonic events were preeminent in providing Eritrea with its present physiography characterized by a gently westward inclined plateau and a sharp, 2000 m high escarpment along the Red Sea coastline.

The sequence of paleoenviroments and events for the northern and central Eritrea since the Neoproterozoic can be outlined as follows:

1. Rifting apart of the Neoproterozoic supercontinent Rodinia about one billion years ago and opening of the Mozambique ocean.
2. Accretion of the new margins (East and West Gondwana) through ocean spreading, crustal growth, island arc- and plume-related magmatic activities.
3. Various phases of continental collisions from c. 650 Ma to c. 550 Ma ago and formation of the East African Orogen within the Gondwana supercontinent.
4. Two tectonically superimposed terranes in the Eritrean East African Orogen: the lower one with amphibolite facies metamorphic rocks, in the upper one greenschist facies metamorphic rocks.
5. Deformation and/or exhumation through low-angle detachments, tectonic escape, suturing, orogenic collapse and unroofing.
6. Denudation of mountain belts and rift basins filled by continental and/or shallow marine sediments with traces of glacial activity (Snow Ball Earth).
7. c. 550 Ma Afro-Arabian Peneplain over the Gondwana Assembly; relative quiescence with very scarce clastic sedimentation occasionally glacigenic till c. 260 Ma.
8. Diffuse intraplate rifting of Gondwana with alluvial, fluvial and deltaic sediments followed at c. 250 Ma by the sea ingression from the Indian Ocean with marine sedimentation lasting 15 Ma.
9. 15 Ma of marine deposition ensued by continental clastic deposition until the end of the Cretaceous.
10. Intense erosion down to the basement marked by the African Surface with notable laterite pedogenesis particularly evident in central Eritrea.
11. Wide areal extension with low relief and low elevation inherited from the Late Cretaceous.
12. Afar plume volcanism (Oligocene Traps) locally overlying basement clastics derived from precursory rifting.
13. Since 20–15 Ma, a regional fault zone linked to enhanced Afar plume impingements generates, from west to east, a wide Eritrean plateau subject to progressive uplift, a diffusely faulted Eritrean escarpment and a coastal lowland.

14. Mio-Pliocene episodes of intense rifting and conglomerate production (Desset Fm.).
15. Late Miocene water circulation restraint up to stalling between the Red Sea and the Gulf of Aden.
16. Extensional phases connected to continuous stretching of the Nubian margin and active spreading since the Pliocene in the Southern Red Sea.
17. Continuous upheaval and slow erosion (c. 30 mm/ka) of the whole plateau and escarpment as confirmed by the occurrence of several ongoing river captures, especially in the easternmost part of the escarpment.

The contrasting physiography and bedrock lithological characteristics produced a variety of modern landscape and landforms. Both old (Proterozoic) and more recent faults exert a substantial control on the river network of Eritrea, whereas the lateritic duricrust acts like hard discontinuity contrasting the landscape incision of the plateau.

The denudation rates calculated with different methods revolve around the value of 30 mm/ka, which may seem a relatively low value for a region that experienced such an impressive uplifting. This value, however, is comparable to that estimated for the present-day global mean (43 mm/ka). The denudation rate of Eritrea is an average value calculated across a long time interval during which river erosion was likely lower at the beginning of the uplifting, when the regional slope was gentle, and progressively increased as the plateau reached its maximum elevation.

Acknowledgements This chapter conveys the results of several field campaigns carried out between 1994 and 2011 in Eritrea and Ethiopia. They were funded by the Italian Ministry for University and Research, the Italian Ministry of Foreign Affairs, MAE Dipartimento per la Cooperazione e la Promozione Culturale, the CNR Italian National Research Council and Florence and Ferrara University.The authors would like to remember the late Piero Bruni and Pietro Passerini both assiduous mentors and substantial contributors in our geological investigations in the field and in the interpretation of the field data.This paper greatly benefited from the scientific support both in the field and in the office of the following colleagues: Mario Sagri, Andrea Albianelli, M. L. Balestrieri, Marco Benvenuti, Vittorio Borselli, Fabio Cozzini, P. Falorni, Milvio Fazzuoli, M. Ghinassi, Marta Marcucci, Mazzini Menotti, Simonetta Monechi, Giovanni Napoleone, Mario Papini, Viviana Reale and Lorenzo Rook.Yosief Libsekal (National Museum of Eritrea, Asmara) is greatly acknowledged for his restless support with the local authorities and in organizing the field campaigns. Alem Kibreab, Tesfamichael Keleta and Tewolde Medin Tecle (Department of Mines, Asmara) and Beraki Woldehaimanot (Asmara University) continuously provided scientific and logistic support.

References

Abbate E, Sagri M (1980) Volcanites of Ethiopian and Somali Plateau and major tectonic lines. Atti Convegni Lincei Roma 47:219–227

Abbate E, Passerini P, Zan L (1995) Strike-slip faults in a rift area: a transect in the Afar Triangle, East Africa. Tectonophysics 241:67–97

Abbate E, Albianelli A, Azzaroli A, Benvenuti M, Tesfamariam B, Bruni P, Cipriani N, Clarke RJ, Ficcarelli G, Macchiarelli R, Napoleone G, Papini M, Rook L, Sagri M, Tecle TM, Torre D, Villa I (1998) A one-million-year-old *Homo* cranium from the Danakil (Afar) Depression in Eritrea. Nature 393:458–460

Abbate E, Balestrieri ML, Bigazzi G (2002) Morphostructural development of the Eritrean rift flank (southern Red Sea) inferred from apatite fission track analysis. J Geophys Res 107(B11):2319

Abbate E, Woldehaimanot B, Bruni P, Falorni P, Papini M, Sagri M, Girmay S, Tecle TM (2004) Geology of the Homo-bearing Pleistocene Dandiero basin (Buia region, Eritrea Danakil Depression). Riv It Paleont Strat 110(Supplement):5–34

Abbate E, Bruni P, Coppa A, Aria D, Ferretti MP, Libsekal Y, Rook L, Sagri M (2012) A new Oligo-Miocene mammal-bearing site from a sedimentary intercalation in the Trap basalts of central Eritrea. Riv It Paleont Strat 118:545–550

Abbate E, Bruni P, Ferretti MP, Delmer C, Laurenzi MA, Hagos M, Bedri O, Rook L, Sagri M, Libsekal Y (2014) The Oligocene intertrappean beds: regional distribution, depositional environments and Afro/Arabian mammal dispersal corridors. J Afr Earth Sc 99:463–489

Abbate E, Bruni P, Sagri M (2015) Geology of Ethiopia: a review and geomorphological perspective. In: Billi P (ed) Landscapes and landforms of Ethiopia. Springer, Heidelberg, pp 33–64

Abebe T, Mazzarini F, Innocenti F, Manetti P (1998) The Yerer-Tullu Wellel volcanotectonic lineament: a transtentional structure in central Ethiopia and the associated magmatic activity. J Afr Earth Sci 26:135–150

Abul-Haggag Y (1961) A contribution to the physiography of Northern Ethiopia. University of London, Athlone Press, 153 pp

Ahnert F (1970) Functional relationship between denudation, relief and uplift in large min-latitude drainage basins. Am J Sci 268:243–263

Albianelli A, Napoleone G (2004) Magnetostratigraphy of the *Homo*-bearing Pleistocene Dandiero Basin (Danakil Depression, Eritrea). Riv It Paleont Strat 110(Supplement):5–34

Alene M, Conti A, Sacchi R, Zuppi G (1999) Stable isotope composition (^{13}C and ^{18}O) of Neoproterozoic limestones and dolomites from Tigrai, northern Ethiopia. Boll Soc Geol Ital 118:611–615

Alene M, Jenkin GRT, Leng MJ, Darbyshire DPF (2006) The Tambien Group, Ethiopia: an early Cryogenian (ca. 800–735 Ma) Neoproterozoic sequence in the Arabian-Nubian Shield. Precambrian Res 147:79–99

Almalki KA, Betts PG, Ailleres L (2014) Episodic sea-floor spreading in the Southern Red Sea. Tectonophysics 617:140–149

Aloisi P (1931) Gli studi petrografici sull'Eritrea, sulla Somalia e sulla Libia. Atti I Congresso, St Colon, Firenze, 12 pp

Al-Subbary AK, Nichols G, Bosence D, Al Kadasi M (1998) Pre-rift doming, peneplanation or subsidence in the southern Red Sea? Evidence from the Medj-zir Formation (Tawilah Group) of western Yemen. https://doi.org/10.1007/978-94-011-4930-3_8

Andersson UB, Ghebreab W, Teklay M (2006) Crustal evolution and metamorphism in east-central Eritrea, south-east Arabian-Nubian Shield. J Afr Earth Sci 44:45–65

Andreatta (1941) Appunti geologico-petrografici su alcuni problemi dello zoccolo cristallino dell'Eritrea. Rend R Acc It 2:166–176

Andrews Deller ME (2003) Locating datable minerals in laterites, using remotely sensing: an example from Eritrea, NE Africa. In: Annual general meeting of the geological remote sensing group, Geological Society of London, Burlington House, Piccadilly, London

Andrews Deller ME (2006) Facies discrimination in laterites using Landsat Thematic Mapper, ASTER and ALI data—Examples from Eritrea and Arabia. Int J Remote Sens 27:2389–2409

Angelucci A, Befani G, Biagi PF, Bono P, Caputo C, Carbone F, Catenacci V, Ciancetti GF, Civitelli G, D'Alessandro L, Girotti O,

La Monaca GB, Lupia Palmieri E, Mariotti G, Matteucci R, Sirna G, Toro B (1981) Geological framework of Tanam, Wusta and Isratu in the Dahlak Islands (Southern Red Sea). Geol Romana 17:345–358
Angelucci A, Matteucci R, Praturlon A (1981) Outline of geology and sedimentary environments of the Dahlak Islands (Southern Red Sea). Boll Soc Geol Ital 99:405–419
Antonielli B, Fidolini F, Righini G (2009) Landsat TM and Quickbird images for geological mapping in the syn-rift lower Dogali formation (Red Sea coast, NE Eritrea). Photointerpr Eur J Appl Remote Sens 3:107–114, 131–134
Assefa G (1981) Gohatsion Formation: a new Lias lithostratigraphic unit from the Abbay River basin, Ethiopia. Geosci J 2:63–88
Avigad D, Stern RJ, Beyth M, Miller N, McWilliams MO (2007) Detrital zircon U-Pb geochronology of Cryogenian diamictites and Lower Paleozoic sandstone in Ethiopia (Tigrai): age constraints on Neoproterozoic glaciation and crustal evolution of the southern Arabian-Nubian Shield. Precambrian Res 154(1–2):88–106
Baldacci L (1891) Osservazioni fatte nella Colonia Eritrea. Mem Descr Carta Geol Ital VI:110 pp, and 1:400,000 Map of Massawa, Keren, Axum e Adigrat
Balestrieri ML, Stuart FM, Persano C, Abbate E, Bigazzi G (2005) Geomorphic development of the escarpment of the Eritrean margin, southern Red Sea from combined apatite fission-track analysis and (U-Th)/He thermochronometry. Earth Planet Sci Lett 231:97–110
Bannert D, Brinckmann J, Käding KC, Kretsch G, Kürsten N, Mayrhofer H (1970) Zur Geologie der Danakil Senke (Nördliches Afar Gebiet), NE Aethiopien. Geol Rund 59:409–443
Barberi F, Santacroce R (1980) The Afar Stratoid Series and the magmatic evolution of East Africa rift system. Bull Soc Geol Fr (7) 22(6):891–899
Barberi F, Borsi S, Ferrara G, Marinelli G, Varet J (1970) Relations between tectonics and magmatology in the northern Danakil depression (Ethiopia). Philos Trans R Soc Lond 267:293–311
Barberi F, Borsi S, Ferrara G, Marinelli G, Santacroce R, Tazieff H, Varet J (1972) Evolution of the Danakil Depression (Afar, Ethiopia) in light of radiometric age determination. J Geol 80:720–729
Berhe SM (1990) Ophiolites in northeast and east Africa: implications for Proterozoic crustal growth. J Soc Geol Lond 147:41–57
Beyth M (1972a) The geology of central and western Tigray. PhD thesis, Reinische Friedrick-Wilhelm University of Bonn, 200 pp
Beyth M (1972b) Paleozoic-Mesozoic sedimentary basin of Mekele outlier, Northern Ethiopia. Am Petr Geol Bull 57:2440–2443
Beyth M (2001) Preliminary indications for Snowball Earth in the East African Orogen. Geol Soc Aust 65 (Abstracts)
Beyth M, Avigad D, Wetzel HU, Matthews A, Berhe SM (2003) Crustal exhumation and indications for Snowball Earth in the East African Orogen: north Ethiopia and East Eritrea. Precambrian Res 123:187–201
Bibolini A (1920) Risultati preliminary delle osservazioni fatten el Nord Est della Colonia Eritrea. Asmara, 25 p
Bibolini A (1921) Sui conglomerati di Rora Bagla e dei Monti Haggar in Colonia Eritrea. Boll Soc Geol It 40:169–176
Bigazzi G, Balestrieri ML, Norelli P, Oddone M, Tecle TM (2004) Fission-track dating of a tephra layer in the Alat Formation of the Dandiero Group (Danakil Depression, Eritrea). Riv It Paleont Strat 110(Supplement):45–49
Billi P, Badri O (2010) Sediment transport of the Blue Nile River at Khartoum. Quatern Int 226:12–22
Bishop P (1985) Southeast Australian late Mesozoic and Cenozoic denudation rates: a test for late Tertiary increase in continental denudation. Geology 13:479–482
Blanford WT (1869) On the geology of a portion of Abyssinia. Q J Geol Soc Lond 25:401–406
Blanford WT (1870) Observations on the geology and zoology of Abyssinia, made during the progress of the British expedition to that country in 1867–68. Macmillan, London, p 487
Bosellini A (1989) The continental margins of Somalia: their structural evolution and sequence stratigraphy. Mem Sci Geol Padova 41:373–458
Bosellini A, Russo A, Assefa G (1995) Il calcare di Antalo nella regione di Macallè (Tigrai, Etiopia settentrionale). Rend Fis Acc Lincei S9(6):253–267
Bosellini A, Russo A, Fantozzi PL, Assefa G, Solomon T (1997) The Mesozoic succession in the Mekele outlier (Tigrai province, Ethiopia). Mem Sci Geol Padova 49:95–116
Bosellini A, Russo A, Schroeder R (1999) Stratigraphic evidence for an Early Aptian sea-level fluctuation: the Graua Limestone of south-eastern Ethiopia. Creta Res 20(6):783–791
Bosellini A, Russo A, Assefa G (2001) The Mesozoic succession of Dire Dawa, Harar Province, Ethiopia. J Afr Earth Sci 32:403–417
Bosworth W, Burke K (2005) Evolution of the Red Sea-Gulf of Aden rift system. In: Post PJ, Rosen NC, Olson DL, Palmes SL, Lyons KT, Newton GB (eds) Petroleum systems of divergent continental margins, SEPM society for sedimentary geology, vol 25, pp 342–372. https://doi.org/10.5724/gcs.05.25.0342
Bosworth W, Huchon P, McClay K (2012) The Red Sea and Gulf of Aden Basins. J Afr Earth Sci 43:334–378
Brinkmann J, Kürsten M (1970) Geological sketch map of the Danakil Depression, scale 1:250,000. Bunderanstalt für Bodenforschungen, Hanover
Bunter MAG, Debretsion T, Woldegiorgis L (1998) New development in the pre-rift prospectivity of the Eritrean Red Sea. J Petr Geol 1 (4):373–400
Burek PJ (1970) Paleomagnetic evidence for an anti-clockwise rotation of the Danakil Alps, Ethiopia. Eos Trans Am Geophys Union Abs 51:271
Burke K, Gunnell Y (2008) The African erosion surface: a continental-scale synthesis of geomorphology, tectonics, and environmental change over the past 180 million years. Geol Soc Am Mem 201:66 pp
Bussert R (2010) Exhumed erosional landforms of the Late Paleozoic glaciation in Northern Ethiopia: indicators of ice-flow direction paleolandscape and regional ice dynamics. Gondwana Res 18:356–369
Bussert R (2014) Depositional environments during the Late Paleozoic ice age (LPIA) in Northern Ethiopia, NE Africa. J Afr Earth Sci 99:38–407
Bussert R, Dawit E (2009) Unexpected diversity: new results on the stratigraphy and sedimentology of Paleozoic and Mesozoic siliciclastic sediments in Northern Ethiopia. Zentr Geol Paläont Teil 1 (2007):181–198
Bussert R, Schrank E (2007) Palynological evidence for a latest Carboniferous-Early Permian glaciations in Northern Ethiopia. J Afr Earth Sci 49:201–210
Carbone F, Matteucci R, Angelucci A (1998) Present-day sedimentation on the carbonate platform of the Dahlak Island, Eritrea. In: Purser BH, Bosence DWJ (eds) Sedimentation and tectonics in Rift Basins: Red Sea-Gulf of Aden, vol 11, pp 523–536
Cecioni G (1981) Precambrian pebbly mudstones in Eritrea, northeastern Ethiopia. In: Hambrey MJ, Harland WB (eds) Earth's pre-Pleistocene glacial record, vol A24. Cambridge University Press, 150
Clynne MA, Duffield WA, Fournier RO, Giorgis LW, Janik CJ, Kahsai G, Lowenstern J, Woldemariam K, Smith JG, Tesfai T (1996) Geology and geothermal potential of Alid volcanic center, Eritrea, Africa. Final Report to the US Agency for International Development, 46 pp

CNR-CNRS (1971) Geological map of the northern Afar, scale 1:500,000. Geotechnip, La Celle St. Claude, France
CNR-CNRS (1975) Geological map of the central and southern Afar, scale 1:500,000. Geotechnip, La Celle St. Claude, France
Coltorti M, Dramis F, Ollier CD (2007) Planation surfaces in Northern Ethiopia. Geomorphology 89:287–296
Coltorti M, Firuzabadi D, Borri A, Fantozzi P, Pieruccini P (2015) Planation surfaces and the long-term geomorphological evolution of Ethiopia. In: Billi P (ed) Landscapes and landforms of Ethiopia. Springer, Heidelberg, pp 117–136
Dainelli G (1943) Geologia dell'Africa Orientale (3 vols. text, 1 vol. maps). Reale Accademia d'Italia, Roma
Dainelli G, Marinelli O (1912) Risultati scientifici di un viaggio nella Colonia Eritrea. Pubblicazioni Regio Istituto di Studi Superiori Pratici e di Perfezionamento Firenze, 601 pp
Davidson A (1983) Reconnaissance geology and geochemistry of parts of Illubabor Kefa, Gemu Gofa and Sidamo, Ethiopia. The Omo River Project, Ministry of Mines and Energy, Ethiopian Institute of Geological Survey Bull, vol 2, 89 pp
Davison I, Al-Kadasi M, Al-Khirbash S, Al Subbary AK, Baker J, Blakey S, Bosence D, Dart C, Heaton R, McClay K, Menzies M, Nichols G, Owen L, Yelland A (1994) Geological evolution of the southeastern Red Sea Rift margin, Republica of Yemen. Geol Soc Am Bull 106:1474–1493
Dawit EL (2010) Adigrat sandstone in Northern and Central Ethiopia: stratigraphy, facies, depositional environments and palynology. PhD thesis, Technical University, Berlin, 166 pp
De Souza Filho CR (1995) Remote sensing and the tectonic evolution of Northern Eritrea. Ph.D. thesis, The Open University, UK, 276 pp
De Souza Filho CR, Drury SA (1998) A Neoproterozoic supra-subduction terrane in northern Eritrea, NE Africa. J Geol Soc Lond 155:55–556
Department of Mines (2009) Mineral potential of the state of Eritrea. Yearly Brochure of Eritrean Mineral Sector, Ministry of Energy and Mines, Asmara, Eritrea
Dow DB, Beyth M, Hailu T (1971) Palaeozoic glacial rocks recently discovered in northern Ethiopia. Geol Mag 168:53–60
Drury SA, Berhe SM (1993) Accretion tectonics in northern Eritrea revealed by remotely sensed imagery. Geol Mag 130(2):177–190
Drury SA, De Souza Filho CR (1998) Neoproterozoic terrane assemblages in Eritrea: review and prospects. J Afr Earth Sci 27:331–348
Drury SA, Kelley SP, Berhe SM, Collier REL, Abraha M (1994) Structures related to Red Sea evolution in northern Eritrea. Tectonics 13:1371–1380
Drury SA, Peart RG, Deller ME (2001) Hydrogeological potential of major fractures in Eritrea. J Afr Earth Sci 32:163–177
Drury SA, Ghebreab W, Andrews Deller ME, Talbot CJ, Berhe SM (2005) A comment on geomorphic development of the escarpment of the Eritrean margin, southern Red Sea from combined apatite fission-track and (U–Th)/He thermochronometry by Balestrieri, M. L. et al. [Earth Planet. Sci. Lett. 231 (2005) 97–110] Earth Planet Sci Lett 242: 428–432
Ebinger CJ, Casey M (2001) Continental breakup in magmatic provinces: an Ethiopian example. Geology 29:527–530
Ebinger CJ, Sleep NH (1998) Cenozoic magmatism through east Africa resulting from impact of a single plume. Nature 395:788–791
ENI (1962–1971) Enciclopedia del Petrolio. ENI, Roma
Francaviglia A (1938) Note sulla costituzione geologica della zona del basso Barca. Giorn Geol 13:23–27
Frazier SB (1970) Adjacent structures of Ethiopia: that portion of the Red Sea coast including Dahlak Kebir Island and the Gulf of Zula. Philos Trans R Soc Lond A 267:131–141
Friedrich AM (2019) Palaeogeological hiatus surface mapping: a tool to visualize vertcal motion of the continents. Geol Mag 156:308–319
Friedrich AM, Bunge HP, Rieger SM, Colli l, Ghelichkhan S, Nerlich R (2018) Stratigraphic framework for the plume mode of mantle convection and the analysis of interregional unconformities on geological maps. Gondwana Res 53:159–188
Fritz H, Abdelsalam M, Ali KA, Bingen B, Collins AS, Fowler AR, Ghebreab W, Hauzenberger CA, Johnson PR, Kusky TM, Macey S, Muhongo RJ, Viola G (2013) Orogen styles in East Africa Orogen: a review of the Neoproterozoic to Cambrian tectonic evolution. J Afr Earth Sci 86:65–106
Garland CR (1980) Geology of the Adigrat area. Geological Survey of Ethiopia, Ministry of Mines, Energy and Water Resources Addis Abeba Mem, vol 1, 51 pp
Getaneh W (2002) Geochemistry provenance and depositional tectonic setting of the Adigrat Sandstone northern Ethiopia. J Afr Earth Sci 35:185–198
Ghebreab W (1996) An outline of major Pan-African lithological assemblages and shear zones in Eritrea: implications for mineral exploration. Afr Geo Rev 3:355–366
Ghebreab W (1998) Tectonics of the Red Sea region reassessed. Earth Sci Rev 45:1–45
Ghebreab W (1999) Tectono-metamorphic history of Neoproterozoic rocks in eastern Eritrea. Precambrian R 98:83–105
Ghebreab W, Talbot CJ (2000) Red Sea extension influenced by Pan-African tectonic grain in eastern Eritrea. J Struct Geol 22:931–946
Ghebreab W, Carter A, Hurford AJ, Talbot CJ (2002) Constraints for timing of extensional tectonics in the western margin of the Red Sea in Eritrea. Earth Planet Sci Lett 200:107–119
Ghebreab W, Talbot CJ, Page L (2005) Time constraints on exhumation of the East African Orogen from field observations and $^{40}Ar/^{39}Ar$ cooling ages of low-angle mylonites in Eritrea. Precambrian Res 139:20–41
Ghebreab W, Kontny A, Greiling RO (2007) Fabric evolution across a discontinuity between lower and upper crustal domains from field microscopic, and anisotropy of magnetic susceptibility studies in central eastern Eritrea, NE Africa. Tectonics 26 TC3015:19 pp
Ghebreab W, Greiling RO, Solomon S (2009) Structural setting of Neoproterozoic mineralization, Asmara district, Eritrea. J Afr Earth Sci 55:219–235
Gherardi D (1951a) Preliminary geological survey of Eritrea. Min Mag 84:265–276
Gherardi D (1951b) Preliminary geological survey of Eritrea. Min Mag 84:338–347
Ghinassi M, Oms O, Papini M, Scarciglia F, Carnevale G, Sani F, Rook L, Delfino M, Pavia M, Libeskal Y, Bondioli I, Coppa A, Frayer DW, Macchiarelli R (2015) An integrated study of the *Homo*-bearing Aalat stratigraphic section (Eritrea): an expanded continental record at the Early-Middle Pleistocene transition. J Afr Earth Sci 112:163–185
Gortani M (1973) La fauna mesocretacea degli Strati di Graua (lavoro compiuto nel 1942). Acc Naz Lincei. Missione geologica dell'Azienda Petroli (AGIP) nella Dancalia Meridionale e negli altopiani Harrarini (1936–1938) 3:1–38
Goudie AS (2005) The drainage of Africa since the Cretaceous. Geomorphology 67:437–456
Hambrey MJ (2007) Glacial sedimentary processes and products: the Ordovician glaciation in Eritrea and Ethiopia, NE Africa. In: Hambrey MJ, Christoffersen P, Glasser NF, Hubbard B. International Association of Sedimentologists. ISBN: 978-1-405-18300
Hamiel Y, Baer G (2016) Crustal deformation associated with the 2011 eruption of the Nabro volcano, Eritrea. Tectonophysics 691:257–262
Hankel O (2013) EarlyJurassic Palynomorphs from the Adigrat Sandstone in well Abre-1, Ogaden, Ethiopia. Sci Study 12 pp
Haq QU, Hardenbol J, Vail PR (1988) Mesozoic and Cenozoic chronostratigraphy and cycles of sea-level changes. In: Wilgus CK

et al (eds) Sea-level changes-an integrated approach. Soc Econom Paleont Mineral Spec Publ 42:71–108
Hoffman PF (2009) A palaeogeographic context for Neoproterozoic glaciation. Palaeogeogr Palaeocl Palaeoeco 277(3–4):158–172, 15:135–156
Hoffman PF, Scharag DP (2002) The snowball Earth hypothesis: testing the limits of global change. Terra Nova 14:129–155
Hoffman PF, Kaufman AJ, Halverson GP, Scharag DP (1998) A Neoproterozoic snowball earth. Science 281:1343–1346
Hughes GW, Varol O, Beydoun ZR (1991) Evidence for Middle Oligocene rifting of the Gulf of Aden and Late Oligocene rifting of the southern Red Sea. Mar Petrol Geol 8:354–358
Hutchinson RW, Engels GG (1970) Tectonic significance of regional geology and evaporate lithofacies in northeastern Ethiopia. R Soc Lond Phil Trans A267:313–329
Johnson PR, Wolhehaimanot B (2003) Development of the Arabian-Nubian Shield: perspectives on accretion and deformation in the northern East African Orogen and the assembly of Gondwana. Geol Soc Lond Spec Publ 206:289–325
Kamen-Kaye J (1978) Permian and Tertiary faunas and paleogeography: Somalia, Kenia, Tanzania, Mozambique and Madagascar, South Africa. J Petrol Geol 1:79–101
Kazmin V (1973) Geological Map of Ethiopia, scale 1:2.000.000. Imperial Ethiopian Government, Geological Survey of Ethiopia, Addis Ababa
Kazmin V (1975) Explanation of the geological map of Ethiopia. Geol Surv Ethiopia Mem 1:1–14
Kenea NH, Ebinger CJ, Ukstins IA, Yirgu G (2001) Late Oligocene volcanism in the southern Red Sea Hills, Sudan. J Geol Soc Lond 158:285–294
Khalil HM (2012) Pliocene–Pleistocene stratigraphy and macrofauna of the Farasan Islands, South East Red Sea, Saudi Arabia. Arab J Geosci 5:1223–1245
Kidane T (2013) Paleomagnetic dating of the Enticho Sandstone at Negash locality (Tigrai Region), Northern Ethiopia: implication for Quaternary remagnetization. SINET Ethiop J Sci 37(1): 31–42
Kidane T (2014) Paleomagnetic dating of Enticho Sandstone at Negash locality (Tigrai region, Northern Ethiopia): implication for quaternary remagnetization. SINET Ethiop J Sci 37(1). ISSN: 0379–2897
Kröner A (1985) Ophiolites and the evolution of tectonic boundaries in the Late Proterozoic Arabian-Nubian Shields of Northeast Africa and Arabia. Precambrian Res 27:277–300
Kröner A, Stern RJ (2004) Pan-African Orogeny. In: Encyclopedia of geology, 12 pp
Kröner A, Greiling R, Reischmann T, MantonW HIM (1991) Evolution of Pan-African island arc assemblages in the southern Red Sea Hills, Sudan, and in southwestern Arabia as exemplified by geochemistry and geochronology. Precambrian Res 53:99–118
Kumpulainen RA (2007) The Ordovician glaciation in Eritrea and Ethiopia, NE Africa. In: Hambrey MJ, Christoffersen P, Glasser NF, Hubbard B (eds) Glacial sedimentary processes and products, pp 321–342
Kumpulainen RA, Uchman A, Woldehaimanot B, Kreuser T, Ghirmay S (2006) Trace fossil evidence from the Adigrat Sandstone for an Ordovician glaciation in Eritrea, NE Africa. J Afr Earth Sci 45:408–420
LeBas MJ, Mohr PA (1968) Feldspathoidal rocks from the Ethiopian Cainozoic volcanic province. Geol Runds 58:273–280
Lewin A, Meinhold G, Hinderer M, Dawit EL, Bussert R (2018) Provenance of sandstones in Ethiopia during Late Ordovician and Carboniferous-Permian Gondwana glaciations: petrography and geochemistry of the Enticho Sandstone and the Edaga Arbi glacials. Sediment Geol 375:188–202
Macchiarelli R, Bondioli L, Chech M, Coppa A, Fiore I, Russom R, Vecchi F, Libsekal Y, Rook L (1998) The Late Early Pleistocene human remains from Buia, Danakil Depression, Eritrea. Riv It Paleont Strat 110:133–144
MacFadyen WA, Cox LR, Brighton AG (1930) The geology of the Farasan Islands, Gizan and Kamaran Island, Red Sea. Part 1—general geology. Geol Mag 68:1–13
Manasse E (1909) Contribuzioni allo studio petrografico della Colonia Eritrea. Nava Siena, 169 pp
Marini A (1938) L'Universo 19:51–65, 131–170
Marini A (1998) Alid volcano in the Colony of Eritrea, translation of Angelo Marini's Il Vulcano Alid nella Colonia Eritrea. USGS open file report 98-218
Martínez-Navarro B, Rook L, Segid A, Yosief D, Ferretti MP, Shoshani J, Tekle TM, Liksekal Y (2004) The large fossil mammals from Buia (Eritrea, Northern Danakil Depression). Riv It Paleont Strat 110:61–88
Martini F, Libsekal Y, Filippi O, Ghebre/her A, Kashay A, Kiros A, Martino G, Okubatsion D, Segid A, Solomon T, Teka Z, Yosief D, Yamane S (2004) Characterization of lithic complexes from Buia (Dandiero Basin, Danakil Depression, Eritrea). Riv It Paleont Strat 110:99–132
McDougall I, Morton WH, Williams AJ (1975) Age and rates of denudation of Trap Series basalts at Blue Nile Gorge, Ethiopia. Nature 254:207–209
Merla G (1963) Missione geologica nell'Etiopia meridionale del Consiglio Nazionale delle Ricerche 1959–1960. Notizie geomorfologiche e geologiche. Giorn Geol 31:1–56
Merla G, Minucci E (1938) Missione geologica nel Tigrai. In: Volume primo: La serie dei terreni. Reale Accademia d'Italia, 363 pp
Merla G, Abbate E, Azzaroli A, Bruni P, Fazzuoli M, Sagri M, Tacconi P (1979) A geological map of the Ethiopia and Somalia and comment with a map of major landformas (scale 1:2,000,000). Cons Naz Ric Roma 95 pp
Mesfin A, Yohannes E (2014) The geology of Northern Danakil Depression and its geothermal significance. In: Proceedings 5th African rift geothermal conference, Arusha, Tanzania, 29–31 Oct 2014, 6 pp
Miller NR, Alene M, Sacchi R, Stern RJ, Conti A, Kröner A, Zuppi G (2003) Significance of the Tambien Group (Tigrai, N. Ethiopia) for snowball earth events in the Arabian Nubian Shield. Precambrian Res 121:263–283
Miller NR, Johnson PR, Stern RJ (2008) Arab J Sci Eng 33(1C):57–77
Miller NR, Avigad D, Stern RJ, Beyth M (2011) The Tambien Group, Northern Ethiopia (Tigre). In: Arnaud E, Halverson GP, Shields-Zhou G (eds) The geological record of Neoproterozoic glaciations. Geol Soc London Mem 36:282 pp
Mohr P (1962) The geology of Ethiopia. University College Addis, Ababa Press, 268 pp
Mohr P (1963) Geological map of Horn of Africa, 1: 2,000,000. Philip and Tracey, London
Mohr P (1966) On erosion surfaces. Bull Geophys 9:40–42
Mohr P (1987) Patterns of faulting in the Ethiopian rift valley. Tectonophysics 143:169–179
Mohr P (1999) The Asmara dike swarm, Eritrean plateau: physical parameters of an off-rift olivine dolerite injection zone. Acta Vulc 11(1):177–181
Mohr P (2009) Africa beckoning: explorers of eastern Africa, its rift valleys and geology, in the 19th and early 20th centuries. Millbrook Nova Press, Tóin An Ghárrain, Corr an Dola, Co. na Gaillime, 192 pp
Montanaro Gallitelli E (1939) Ricerche sull'età di alcuni depositi insulari e marginali del Mar Rosso (isole Daalac), penisola di Buri, dintorni di Massaua fino al Sahel. Palaeontogr Ital 39(9):215–258
Montanaro Gallitelli E (1973) Ricerche eseguite negli anni 1941–42 sopra coralli delle scogliere emerse di Massaua e Gibuti. In: Missione geologica dell'Azienda Petroli (AGIP) nella Dancalia

meridionale e negli altopiani harrarini (1936–1938). Atti Acc Naz Lincei 4:445–520
Morley C (1988) Variable extension in Lake Tanganyika. Tectonics 7:785–801
Moucha R, Forte AM (2011) Changes in African topography driven by mantle convection. Nat Geosci 4:707–712
Overstreet WC, Stoeser DB, Overstreet EF, Goudarzi GH (1977) Tertiary laterite of the As Sarat Mountains, Asir Province, Kingdom of Saudi Arabia. Min Res Bull Saudi Arabia Direct Gen Min Res 21:1–30
Park Y, Swanson-Hysell NL, MacLennan SA, Maloof AC, Gebreselassie M, Tremblay MM, Schoene B, Alene M, Anttila ESC, Tesema T, Haileab B (2020) The lead-up to the Sturtian Snowball Earth: neoproterozoic chemostratigraphy time-calibrated by the Tambien Group of Ethiopia. Bull Geol Soc Am 75:1119–1149
Perelló J, Brockway H, García A (2020) A minimum Thanetian (Paleocene) age for the African Surface in the Eritrean highlands, Northeast Africa. J Afr Earth Sci 164:103782
Persano C, Stuart FM, Bishop P, Barfod DN (2002) Apatite (U–Th)/He age constraints on the development of the Great Escarpment on the southeastern Australian passive margin. Earth Planet Sci Lett 200:79–90
Pfeiffer AM, Finnegan NJ, Willenbring JK (2017) Sediment supply controls equilibrium channel geometry in gravel rivers. Proc Natl Acad Sci USA 114(13):3346–3351. https://doi.org/10.1073/pnas.1612907114
Pik RD, Coulon C, Yirgu G, Hofmann C, Ayalew D (1998) The northwestern Ethiopian plateau flood basalts: classification and spatial distribution of magma types. J Volcanol Geotherm Res 81:91–111
Pik R, Marty B, Carignan J, Lavé J (2003) Stability of the Upper Nile drainage network (Ethiopia) deduced from (U–Th)/He thermochronometry: implications for uplift and erosion of the Afar plume dome. Earth Planet Sci Lett 215:73–88
Porro C (1936) Esplorazione geologica della zona costiera Eritrea a Nord di Massawa. Mem R Acc Torino 78:35–72
Prior GT (1900) On Aegirite and Riebeckite Anorthoclase rocks related to the "Grorudite-Tinguaite" series, from the neighborhood of Aodowa and Axuma, Abyssinia. Min Mag 12:255–273
Purcell PG (2018) Re-imagining the development of the East African Rift. Petrol Geosci 24:21–40
Rooney TO (2017) The Cenozoic magmatism of East-Africa: part I—flood basalts and pulsed magmatism. Lithos 286–287:264–301
Röser HA (1975) A detailed magnetic survey of the southern Red Sea. Geol Jahrb 13:131–153
Rüppell WE (1834) Skizze der geologischen Formation Abyssiniens. Abh Mus Senckenb 1:286–288
Russo A, Getaneh A, Atnafu B (1994) Sedimentary evolution of the Abbai River (Blue Nile) Basin, Ethiopia. N J B Geol Palaeo Mh 5:291–308
Ruxton BP, McDougall I (1967) Denudation rates in northeast Papua from potassium-argon dating of lavas. Am J Sci 265(7):545–561. https://doi.org/10.2475/ajs.265.7.545
Sacchi R, Alene M, Barbieri M, Conti A (2007) On the Paleozoic Tillite of the Adigrat Group (Tigrai, Ethiopia). Period Mineral 76:241–251
Sagri M, Abbate E, Azzaroli A, Balestrieri ML, Benvenuti M, Bruni P, Fazzuoli M, Ficcarelli G, Marcucci M, Papini M, Pavia G, RealeV, Rook L, Tecle TM (1998) New data on the Jurassic and Neogene sedimentation in the Danakil Horst and Northern Afar Depression, Eritrea. In: Crasquin-Soleau S, Barrier E (eds) Peri-Tethys Mem 3: stratigraphy and evolution of Peri-Tethyan platforms. Mém Muséum Nat Hist Nat Paris 177:193–214
Savoyat F, Shiferaw A, Balcha J (1989) Petroleum exploration in the Ethiopian Red Sea. J Petrol Geol 12:187–204
Saxena GN, Getaneh A (1983) New evidence on the age of the glacial rocks of northern Ethiopia. Geol Mag 120:549–554
Selli R (1973) Molluschi quaternari di Massaua e di Gibuti, In: Missione geologica dell'azienda generale Italiana petroli (AGIP) nella Dancalia meridionale e sugli altipiani Hararini (1936–1938). Vol. 4: Documentazione paleontologica, parte seconda. Accademia Nazionale dei Lincei, Roma, pp 153–444
Sembroni A, Faccenna C, Becker TW, Molin P, Abebe B (2016) Longterm, deep-mantle support of the Ethiopia-Yemen Plateau. Tectonics 35:469–488
Sepulchre P, Ramstein G, Fluteau F, Schuster M, Tiercelin JJ, Brunet M (2006) Tectonic uplift and Eastern Africa aridification. Science 313:1419–1423
Shoshani J, Walter RC, Abraha M, Berhe S, Tassy O, Sanders WJ, Marchant GH, Libsekal Y, Ghirmai T, Zinner D (2006) A proboscidean from the late Oligocene of Eritrea, a "missing link" between early Elephantiformes and Elephantimorpha, and biogeographic implications. PNAS 1 Fr 6:925–933
Sklar LD, Dietrich WE (2008) Implications of the saltation-abrasion bedrock incision model for steady-state river longitudinal profile relief and concavity. Erath Surf Process Landforms 33:1129–1151. https://doi.org/10.1002/Esp.1689
Solomon S, Ghebreab W (2006) Lineament characterization and their tectonic significance using Landsat TM data and field studies in the central highlands of Eritrea. J Afr Earth Sci 46:371–378
Stefanini G (1933) Carta Geologica dell'Eritrea, della Somalia e dell'Etiopia alla scala di 1:2.000.000 con Note illustrative. Istituto Geografico Militare, Firenze, p 195
Stern RJ (1994) Arc assembly and continental collision in the Neoproterozoic East African orogen: implications for the consolidation of Gondwana. Annu Rev Earth Planet Sci 22:119–151
Stern RJ, Johnson P (2010) Continental lithosphere of the Arabian plate; a geologic, petrologic, and geophysical synthesis. Earth Sci Rev 101:29–67
Stern RJ, Miller E (2019) Neoproterozoic glaciation-snowball earth hypothesis. Encycl Geol. https://doi.org/10.1016/B978-0-12-409548-9.12107-4
Sultan M, Becker R, Arvidson RE, Sore P, Stern RJ, El-Alfi Z, Guinnes EA (1992) Nature of the Red Sea crust, a controversy revisited. Geology 20:593–596
Summerfield MA (1991) Global geomorphology: an introduction to the study of landform. John Wiley and Sons Inc. New York, 525 pp
Sutcliffe JV, Parks YP (1999) The hydrology of the Nile. IAHS Special Publication 5, IAHS, Wallingford, UK, 179 pp
Tadesse T (1996) Structure across a possible intra-oceanic suture zone in the low-grade Pan-African rocks of northern Ethiopia. J Afr Earth Sci 23:375–381
Talbot CJ, Ghebreab W (1997) Red Sea detachment and basement core complexes in Eritrea. Geology 25:655–658
Tefera M, Chernet T, Haro W (1997) Geological map of Ethiopia, scale 1:2,000,000. Ministry of Mines and Energy, Geological Survey of Ethiopia, Addis Ababa
Teklay M (1997) Petrology, geochemistry and geochronology of neoproterozoic magmatic arc rocks from Eritrea: implications for crustal evolution in the Southern Nubian Shield. Department of Mines, Eritrea, Memoir 1, 125 pp
Teklay M (2006) Neoproterozoic arc-back-arc system analog to modern arc-back-arc systems: evidence from the tholeiite-boninite association, serpentinite mudflows and across-arc geochemical trends in Eritrea, southern Arabian-Nubian shield. Precambrian Res 145(1–2):81–92
Teklay M, Hofmann AW, Schwarz W, Trieloff M (2003) $^{40}Ar/^{39}Ar$ whole rock dating of low-Ti tholeiitic and high-Ti alkaline flood basalts from Eritrea: implications for the duration of volcanism at

the Afro-Arabian volcanic province. In: Yemeni scientific research foundation, the science conference 2003 abstract volume

Teklay M, Asmerom Y, Toulkeridis T (2005) Geochemical and Sr-Nd isotope ratio in Cenozoic basalts from Eritrea: evidence for temporal evolution from low Ti tholeiitic to high-Ti alkaline basalts in Afro-Arabian Continental Flood Basalt Province. Per Mineral 74 (3):167–182

Tesfaye S, Ghebreab W (2013) Simple shear detachment fault system and marginal grabens in the southernmost Red Sea rift. Tectonophysics 608:1268–1279

Verri P (1909) Contributo allo studio geografico della Colonia Eritrea. Boll Soc Geogr Ital 46:251–301 (with 1:1,500,000 geological map)

Vialli V (1966) Sul rinvenimento di Dinoterio (*Deinotherium* cf. Hobley Andrews) nelle ligniti di Adi Ugri (Eritrea). Giornale Geol 2 (33):447–458

Vinassa de Regny P (1921–1924) Dancalia. Alfieri e Lacroix, Roma, 140 pp (with a 1:500,000 geological map)

Vinassa de Regny P (1930) La geologia delle Alpi Dancale. Boll Soc Geol Ital 50:1–24

Wolde B (1989) Cenozoic volcanism and rift development in Ethiopia. J Afr Earth Sci 8:99–105

Woldehaimanot B (1995) Structural geology and geochemistry of the Proterozoic Adhoba and Adola belts (Eritrea and Ethiopia). Ph.D. dissertation, Giessener Geol Schr 54:218 pp

Woldehaimanot B (2000) Tectonic setting and geochemical characterization of Neoproterozoic volcanics and granitoids from the Adoba belt, northern Eritrea. J Afr Earth Sci 21:459–476

Wolfenden E, Ebinger C, Yirgu G, Renne P (2005) Evolution of a volcanic rifted margin: Southern Red Sea, Ethiopia. Bull Seism Soc Am 102:439–458

Wolman MG, Miller JP (1960) Magnitude and frequency of forces in geomorphic process. J Geol 68:54–74

Yemane K, Taieb M, Faure H (1987) Limnogeologic studies on an intertrappean continental deposit from the northern Ethiopian Plateau (37°03′E, 12°25′N). J Afr Earth Sci 45:156–161

Zanettin B, Bellieni G, Justin Visentin E, Haile T (1999) The volcanic rocks of the Eritrean plateau: stratigraphy and evolution. Acta Vulc 11(1):183–193

Zanettin B, Bellieni G, Justin Visentin E (2006a) Stratigraphy and evolution of the trachy-rhyolitic volcanism of the Senafe area (Eastern Eritrean Plateau). J Afr Earth Sci 45:478–488

Zanettin B, Bellieni G, Justin Visentin E (2006b) New radiometric age of volcanic rocks in the central Eritrean plateau (from Asmara to Adi Quala): considerations on stratigraphy and correlations. J Afr Earth Sci 45:156–161

Zeyen H, Volker F, Wehrle V, Fuchs K, Sobolev SV, Zeyen RA (1997) Styles of continental rifting: crust-mantle detachment and mantle plumes. Tectonophysics 278:329–352

Zhang L, Yang D, Zhu D (2003) Denudation rates of crystalline rock in the Huangling anticline of the Three Gorges of the Yangtze River in China since the Eocene. Sci China Ser D Earth Sci 46:928–940. https://doi.org/10.1007/BF02991339

Zuffardi Comerci R (1936) Corallari neogenici del Sahel. Atti R Acc Torino 71:1–15

Zwaan F, Corti G, Keir D, Sani F (2020) A review of tectonic models for the rifted margin of Afar: implications for continental break-up and passive margin formation. J Afr Earth Sci 164:103649

Geomorphology of Afar

3

Jacques Varet

Abstract

This chapter provides a wide overview of the geomorphology of the Afar triangle (which extends over SE Eritrea, Djibouti and NE Ethiopia), based on the geological formations encountered on this wide depression floor and its margins. It shows the diversity of volcanic tectonic and sedimentary landforms resulting from the successive phases of rifting having affected the area until the present active oceanic floor spreading. If erosion takes place along the escarpments bordering the Afar depression, the landscape remains essentially determined by normal faulting and recent volcanic activity on the Afar floor thanks to both the arid climate and persistent telluric activity. Of course, such a depression will also tend to be filled by detrital colluvial, alluvial, lacustrine and eventually also marine sediments. The picture described by the author results from 50 years of geological observations. No doubt that landforms will continue to evolve along this active spreading plate margin (with a spreading rate of 2 cm a year) where an ocean floor can exceptionally be observed at the free-air surface within the African continent.

Keywords

Afar • Danakil • Ethiopia • Djibouti • Eritrea • Red Sea—Aden oceanic ridge • Depression • Desert • Below sea level • Volcanic • Tectonic • Geothermal • Geodynamic • Active • Telluric • Spreading • Rift • Fault • Open fissure • Basalt • Salt plain • Awash basin

J. Varet (✉)
22 rue du Bœuf St Paterne, 45000 Orleans, France
e-mail: j.varet@geo2d.com
URL: http://www.geo2d.com

3.1 Introduction

Afar is a striking geographic feature in the African continent: a 200,000 km^2 wide triangular depression with a floor ranging from 157 m bsl to the east (Lake Asal, Djibouti) and 120 m bsl to the north (great salt plain, Afar Regional State, Ethiopia) up to 800 m asl to the south (Fig. 3.1). At an average altitude of 200 m asl—except for a few elevated volcanoes like Nabro in Eritrea (2218 m asl) or Mousa Ali in Djibouti (2028 m asl)—it is surrounded by the plateaus of Nubia, Somalia and Yemen with steep escarpments and contrasting altitudes ranging from 2500 to 4500 m asl (Fig. 3.1). Its arid climate and intense telluric[1] activity have produced remarkable landscapes and landforms, highlighted by erosion, showing a variety of volcanic and tectonic features expressing the specific geodynamic character of the area (Tazieff et al. 1970; Barberi et al. 1972a, b; Barberi and Varet 1975a, b; Varet 2008): a unique place on Earth where two oceanic ridges (Red Sea and Gulf of Aden) meet a continental rift (the East African Rift Valley).

3.2 A Toponymy Expressing Landscapes and Landforms

The Afar triangle, as seen in Fig. 3.1, extends over three countries (Ethiopia, Eritrea and Djibouti) and corresponds with a well-defined human entity. This arid area is populated by pastoralist communities over the whole region, even the most unexpected parts. Though more densely settled along rivers or wetlands, as along the Awash River basin in southern Afar or in the marginal grabens found at the foot of the Nubian escarpment in north-western Afar, isolated settlements may be found anywhere, including the highest mountains, isolated craters or apparently rugged plains and bared basaltic surfaces.

[1] Telluric activity includes: volcanic, tectonic and hydrothermal.

P. Billi (ed.), *Landscapes and Landforms of the Horn of Africa*, World Geomorphological Landscapes,
https://doi.org/10.1007/978-3-031-05487-7_3

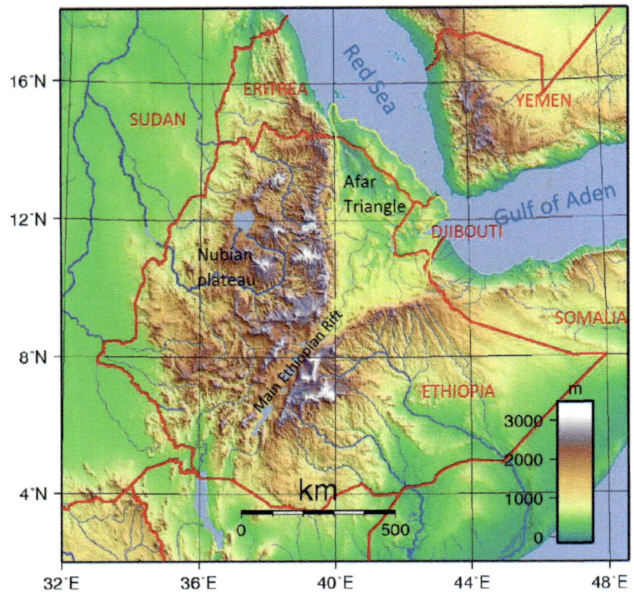

Fig. 3.1 Afar triangle, a depressed area surrounded by the Nubian, Yemen and Somalian plateaus, in the area of junction of the Gulf of Aden, the Red Sea and the Main East African Rift Valley, as seen—enlightened in yellow—on this relief map, also showing the political boundaries (in red) of concerned countries: Ethiopia, Eritrea, Djibouti, Somalia and Yemen. Apart from the rather arid Afar endoreic basins, observe the mainly diverging flow pattern of the rivers (towards Mediterranean and Indian ocean basins)

Afar people share the same language, and all places are given a name which provides information on its landform, colour or even geological nature. Table 3.1 provides a selection of the most common Afar words of use in the toponomy, and Table 3.2 lists some geologically useful terms with their English translation (source: E. Chedeville in Tazieff et al. 1970, completed by Varet 2018a, b).

3.3 Afar Margins

Afar margins are characterized by steep slopes, crowned with sharp vertical sections, formed by intense normal faulting cutting through the pre-Miocene formations: the metamorphic basement, its Mesozoic sedimentary cover and the basaltic traps of the plateau (dominantly basaltic flows with minor silicic tuffs) (Black et al. 1972). Well-marked along the Nubian and Somalian escarpments to the west and south, where two kinds of faulting are observed (with or without marginal graben, as sketched on Fig. 3.2), the relief contrasts are less marked in the north-east (Danakil Alps) and the south-east (Aisha—Ali Sabieh Horst).

These faulted margins are the areas in Afar where erosion and consequent sedimentary features are observed, contrasting with the rest of the depression where "telluric" landforms remained almost intact. This is for three reasons:

Table 3.1 Translation of a few Afar words of use in the toponymy (from Varet 2018a, b)

English	Afar	English	Afar
Nouns		Adjectives	
Mountain, massif	'Ale	Large	Kadda
Mount, hill	Kôma	Small	'Unda
Rock, small rocky hill or plateau	Dâ	Black	Data
Large rocks	Deet	White	Ado
Plateau	Râsa	Red	'Asa
Ridge	Gêra	Yellow	Hurud
Summit	San	Green	Ar'Dar
Cape, point	Damum		
Col, pass	Dâba		
Hole	Bôdu		
Fire	Kira		
Plain	Bahari		
Powder	Bodo		
Valley, Wadi	Da'ar		
Small valley	Dabba		
Water, water point	Lê		
Well	Ela		
Shallow well	Buyyi		
Lake, sea	Bad		

Table 3.2 Selected useful Afar geological terms with their English translation (from Varet 2018a; b).

Geological terms	
Earth	Baloo
Basalts (black rocks)	Dâta Deet
Rhyolites (red rocks)	'Asa Deet
Scoriaceous lava flow ("Aa")	Manda
Smooth lava flow (Pa Hoe Hoe)	Rasa
Cooked earth	Alayta
Fault	Andidou
Earthquake	Baloo Anguaui
Open fissure	Adale Andidou
Necked clay plain	Bôda
Pasture land	Affara Dara
Evaporitic plain	Koubi Adoul
Diatomite (white powder)	Ado Bodo
Travertine	Diikiil
Hot spring	Ni'iI Lê
Fumarole, steam vent	Boïna
Eruption	Badok Tawle
Smoke	'Irta
Intermittent hot spring ("coughing water")	Kahouh Ye
Salty water	Asbo lê
Water spring	Geda Lê
Cold water	Da'Him Lê
Putrefied lake	Abhe Bad
Sea (bad water)	Ad'He bad
Camel hill	Gali Koma
Coloured mountain	Bora'le Ale
Stomach like mountain	Ale Bagu
Long mouth	Afdera
Huge (mountain)	Dabbahu

- The tectonic activity continuously affecting these escarpments since 25 My (Barberi and Varet 1975a, b);
- The high rainfalls on the plateaus with consequent intense erosion along the escarpments (contrasting with the dry Afar floor);
- The older age of the geological units found outside Afar and along its sides, as besides the pre-Mesozoic units, volcanic units of Miocene age are exposed along these margins (Fig. 3.3). This is evident, especially in the southern half of Afar (Mille area to the west, North-Tadjourah and Ali Sabieh areas to the east) where large surfaces are covered by rhyolitic and basaltic units ranging in age from 14 to 4 My.

As a result, the typical landforms in these areas are faulted and tilted blocks, deeply eroded along the foothill (Fig. 3.4), surrounded by detrital terraces (coarse grained scree slope and alluvial deposits) inclined towards Afar floor (Figs. 3.4 and 3.5 and the former eventually affected by successive cycles of normal faulting and erosion (as seen in Fig. 3.5, the red beds, tilted and faulted, are covered by grey layers, both part of the polychromatic formation). The Miocene volcanic units in SW and SE Afar (Figs. 3.6, 3.7 and 3.8) show geomorphic features similar to those of the pre-Mesozoic units in northern Afar, but the former are frequently partly hidden by the more important lacustrine and detritic sedimentation resulting from the Awash endoreic basin.

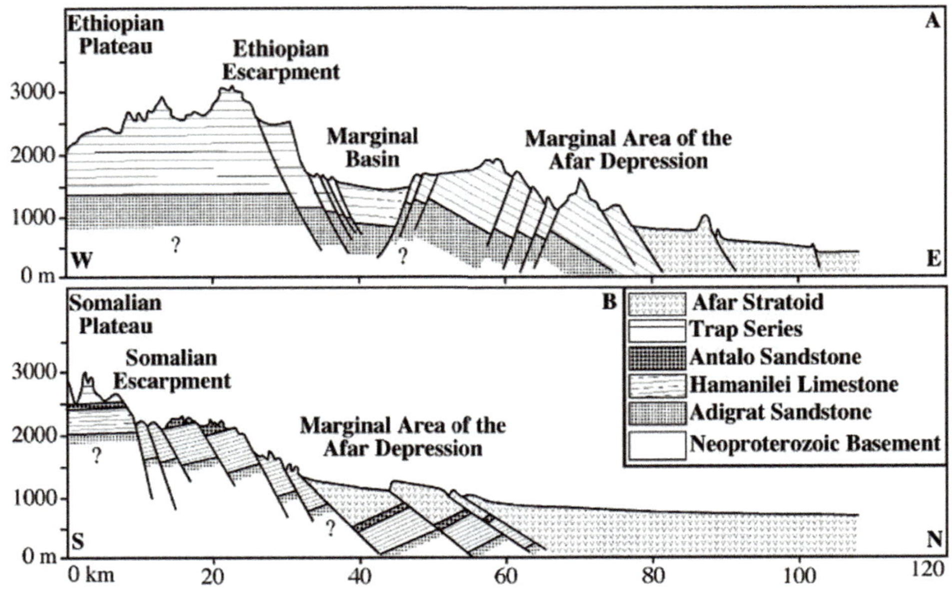

Fig. 3.2 Stylized synthetic geological sections illustrating the two kind of tectonics characterising the Nubian (upper section) and Somalian (lower section) escarpments, according to Beyene and Abdelsalam (2005). Note that, if marginal graben are frequently present, the tectonic style can be rather more complex. As observed in Fig. 8.4, the faults are dipping towards the depression and the blocks are dipping towards the plateau (as in the lower figure) in the lowest compartments along the western margin in Northern Afar

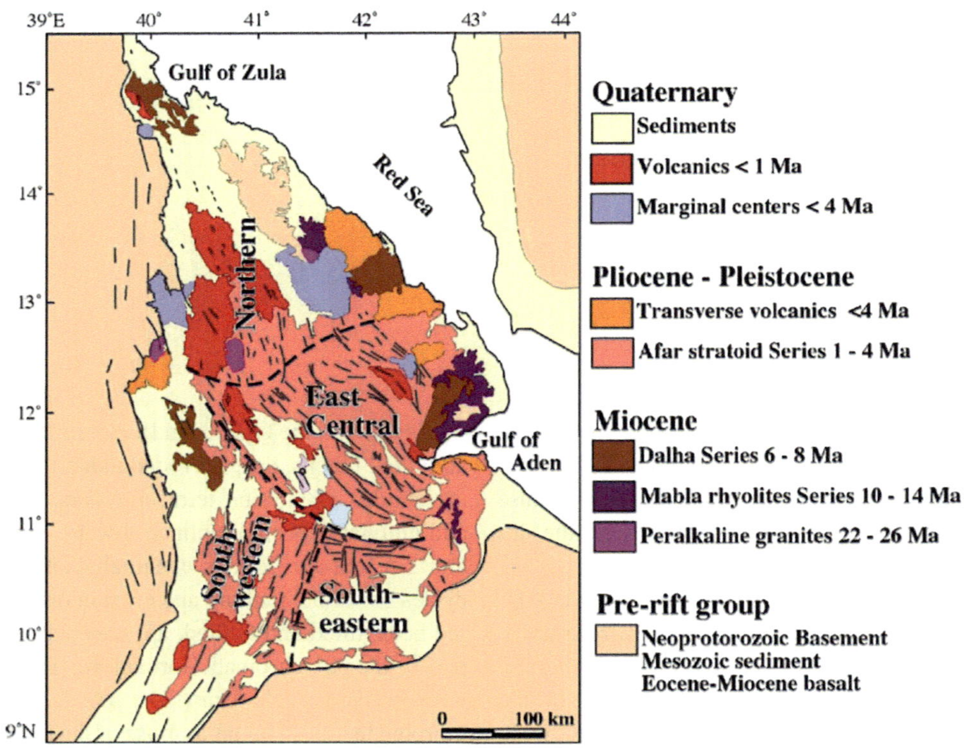

Fig. 3.3 Geological sketch map of Afar (after Varet 1978, Beyene and Abedsalam 2005, modified) showing the location of the volcanic units of Afar (from Miocene to present). Miocene units (earlier continental rift type) are located along both sides of the now stabilized margins of eastern and western Afar. Most of the Afar floor is covered by the dominantly basaltic stratoid series. During the last million year, the activity concentrated in axial ranges, as well as in marginal centres and transverse units. The names of the axial ranges are in black, whereas the marginal central volcanoes and transverse ranges are in red. The limits of south-eastern and south-western Afar are drawn with dashed lines. The Gulf of Tadjourah is the western extension of the Gulf of Aden penetrating into the African continent. It ends in the Ghoubbet nearly closed sea, part of the Asal-Ghoubbet rift, emerged between Lake Asal and the Goubbet Gulf. The blue rectangle show location of map Fig. 3.24, the green rectangle of map Fig. 3.32

Fig. 3.4 At the bottom of the escarpment, while approaching the great salt plain, the basement formation and its Mesozoic cover is affected by normal faults, of Red Sea direction (NNE) and block are tilted with the Mesozoic sedimentary sequence dipping west. The strongly tilted block is (**a**) surrounded by tertiary detrital sediments (**b**). (Photograph taken from the Mekele-Dallol road, looking north—13°54′ N, 40°08′ E—by Varet 2013)

Fig. 3.5 View—13°55′ N, 40°10′ E—of the polychromatic formation on the western flank of the Northern Afar depression, at the foot of the Ethiopian Escarpment (visible in the back). It is clearly made of several sub-units (B, yellow colour, and covering C, red colour), the eldest (red series) successively faulted and eroded (photography J. Varet 2013)

North of the Gulf of Tadjourah (in Djibouti Republic), the Dalha basalt pile (8–4 My) is the thickest. It discordantly covers the faulted and eroded Mabla rhyolites (14–10 My) (Fig. 3.9) (Marinelli & Varet, 1973). The area is lifted up north of the Ghoubbet rift, normally faulted and actively eroded towards this marine Gulf (Fig. 3.8). The basalts of the gulf (3–1 My) cover in discordance the tilted erosion terraces (Richard & Varet, 1979). Similar formations of the

Fig. 3.6 Typical dissected landscape of the Eastern Afar margin, in the northern part of the Djibouti Republic. Mabla rhyolites **a**, deeply eroded and are partly covered by the Dalha flood basalts **b** contrasting with their flat-lying surface marking the horizon to the NW (and slightly dipping in that direction). Circa 11°53′ N, 48°47′ E, looking North. From Varet (2017)

Fig. 3.7 On the western, central Afar margin rhyolite flows and domesof the Mabla formation are also faulted, tilted and eroded (**a**), and later covered by basalts of the Dlaha formation (**b**). Photography taken, looking North, at the foot of the Nubian escarpment SW of Digdiga (Sullu Adu massif). Later detrital terraces are observed in the front Circa 11°32′ N, 40°31′ N. *Photography* Varet (2015)

same age are found along the Nubian foothill in western-central Afar. They are therefore thought to characterize the early stage of rifting in Afar and in the Red Sea, until then of continental type.

A particular geological feature observed in Central Afar and surrounding margins are peralkaline granites. They intruded in the trap basalts of the plateau, were dissected by normal faults, all dated between 20 and 25 My, and considered as marking the initial stage of formation of Afar (Barberi and Varet 1975a; b). These white angular reliefs are well visible as they contrast with the surrounding landscape (Fig. 3.10) and may represent the apex of a large pluton which occupied the central part of the Afar triangle, i.e. the original Afar Plume (Varet 2018a; b).

Fig. 3.8 Dalha basalt piles (8–6 My) in Randa area—circa 11°45′ N, 42°34′ E—(Nothern part of Djibouti Republic), down-faulted to the south and actively eroded towards the Ghoubbet marine gulf (western extremity of the Gulf of Tadjourah)

3.4 Afar Floor: The Basaltic Stratoid Series

Most part of the Afar floor is made of a typical geological unit: the Afar stratoid series (as seen in Fig. 3.3). This is characterized by a pile of basaltic flows that reach up to 1000 m in thickness (in the highest visible outcrop, along the Gamari Fault in Central Afar), but very likely it reaches even higher thickness under the Afar floor. This unit is visible all over the Afar floor, although is more limited in the north where younger volcanic units dominate (Fig. 3.4). Originally emitted as sub-horizontal basaltic flows, as it can be observed on the outcropping surface of this unit now forming large plateaus, it was discontinuously affected by normal faulting that transformed the original surface into a horst-and-graben topography and eventually tilted the blocks (Fig. 3.11). As a result, the lava piles observed at the bottom of the fault-generated escarpments are frequently tilted, with an angle increasing with depth.

This unit certainly results from fissure-type basaltic emission, although it is quite unusual to observe the feeding dikes (see Varet 2018a; b for explanation). Locally, two or more directions of faults are crossing each other. As a result, diamond-shaped blocks are formed (Fig. 3.12). Silicic volcanic units are also found in such tectonic environments, topping the basaltic stratoid series, either as well-defined

Fig. 3.9 Satellite image view of the Mabla rhyolites of Miocene age (10–14 Ma) forming the Eastern Afar margin in the North-Tadjourah block, Djibouti Republic. Observe the dominantly NNW-trending faults and domes that were deeply eroded (A) before being invaded by the Dalha basalts 6–8 Ma), They were themselves later also eroded, an forming inselberg (**b**) and Mesas (**c**) of which is observed along the Red Sea margin and visible on the eastern side of the image (from Varet 2017)

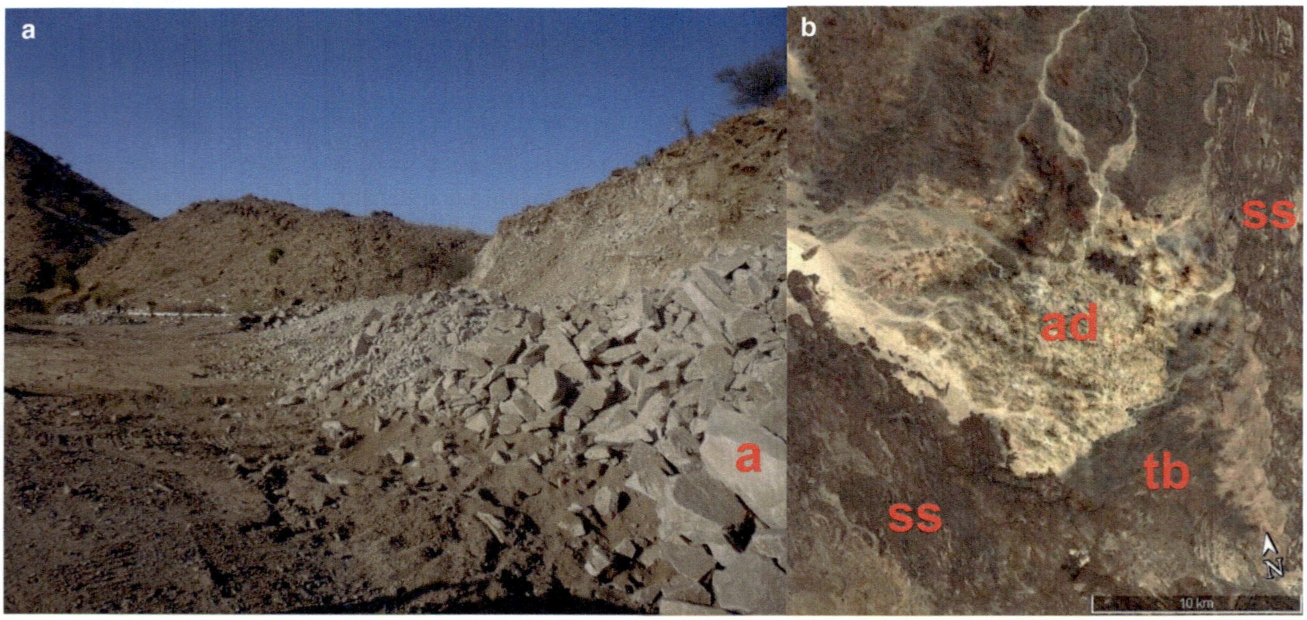

Fig. 3.10 Left (**a**) is a view—12°31′ N, 40°01′ E—of Limmo peralkaline granite of Miocene age (white: a) intruded in the altered, tilted plateau basalts (brown) along the lower part of the Nubian escarpment (*Photo* Varet 2016). Right (**b**) is a satellite image showing the Affara Dara (ad) granitic body—12°26–30′ N, 40°46–53′ E—outcropping in north-central Afar affected by WNW-ESE and NNE-SSW tectonics, intruded in older trap basalts (tb) and covered in discordance by the faulted stratoid series (ss)

Fig. 3.11 Aerial view of the stratoid effected by distensive tectonics: normal faults (**a**) and open fissures (**b**) in central-northern Afar (Varet 1970)

Fig. 3.12 Double faulting affecting the stratoid series—12°12′ N, 41°11′ E—producing blocks of diamond shape central-eastern Afar (Varet 2017)

Fig. 3.13 Ela silicic volcanic centre—12°28–32′ N, 41°10–14′ E—in the upper part of the stratoid series, central Afar (Varet 2016)

Fig. 3.14 Faulted statoid series in central-eastern Afar—12°10′ N, 41°11′ E—showing the rather un-eroded resulting relief. Fallen blocks at the foot of the fault mainly result from seismic activity (*Photo* Varet 2016)

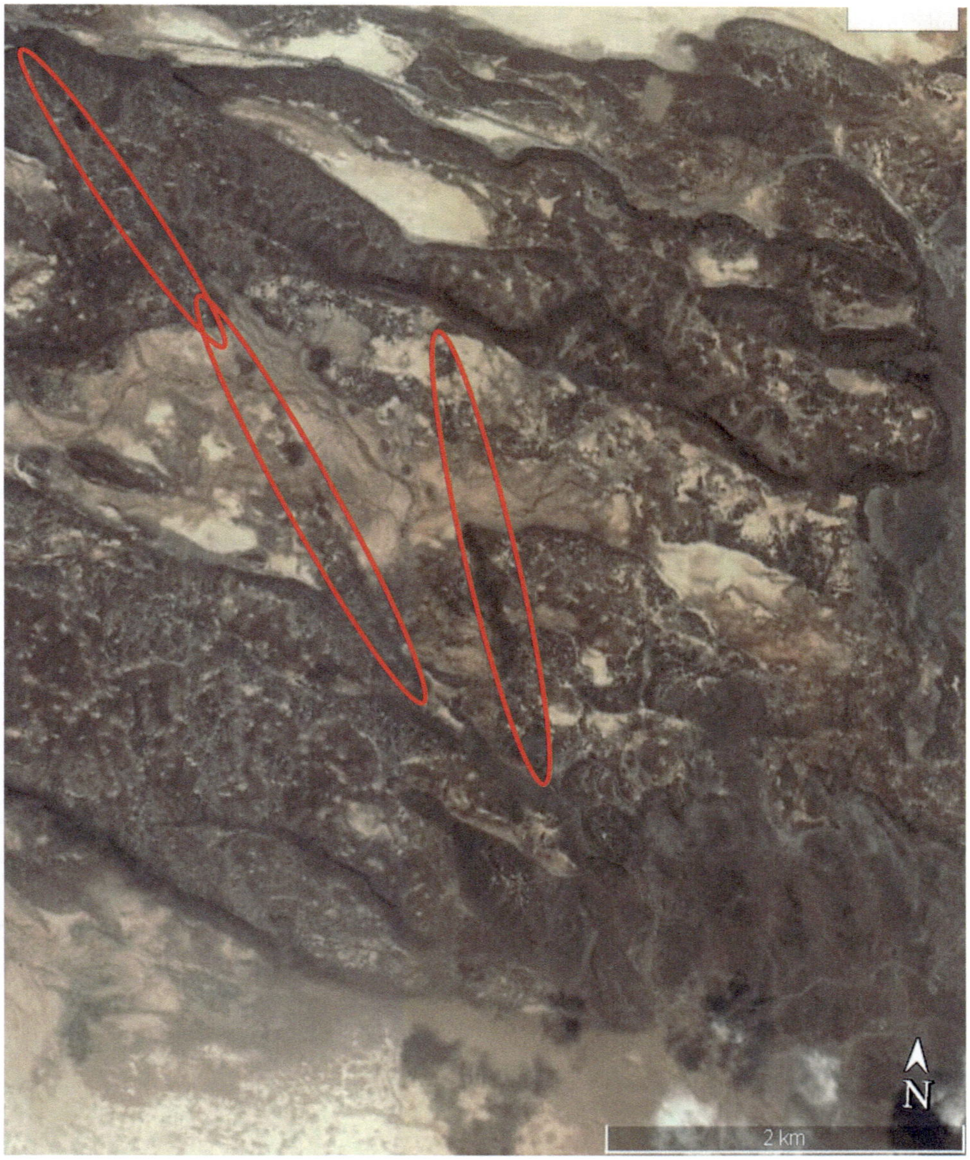

Fig. 3.15 Example of alignments of basaltic spatter cones along emissive fissures of NNW direction visible at the surface of the stratoid series (underlined by red ellipses), south from Ghoubbet—11°24′ N, 42° 42′ E—(images from Google Earth Pro in Varet 2017)

reliefs, some of high elevation (Fig. 3.13) or within fault escarpments, as thick yellowish layers interbedded in the upper part of the dark basaltic pile (Fig. 3.14). Despite extensive sampling and apparent older ages of the lower, hydrothermally altered basalts at the bottom of the lava piles, it appears that the oldest ages found in this unit do not exceed 3.5 My. Hence, the whole Afar floor has rather recent morphology, with the surface of the lava plateaus and fault scarps nearly un-eroded. Fallen blocks, with poorly altered surfaces, sharp corners and with a rather fresh shape found at the bottom of the fault scarps generally results from later seismic activity rather than from erosion (Fig. 3.14). This age (3.5 My) corresponds with the formation of the Gulf of Tadjourah, which marks the western penetration of the Gulf of Aden inside the African continent (Richard and Varet 1975), and with the formation of the axial trough of the Red Sea, i.e. the age of the initiation of the Afar oceanic floor along the eastern part of the Afro-Arabic plate margin. Although the corresponding spreading axes are ill-defined, the Afar stratoid series can be regarded as the emerged equivalent of a basaltic oceanic floor (Barberi and Varet 1977; Varet 2018a, b).

Fig. 3.16 Alol and Sakalol graben as seen from the northern fault scarp—12°04′ N, 42°15′ E—(looking SE) green area are wet zones thanks to hot springs emerging from border faults. The floor is filled with eolian, lacustrine and evaporitic sediments. Basaltic blocks offer a typical view of the surface of the stratoid series. *Photo* Varet (2012)

The age of the stratoid series top is less than 1 My old, and locally, it may be even much younger, as witnessed in places by remnants of spatter ramparts (Fig. 3.15) which may be considered as the last episodes in the formation of the Afar stratoid series. However, most of the present-day volcanic activity concentrates in specific volcanic units that are described below (Sects. 3.4, 3.5 and 3.6).

The grabens (Fig. 3.16), formed by successive extension and normal faulting identified bookshelf faulting and block rotations (Manighetti et al. 1998, 2001), were later filled by sediments of fluvio-lacustrine and eolian origin, as described in Sect. 3.6.

3.5 Dominantly Silicic Stratovolcanoes of Afar Margins and MER Northern End

Another kind of volcanic relief of Afar consists of large, central stratovolcanoes located along its margins and covering the underlying faulted basement, partly made of older volcanics. These are dominantly silicic in composition, with a substantial proportion of pyroclastic products (ignimbrites and pumice falls). They generally have summit calderas, commonly showing poly-phase eruptions and post-caldera units, at the summit, flank or foot of the volcanoes. Extensive

Fig. 3.17 Wide flat plain north of Ma'Alalta dominantly silicic stratovolcano (forming the horizon line) resulting from a huge ignimbrite eruption having covered the whole area −13°08′ N, 40°15′ E—(*Photo* Varet 2017)

Fig. 3.18 Ma'alalta volcano (coordinates: 13°01′ N, 40°11′ E; location on Figs. 3.3 and 3.11) on the western border of Afar depression, seen from south (Google Earth oblique image). The double rimmed caldera toping the stratovolcano and elongated in a transverse direction (NE-SW) is visible as well as post-caldera obsidian and pumice domes and viscous flows, and as well as basaltic flows (black) north and south. The bottom of the scarp of the Pre-Mesozoic basement is visible in the SW corner

ignimbrites and ash fall cover and flatten all the surrounding relief (Fig. 3.17). Basaltic and intermediate lava flows and domes are also observed, either preceding or post-dating the pyroclastic events (Fig. 3.18). The intersection of transverse faulting (NE-SW) with the regional (NNW-SSE) crustal spreading facilitated the development of these central magmatic units. The shape and orientation of the calderas reflects these structural controls. Geochemical studies confirm the interference of continental crust components in the petrogenesis of these silicic products (Barberi et al. 1970).

Fig. 3.19 Google Earth satellite image of the ash plume emitted from Nabro volcano (part of the Bidu massif including two other stratovolcanoes with calderas)—located in Eritrea near the Ethiopian border on the eastern Afar margin (13°22′ N, 41°42′ E. see Fig. 3.3)—during the June 2011 eruption

These units are generally hydrothermally active, with fumaroles and hot grounds found in the calderas and along faults affecting the flanks of the volcanoes. On Eastern Afar Margin, the Bidu massif is particularly spectacular with three large calderas aligned along a NE-SW direction, one of which—called Nabro—produced a large ash eruption in 2011 (Fig. 3.19), with lava flows and pumice falls affecting the surroundings, causing fatalities and property loss (Goitom et al. 2015).

In northern Afar, the Alid dome-volcano (Eritrea) (see also Chaps. 5 and 6 of this book) displays very similar characteristics. Despite being located along the rift axis showing elsewhere an oceanic affinity, the close presence of the basement in this narrowest part of the Afar triangle fully explains this particularity.[2] The extension process in this apex of the Afar rift is very limited, even less than along the MER, as shown by GPS measurements (Calais 2016).

These central rhyolitic stratovolcanoes located in sialic crustal environment show some striking similarities with the volcanic centres found along the Main Ethiopian Rift (MER), including the silicic volcanoes located along its northern portion that characterize southern Afar (Gabilema, Yangudi, Ayelu, Aabida, Dofan, Fantale, Fig. 3.20).

Although slightly older and no more active, Mousa Ali volcano, the highest summit in Afar (2021 m) that marks the triple border between Ethiopia, Eritrea and Djibouti, can be assigned to the same category (De Fino et al. 1973a, b). Compared with others, the caldera is not so well-developed (although clearly expressed by the truncated morphology of this large cone, as seen in Fig. 3.21) and pyroclatic products are less abundant. Thick silicic lava flows and domes dominate, and the volcano, therefore, display steeper slopes than its equivalents where pyroclatites dominate.

---

[2] This strato-volcano and dome even show outcrops of pre-Mesozoic basement reworked in the volcanic unit (Varet, 2018a, b, p.214; Duffield et al. 1997).

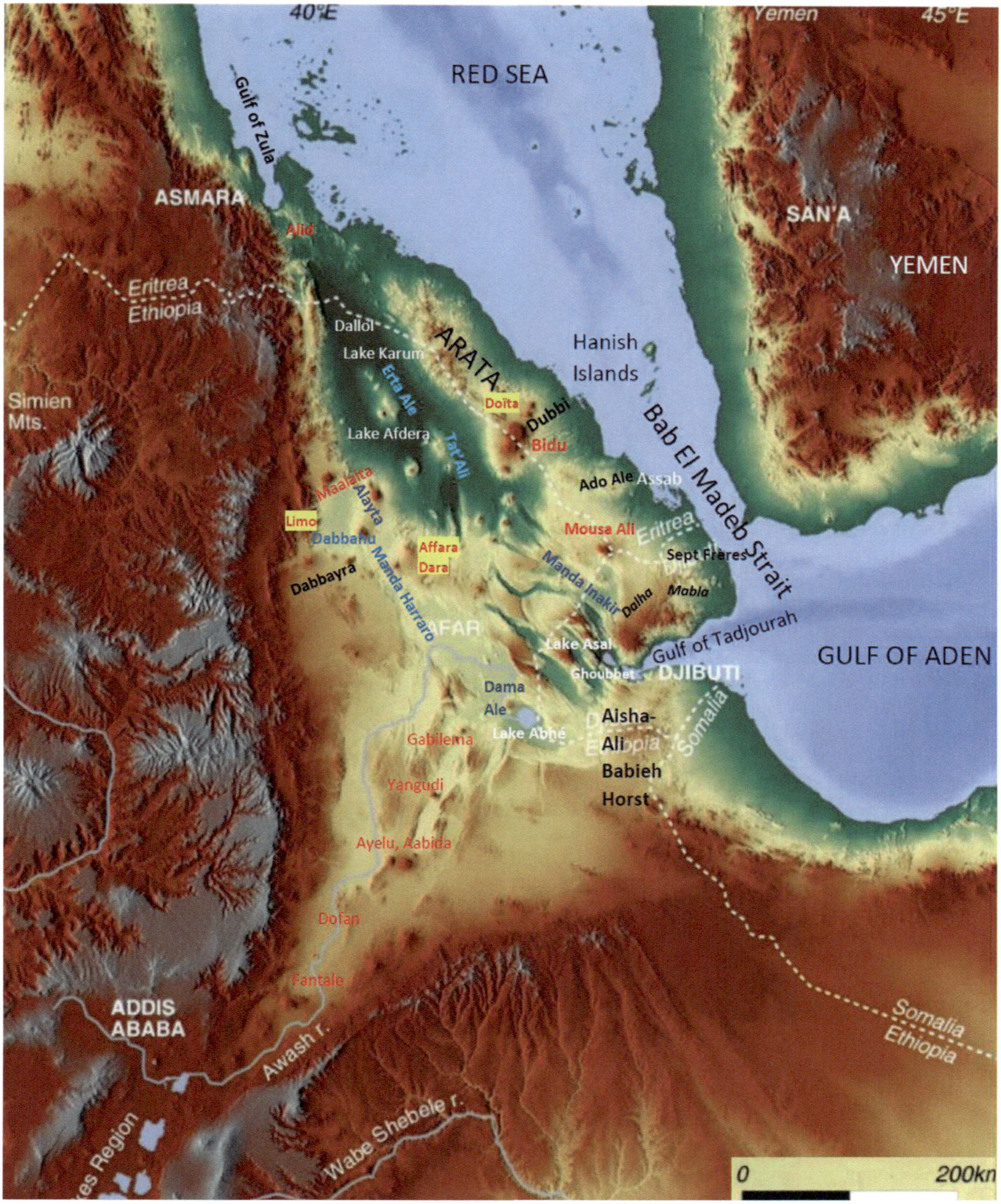

Fig. 3.20 Relief map of the Afar triangle and surrounding plateaus, with the toponymy of the main units cited in the text. NNW-SSE NW–SE-trending axial ranges (dominantly basaltic spreading segments) in blue. Central silicic volcanoes in red. Lakes and Gulf of Ghoubbet in white. Granitic massifs in red over yellow. Transverse ranges of the margins in thick black. Location of geological formation's names in italics

Fig. 3.21 View of Moussa Ali volcano—12°28′ N, 42°24′ E—as seen from its southern foot. The truncated neck indicates the summit caldera (*Photo* Varet 2012)

3.6 Transverse Basaltic Alignments

Less spectacular but characterized by specific landforms are the transverse basaltic alignments observed along the Afar margins. These are scoria and spatter cones emitted along fissures oriented E-W and/or NE-SW. Unlike the lavas found on the Afar floor, which have transitional compositions with tholeiitic affinity, these are typically alkali basalts. Peridotite nodules (mantle xenoliths) and gabbro cumulates (synmagmatic) are frequently found in these scoria and lava and led Barberi et al. (1974a, 1974b) to consider these features as analogues to the oceanic fracture zones yet characterized by peridotites and alkali basalts contrasting with the tholeiites of the Mid-Oceanic Ridges (MOR). Studied by Otonello et al. (1978), these peridotite xenoliths provided information on the composition of the upper mantle.

The most productive areas—and spectacular landforms—are in the middle of Afar on both sides of the depression: Dubbi (Fig. 3.3) in the southern Danakil Alps and along the Red Sea (in Eritrea) and Dabbayra at the foot of the Nubian escarpment (Fig. 3.22). Both are characterized by important basaltic emissions forming embryonic shield volcanoes built along NE-SW feeding fissures. They are considered by Varet (2018a; b) to be controlled by a major fracture zone that extends towards the Red Sea in the Anish Islands, which are also built along feeding fissures of the same direction. This also marks the end of the Red Sea axial trough and the displacement of the oceanic spreading within Afar, with the spectacular Manda Harraro range identified as one of the major spreading segments in Afar by Barberi and Varet (1977), confirmed by the 2005–2010 Dabbahu event (see Sect. 3.4). This can be considered as equivalent to Oceanic Fracture Zones (OFZ) as it also corresponds to a major discontinuity of the Nubian escarpment, like the one observed further north at the level of Ma'Alalta (see Fig. 3.3).

Further south, several such transverse alignments are observed along the stabilized margin of Afar along the Red Sea. All are trending E-W with no really visible associated tectonic features except the alignment of the scoria cones. The most spectacular is the Ado Ale range near Assab in Eritrea where two such alignments of scoria cones along distinct feeding fissures can be observed (Fig. 3.23). East of Moussa Ali, another similar alignment is observed in Eritrea. In Djibouti, the Sawäbi (Sept Frères islands) alignment is made of hyaloclastite cones crossing through the Bab-El-Mandeb in the same E-W direction (Fig. 3.2).

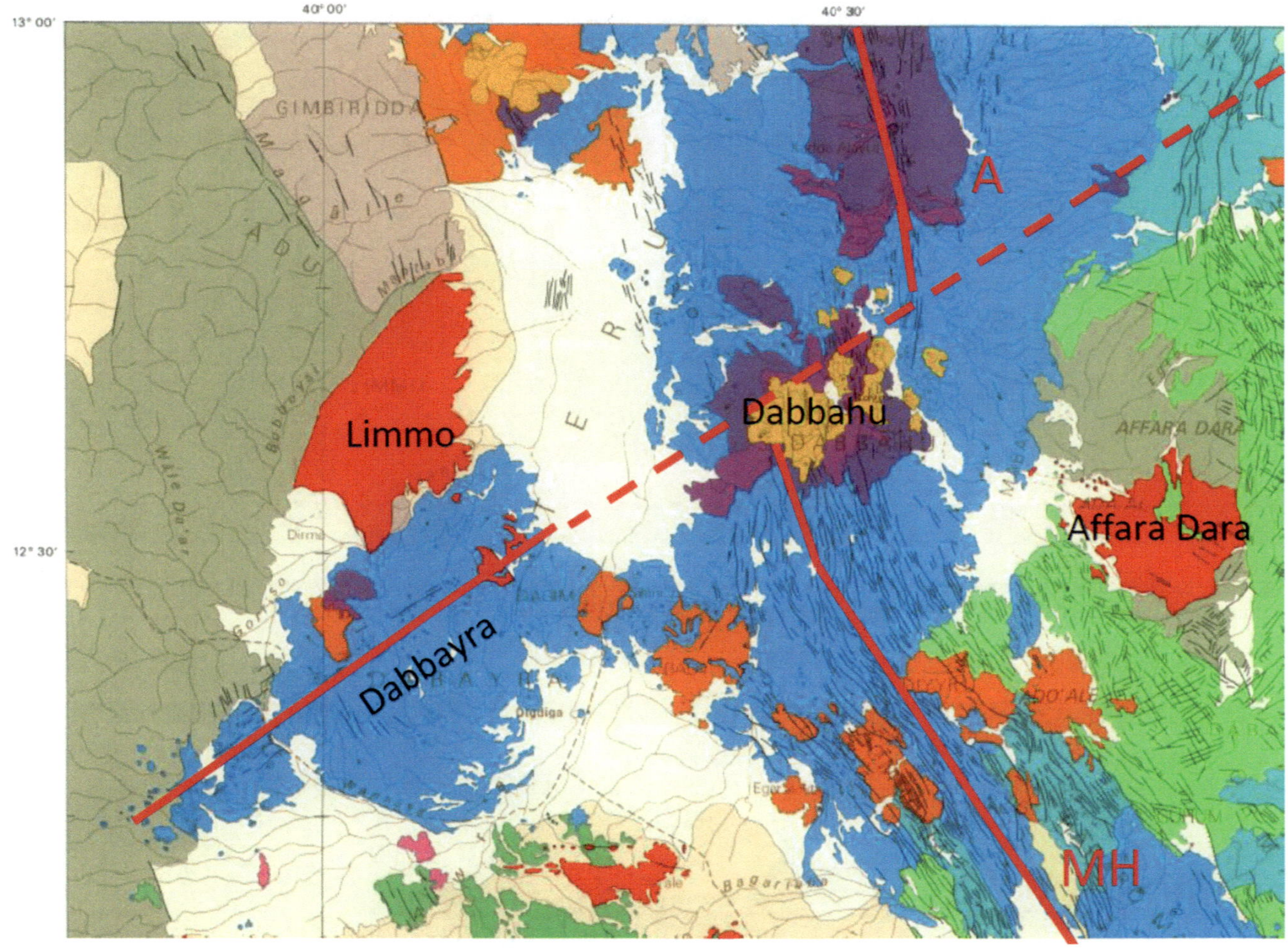

Fig. 3.22 Geological map (from Varet 1975, see location in Fig. 3.3) showing the Dabbayra shield volcano (D) built along transverse faults of NE-SW direction, also marking a shift in the Nubian escarpment (equivalent to Oceanic Fracture Zones), and the Transform Fault Zone (TFZ), explaining accommodating the shift between Alayta (A) and Manda Harraro (MH) axial ranges, both bent. This discontinuity cuts also mid-way between the two Miocene peralkaline granites of Limmo and Affara Dara and extends NE through Afar up to Bidu, Dubbi and Hanish islands in the Red Sea. Pre-Miocene basement in beige, Miocene granite in red, Dalha basalts in dark green, stratoid basalts in pale green and associated rhyolitic centres in orange, Dabbahu volcanic centre developed on this TFZ, with a succession of recent basaltic flow (in blue), lava flows of intermediate composition (mugearites, trachytes) in violet and peralkaline obsidians (in yellow) toping the shield

3.7 Axial Ranges (Active Rift Axis)

In Afar, volcanic activity developed also in submarine (northern Afar and Djibouti) and sub-lacustrine environments (central Afar mainly). This resulted in very specific landforms that are usually found in submarine environments like guyots (Fig. 3.24, hyaloclastite ash rings (Fig. 3.25a), pillow lavas and lava flow surfaces with small pavements (Fig. 3.25b) Bonatti et al. 1979; Bonatti & Tazieff, 1980. These volcanic landforms, unusual at the earth surface, are commonly observed at the axial ranges initial stage of activity (Fig. 3.26).

Axial volcanic ranges are the most spectacular landforms in terms of volcanology and geodynamic significance. These are active spreading segments—of mid-oceanic ridge (MOR) type even if dominantly emitted in free-air environment—exposed at the surface on the Afar floor. Seven such ranges were identified, all trending in a NNW-SSE or NW–SE direction (Barberi and Varet 1975a; b). They are dominated by

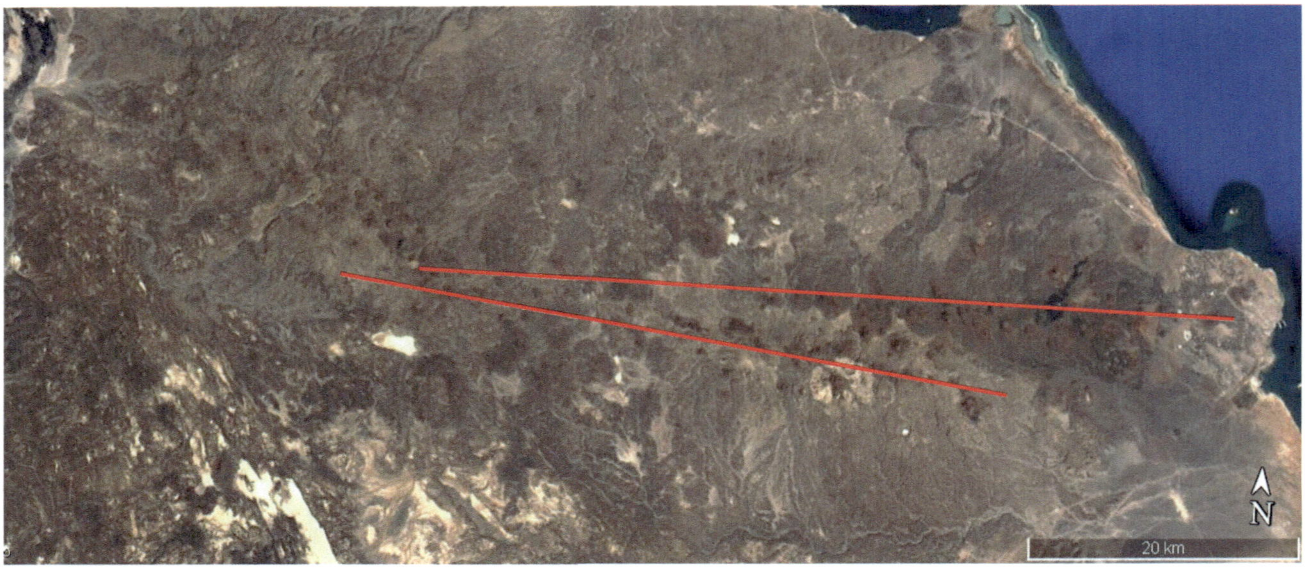

Fig. 3.23 Satellite image of Ado Ale range west from Assab port in Eritrea showing a double alignment of scoria cones (underlined in thin red) emitting flows on both sides of the E–W feeding fissures −13°00′, 42°09–43′ E—(from Google Earth Pro)

Fig. 3.24 Asmara guyot—12°14′ N, 41°32′ E—a sublacustrine hyaloclastite cone in the Awash lower basin, SW of Dama Ale (**a** and **b**: land and air photos coll. G. Marinelli, Pisa Univ. and (**c**) satellite image from Google Earth Pro). Observe the NW–SE fault affecting the volcano

basaltic eruption of fissural type emitted from dikes along open faults and fissures that develop along the range axis. On both sides of the axis, a symmetrical graben is formed with lavas becoming older with distance from the rift axis (Fig. 3.29).

With time, a shield volcano, elongated along the same direction, may form by accumulation of lava issued from the same axial fissure, eventually dissected by later re-opening of the rift axis (Fig. 3.7). In a few cases (e.g. Alayta and Manda Inakir), spreading developed with time at the foot of the shield, along its eastern flank.

In some cases, when the spreading rate decreases, as well observed in the northern part of the range, the shield may evolve in a cumulo-volcano by accumulation of more trachytic and rhyolitic viscous lavas. These lava types are always quantitatively limited in these environments dominated by basaltic lava flows (Fig. 3.27).

The Erta Ale range (Fig. 3.28) offers the most complete view of such volcanic and petrologic evolution and was described in detail by Barberi and Varet (1972 & 1970). The Erta Ale shield volcano is topped by an elongated elliptic

Fig. 3.25 **a** Oblique aerial view—13°54′ N, 40°27′ E—of Catherine volcano seen from north, a submarine hyaloclastite cone later affected by a NNW-trending open fissure that emitted a subaerial scoria cone on its southern flank. The lake is filled by hot springs (photo Coll. G. Marinelli, Pisa University). (B) paving stone basaltic lava flow submarine surface, to the south of Erta Ale range in the vicinity of lake Afrera (*Photo* Varet 2015)

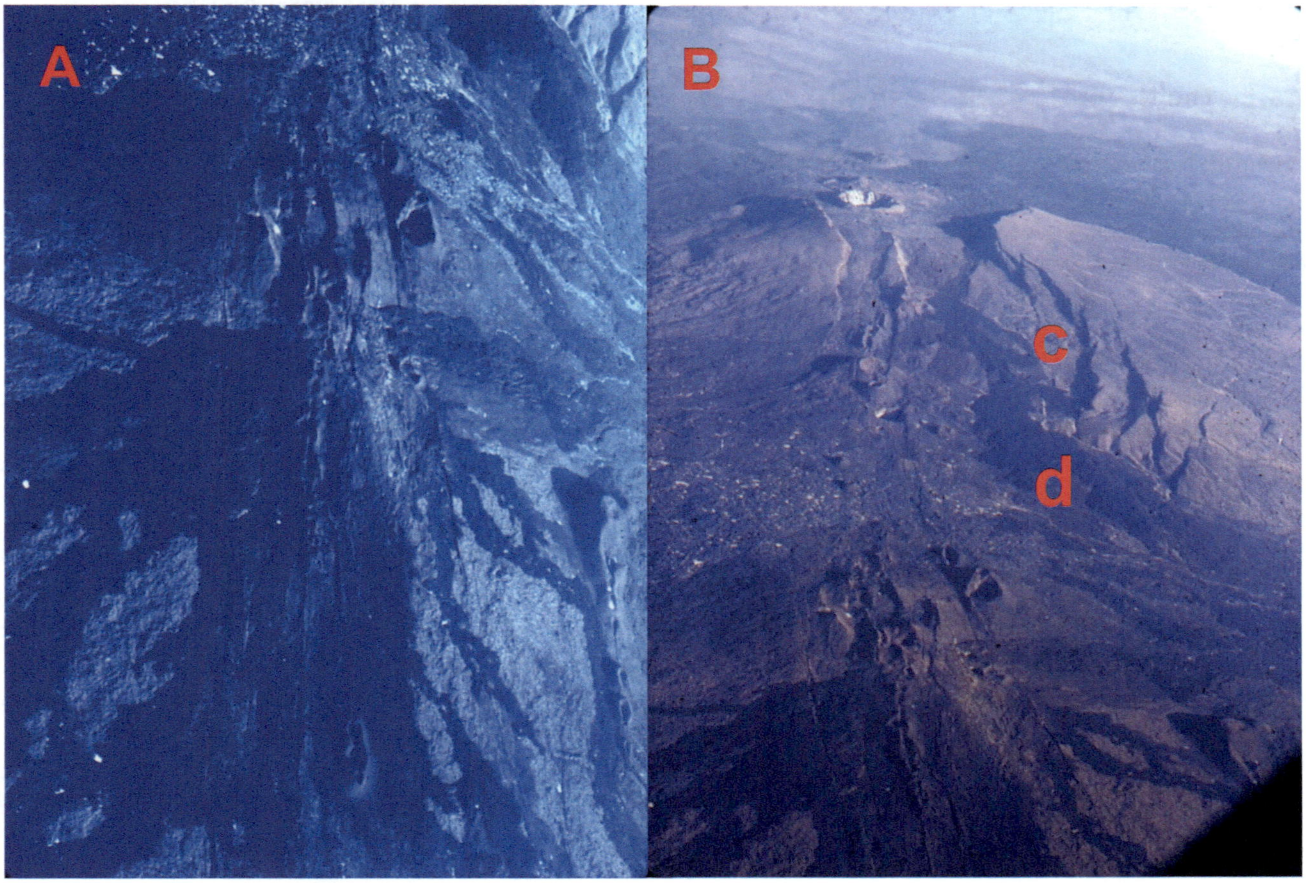

Fig. 3.26 Haili Gub spreading segment (13°30′ N, 40°42′ E—13°21′ N, 40°47′ E; south Erta Ale range): (**a**) aerial photograph of the open emissive fissures with spatter ramparts and (**b**) oblique air photograph of the former shield affected by ring faults (**c**) posterior to inflation and axial rift collapse, with later emission of a trachyte flow (**d**)

Fig. 3.27 Alu—Dalafilla rhyolitic cumulo-volcano, in the northern part of the Erta Ale range—13°47′ N 40°43′ E—seen from west. The small transverse graben affecting the older shield volcano is visible as well as the ring fissures emissive of coria scones and spatter ramparts. (photograph Smithsonian Institution, global volcanism programme)

caldera where two lava lakes have been permanently active for at least the last 50 years (Fig. 3.29). Further south, another elliptic caldera also contained a lava lake in 2017 and early 2018 (Fig. 3.30).[3] The lava flows typically display surfaces of "Aa" (scoriaceous); "Pa Hoe Hoe" (smooth) or "cordées" types (Fig. 3.31). Lava channel and lava tubes are also common, sometimes revealed by block falls (Fig. 3.32). A common landform in the summit part is also the hornitos (Fig. 3.33).

Further south, the Erta Ale range splits into two parallel axial ranges: Alayta to the west and Tat'Ali to the east, as seen in Figs. 3.3 and 3.28. This split occurs at the level of Lake Afdera which in a NE-SW direction. It was interpreted by Illsley-Kemp et al. (2018) as a proto-transform fault, a view based on the strike-slip motion observed from seismic studies. Bonatti et al. (2017) could show from bathymetric studies that the bottom of the lake is at -190 m below sea level and is still presently sinking, so that the lakes bottom can be considered the equivalent of the "nodal deep" of a slow-slip oceanic transform.[4] Alayta is composed of two parts, a wide shield volcano elongated NNW-SSE affected by an axial graben with a summit crater (Fig. 3.34) and an even wider basaltic fissural lava field, dominantly of Aa type, which corresponds to an eruption which occurred in 1908 according to historic observations (Fig. 3.35) along N–S-trending open fissures. Tat'Ali is a smaller axial range which developed along fissures of the same NNW-SSE direction on the SE side of

[3] Since 2019, as a consequence of this important basaltic eruption, the level of the lava lakes did sink in the Erta Ale northern caldera, so that the volcanic activity at the lava surface was no more visible in the last months (until December 2019 at least).

[4] This discontinuity in the axial ranges was probably inherited from the discontinuity in the Nubian escarpment observed at the level of Ma'alalta volcano (see Fig. 8.3 and 8.32).

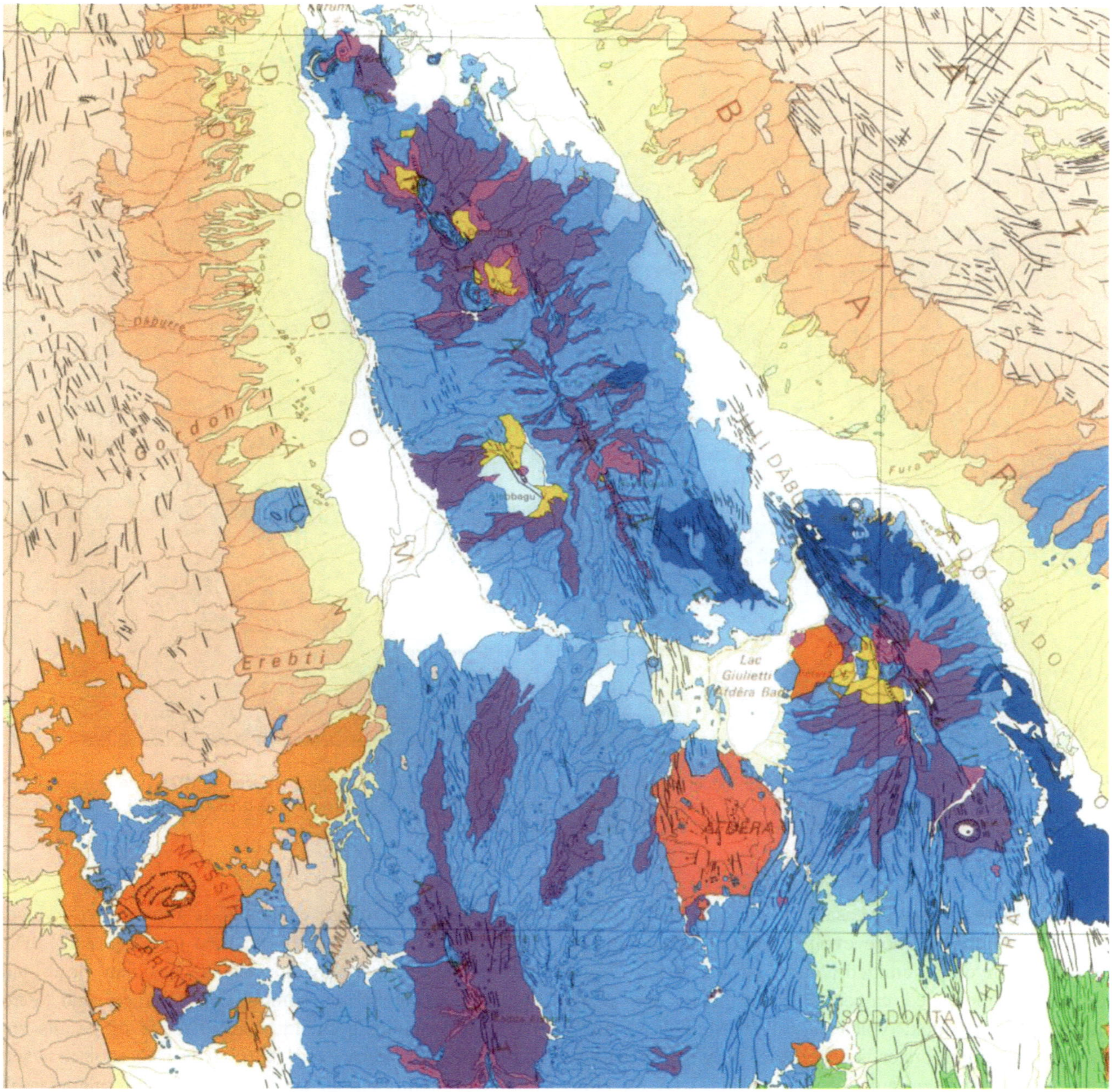

Fig. 3.28 Erta Ale range located in its northern Afar geological context, extracted from the geological map of Northern Afar at 1/500.000 scale (CNR-CNRS Afar Team 1973). See location on Fig. 3.3, along the axis of the salt plain resulting from the drying of the Danakil Sea, whereas the faulted pre-Mesozoic basement, the polychrome tertiary detritic series and the coral reefs of the former Danakil Sea (yellow) are visible on both sides. The lava sequence of Erta Ale is made of olivine basalts (blue, deep blue when picritic), andesine basalts and ferrobasalts (violet) dark trachytes (pink), and silicic trachytes and rhyolites (yellow) in the most evolved volcanic units (to the north of the range, and also at Ale Bagu, a lateral volcano displaying also phyroclatic-hyaloclatic products (pale blue). This map also shows how the range ends south, at the level of lake Afdera, where it splits into two: Tat'Ali SE and Alayta SW. A discontinuity inherited from the structure of the basement, showing a parallel shift at the level of Ma'alalta silicic (orange) volcano (called Pierre Provost before we could establish the Afar toponymy)

Fig. 3.29 View from the north of the Erta Ale caldera—13°36′ N, 40°40′ E—toping the shield volcano, as seen in 1975. The two active lava lakes are visible in the front caldera, as well as the flat surface of the southern one. The wide elliptic sing is also visible as well as the Haili Gub shield and axial graben. All are aligned along the same NNW rift axis (*Photo* J. Varet)

Lake Afrera. It is made of basaltic fissural lava fields to the north and south and more evolved lavas (trachytes up to peralkaline rhyolites) emitted along a summit axial rift and forming an elongated caldera. To the south of Tat'Ali range, a wide shield volcano, called Mat'Ala, with a spectacular summit crater is observed. Fairly recent activity developed along open faults of N-S to NNE-SSW direction on the western side of this shield.

It appears that, to the south of Alayta and Tat'Ali, another major discontinuity affects the Afar triangle along a NE-SW (transverse) direction that is not directly expressed at the surface on the Afar floor itself (where it, however, marks major discontinuities) but is visible along the margins as both volcanic units of Dabbayra (West) and Dubbi-Bidu (East) are clearly fed by dikes and controlled by faults of transverse direction (as seen in Sect. 3.4). Dabbahu volcano[5] is located at this level, at the place where the Alayta axial range ends and is replaced by Manda Harraro range further south and west (see Fig. 3.22). This unit is characterized by spectacular obsidian flows and domes of pantelleritic composition (Barberi and Varet 1975a; b) topping a shield volcano made of basalts and intermediate products (Fig. 3.36). It is also a place where important hydrothermal activity occurs with innumerable steam vents frequently engineered by Afar people to condensate the water. As a whole, this major transverse structure crossing through Afar and the Red Sea can be considered as an incipient Oceanic Fracture Zone (Varet 2018a; b), allowing the MOR axis of the Red Sea to shift into Afar where it connects with the Aden Ridge through the Gulf of Tadjourah, whereas the Bab-El-Madeb strait is unactive (Barberi and Varet 1977).

Manda Harraro range is a less spectacular volcanic landform at first sight, as it does not show the typical shape expected for a volcano. The range is made of flat-lying flood basalts emitted along open fissures of NNW-SSE direction, over a length of more than 70 km and a width that vary from a few up to 10 km (Fig. 3.37). In details,

[5] Earlier called "Boina" (meaning steam vent) before we established the Afar toponymy with the help of E.Chedeville.

Fig. 3.30 After a rise of the lava in the central pit, with consequent new overflows from the caldera to the south, new fissures of NNW-SSE direction opened up on 21 January 2017 in the large southern caldera, spilling large amounts of lava. Image captured by the Operational Land Imager (OLI) sensor on Landsat 8 on 26 January 2017. Infrared hot spots show a lava flow and a new lava lake aligned along the same open fissure, a tectonic event cutting through the caldera that allowed to spread Aa flows (black colour) over large parts of that caldera floor with overflow to the NE, to the West and to the South. *Source*: https://landsat.visibleearth.nasa.gov/view.php?id=89544

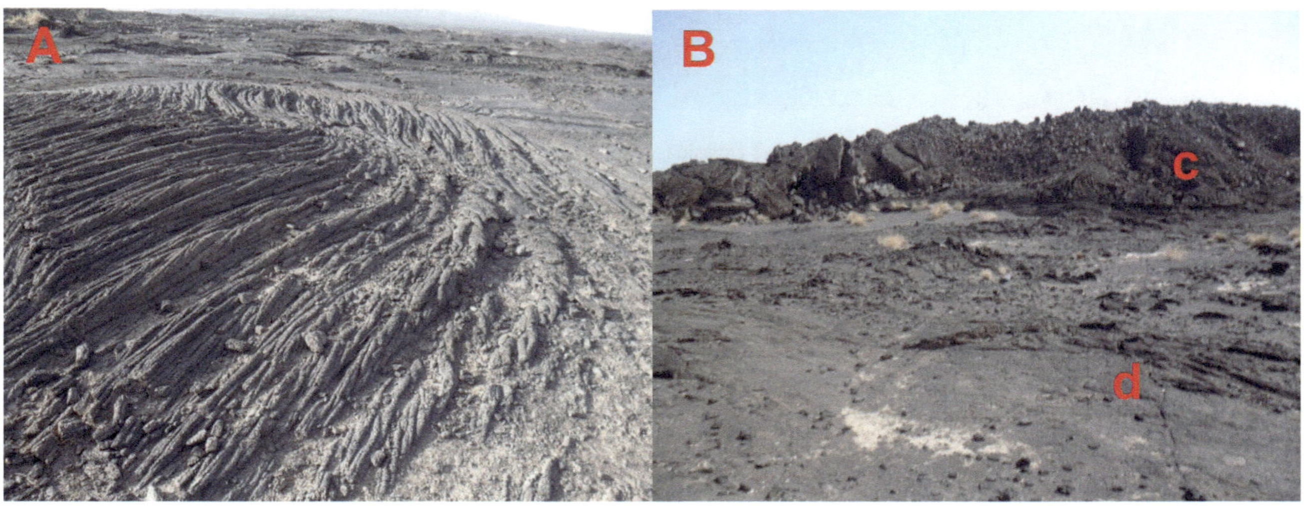

Fig. 3.31 Three kinds of lava flow surfaces: "lave cordée" (**a** left) and Aa flow (Scoriceous, **c**) covering a pa hoe hoe (smooth, **d**) flow (**b**, right), as seen on the western flank of Erta Ale shield—13°34′ N, 40°36′ E—(*Photos* Varet 2013)

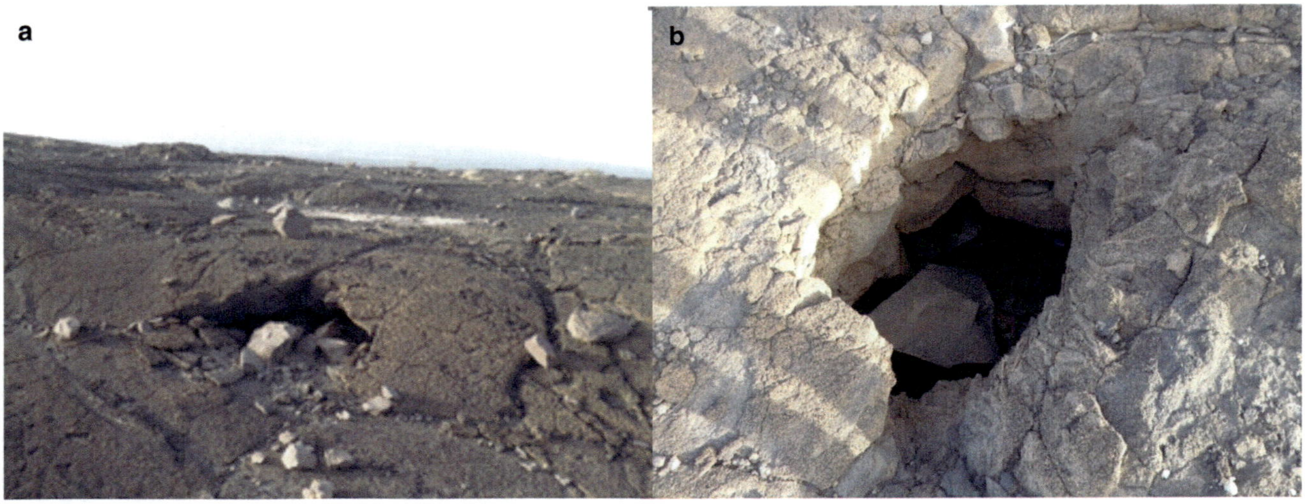

Fig. 3.32 Angular blocks projected by Erta Ale crater having smashed the surface of the lava flows revealing lava tunnels on the western flank—13°35′ N, 40°39′ E—(*Photos* **a** and **b**, Varet 2013)

Fig. 3.33 Detailed view of one of the fixed hornitos, site of emission of basaltic lava tubes, on the southern flank of Erta Ale volcano—13° 35′ N, 40°39′ E—(*Photo* Varet 2013)

smaller structures like shield volcanoes and various types of craters (some of them rectangular) are observed, the most recent being aligned along the median axis. This is also the range that shows the most tholeiitic kind of basalts and the less extensive crystal fractionation processes (lack of differentiated products). In the period 2005–2010, it was affected by an important seismovolcanic crisis that resulted in the opening of a 8 m wide and 70 km long trench along the axial graben Ayele et al. 2007 & 2009. It was later filled by successive basaltic dikes, with only rare emissions at the surface (Fig. 3.38). The area was extensively studied during this event (see, among other: Yirgu et al. 2006; Grandin et al. 2009; Keir et al. 2011; Ebinger et al. 2010; Ferguson et al. 2013) allowing to uncover the mechanisms of accretion occurring in the Afar rift along these axial ranges. The resultant landform is a flat-lying surface covered by recent basaltic lava flows and crossed by normal faults open fissures,[6] topped by spatter ramparts and craters aligned along the axis.

The Manda Harraro range ends on the western side of the Tendaho graben that extends further south-east down to Lake Abhe at the border with the Republic of Djibouti. A few volcanic units are observed in this plain, the most extensive being Gabillema where sub-E-W faults and fissures cross the NNE-SSW-trending axis of the Main Ethiopian Rift (MER) that ends there while reaching the Red Sea—Gulf of Aden spreading axis. Dama Ale is a circular shield volcano displaying the entire sequence from basalts to rhyolites, with a wide crater at the top where a few pyroclastic products are observed (Fig. 3.39). Its shape and structure as well as its composition result from the intersection of the NW–SE-trending faults marking the extremity of the Tendaho graben with the northern tip of the—here dying—NNE-SSW MER. This volcano was probably active in recent historic time as shown by its recent crater and dried lava lake.

Further west, the Manda Inakir range—crossing the Ethiopian-Djibouti border—also displays the typical characteristics of an axial range, with a basaltic shield (Inakir) to

[6] Open fissures are not just surface manifestations but typical tectonic features of crustal extension (oceanic spreading), well documented in the litterautre in rift zones, as in Iceland where the Silfra fissure opened in 1789, due to the produce fissure eruption. Th e most recent were observed in Afar at Asal in 1978 and Manda Harraro in 2010.

Fig. 3.34 Alayta shield volcano—13°30′ N, 40°37′ E—seen from south. In front, the basaltic lava fields are affected by open fissures, well visible on both sides of the image. The axial graben is detached on the horizon, with recent basaltic flows emitted along the NNW rift axis (*Photo* coll. G. Marinelli, Univ. Pisa)

the south and an active rift (Manda) to the north (Fig. 3.40). Local observations reveal that Manda Inakir was active in 1928–1929 (Audin et al. 1990). This range shows analogous volcano-tectonic (symmetrical normal faults and open fissures along the axis) and magmatic character with other Afar axial ranges (De Fino et al. 1973a, b).

The south-westernmost axial range, located in Djibouti Republic, between the Ghoubbet marine pass and Asal Lake (157 m bsl), is called Asal (Fig. 3.41). It is a "textbook" axial range with a well-marked volcanic axial rift located along the axis of a graben itself dissecting a former shield volcano. Asal is the last emerged segment, as it extends under the sea along the Ghoubbet rift axis, whereas the next axis of spreading is located further east along the bottom of the Gulf of Tadjourah, where bathymetric and seismic studies allowed to map this westernmost extension of the Aden Mid-Oceanic Ridge (MOR). Asal was chosen in 1970 as a site for measuring the extension between the Arabic and the African plates on the ground. For five years, regular laser measurements did not provide any sign of motion, until the Ardoukoba seismic and volcanic event occurred in 1978 Demange et al. (1980). A basaltic eruption was observed along the rift axis and open fissures, as much as 2 m wide, originated within the axis of Asal and Ghoubbet rifts over a length of 40 km (Fig. 3.42), whereas normal faults also accommodated extension on both sides, particularly along the "petit rift" located north. This is also the axis along which the main sea water flow from Ghoubbet Gulf to Lake Asal occurs. Continuous seismological studies allowed to show that a magma chamber located at a depth of 7 to 2 km underneath the Fiale caldera was filled with basalts during the period 1988–2000 (Fig. 3.43). This is the site selected by the World Bank for exploration drilling in order to develop a geothermal field for electricity production. This is, however, risky given that the opening of the rift facilitated the groundflow of sea water along this axis also, hence cooling the geothermal reservoir (Varet 2014).

3.8 Sedimentary Plains and Lakes

Although volcanic and tectonic landforms dominate the landscape in Afar, sedimentary plains also need to be mentioned. In these areas, most of human activities are located

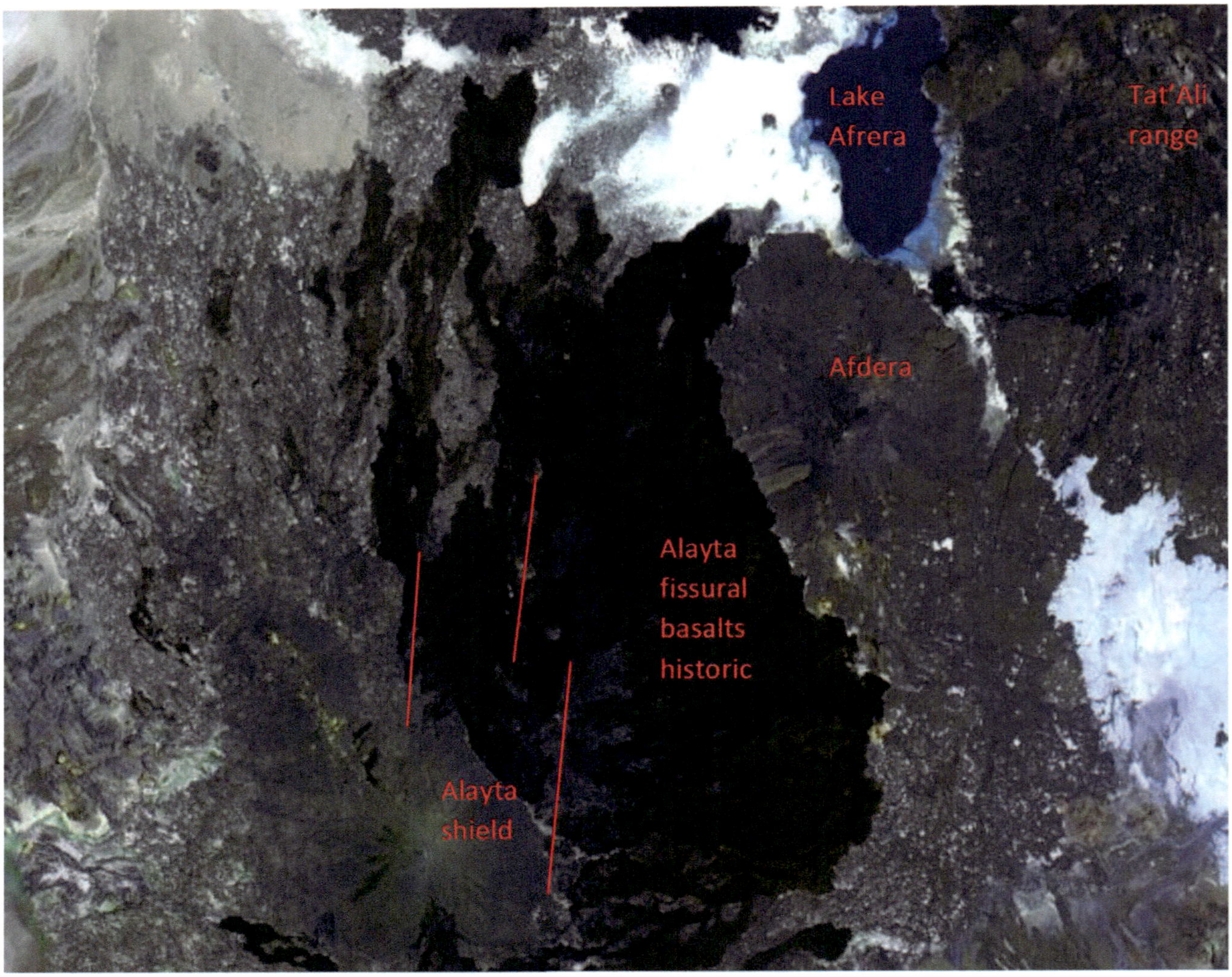

Fig. 3.35 Alayta shield volcano (greenish colour, SW side), with series of very recent craters aligned along the NNW-trending axis of the basaltic-to-trachytic shield (seen in Fig. 3.36). Very fresh Aa basaltic lava flows (dark black colour) resulting from a partly historic eruption (1908) emitted from sub-N-S-trending fissures along the east side (central part of the image) laps up against the western flank of Afdera volcano, immediately south of Lake Afrera

and tend to develop, whether by a traditional (pastoralist activity) or a more recent (roads, cities, mineral extractions, plantations, etc.) approach. They are also the places where the most of ancient human fossils were found. All sedimentary plains in Afar result directly or indirectly from tectonic processes.

In northern Afar and along the Nubian and Somalian escarpments, sedimentary plains result from the erosion of the surrounding faulted plateaus, with detrital terraces sloping towards the Afar depression. At the foot of these terraces, flat-lying fluvio-lacustrine or aeolian sediments are generally made of fine particles and the water table allows

Fig. 3.36 Large pantelleritic obsidian flows toping the Dabbahu volcano, 1307 m—12°35′ N, 40°28′ E—(Google Earth image)

for the development of green areas, as observed in Dodom or Teru (Fig. 3.44).

In the northernmost part of northern Afar (also called Danakil), the lower surface (120 m bsl) is occupied by a thick salt formation that extends over 120 km long and 20–50 km wide (Fig. 3.45). Salt domes and hydrothermal activity affect this otherwise monotonous area, the surface of which is reworked every year. The salt surface is dissolved during the rainy season, and recrystallization occurs during the dominant dry season. Salt is exploited by hand cutting of rectangular blocks and then attached and carried away on camel backs (Fig. 3.46). Potash mining companies are now ready to exploit potash around Dallol (Ethiopia) where sylvite beds are exposed and found at depth (Fig. 3.47). At Afdera, salt extraction was recently improved by evaporation of the lake water in ponds constructed around the shores of the lake (Fig. 3.48), but thus negatively affecting the ecosystem along the lake shores.

It is worth noticing that the former Danakil seashores are well visible all around the northern Afar, with coral reefs and

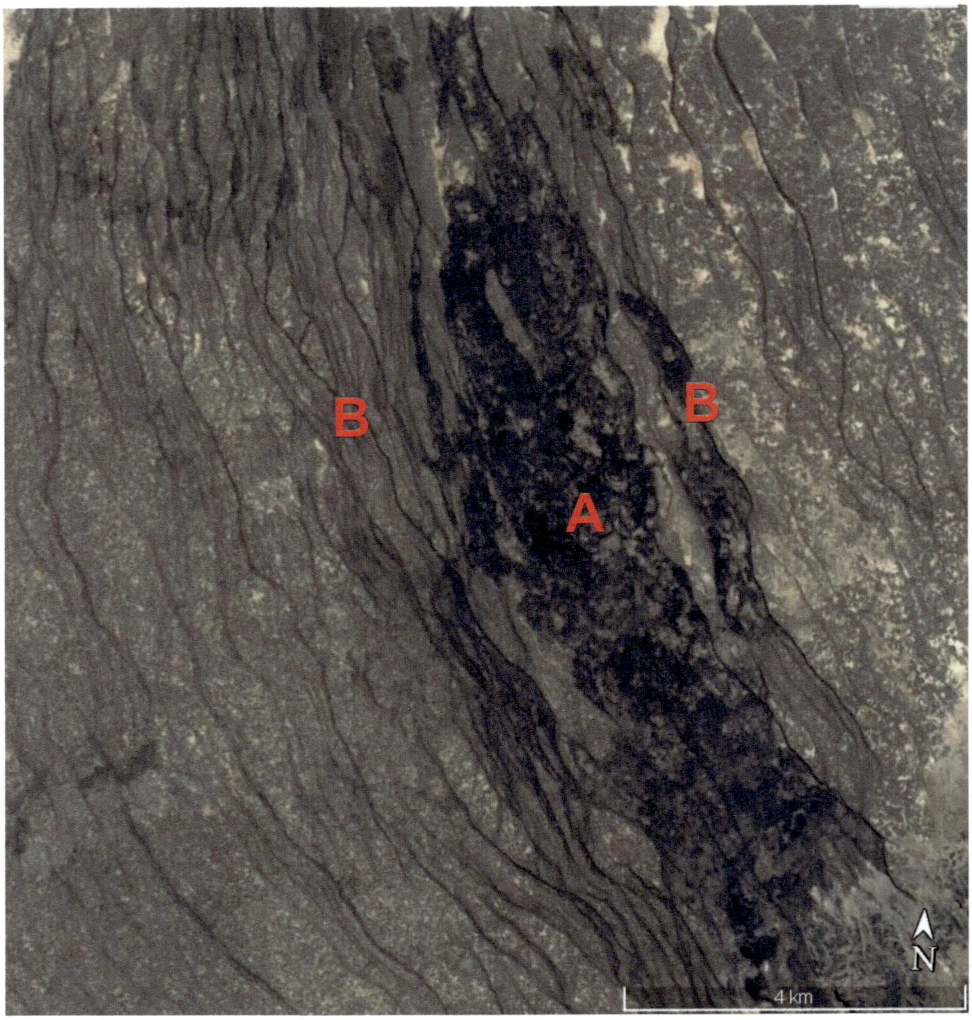

Fig. 3.37 Satellite view (Google Earth Pro) of the axial graben of the northern part of Manda Harraro range (see location on Fig. 3.3). Fissural basalts are emitted from open fissures (**a**), and surrounding symmetrical normal faults affecting recent flows on both sides (**b**) show an intense tectonic activity contemporaneous with basalt emissions along the same faults

former beaches with flat urchins seated on the detrital terraces of the polychromatic series (Fig. 3.49). Human tools were dated to 120,000 years thanks to associated corals (Tazieff et al. 1970, 1972).

In southern Afar, no trace of marine environment was found, and the lacustrine environment appeared to have been suitable for early human settlements. The Hadar unit (western tributaries of Awash River) was extensively studied during the last 50 years, as it displays important fossil sequences. Following Taïeb et al. (1976), important mammals' bones (hippo, elephant, rhinoceros, etc.) are present and the species *Australopitecus Afarensis* was dated 3.2 Ma, called "Lucy" by the US members of the team and "Dinkenesh" by the Ethiopian counterpart. Earlier, *Ardipithecus Kabbada* and *Ardipithecus Ramidus* were identified, underlying the Hadar formation and *Homo*

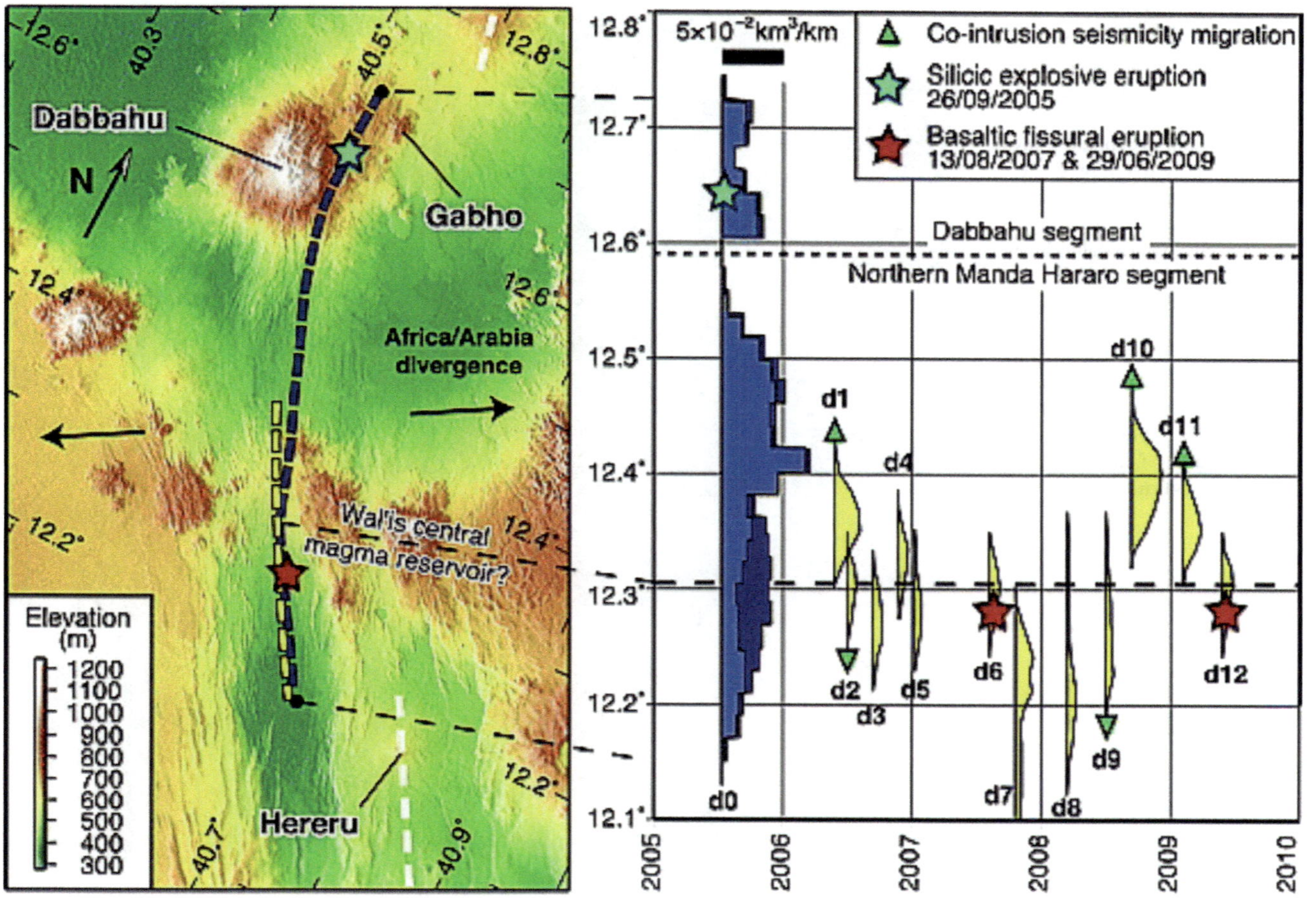

Fig. 3.38 Digital elevation model (DEM) of the Manda Harraro–Dabbahu rift (map, right). Thick dashed lines indicate the location of dike intrusions (blue, September 2005 mega-dike; yellow, 2006–2009 dikes); white dashed lines show the location of neighbouring Alayta and Southern Manda Harraro axial ranges. Distribution of dike intrusions as a function of time and latitude; thickness of the coloured curves indicates dike volume per unit distance along the dike. Green triangles show the direction of the migration of seismicity coeval to dike injection. Stars indicate eruptions (Grandin et al. 2009)

Erectus was also discovered in the upper fluvial formation called Basidima (2.7 Ma to present) and in the Samoti plain in the Eritrean Danakil (Abbate et al. 1998) (see also Chap. 6 of this book). Afar became then known worldwide as the "birth place of mankind", although this has been disputed by other places in Africa since Bruxelles (2006) studies.

All over Afar, smaller plains are filling the graben or half-graben floors with aeolian sediments and earlier lacustrine sediments deposited in the Plio-Pleistocene phases when Afar was subjected to more humid condition, during which wide lakes covered these depressions (Fig. 3.50. Pulverulent limestones—well expressed in the Sodonta plain (Fig. 3.51)—and diatomite beds and gypsum deposits—found in many

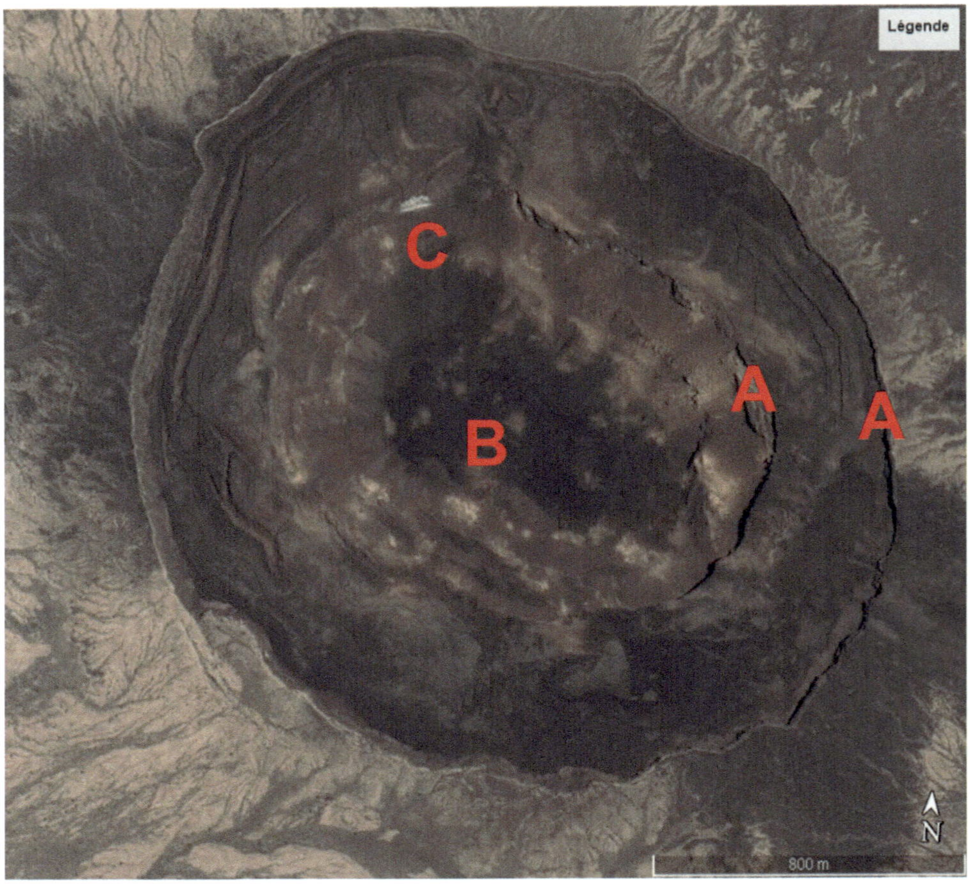

Fig. 3.39 Satellite image (Google earth Pro) of the summit part of Dama Ale volcano, 909 m, 12°16′ N, 41°38′ E; showing the double-rimed crater (**a** and **a′**), both occupied by lava lakes that partly overflowed the crater rim and consequently sunk. The floor of the crater was subject to later event, with basaltic lava flows and hornitos (**b**). The latest eruption was reported to occur in 1631. Fumaroles are observed in several places, in particular along the inner crater walls (**c**). Hot springs also occur at the foot of the volcano, notably along the NE-SW-trending faults around Lake Abhe (240 m, south)

grabens in Djibouti are typical remnants of these humid periods Gasse & Gasse et al. (1977, 1989, 1978, 1980).

The case of Asal Lake is of particular interest. It was flowing from inland to the sea in Ghoubbet over the Asal shield volcano during the last wet episode, whereas the sea water now flow from Ghoubbet to lake Asal through the open fissures and faults affecting the rift axis (Ardoukoba and "Petit Rift"). Asal Lake is saturated in salt, with evaporation compensating the sea water inflow.

3.9 Hydrothermal, Tectonic and Volcanic Activity; Geothermal Perspectives

The region is tectonically and volcanically active, subject to extensive normal faulting affecting the whole Afar floor, with the noticeable exception of the area bounding the Red Sea (which happen to more or less correspond with the Eritrean border), which appears to have stabilized in the last 5 My along the Bab-El-Mandeb strait. In the norther part, the Arata Block (or Danakil Alps[7]) acts as a microplate subject to anti-clockwise rotation as it is attached to the Nubian plate north and to the Arabian plate south. But the southern Afar Red Sea side (in southern Eritrea and down to the northern shore of the Gulf of Tadjourah in northern Djibouti Republic) acts since 5My as an accretion of the Arabic plate (Fig. 3.52).

In Afar, hydrothermal activity, mainly in the form of fumaroles and steam vents, is observed in many places. Quite widespread along the axis of the volcanic ranges and within caldera walls of the central volcanoes, they also occur along open fissures affecting the stratoid series or other geological formations, like sediments in the Tendaho-Gobaad graben. These are in fact the most ubiquitous form of water emergence in Afar, and therefore, these sites—called Boïna in Afar language—are frequently engineered by local tribesmen in order to condensate the steam and produce liquid water that is collected in underground basins made of red clay resulting from the hydrothermal alteration of the basalts (Fig. 3.53).

Hot springs are more frequent along the faults at the bottom of the Nubian escarpment where it affects pre-Mesozoic basement (emerging from tilted Jurassic limestones, Fig. 3.54) and in the Awash River basin (where it emerges from recent quaternary basalts Fig. 3.55. The

[7] The hills located between the Red Sea and the Danakil Sea, forming a "microplate".

Fig. 3.40 Satellite image showing the more recent nature (black Aa flows, partly historic) of the northern half of the range (Manda) with respect to Inakir (**b**) now extinct shield (source: Google Earth Pro). This unit (see location on Fig. 3.3) cross the Ethio-Djibouti border and is crossed by the road to Assab. Moussa Ali volcano is partly seen on the NE corner (**c**)

Fig. 3.41 Symmetrical rift structure of Asal rift, as seen from Lake Asal—11°39′ N, 41°22′ E—(154 m bsl). *Photo* Varet

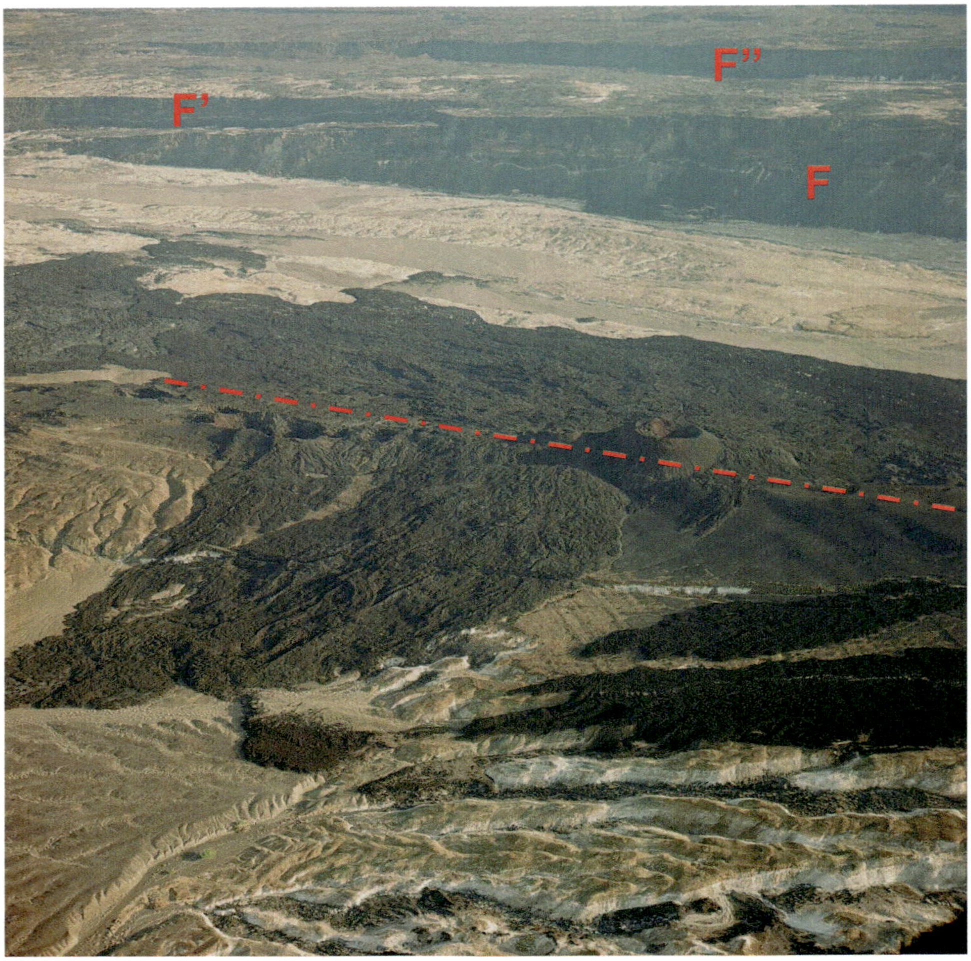

Fig. 3.42 General view of the Ardoukoba—11°35′ N, 42°29′ E—eruptive fissure (including the 1978 eruption) along the axis of Asal rift seen from south. The faults bordering the inner graben north (F) are seen in the back. Observe the lava covering the 500 years old lacustrine deposits, easily eroded and frequently faulted

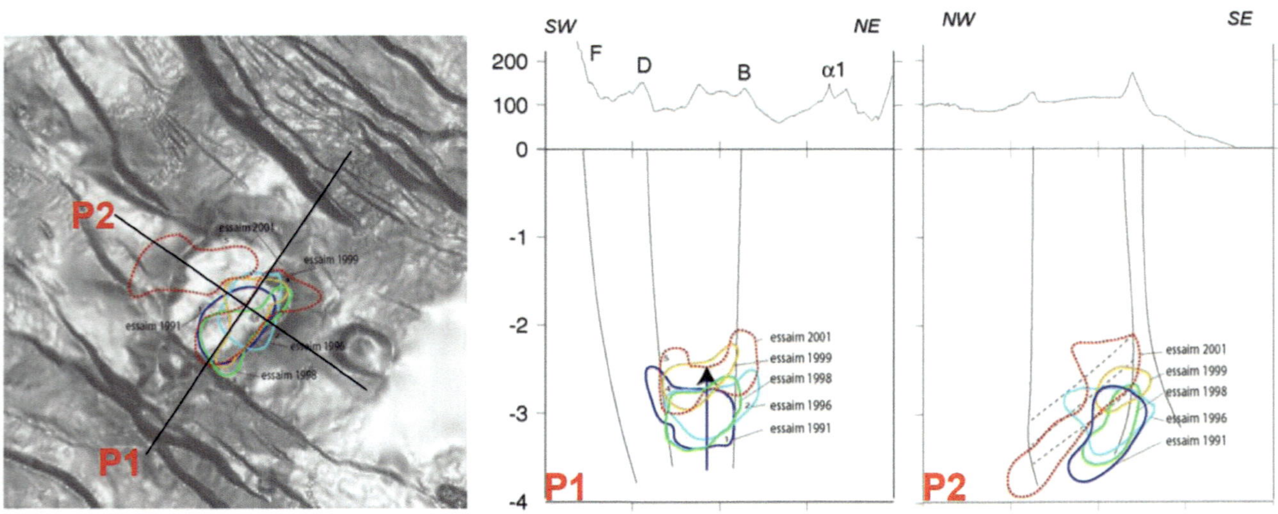

Fig. 3.43 Detailed mapping of the chronology of the main seismic events having occurred at Fialé caldera in the period 1991–2001 showing a general movement of rise of the seismic activity which can be interpreted as a rise of the magma chamber, the upper part of which having migrated from 3.5 to 2.0 km depth with a displacement of the events towards NW where the uplift was also observed from radar interferometry (after Doubre et al. 2007)

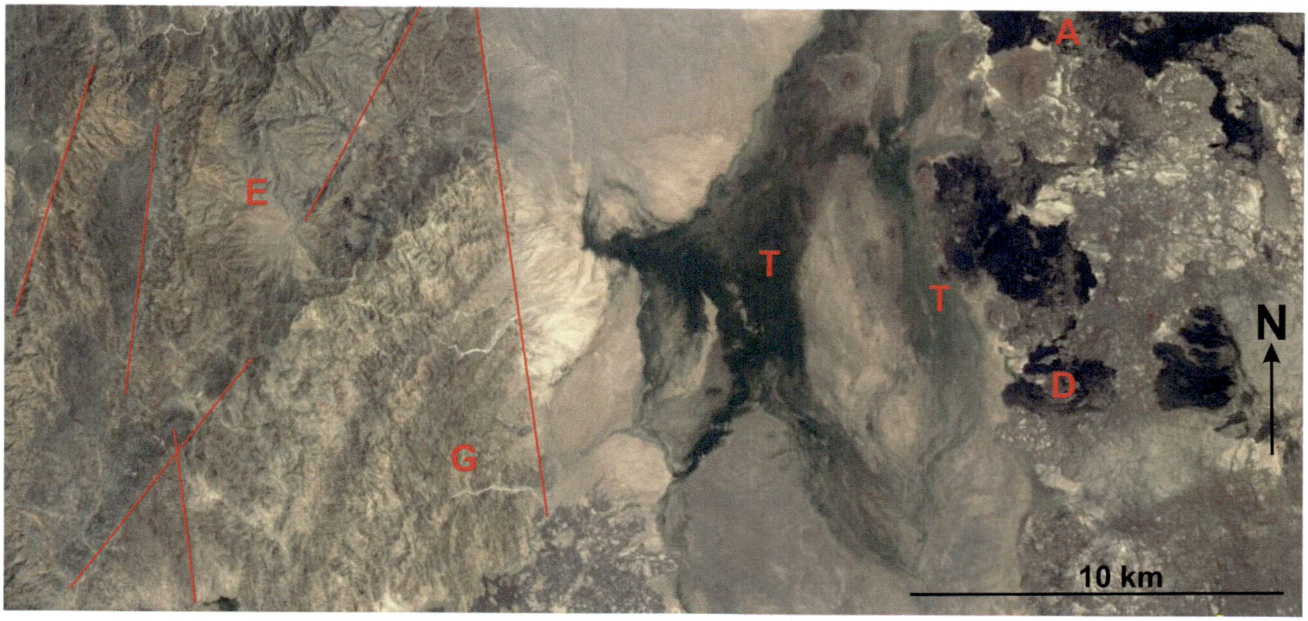

Fig. 3.44 Satellite image of the recent quaternary Teru sedimentary plain (T), a green, fertile area between the foot of the faulted and eroded pre-Mesozoic Ethiopian escarpment (E), with the Limmo Miocene granite (G) in yellow to the west and the recent volcanic ranges of Alayta (A) and Dabbahu (D) to the east.

most spectacular hydrothermal phenomena are observed in and around Dallol (Northern Afar State, Ethiopia) where hot springs and steam vents produced inversed stalagmite by dissolving the salt leaving colourful deposits due to variable degree of oxidation of dissolved metallic sulphides (Fig. 3.56). Phreatic explosion craters are also observed in this environment (Figs. 3.57 and 3.58).

As a whole, Afar appears as one of the most promising areas in the world for geothermal energy production, as it is the only place—together with Iceland—where an oceanic ridge is emerged at the surface allowing for exceptionally high heat flows (up to 10 time the continental average, i.e. of the order of 1 MWe per km^2). Several sites are presently being investigated and subject to exploration drilling, like Tendaho (central Afar), Fantale (southern Afar) and Asal (Eastern Afar, Djibouti Republic). Other sites are being considered at North Ghoubbet (Djibouti Republic), Dama Ale and Abhé lake (Varet 2018a, b), Dofan (southern Afar),

Fig. 3.45 View of the salt plain in Northern Afar (14°08′ N, 40°20′ E); along the horizon, the salt is cut in rectangular pieces and carried by caravans of camels carrying the salt to the Makalé salt market on the plateau. The surface is particularly fresh, being recrystallized every year after the rainy season had dissolved the surface, regenerated by deposition due to intense evaporation along the otherwise all-year-long arid climate (*Photo* Varet 2013)

Table 3.3 Record of known historical volcanic eruption having occurred in Afar

Date	Volcanic range or volcano	Kind of eruption
1400	Dubbi	Basaltic flow reaching the sea
1631	Dama Ale	Summit eruption: lava lake overflow
1861	Dubbi	Basaltic flows and 17 scoria cones
1907	Alayta	Basaltic fissural flow
1928–1929	Manda Inakir	Basaltic fissural flow
1967-present	Erta Ale	Permanent lava lakes with periodical overflows
1978	Asal	Scoria cones and few basaltic flows along axis
2005	Dabbahu	Small pumice eruption
2007, 2009	Manda Harraro	Scoria and spatter cones with limited flows
2008	Dala Filla	Mugearite (?) flow
2011	Nabro	Cinder and pumice eruption; lava flows
2017-present	S. Erta Ale	Fissural eruption and episodic lava lakes

Fig. 3.46 The salt is exploited manually by digging along the hexagonal contraction fissures (**a**) and detaching a salt slate along a weaker, slightly clayed horizon (**b**). The salt slates are then cut in regular rectangles, cleaned by pealing the clayed surfaces (**c**), and packed (**d**) for transport by camels or donkeys (*Photos* Varet 2015).

Fig. 3.47 Detailed view of layered halite-sylvite deposits, as observed on eroded cliffs in the SW part of the Dallol dome (14° 14′ N, 40°17′ E; *Photo* Varet 2015)

Fig. 3.48 Workers on duty extracting the salt from a pound along Lake Afrera shoreline. 13°13′ N, 40°52′ E. Observe the palm trees that developed along the shoreline in the hot spring environment, now victims of the salinization of the water table due to the development of these salt pounds (*Photo* Varet 2015)

Fig. 3.49 Outcrop (**a**) of oolitic limestone—14°06′ N, 40°13′ E—from the upper sequence of the marine littoral belt surrounding the Danakil Sea (125 ky), with abundant corals, oysters (**b**) and flat urchins in the bioherms of the medium part of the sequence

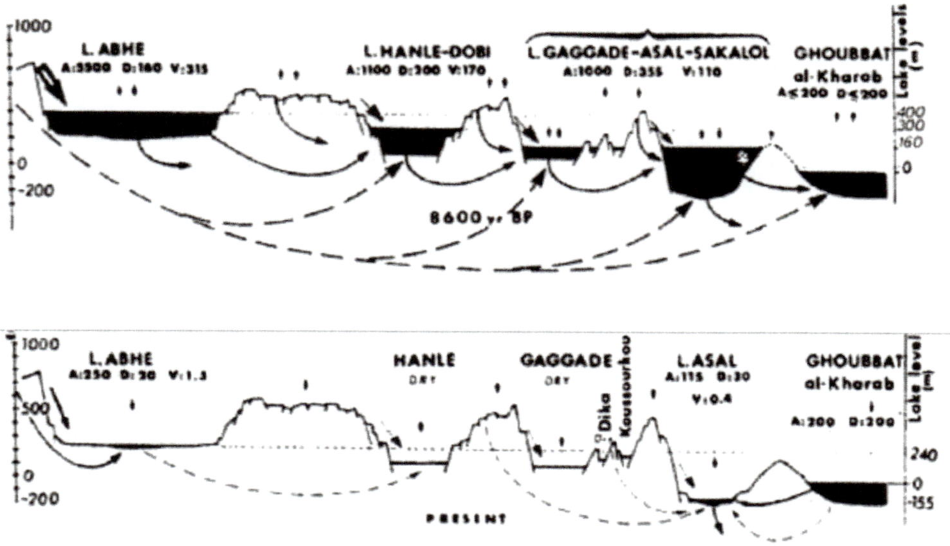

Fig. 3.50 Lacustrine deposits are found in most Afar grabens resulting from the past 8.500 years to present water levels changes, inducing surface and subsurface circulations from EW to NE in western Afar, Djibouti (after Audin et al. 2001)

Fig. 3.51 Sodonta plain with a surface of 450 km^2, covered with pulverulent limestones and gypsum sediments, extends SE from Afdera (seen in the back). *Photo* Varet. 12°47′ N, 41°02′ E

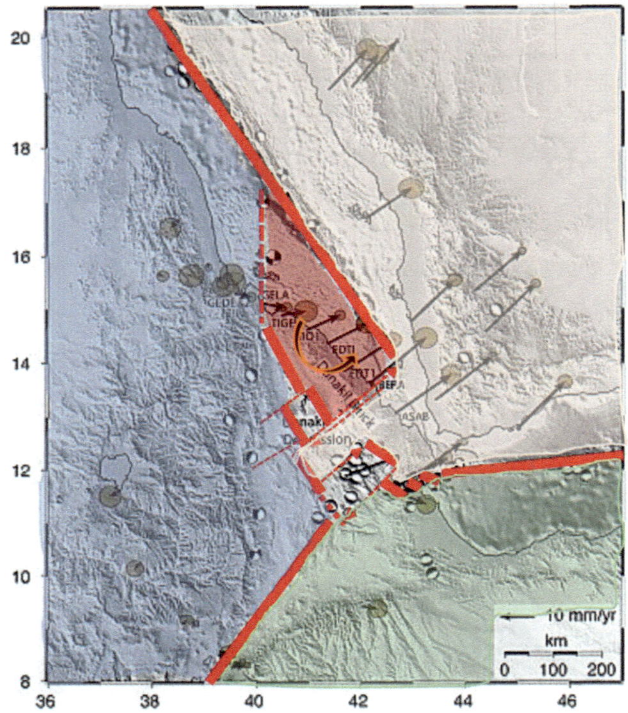

Fig. 3.52 Map of the Afar Triple Junction showing GPS velocities and 95% confidence ellipses with respect to Nubia. Focal mechanisms are from Harvard catalogue, 1976–2009; topography and bathymetry from SRTM30 PLUS (McKlusky et al. 2010). Author's interpretation of active plate boundaries, with axial ranges acting as spreading segments (in thick red), transform fault zones (dotted arrows) and fracture zones (dotted red line), and resulting major plate boundaries in beige (Arabia), green (Somalia), blue (Nubia) and ping (rotatin Arata block. The plates are affected by intense deformations (extension, rotation, etc) in Afar

Dallol (N Afar in relation with the potash extraction), Alid (Eritrea). Dabbahu (north-western Afar) and Tat'Ali (north-eastern Afar) are also considered by a recently formed AGAP (Afar community based geothermal development company, Nabro et al. 2016; Gardo and Varet 2018a; b) (Table 3.3).

3.10 Conclusion

Thanks to its arid climate and active geodynamic environment, Afar displays a variety of volcanic and tectonic landforms that reflect its specific geology, where an oceanic crust is being created at the earth surface, a situation that usually occurs in deep oceans floors.

As a whole, Afar appears as a real "textbook" for this type of geomorphology, dominated by active faulting, basaltic volcanism and associated hydrothermal activity.

In addition to this incursion in the usually submarine ocean floor, Afar also offers a view of the successive stages of evolution from a continental type of rifting to an oceanic rift.

The chapter presents the situation as observed and interpreted to date (and benefitted from 50 years observations in the period 1967–2017), but—as the process is continuing—the active nature of the area will allow to witness changing landforms in the future and will provide new opportunities to further observe unique active spreading with associated fissuring, faulting, diking, basaltic flooding and volcano building, as well as hydrothermal emergences and surface sedimentary filling.

The area being affected by climate change and geothermal energy will hopefully offer a resilient option for the Afar communities. A unique socio-economic target for geo-scientists.

Fig. 3.53 Artisanal system of water condensation from the geothermal steam in Afar: the emissive fissure is cleaned by removing the red clay found in the conduit resulting from hydrothermal alteration of the basalt, which is used to impermeabilize a reservoir in which the drops of condensed water will fall. A dry wall is build surrounding the emission centre allowing to maintain branches of acacias that will ensure the condensation. When needed, the water is then extracted from the reservoir using a recipient attached to a rope. *Photo* Varet (2016). 12°30′ N, 40° 26’E

Fig. 3.54 Thermal spring (72 °C) emerging from faulted Jurassic limestones in the Erepti river—13°16′ N, 40°09′ E—and producing carbonate travertine deposits at the foot of the Nubian escarpment. *Photo* Varet (2016)

Fig. 3.55 Hot springs emerging from a normal fault, north of Fantale. *Photo* Varet (2015)

Fig. 3.56 Hydrothermal manifestations producing inversed stalactites (**a**) and salt concretions (**b**) in Dallol—14°14′ N, 40°17′ E. Photos Varet

Fig. 3.57 Phreatic explosion crater—14°14′ N, 40°18′ E—which appeared in 2012 with still active gas emission (Varet 2013)

Fig. 3.58 As'Ale phreatic explosion crater ring along the axis of the salt plain—14°07′ N, 40°20′ E—north of Erta Ale range (*Photo* Marinelli 2011)

References

Abbate E, Albianelli A, Azzaroli A, Benvenutu M, Tesfamatiam B, Bruni P, Cipriani N, Clarke RJ, Ficcarelli G, Napoleone G, Papini M, Rook L, Sagri M, Tecle TM, Torre D, Villa I (1998) A one-million-year-old *Homo* cranium from the Danakil (Afar) depression of Eritrea. Nature 393:458–460

Audin J, Vellutini PJ, Coulon C, Piguet P, Vincent J (1990) The 1928–1929 eruption of Kammourta volcano—Evidence of tectonomagmatic activity in the Manda-Inakir rift and comparison with the Asal Rift, Afar depression, Republic of Djibuti. Bull Volcanol 52:551–561

Audin L, Manighetti I, Tapponnier P, Métivier F, Jacques E, Huchon P (2001) Fault propagation and climatic control of sedimentation on the Ghoubbet rift floor: insights from the Tadjouraden cruise in the western gulf of Aden. Geophys J Int 144:391–413

Ayele A et al (2007) The volcano-seismic crisis in Afar, Ethiopia, starting September 2005, *Earth Planet*. Sci Lett 255:177–187

Ayele A, Keir D, Ebinger C, Wright TJ, Stuart GW, Buck WR, Jacques E, Ogubazghi G, Sholan J (2009) September 2005 mega-dike emplacement in the Manda-Harraro nascent oceanic rift (Afar depression) (PDF file). Geophys Res Lett vol 36

Barberi F, Borsi S, Ferrara G, Marinelli G, Varet J (1970) Relations between tectonics and magmatology in the northern Danakil Depression (Ethiopia). Philosophical *Trans. Royal Soc. London*, A267, 293 - 311.

Barberi F, Tazieff H, Varet J (1972a) Volcanism in the Afar depression: its tectonic and magmatic significance. Tectonophysics 15:19–29

Barberi F, Borsi S, Ferrara G, Marinelli G, Santcroce R, Tazief H, Varet J (1972b) Evolution of the Danakil Depression (Afar Ethiopia) in light of radiometric age determination. J Geol 80:720–729

Barberi F., Santacroce R., Ferrara G. & Varet J. (1974a) - Evolution of Afar in light of K/Ar age determinations. *Afar Symposium, Bad Bergzabern*, April 1–6 (abstract).

Barberi F, Bonatti E, Marinelli G, Varet J (1974b) Transverse tectonics during the split of a continent data from the Afar Rift. Tectonophysics 23:17–29

Barberi F, Ferrara G, Santacroce R, Treuil M, Varet J (1975) A transitional basalt-pantellerite sequence of fractional crystallisation, the Boina centre (Afar rift, Ethiopia). J Petrol 16:22–56

Barberi F, Varet J (1970) The Erta `Ale volcanic range. Bull Volc 34:848–917

Barberi F, Varet J (1972) Carte volcanologique et pétrologique de la chaine de l'Erta'Ale (Afar, Ethiopie) au 1/100.000 Géotechnip, La Celle-Saint-Cloud

Barberi F, Varet J (1975a) Volcanological research in afar (L. R. Wager prize summary lecture). Bull Volc 39(2):166–174

Barberi F, Varet J (1975b) Recent volcanic units of Afar and their structural significance. In: Pilger A, Rosler A (eds) Afar depression of Ethiopia, proceedings of an international symposium on the Afar region and rift related problems, Bad Bergzabren, Germany, 1974, vol 1, E. Schweizerbart'sche Verlagsbuchhandlung, Stuttgart, Germany, pp 174–178

Barberi F, Varet J (1977) Volcanism in Afar: small-scale plate tectonic implications. Bull Geol Soc Amer 88:1251–1266

Beyene A, Abdelsalam M (2005) Tectonics of the Afar depression: A review and synthesis. J African Earth Sci 41:41–59

Black R, Morton WH, Varet J (1972) New data on Afar tectonics. Nat Phys Sci 240:170–173

Bonatti E, Emiliani C, Ostlund G, Rydell H (1979) Final dissecation of the Afar rift, ethiopia. Science 172:468–469

Bonatti E, Gasperini E, Vigliotti L, Lupi L, Vaselli O, Polonia A, Gasperini L (2017) Lake Afrera, a structural depression in the northern Afar Rift (Red Sea). Heliyon 3(5):e00301

Bonatti E, Tazieff H (1980) Submarine volcanoes in the Afar rift. Amer Geoph Un Gen Meeting. AGU Trans

Bruxelles L (2006) A la recherche d'une nouvelle pièce du berceau de l'humanité en Afrique Australe. Géosciences 21:44–49

Calais E (2016) Le rift est-africain, laboratoire de la rupture continentale. Geosciences 21:28–32 (2016)

Civetta L, De Fino M, La Volpe L, Lirer L (1974) Geochemical trends in the alkali basaltic suite of the Assab range (Ethiopia). Chem Geol 13:149–162

CNR-CNRS Afar Team (1973) Geology of northern Afar (Ethiopia). Rev Geogr Phys Geol Dyn 15:443–490

De Fino M, La Volpe L, Lirer L, Varet J (1973a) Geology and petrology of Manda-Inakir range and Moussa Alli volcano, central eastern Afar (Ethiopia and T.F.A.I.). Rev Geog Phys Géol Dyn 15 (2):373–386

De Fino M, La Volpe L, Lirer L (1973b) Volcanology and petrology of the Assab range (Ethiopia). Bull Volc 37-1:1–16

Demange J, Stieltjes L, Varet J (1980) L'éruption d'*Asal* de novembre 1978. Bull Soc Géol France 12(6):837–843

Doubre C, Manighetti I, Dorbath C, Dorbath L, Jacques E, Delmond J-C (2007) Crustal structure and magmato-tectonic processes in an active rift (Asal-Ghoubbet, Afar, East Africa): 1. Insights from a 5-month seismological experiment. J Geophys Res 112

Duffield WA, Bullen TD, Clynne MA, Fournier RO, Janik CJ, Lanphere MA, Lowenstern J, Smith JG, Giorgis L, Kahsai G, Mariam K, Tesfai T (1997) Geothermal potential of the Alid volcanic center, danakil depression, eritrea. US Geological Survey Open File Report 97–291

Ebinger C, Ayele A, Keir D, Rowland J, Yirgu G, Wright T, Belachew M, Hamling I (2010) Insights into extensional processes during magma assisted rifting: Evidence from aligned scoria cones. Ann Rev Earth Planet Sci 38:437–464

Ferguson DJ, Maclennan J, Bastow ID, Pyle D, Jones SM, Keir D, Blundy JD, Plank T, Yirgu G (2013) Melting during late-stage rifting in Afar is hot and deep. Nature 499(7456):70–73

Gardo IA, Varet J (2018a) Dabbahu (Teru Woreda) in Northern Afar. A major Ethiopian Geothermal Site Leased by AGAP. In: Proceedings, 7th African rift geothermal conference Kigali, Rwanda 31st Oct 2nd November 2018a. p 14

Gardo IA, Varet J (2018b) Ta'Ali geothermal site, Afdera Woreda, Northern Afar, Ethiopia. In: Proceedings, 7th African rift geothermal conference Kigali, Rwanda 31st Oct 2nd Nov 2018b. p 14

Gasse F (1977) Evolution of lake abhe (Ethiopia and TFAI), from 70,000 b.p. Nature 265:42–45

Gasse F, Fontes JC (1989) Palaeoenvironments and palaeohydrology of a tropical closed lake (Lake Asal, Djibouti) since 10,000 yr B. P. Palaeogeogr Palaeoclimatol Palaeoecol 69:67–102

Gasse F, Street A (1978) Late quaternary lake-level fluctuations and environments of the northern Rift Valley and Afar region (Ethiopia and Djibouti). Palaeogeogr Palaeoclimatol Palaeoecol 24:279–325

Gasse F, Richard O, Robbe D, Rognon P, Williams MAJ (1980) Evolution tectonique et climatique de l'Afar central d'apres les sediments plio-pleistocenes. Bull Soc Geol Fr 6:987–1001

Grandin R, De Chabelier J-B, Socquet A, Binet R (2009) Evènement en frontière de plaque : le cas du Manda Hararo (Ethiopie). Géosciences 9:20

Goitom B, Oppenheimer C, Hammond JO, Grandin R, Barnie T, Donovan, A (2015) First recorded eruption of Nabro volcano, Eritrea, 2011. Bull Volcanol 77(10):85

Haga AO, Youssouf SK, Varet J (2012a) The Manda-Inakir geothermal prospect area, Djibouti Republic. In: Proceedings of the 4th African rift geothermal conference. Nairobi, Kenya, 21–23 Nov 2012a, p 7

Houmed AM, Haga AO, Abdilahi S, Varet J (2012b) The Asal geothermal site, Djibouti Republic (model update, 2012b). In:

Proceedings of the 4th African rift geothermal conference. Nairobi, Kenya, 21–23 Nov 2012b, p 9

Houmed AM, Haga AO, Varet J (2012c) A revised approach to the Hanlé-Gaggadé (Djibouti Republic): the Garabbayis geothermal site. In: Proceedings of the 4th African rift geothermal conference. Nairobi, Kenya, 21–23 Nov 2012c, p 7

Houmed AM, Haga AO, Abdilahi S, Varet J (2012d) Nord-Ghoubbet geothermal site, Djibouti Republic. In: Proceedings of the 4th African rift geothermal conference. Nairobi, Kenya, 21–23 Nov 2012d, p 7

Illsley-Kemp F, Bull JM, Keir D, Gerya TV, Pagli C, Gernon T, Ayele A, Goitom B, Hammond J, Kendall JM (2018) Initiation of a proto-transform fault prior to seafloor spreading. Geochem Geophys Geosyst Dec 2018

Keir D et al (2011) Mapping the evolving strain field during continental breakup from crustal anisotropy in the Afar depression. Nat Comm 2:285

Manighetti I, Tapponnier P, Gillot P-Y, Jacques E, Courtillot V, Armijo R, Ruegg J-C, King G (1998) Propagation of rifting along the Arabia-Somalia plate boundary: Into Afar. J Geophys Res 103:4947–4974

Manighetti, I, King GCP, Gaudemer Y, Scholtz CH, Doubre C (2001) Slip accumulation and lateral propagation of active normal faults in Afar. J Geophys Res 106(13):13667–13696

Marinelli G (1971) La province géothermique de la dépression Danakil. Ann Mines, mai 1971

Marinelli G, Varet J (1973) Structure et évolution du Sud du "horst Danakil" (TFAI et Ethiopie). C.R. Acad Sci (D) 276:1119–1122

Nabro A, Gardo IA, Varet J, & Onyango S (2016) Community-based geothermal development perspective in Afar: a new player Afar Geothermal Development Company (AGAPI) Proceedings, 6th African Rift Geothermal Conference, Addis Ababa, Ethiopia, 2nd – 4thNovember 2016

Ottonello G, Piccardo GB, Joron JL, Treuil M (1978) Evolution of the upper mantle under the Assab region (Ethiopia): suggestions from petrology and geochemistry of tectonic ultramafic xenoliths and host basaltic lavas. Geol Rundsch 67(2):547–575

McClusky S, Reilinger R, Ogubazghi G, Amleson A, Healeb B, Vernant P et al (2010) Kinematics of the southern Red Sea-Afar Triple Junction and implications for plate dynamics. Geophys Res Lett 37:L05301

Richard O, Varet J (1979) Study of the transition from deep oceanic to emerged rift zone : gulf of Tadjoura (Republic of Djibouti). Int Symp Geodyn Evols Afro-Arabian System, Roma

Taieb M, Johanson DC, Coppens Y, Aronson JL (1976) Geological and paleontological background of Hadar hominid site, Afar, Ethiopia. Nature 260:289–293

Tazieff H, Barberi F, Borsi S, Ferrara G, Marinelli G, Varet J (1970) Relationships between tectonics and magmatology in the Norther Afar (or Danakil) depression. Phil Trans Royal Soc London A 267:293–311

Tazieff H, Barberi F, Giglia G, Varet J (1972) Tectonic significance of the Afar (or Danakil) depression. Nature 235:144–147

Varet J (1975) Carte géologique de l'Afar central et méridional, CNR-CNRS, 1/500,000 Géotechnip

Varet J (2014) Asal-Fialé geothermal field (Djibouti republic): A new interpretation for a geothermal reservoir in an actively spreading rift segment. In: Proceedings 5th African rift geothermal conference. Arusha, Tanzania, 29–31 Oct 2014

Varet J (2018a) Geology of Afar (East Africa) ed. Springer. 1st ed., 397 p., 377 illus. http://www.springer.com/in/book/9783319608631

Varet J (2018b) Geothermal resource along the border: The Ethiopia-Djibouti case. In: Proceedings, 7th African rift geothermal conference Kigali, Rwanda 31st Oct—2nd Nov 2018b, p 14

Yirgu G, Ayalew D, Ayele A (2006) Recent seismovolcanic crisis in northern Afar, Ethiopia. EOS Trans Am Geophys Union 87 (33):325–329

Geomorphology of Eritrean River Systems

4

Paolo Billi

Abstract

Eritrea is a mountainous country. Its physiography and landscape are the results of uninterrupted uplift and faulting in the Cenozoic. Both the highlands and the rift escarpment are characterised by a high relief ratio, the occurrence of deeply incised rivers systems and rugged geomorphology. Rivers in the headwaters are incised into bedrock or may have the streambed covered by a thin accumulation of bed material. Notwithstanding the steep gradient, several bedrock channels with alluvium show a meandering pattern. Such meandering pattern is unexpected and poorly studied in a mountain environment. A relatively small set of meander parameters was measured to investigate processes generating meanders in a steep bedrock channel and to compare the bedrock curves geometry to lowland, alluvial meandering rivers. The data showed that both steep bedrock and flatter alluvial meanders share very similar geomorphic characteristics. Though it is generally recognised that tectonics may have influenced the shape of the longitudinal profile of a river, few studies were conducted on this topic in an area with a very high uplift rate such as Eritrea. The downstream variation of the stream gradient index (SGI) was obtained from Google Earth for some rivers representative of different physiographic areas of Eritrea. The data indicate that the SGI is sensitive to recent tectonics, whereas past uplift is hardly recorded in the river profiles, which may have reached equilibrium condition in a shorter time than expected.

Keywords

River profile • River network • River capture • Stream gradient index • Dryland • Eritrea

P. Billi (✉)
International Platform for Dryland Research and Education, Tottori University, Tottori, Japan
e-mail: bli@unife.it

4.1 Introduction

The large-scale physiography of Eritrea consists mainly of a plateau gently inclined to the west, a sharp escarpment and a coastal lowland in the east. This setting results from uplift associated with the doming of the Arabian–Ethiopian region (Merla et al. 1979) and the formation of the Red Sea as the northern trough of the Afar triple point. In Eritrea, uplift pushed up the basement rocks to an elevation of 2500 m asl and created the present-day relief. The doming, associated with extensive faulting, began in the Oligocene, and it is the result of a continuous mantle plume activity under the Danakil crust (Merla et al. 1979). The trough formation was simultaneous with the doming or followed shortly after (Drury et al 1994). Uplifting and faulting continued until the last Pleistocene phase of the great uplift of the Ethiopian region (Merla et al. 1979). Eritrea has therefore a very complex geological structure formed by a long history of tectonic events (see also Chap. 2). These crustal processes have been active till recent creating and affecting the development of the modern landscape and drainage systems (Fig. 4.1).

River profiles and patterns are strongly influenced by tectonic structures, vertical movements and rock resistance to erosion, but at the same time, according to a few authors (e.g. Hack 1973; Kirby and Whipple 2001; Keller and Pinter 2002; Roberts and White 2010; Gaidzik and Ramírez-Herrera 2017), the characteristics of drainage networks and longitudinal river profiles may help in understanding temporal changes in surface uplift and, therefore, to a larger extent, the origin of the regional landscape as a response to tectonic events. Although most of the studies used a theoretical approach or assumed particular conditions, such as bedrock homogeneity, to investigate the response of rivers to variations in rock uplift rate (Wobus et al. 2006), there is still some uncertainty about the information we can get from the analysis of river profiles and no method has yet proven to perform better in providing tectonic information. The use of

P. Billi (ed.), *Landscapes and Landforms of the Horn of Africa*, World Geomorphological Landscapes,
https://doi.org/10.1007/978-3-031-05487-7_4

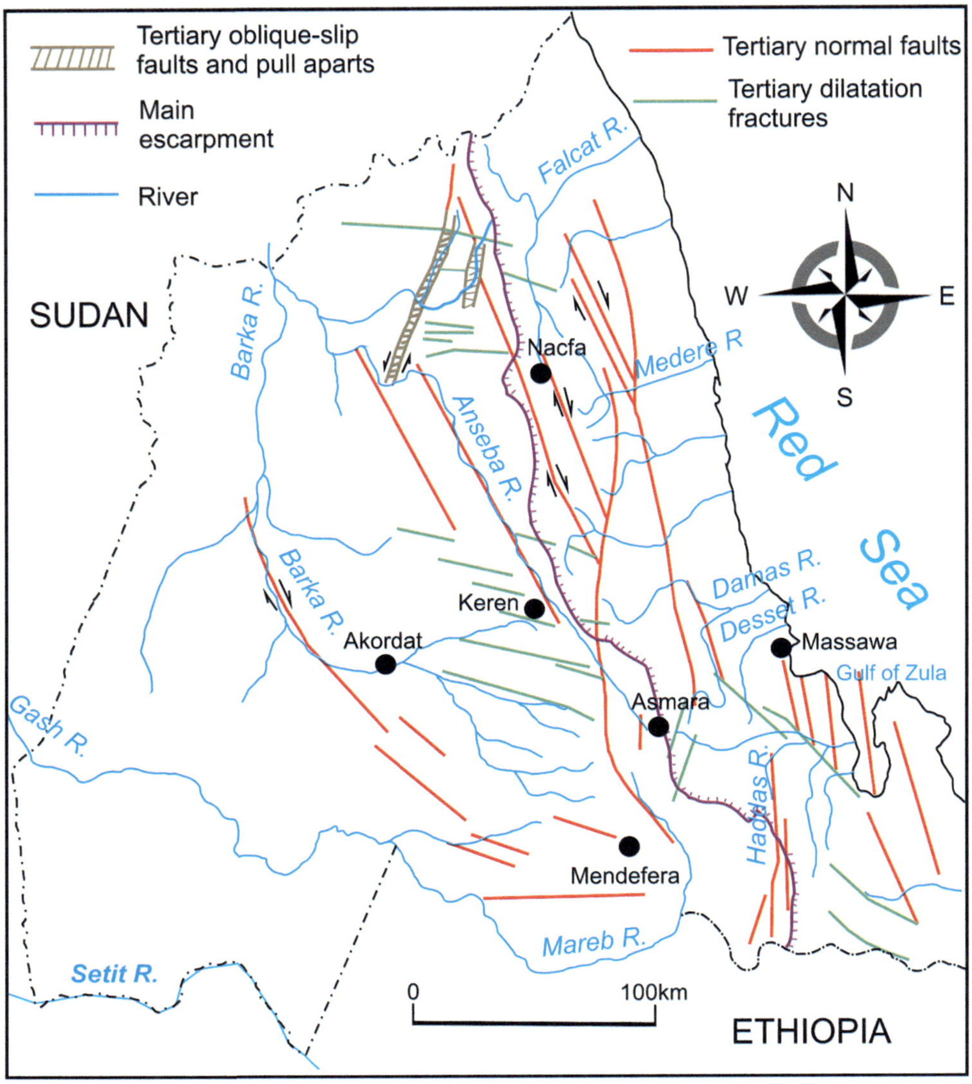

Fig. 4.1 Main structural features and drainage systems of Eritrea (modified from Drury et al. 1994)

simple methods, such as the stream gradient index of Hack (1973), can provide an insight into the recent tectonics and, at least, qualitative information on the influence of tectonics on river channels. For the largest part, Eritrea is underlain by Proterozoic and Paleozoic basement rocks which consist of granites and high- to low-grade metamorphic rocks, i.e. gneisses and schists, whose resistance to erosion is variable but they underlie large areas with uniform characteristics, thus limiting the influence of lithological factor in interpreting river profile data (Demoulin et al. 2017) only to specific cases.

Commonly, a simple model of river longitudinal profile is of the type (Knighton 1998):

$$H = H_0 e^{-kL} \qquad (4.1)$$

in which H and H_0 are the elevation at a given distance L and $L = 0$, respectively, and k is a constant. Morris and Williams (1997) showed by a theoretical model that streams have an exponential profile if bedload undergoes either abrasion or hydraulic sorting, but if sediment is subjected to both processes (which is the most common situation in nature), only short reaches have an exponential profile. Snow and Slingerland (1987) used a numerical simulation to demonstrate that both exponential, power and logarithmic function can fit well river profiles, depending on which variable has the main influence on the profile form. According to these authors, rivers characterised by a downstream increase in discharge and sediment load have a power function profile. If bed material declines in size very quickly downstream, the profile best approximation function will be logarithmic, whereas an exponential profile will be achieved when bed material grain size decreases downstream under low sediment transport conditions, the other controlling variables being almost constant. Though most of river profiles, no matter the approximation function, tend to develop a concave upward shape, tectonic activity or the occurrence of bedrock with marked different hardness may produce convex profiles or profiles with convex portions. These are considered by many authors an exception with respect to the "normal" concave upward profile (Knighton 1998). Several studies (e.g. Knighton 1998, Table 5.11 therein) confirmed that streambed gradient is a function of both discharge and bed material grain size. However, under dynamically and relatively fast changing conditions, such as tectonic uplift, if we assume that runoff does not change significantly, an increased rate of sediment supply may be locally balanced by an increase in the transport of finer particle, which results in an increase of both bed material grain size and roughness, thus temporarily decreases the erosive capacity of the river until new equilibrium conditions (namely a local increase in streambed gradient) are achieved.

Most rivers achieve their concave upward profile as a result of downstream increase in discharge and grain size fining. Eritrean River systems, however, are presently dry for most of the time, especially in their mountain portions, and erratic rainfall may not cover the entire surface of the catchment. Low discharges are also decreasing in the downstream reaches due to infiltration. Though in the early Holocene the Eritrean highlands may have experienced a short interval of precipitation higher than today, the current conditions of low-volume flow may have delayed the development of concave upward stream profiles, thus preserving evidence of recent tectonic deformations and implications on the river pattern and channel morphology. In fact, though the profile of a river may also reflect the longer-term uplift (Phillips and Lutz 2008), the most recent uplift and horizontal deformations, faulting, graben formation and climate have more influence on the river pattern and on specific channel features and processes such as captures, alluvial fans and channel morphology.

Eritrea is a mountainous country characterised by a high relief ratio. Most of the rivers are steep and incised into bedrock, and only few examples of alluvial reaches are present. Typical mountain streams are straight, with no or little bed material, confined by bedrock valley sides, and with no or a very small, discontinuous alluvial plain. A closer observation of the mountainous areas of Eritrea reveals that steep bedrock streams with a thin streambed alluvium and a meandering pattern are very common. Very few studies dealt with mountain stream geomorphology (see Wohl 2013, for a comprehensive review) and even fewer focused on high-gradient, bedrock-incised meandering channels (Fairbridge 1993; Hovius and Stark 2001; Limaye and Lamb 2014). Classical studies on mountain rivers (e.g. Montgomery and Buffington 1997; Montgomery and Gran 2001; Buffington and Montgomery 2013) neglected the occurrence of steep bedrock meandering channels, even though they are very common in semi-arid mountains. Hovius and Stark (2001) in their study on Taiwan bedrock meandering channels concentrated mainly on lateral erosion and concluded that in mountain areas active river meandering can be a purely erosional process. The thesis supported in this study is that bedrock meanders may be formed by differences in boundary rock resistance, but also other hydraulic factors such as the uniform distribution of energy expenditure are necessarily involved to give way to fully developed meanders.

Unfortunately, due to decades of independence war and international conflicts, lack of budget and inadequate maintenance of flow gauges, no hydrological data is available for the rivers of Eritrea (Alemngus et al. 2017). Therefore, river flow data analysis, as a complementary information to the river landscape and channel morphology, was not possible. Therefore, the main research aims of this chapter are: (1) to investigate the generating conditions of bedrock meandering channels through the analysis of their main geomorphic parameters and processes, and to verify their similarities and differences with low-gradient, free-meandering alluvial rivers; (2) to investigate the relationship between longitudinal profile characteristics and recent tectonic in respect to a few rivers representative of the main physiographic zones of Eritrea.

4.2 Study Area

Eritrea can be subdivided into three main physiographic regions: the western lowlands, the plateau and the eastern coastal lowlands, separated by the main escarpment (Figs. 4.1, 4.2 and 4.3). The plateau is gently inclined to the west. In the relief close to the eastern plateau margin, the higher mountains (between 2000 and 2500 m asl) are found, with the highest peak (Mt. Soira) reaching an elevation of 3018 m asl. The plateau declines towards the western lowlands down to elevations around 500 m asl close to the Sudanese border and around 250–300 m asl in the Barka River valley bottom (Figs. 4.1 and 4.2).

Neoproterozoic and Paleozoic rocks underlie the plateau which is the largest physiographic region of Eritrea and takes up the largest proportion of Eritrea territory (see Chap. 2 for more details about the geology of Eritrea) (Fig. 4.4).

In the northern portions, the coastal belt includes an inner, detached mountain range, with peaks that seldom exceed the elevation of 1000 m asl, and some isolated groups of hills, commonly less than 300 m asl high. This area is underlain by Neoproterozoic basement rocks and, subordinately, by recent (Neogene) sedimentary rocks. In the southern portion, the Danakil Alps range in height between 500 and 1000 m asl. They consist mainly of Mesozoic sedimentary rocks and separate the Danakil depression from the Red Sea, whereas in the southernmost portion, higher elevations are reached by Pleistocene and contemporary volcanoes (e.g. the Nabro volcano 2218 m asl.).

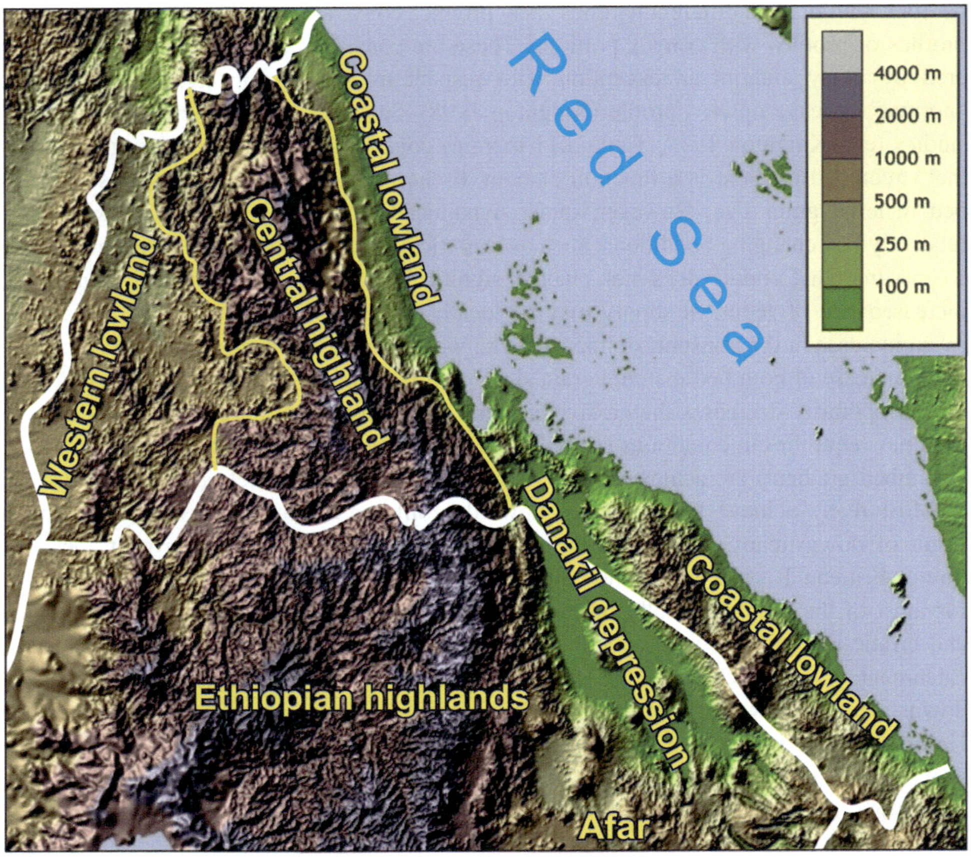

Fig. 4.2 Main physiographic regions of Eritrea

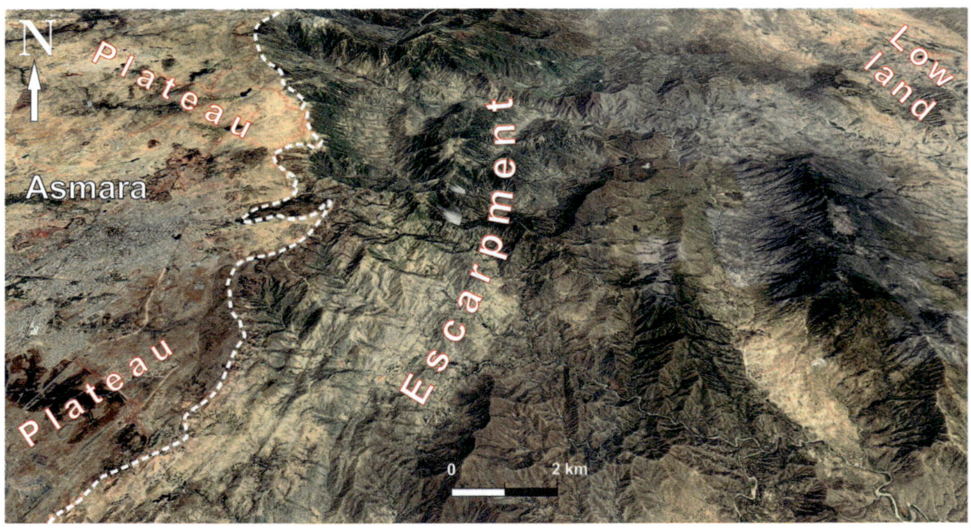

Fig. 4.3 Transverse view of the plateau escarpment near Asmara

A peculiar, small physiographic area is the Eritrean portion of the Danakil depression. For the largest part, it is a structural trough (in some places, close to the Ethiopian border, the ground surface is about 120 m below sea level), bound by large volcanoes (e.g. the Alid, 904 m asl) and metamorphic basement rock on the eastern and the western side, respectively, and filled with recent fluvio-lacustrine and aeolian sediments (see Chaps. 3 and 5).

The climate of Eritrea is largely influenced by its contrasting physiography. Temperatures are lower, and precipitation is higher in the central highlands (e.g. in Asmara average monthly low temperature ranges from about 6 to 12 °C and the maximum temperature ranges between 22 and 26 °C; annual precipitation is 520 mm). In the low-lying lands close to the Sudan border, temperatures are higher (e.g. the average annual temperature at Teseney is 28.6 °C) and

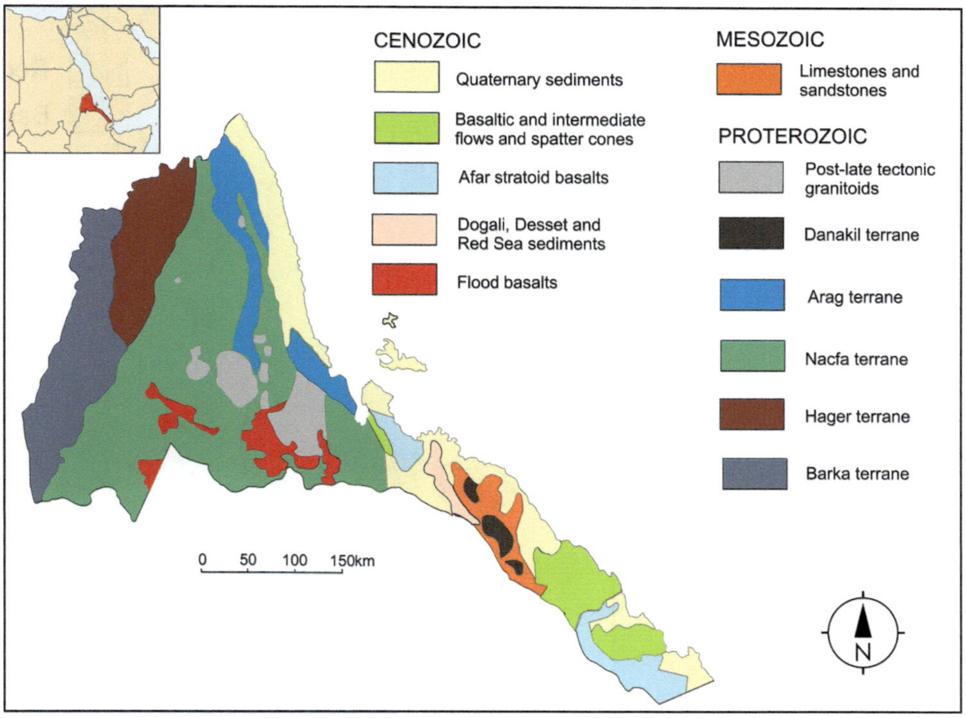

Fig. 4.4 Simplified geological map of Eritrea (based mainly on Drury and De Souza Filho 1998). The Barka terrane includes mainly gneisses, amphibolites and paraschists; the Hager terrane includes meta-igneous and meta-sedimentary rocks; the Nacfa terrane includes granites, gneisses, migmatitic amphibolites, and paraschists; the Arag terrane includes gneisses, paraschists and granitoid masses and the Danakil terrane consists of phyllites, slates, greywackes, metavolcanics and metacarbonates

precipitation is less (e.g. 341 mm yr^{-1} at Teseney) than in the central highlands. In the coastal lowlands, temperature is the highest (e.g. mean annual temperature at Assab is 30.9 °C and 29.4 °C at Massawa) and annual precipitation is the lowest (e.g. 50 mm at Assab and 115 mm at Massawa) (see Chap. 1). Very few climate data are available for the Eritrean portion of the Danakil, but likely the highest temperature and the lowest rainfall of the country are expected to be recorded there.

Large rivers, whose catchments are entirely contained within Eritrea, are very few. Though they have their headwaters in Eritrea, all the larger rivers exit the country and flow beyond the Sudanese border. Six main drainage areas can be recognized (Fig. 4.5): the Red Sea basins; the Barka–Anseba river systems; the Mareb–Gash; the Setit–Tekeze Eritrean tributaries; the Danakil basin; and a few small rivers flowing to Sudan in the western part of Eritrea. Though before the 1930s some gauging station were operating, most of them have been destroyed during the state of war that lasted more than two decades, and, even if recently some flow gauges were reactivated (Fig. 4.6), in practice no flow data are available for the Eritrean rivers (Alemngus et al. 2017). In Table 4.1, some data provided by Alemngus et al. (2017) regarding the drainage basin area and annual runoff are reported.

The largest rivers of Eritrea are the Setit and the Mareb. Setit is the name used in Eritrea to indicate the lower reach of the Tekeze River, which marks the boundary between Eritrea and Ethiopia. In its upper reaches, the Tekeze has some flow throughout the year (Billi et al. 2015), but in the Setit reach discharge is very seasonal as water flows only for no more than three months a year. In the dry months, however, pools with water supplied by subsurface flow may be present in places (Bertarelli 1929).

The source of the Mareb River is located on the Emba Tekera, a basalt trap mountain (2563 m asl) a few kilometres south of Asmara. In its upper reaches, the river is almost permanent (sometimes, also in the upper reaches the river may dry up in a few reaches), though commonly shows rather low flow (Fig. 4.7) for the most part of the year. In the downstream reaches of the lowlands, approaching the Sudanese border, the flow becomes very seasonal and intermittent. Beyond the border, the river takes the name of Gash and becomes totally ephemeral, i.e. dry for the largest part of the year, with water flowing only in response to wetter than usual rainy season (typically from June to October). High flows in the lower reaches of the Mareb range from 50 to 500 m^3s^{-1} (Bertarelli 1929).

The Barka River is the largest one flowing to the Red Sea. It is about 640 km long, 430 km of which are within Eritrea. Its source is located in the mountains west of Decorassi, about 50 km SW of Asmara, i.e. in the north-western side of the same mountain range, where the Mareb originates, and its outlet is in Sudan near the town of Tokar. In the intermediate reaches, the Barka River flow is intermittent (Fig. 4.8) and becomes more conspicuous only in response to intense and prolonged precipitation. Its main tributary, the Anseba River (350 km long), is characterized by more persistent

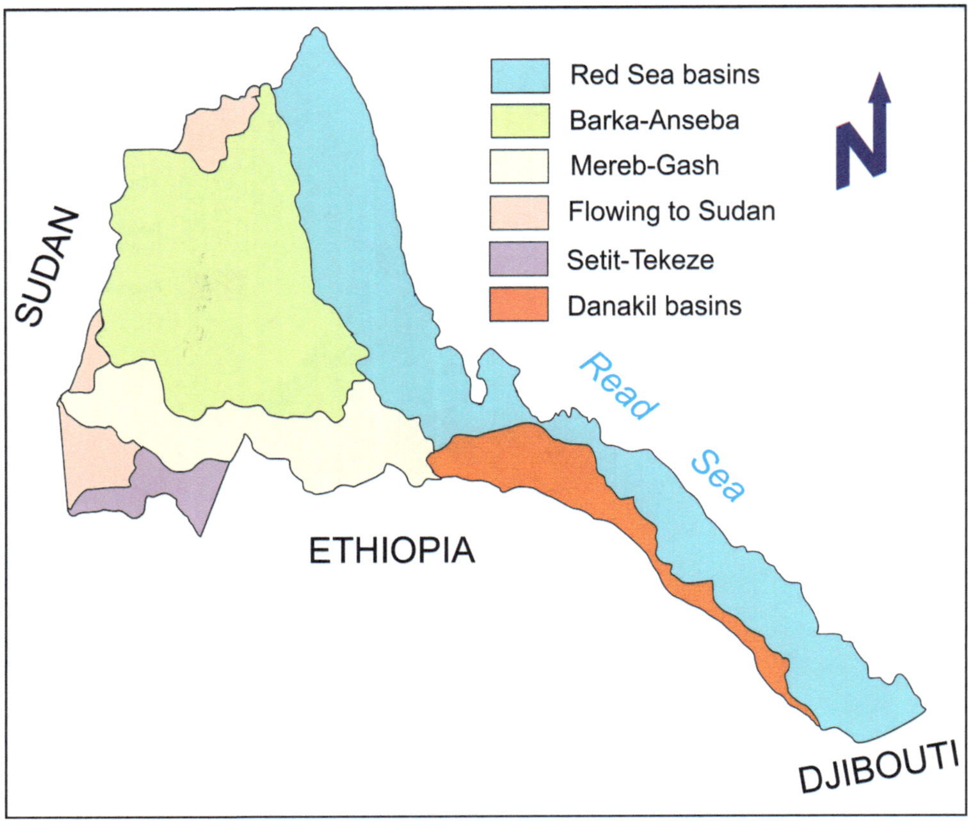

Fig. 4.5 Main drainage basins of Eritrea

Fig. 4.6 Recent flow gauge on the Mareb River, a few kilometres from Asmara

flow than the Barka (Bertarelli 1929). Its source is located a few kilometres east of Asmara, and the river flows through the capital city (Fig. 4.1).

The Red Sea basin (Figs. 4.1 and 4.5) includes several short rivers. Longer rivers are found in the northern portion, i.e. from the border with Sudan to the Gulf of Zula. They are

Table 4.1 Area and annual runoff of the main Eritrean drainage basins (data from Alemngus et al. 2017)

Basin	Area (km^2)		Annual runoff (10^6 m^3)
	Total	Within Eritrea	
Red Sea	44,376	44,376	961
Barka–Anseba	66,200	41,920	932
Mareb–Gash	23,176	17,256	1423
Setit–Tekeze	68,255	7,292	6230
Danakil	10,485	8,905	422

Fig. 4.7 Water pond in an upstream bedrock reach of the Mareb River, a few kilometres south of Asmara

Fig. 4.8 Dry bed of the Barka River near Aderde

80–140 km long and have their headwaters in the main escarpment. These rivers have some intermittent and/or seasonal flow in the upper reaches, but in their coastal reaches, they are ephemeral and dry for most of the time. The rivers of the southern portion of the Red Sea basin are very short, and water flow is very sporadic and occasional. Larger rivers are present in the Danakil basin. The longest, the Aghir-Endeli river, has its source in the Senafe basin, and after 130 km, it reaches, with a permanent flow, the salt plain in the Ethiopian Danakil depression (Bertarelli 1929). Another relatively long river (60 km) is the Dandero, which originates in the Soira mountains (Mt. Soira, 3018 m asl is the highest mountain in Eritrea) and ends up by evaporation and infiltration in the sandy Samoti plain. The Dandero is a typical ephemeral stream, in which water flows only in response to heavy local rains or prolonged precipitation in the headwaters (see Chap. 6).

4.3 Channel Morphology

The majority of Eritrean rivers are bedrock mountain streams, and very few reaches show alluvial characteristics. In Eritrea, mountain streams typically have no flood plain and occupy the very narrow bottoms of V-shaped valleys. These rivers are ephemeral and have flow only in response to intense rainfalls. The climate of Eritrea, in fact, ranges from arid to sub-humid, with the highest annual precipitation around 800–820 mm recorded in the central and southern highlands and desert-like conditions observed along the coast, in the western lowlands and in the Danakil, where annual precipitation ranges from 100 to 250 mm. Mountain streams are commonly straight and have the channel incised directly into bedrock as the flow transport capacity exceeds the sediment supply. Due to steep gradient and lateral confinement, unit stream power is typically one order of magnitude higher than in alluvial channels (Bravard and Petit 2009). This kind of steep bedrock streams typifies the lower order channels, and these are found in the upper portions of the headwater areas (Fig. 4.9). They are equivalent to the *montane bedrock channels* of Sutfin et al. (2014) classification of ephemeral channels in mountain environments (SBC). Higher order streams have lower gradient (0.05–0.02) and are confined by valley sides, flowing within very narrow valley bottoms, which are filled with a thin layer of coarse-grained alluvial deposit reworked by braided stream flows. Unlike steep bedrock streams of lower order, braided rivers confined in these valley bottoms seem to supply sediment supply in quantities equal to transport capacity and give way to an accumulation of coarse sediment that mantles the valley bottom (Fig. 4.9). These streams are equivalent to the *bedrock with alluvium channels* (BAC) of Sutfin et al. (2014). In fact, they contain a persistent bed of active alluvium with braided channels, and in some reaches a very narrow flood plain, narrower than the active channel width, may be present (Fig. 4.10). These rivers are very common all over Eritrea and show a main straight channel incised into alluvial deposits with dry braided channels on top. (Fig. 4.11) or a meandering channel pattern (Fig. 4.12). The meandering channels have variable sinuosity (sinuosity is the river length to the valley axis length ratio), ranging from 1.2 to 1.7 and are entrenched into bedrock with no flood plain or just a few rare patches in some places. In bedrock meandering rivers (BAC), typical features of alluvial meandering rivers such as lateral migration, point bars, chute cutoffs, point bar scrolls can be commonly observed (Fig. 4.13).

Very few examples of purely alluvial channels are present in Eritrea. These include: (1) alluvial meandering reaches (AMR) in the lowlands (Fig. 4.14); (2) braided reaches (Fig. 4.15); (3) straight, single thread channels (Fig. 4.15); and (4) distributary reaches within deltas on the coastal plain, terminal fans (Fig. 4.16), and the inland closed basins (Fig. 4.17). Alluvial meandering rivers (AMR) flow within a narrow flood plain entrenched within bedrock sides or within consolidated alluvium (Fig. 4.14). These rivers are similar to the *incised alluvium channels* of Sutfin et al. (2014) classification. They have, in fact, a distinctive flood plain, and channel width is on average one third of the flood plain width (Table 4.2).

In order to investigate the geomorphic characteristics of bedrock confined meandering and alluvial meandering channels, some representative reaches were selected. In Eritrea, there are very few (probably no more than five) undisturbed alluvial meandering channels (AMR), whereas very many bedrock meandering reaches (BAC) were easily identified and 15 of them were randomly selected for comparison. The author is well aware that five reaches, though representative of different regions of the country, are not enough for any statistical validation of data analysis; however, the occurrence of very few alluvial meandering reaches is imposing this limitation. Nevertheless, some interesting considerations can be made and may stimulate further studies on a so poorly investigated topic such as meandering rivers in mountainous areas (Johnson and Finnegan 2015; Lymaye and Lamb 2014).

In Table 4.1, data of the main geomorphic parameters of both alluvial and bedrock meandering rivers are reported. The data were obtained from Google Earth satellite images and were measured following the same methods used by Billi et al. (2018) in their study of ephemeral meandering streams.

The geomorphic data show that the confined alluvial valleys are rather narrow as the average width of meandering channels is about one third of the alluvial plain. Bedrock meandering streams (BAC) are one order of magnitude

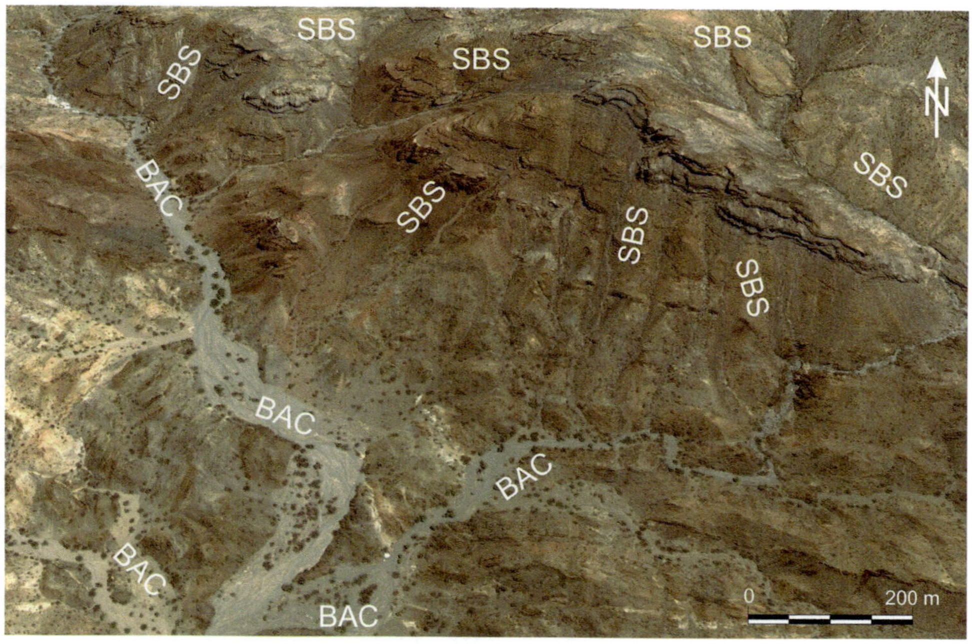

Fig. 4.9 Main bedrock mountain streams of Eritrea: SBS = steep bedrock channel; BAC = bedrock with alluvium channels (Sutfin et al. 2014). Northern highland, close to eastern margin, north of Nacfa

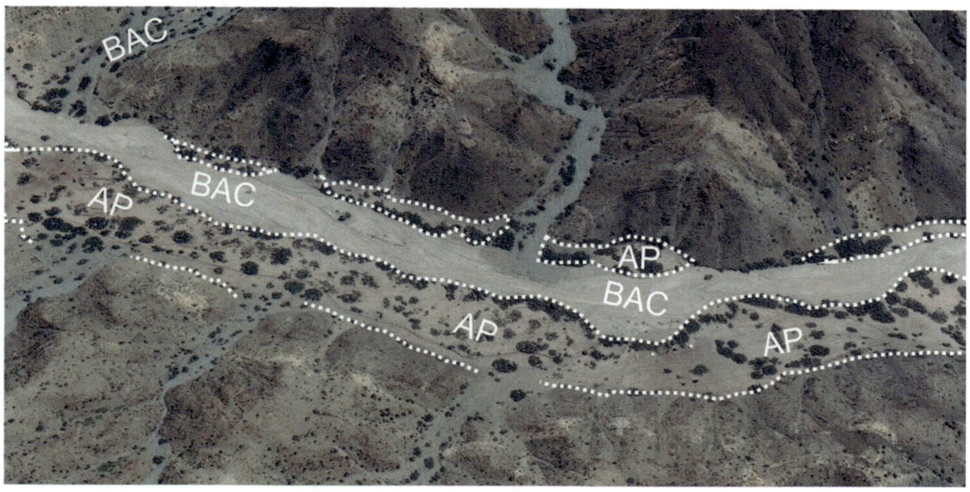

Fig. 4.10 Close-up of bedrock rivers with alluvium channels (BAC) and a small flood plain (AP) (Northern Escarpment, North-East of Nacfa)

steeper and have a higher sinuosity than alluvial meandering channels (AMR), whereas the average radius of curvature/channel width ratio (R_c/W) is almost the same (1.67 and 1.68, respectively; range 0.88–3.10) (Table 4.2). Only three values of the R_c/W ratio lie within the 2–3 range expected from theoretical analysis on the basis of minimum flow resistance increase observed in experiments in pipes and field data (Leopold and Wolman 1960; Bagnold 1960). In this study on bedrock with alluvium and incised in alluvium streams, most of the data (80%) fall within the 1–2 range (Table 4.2) as observed by Billi et al. (2018) for ephemeral, free-meandering rivers, though with a lower frequency (45%) (Fig. 4.18).

According to Langbein and Leopold (1966), meander geometry is best approximated by a sine-generated curve and the radius of curvature can be obtained from the following relationship:

$$R_c = L_w S^{1.5} / 13(S - 1)^{0.5} \tag{4.2}$$

in which L_w is wavelength and S is sinuosity.

Figure 4.19 confirms that the radius of curvature of both upland and lowland BAC and AMR channels can be also predicted by Eq. (4.2) with high accuracy ($R^2 = 0.97$). Average measured and predicted R_c are 118 m and 119 m and 639 m and 620 m for BAC and AMR channels, respectively.

In AMR channels, 98% of wavelength (L_w) variability is explained by channel width (W), but such correlation is poorer for BAC channels ($R^2 = 0.58$) (Fig. 4.20). The average wavelength to channel width ratio is 8.3 and 6.8 for

Fig. 4.11 Braided reach of bedrock river with alluvium (BAC) (northern escarpment, south-east of Nacfa)

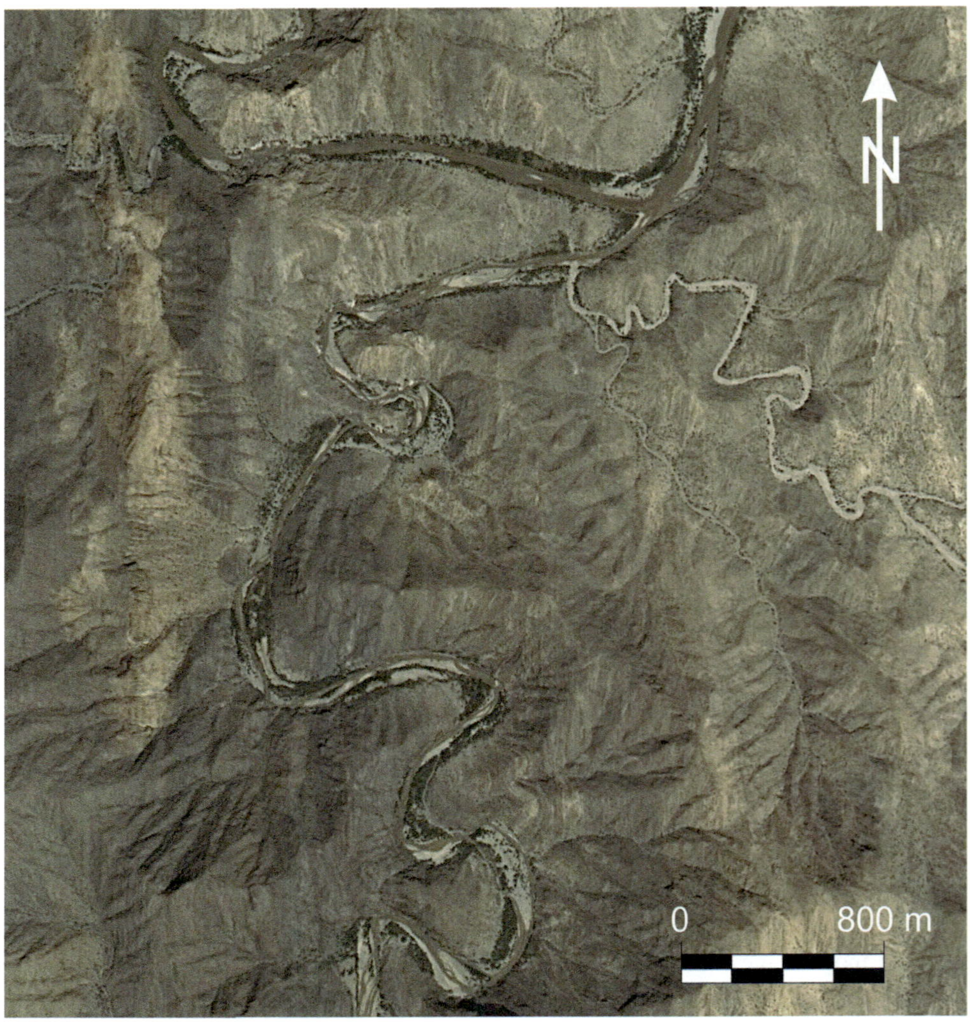

Fig. 4.12 Example of a meandering bedrock channel with an alluvium (BAC) (northern escarpment, south of Alghiena)

BAC and AMR channels, respectively (Table 4.2), whereas Billi et al. (2018) reported a higher value of about 16 for both alluvial ephemeral and perennial streams. Leopold and Wolman (1960) reported an L_w/W ratio of 10, considering both perennial and ephemeral rivers, and Yalin (1971) obtained a value of 6.3.

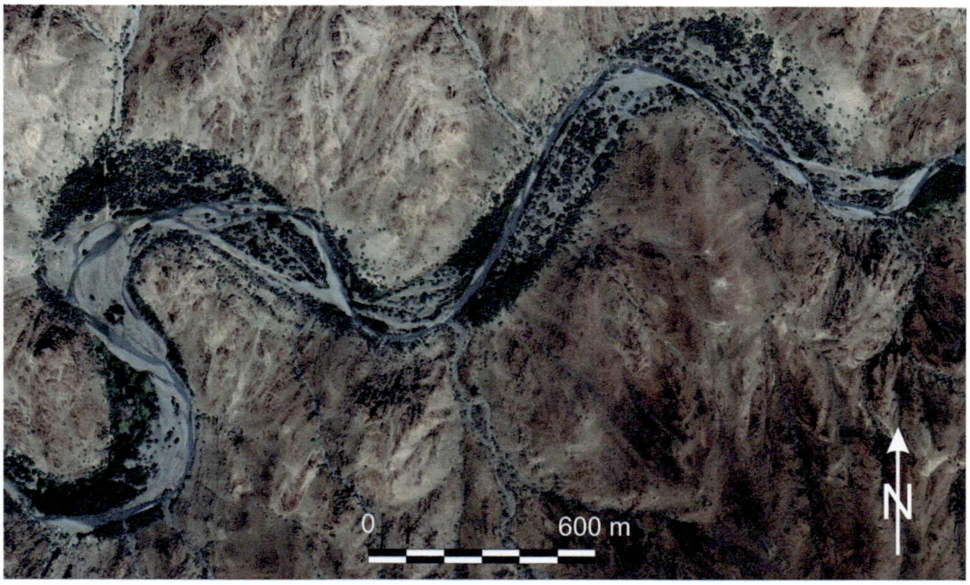

Fig. 4.13 Close-up of a meandering bedrock channel with alluvium. Typical features of meandering rivers such as point bars (P), meander migration (M), chute cutoffs (C), point bars scrolls (S) can be seen (Google Earth coordinates: 17° 20′ 18″ N; 37° 51′ 11″ E)

Fig. 4.14 Anseba River is one of the very few examples of alluvial meandering rivers with a confined flood plain (AMR) in Eritrea (Google Earth coordinates: 16° 30′ 10″ N; 38° 06? 42″ E)

Fig. 4.15 Alluvial braided river **a** turning into a straight, single thread reach **b** in the coastal belt. (Google Earth coordinates: 17° 23′ 16″ N; 38° 32? 20″ E)

Fig. 4.16 Distributary reaches of a delta near the town of Tio

Wavelength for both BAC and AMR meanders is well correlated with the radius of curvature (R^2 = 0.91 and 0.95, respectively) (Fig. 4.21). This and the results obtained by Eq. (4.2) confirm that in Eritrea meander geometry of both mountain steep bedrock streams (BAC) and lowland channels incised in alluvium (AMR) is similar to that of perennial, free-meandering alluvial rivers.

4.4 Geological Structure and River Pattern

The drainage pattern of the majority of Eritrean rivers is substantially influenced by bedrock structure and its tectonic evolution (Fig. 4.1). The river pattern is commonly set on and driven by faults and affected by recent uplift. Several studies explored the possibility to develop a quantitative method to assess the uplift rate of a region from the river profile characteristics (e.g. Hack 1973; Merritt and Vincent 1989; Wobus et al. 2006; El Hamdouni et al. 2008; Troiani and Della Seta 2008; Roberts and White 2010; Sembroni and Molin 2018). These authors used different approaches, but the most common ones include the river longitudinal profile analysis and/or the longitudinal variation of stream gradient by means of the stream gradient index (SGI) developed by Hack (1973). The SGI is defined as:

$$\mathrm{SGI} = (\mathrm{DH}/\mathrm{l_r})\mathrm{L_p} \quad (4.3)$$

in which ΔH is the change in altitude, l_r is the reach length and L_p is the horizontal length from the divide to the mid-point of the reach.

Wobus et al. (2006) warned about the appropriate scale of observation for digital elevation model (DEM) analysis, but Finlayson and Montgomery (2003) were able to extract useful information from 30 arc second ($\sim$1 km) GTOPO30

Fig. 4.17 Inland distributary systems in the Danakil depression (Google Earth coordinates: 14° 24′ 08″ N; 40° 24′ 35″ E)

data. In this study, unfortunately, no DEM was available so basic data of Eq. (4.3) for a few rivers, representative of different areas of Eritrea, were obtained directly from Google Earth images by measuring the elevation difference at the extremes of consecutive reaches of variable length.

As it is well known, in Google Earth, the accuracy of horizontal distances is acceptable and can be easily improved by averaging repeated measurements. The accuracy of elevation data, instead, is rather low. In order to improve the quality of elevation data, 4–5 points at each extremity of a short reach with uniform channel gradient and morphology were selected and their elevation was averaged. In the upstream, steeper river portions, the reach lengths were shorter, given higher accuracy in gradient measurement associated with larger altitude differences of the reach extremes. In lowland reaches, the lower gradients resulted in small differences in elevation and required to consider longer reaches to maintain a sufficient difference in elevation that could reduce the intrinsic error of Google Earth elevation data. Unavoidably, this procedure implied reaches of slightly different length, ranging from 500 m (in the headwaters) to 5000 m in the lower reaches. The latter, however, occur mainly in the coastal belt of the easternmost lowlands that are very little or not at all influenced by uplift.

According to a few authors (e.g. Keller and Pinter 2002; El Hamdouni et al. 2008), high values of the SGI are characteristic of river reaches set on resistant rocks or affected by active uplift, whereas lower values are measured on reaches parallel to strike-slip faults. The rivers selected for longitudinal profile analysis (Barka, Anseba, Khomib and Mareb) are set on the uplifted Neoproterozoic and Paleozoic basement rocks. These consist of low-grade metamorphic rocks, which underlie the largest proportion of the Eritrean territory and include slates, schists and metavolcanics, high-grade metamorphic rocks (gneisses, amphibolites and quartzites), which outcrop in a narrow belt parallel to the main escarpment and granitoid plutons (Merla et al. 1979) (Fig. 4.4). These rocks are different in terms of mineralogical composition, petrographic structure and age, but at large scale, they can be considered rather uniform in terms of hardness and response to linear river erosion, thus reducing the effect of lithology on the variability of SGI values (Demoulin et al. 2017). Within profile segments with homogeneous lithology, therefore, the changes in SGI are expected to be mainly influenced by recent crustal deformations. Using the same data, also the classical longitudinal profiles of the same rivers were constructed and analysed.

Table 4.2 Main geomorphic data of the study meandering channels

	L (m)	S	J (m m^{-1})	W (m)	L_w (m)	R_c (m)	R_c/W	L_w/W	W/W_{fp}
Alluvial meandering rivers (AMR)									
	52,522	1.12	0.0014	465	3859	866	1.86	8.3	0.32
	58,181	1.14	0.0017	692	4709	1209	1.75	6.8	0.48
	28,055	1.34	0.0031	229	1459	390	1.70	6.4	0.24
	25,745	1.17	0.0042	274	1886	604	2.21	6.9	0.36
	11,486	1.46	0.0039	144	844	127	0.88	5.9	
Mean	*35,198*	*1.25*	*0.0029*	*361*	*2551*	*639*	*1.68*	*6.8*	*0.35*
Bedrock channels with alluvium (BAC)									
	14,494	1.34	0.0143	185	1162	218	1.18	6.3	
	8067	1.65	0.0172	60	431	90	1.51	7.2	
	13,528	1.65	0.0102	117	885	212	1.81	7.6	
	15,108	1.59	0.0035	81	1146	199	2.46	14.2	
	4275	1.54	0.0131	66	414	80	1.22	6.3	
	5622	1.75	0.0156	65	496	68	1.04	7.6	
	7165	1.86	0.0236	48	404	91	1.92	8.5	
	4087	1.54	0.0323	44	215	51	1.16	4.9	
	4354	1.50	0.0165	78	567	118	1.50	7.3	
	7557	1.73	0.0180	67	424	86	1.29	6.3	
	6363	1.56	0.0069	23	379	72	3.10	16.4	
	8250	1.32	0.0254	98	674	169	1.73	6.9	
	6437	1.29	0.0174	59	625	129	2.18	10.6	
	3987	1.53	0.0261	52	334	70	1.34	6.4	
	16,317	1.30	0.0153	166	1183	265	1.59	7.1	
Mean	*8374*	*1.54*	*0.0172*	*74*	*583*	*118*	*1.67*	*8.3*	

L river reach length; S sinuosity; J streambed gradient; W average channel width; L_w meander wavelength; R_c meander loop radius of curvature; W_{fp} average width of flood plain

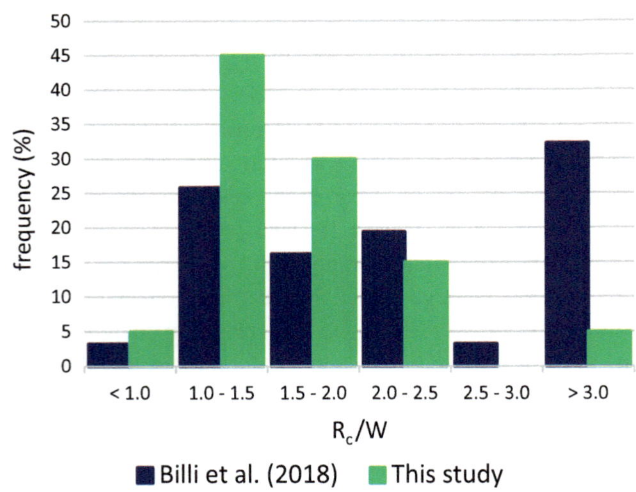

Fig. 4.18 Frequency distribution of the curvature ratio (radius of curvature/channel width). The majority of the values are less 2, rather than within the 2–3 range as observed in experiments in pipes and derived from field data by Leopold and Wolman (1960) and Bagnold (1960)

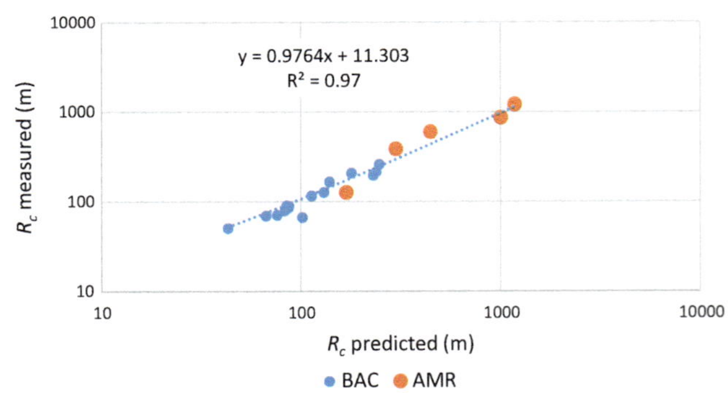

Fig. 4.19 Comparison between the measured radii of curvature and values predicted by Langbein and Leopold (1966) equation (Eq. 4.2)

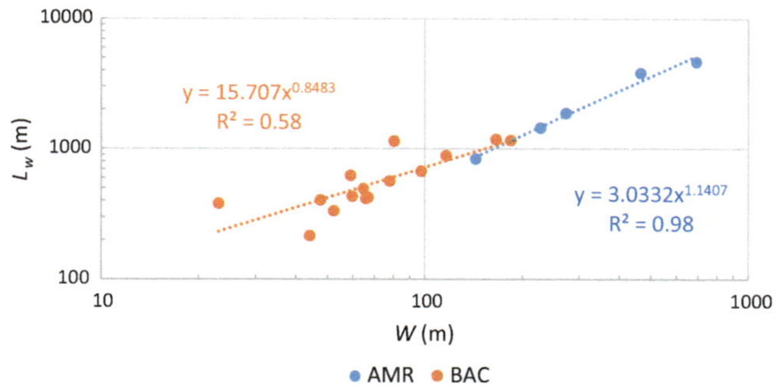

Fig. 4.20 Plot diagram of channel width (W) versus meander wavelength (L_w)

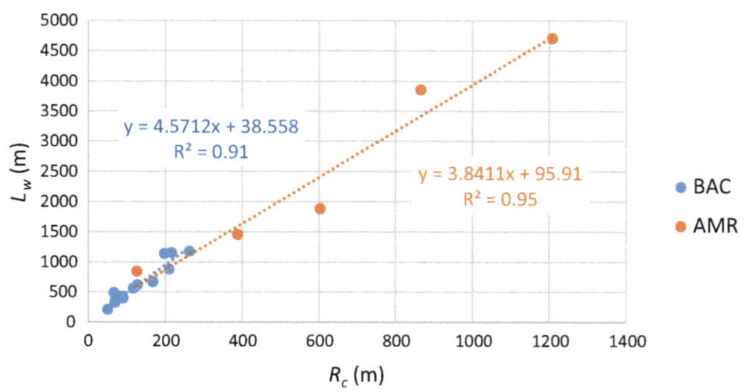

Fig. 4.21 Plot diagram of radius of curvature (R_c) versus meander wavelength (L_w) for both BAC and AMR channels of Eritrea

4.5 Longitudinal River Profile and Tectonics

4.5.1 Energy Gradient and Longitudinal Profile

Field observation suggests that the longitudinal profile of most rivers typically has a smooth concave form, which reflects the downstream distribution of the energy gradient. In natural river systems, the energy gradient depends on many factors, the most important of which are water discharge, sediment size and geo-lithological constraints (Leopold and Langbein 1962). In the upstream reaches of a river, discharge is low and the sediment size is coarse (high roughness). By contrast, in the downstream reaches discharge is large and the sediment is fine (low roughness). Therefore, also the energy gradient decreases downstream. A gradually varied flow is the most common flow condition of natural rivers as it is influenced by streambed slope and friction and there are no abrupt changes (e.g. such as those imposed by engineering works). Therefore, for natural rivers streambed slope is commonly considered as a good approximation of the energy gradient (Singh 2014). On the base of these assumptions, it is evident that the longitudinal variation of the SGI (Hack 1973) is a simplified representation of the downstream variation of the energy gradient.

Theoretically, in a graded river, stream power, which is basically the product of channel slope and discharge, should be constant along the river since the increase in discharge is compensated by the decrease in channel slope. Assuming that stream length is proportional to discharge, also the SGI should be constant, whereas its variability can be taken as an expression of the deviation from the theoretical concave profile whenever changes in bedrock hardness or crustal deformation occur (Phillips and Lutz 2008). Time is an important factor, too. Leopold and Langbein (1962) suggested that random processes like the formation of a drainage network may attain their characteristics rather quickly and Phillips and Lutz (2008) indicated the time span required to produce a profile without significant convexities for >1.3 Ma, though it is not clear if these latter authors assumed stable boundary conditions.

In this study, four rivers draining the Eritrean plateau (Barka, Anseba, Khomib and Mareb), three rivers draining the coastal belt (Falcat, Medere and Damas) and one river draining the main escarpment and ending up in the Danakil depression (Dandero) were selected (Fig. 4.1) to analyse the geologic and tectonic control on the characteristics of longitudinal profiles.

4.5.2 Stream Gradient Index (SGI)

4.5.2.1 Plateau Rivers

The variation of the SGI with distance from headwater divide for the highland rivers is plotted in Fig. 4.22. The Barka and Anseba (which is the main tributary of the Barka River) have a very similar diagram. These two rivers, in fact, have their sources in the flat plateau around Asmara, and beyond Akordat, the Barka flows to the NNW, and it is almost parallel to the Anseba. Both rivers show a very pronounced peak of SGI (about 8500 and 5100 gradient*m, respectively), which corresponds to the western slope of the Asmara plateau. This tableland is a planation surface underlain by Neoproterozoic meta-volcanic rocks (mainly meta-andesite) and some small granitoid plutons (NW of Asmara) (Ghebreab et al. 2005; 2009), and it is covered, especially in its southern portion, by a lateritic crust, which is rather resistant to erosion forming low, flat-topped hills (Abul-Haggag 1961) (Fig. 4.23). Around Asmara, the resistant tableland is bounded to the east by the Central Steep Belt (CSB) (Drury and De Souza Filho 1998) and separated from it by a shear zone (Drury et al 1994; Ghebreab et al. 2009) (Figs. 4.1 and 4.10). According to Drury and De Souza Filho (1998), the CSB is a regional horst of lower crustal material consisting of deformed granitic rocks, gneisses and migmatites, which is bounded westward by late granites, crossed by the Barka and north-westward by early

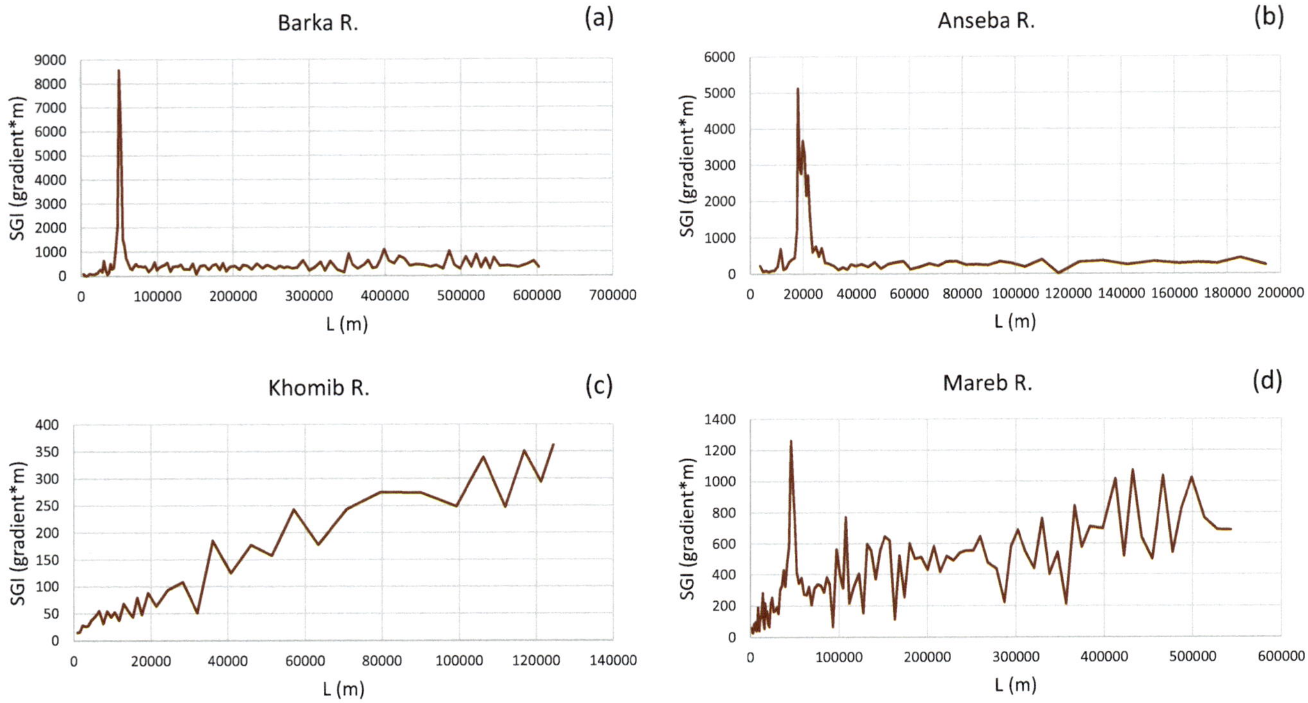

Fig. 4.22 Plot diagram of the stream gradient index (SGI) (Hack 1973) with an increasing distance from divide (L) for the four rivers representative of the highlands

granitoids, crossed by the Anseba (Drury and De Souza Filho 1998) (Fig. 4.24).

The SGI peaks of the Barka and Anseba rivers occur in the CSB, whereas in the Barka downstream reaches as far as the Sudanese border the SGI takes much lower values around 500 gradient*m and values around 300 gradient*m in the Anseba as far as the confluence with the Barka. Such relatively low values are probably associated with the downstream reaches of both rivers flowing parallel to Tertiary normal faults (Drury et al 1994; Keller and Pinter 2002; El Hamdouni et al. 2008). The combination of contrasting physiography and exhumed geomorphology between the Asmara plateau and the western slope and the occurrence of marked crustal deformations and Tertiary faults may substantially affect the SGI values. The Tertiary faults are crossed almost perpendicular by both rivers in their upstream reaches, but in the lower reaches, they flow parallel to them. This setting may also account for the contrasting high peaks of SGI close to the headwaters and much lower values of SGI in the downstream reaches, which are far from the most tectonically active area. This general setting is confirmed by the SGI diagram of the Khomib River (Figs. 4.1 and 4.22), a left tributary of the Barka, which originates from low elevated (around 800 m asl) hills in the western lowland of Eritrea and, after about 120 km, joins the Barka at 16°12″ latitude North. The SGI of this river is very low (<100 gradient*m) in the headwater reaches, but tends to increase downstream reaching the maximum values (around 320 gradient*m) as the river approaches the confluence with the Barka. The value of 300 gradient*m is considered to reflect low to intermediate tectonic activity (El Hamdouni et al. 2008 in Mahmood and Gloaguen 2012). For the first 80 km, the Khomib River flows to the north on a gently inclined surface, then it turns to NE and skims the southern edge of a sheath fold consisting mainly of Neoproterozoic gneisses (Drury and De Souza Filho 1998) witnessing a Precambrian shear zone (Drury et al. 1994). A few kilometres upstream of the confluence with the Barka, the Khomib crosses almost perpendicularly a Tertiary strike-slip fault (Drury et al 1994) (Fig. 4.1). Moreover, the modern Khomib seems to result from a river capture (Fig. 4.25). Upstream of the turn to NE, the river course could be ideally prolonged to join the Lokayeb River, which does not change its northward direction as far as its confluence with the Barka. The Khomib reach upstream and around the right turn is 0–100 m higher than the Lokayeb River reach upstream of the ideal confluence of the Khomib. Though old (Precambrian) South-North lineaments can be supposed in this area (Drury et al. 1994), more recent major strike-slip and transverse faults may have favoured the capture, thus resulting in higher than average SGI values in the most downstream reach of the Khomib.

Though the average SGI of the Mareb River is lower (408 gradient*m) than that of the Barka (580 gradient*m) and of

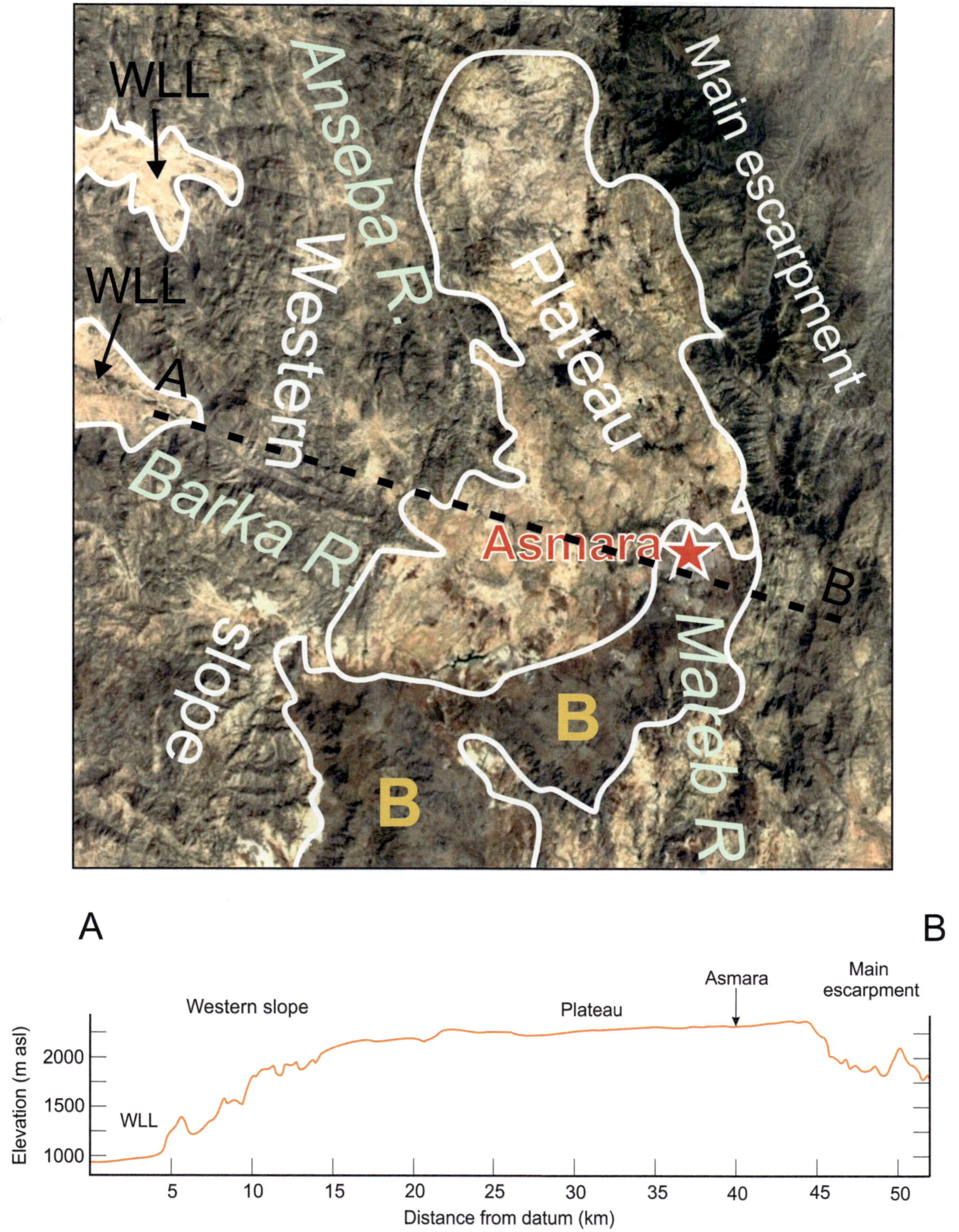

Fig. 4.23 Plateau of Asmara from satellite image and cross section. WLL = western lowlands: B = Oligocene basalts (Ghebreab et al. 2009)

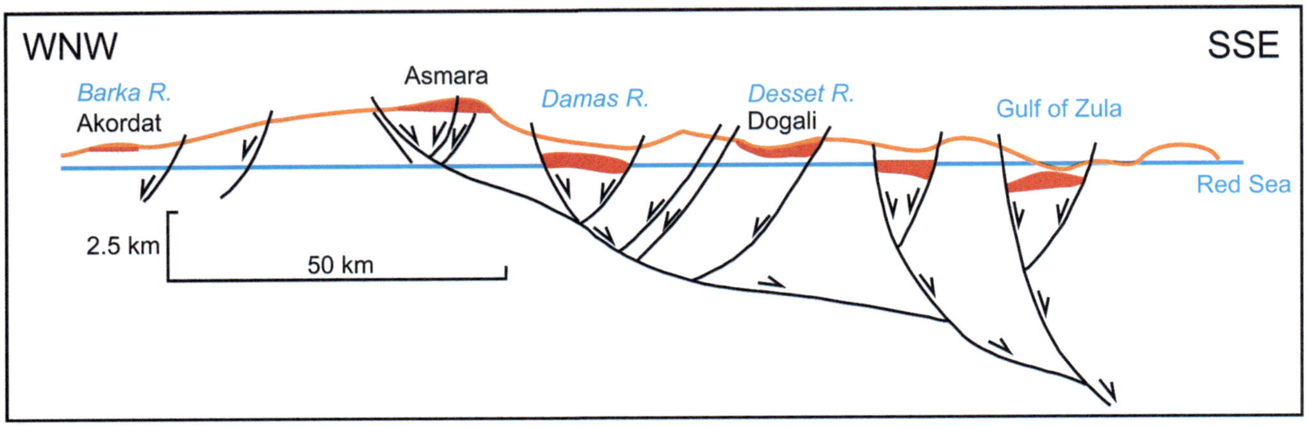

Fig. 4.24 Cross section from Akordat to the Gulf of Zula. The thick red parts are Cenozoic basalts resting on laterite (modified from Drury et al. 1994)

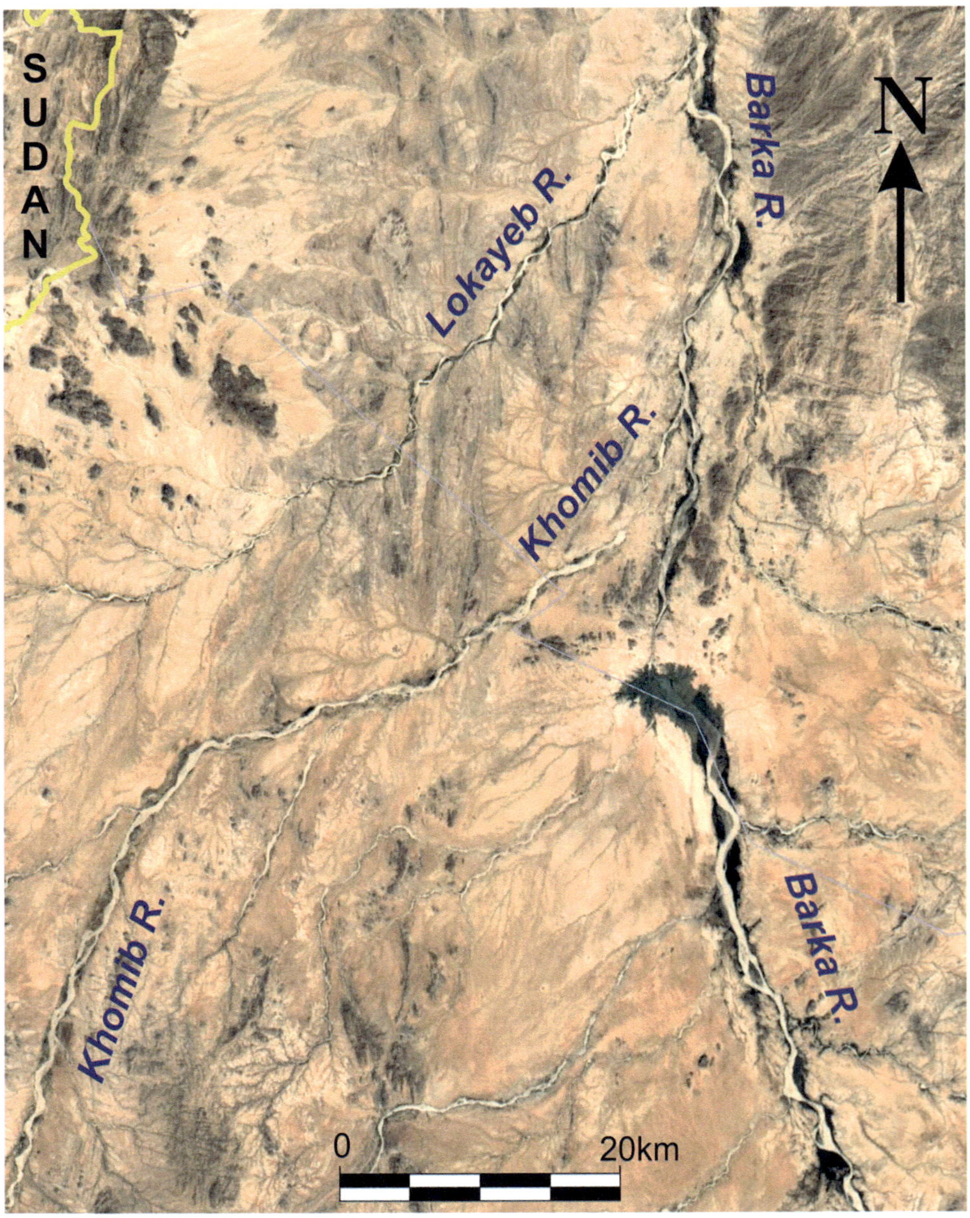

Fig. 4.25 Khomib and Lokayeb rivers, left tributaries of the Barka. Topographic evidence suggests that in the recent past the upper part of the modern Khomib was actually the upstream reach of Lokayeb that was later captured by the lower Khomib

the Anseba (647 gradient*m), the most upstream reach of about 80 km shows a SGI pattern characterized by initial very low values (around 50 gradient*m) followed by a marked peak of 1263 gradient*m, similar to those observed in the Barka and Anseba. Beyond this peak, however, the Mareb SGI shows a downstream increasing pattern, from about 400 gradient*m up to values oscillating between 500 and 1000 gradient*m. The Mareb originates near Asmara, and the initial low values of the SGI can be associated with the flat morphology of the Asmara plateau. The SGI peak occurs where the river crosses, almost perpendicularly, an intensely sheared area characterised by marked SSW-NNE lineations (Drury and De Souza Filho 1998, Fig. 4.26). In this reach, the river has its greatest gradients, is entrenched and carved into bedrock, and lithological changes occur within a short distance (Drury and De Souza Filho 1998). If such combination of crustal deformation and lithological changes can account for the SGI peak, the explanation for the downstream increasing trend of SGI (Fig. 4.22) is not straightforward. The river, after running for about 100 km southward, turns westward and after further 70–80 km turns to NW as far as the Sudanese border, though some southward changes (e.g. south of Barentu) are present as well. The Tertiary uplift, associated with the doming and rifting of the East African Orogen, tends to decrease in intensity and effects with an increasing distance from the main rifting zones, and the Mareb is flowing just westward, so the increase in SGI recorded as the river approaches the Sudanese flat lowland seems a contradiction. Merla et al. (1979) included in their map of major landforms a rectangular fault slope almost parallel to the Sudanese border (Fig. 4.27). These authors estimate this normal fault to be old enough (though they do not say how much old) to have been almost completely obliterated by erosion. Therefore, any effect of this fault on the increasing trend of the SGI has to be reasonably excluded.

In its middle to lower reaches, as far as the Sudanese border, the Mareb flows westward and crosses different terranes (Drury and De Souza Filho 1998), consisting of different Neoproterozoic rocks, and a large transpressional belt (Ghebreab et al. 2009), both oriented SSW-NNE. Such crustal deformations are 800–600 Ma old, and there is no evidence of recent tectonic activity in the western part of Eritrea, where rocks are not harder than in the eastern part crossed by the Mareb (Abbate, personal communication). These data contradict the downstream increase of SGI observed in the Mareb. By contrast, a recent uplift seems to be confirmed by the convex profile in a distance to elevation semi-log diagram (Fig. 4.28), in which the riverbed profile portion above the interpolating line indicates a disequilibrium river which has not yet attained the theoretical equilibrium line, provided tectonic and climatic stability will be maintained in the future. A higher than expected profile portion, similarly to high SGI values, is considered characteristic of streams undergoing rapid uplift (e.g. Merrits and Vincent 1989; Keller and Pinter 2002).

4.5.2.2 Coastal Belt Rivers

A similar pattern of downstream increase of SGI was found also for two coastal rivers (the Falcat and the Medere) (Figs. 4.1 and 4.29) and the Danakil River (the Dandero) (Fig. 4.29). The third coastal river, the Damas (Fig. 4.1), instead, shows a decreasing trend of SGI (Fig. 4.29). In the Falcat, Medere and Dandero rivers, close to the upstream divide, the values of the SGI are relatively low (<400 gradient*m) and tend to grow up to 800–1000 gradient*m along a downstream a distance of 40–60 km from the source. Beyond this distance the SGI trends became almost horizontal and the index oscillates around a mean value of 600, 800 and 1000 gradient*m in the Falcat, Medere and Dandero rivers, respectively. These SGI data, characterized by relatively low values in the main escarpment reaches are unexpected, but may indicate that the effects of the great Tertiary uplift, associated with the doming and rifting of the area, were already cancelled from the river profile by the long-lasting action of river erosion. Similarly, to the Mareb, in fact, these rivers have also convex longitudinal profiles in a semi-log diagram of elevation vs longitudinal distance. By contrast, according to Drury et al. (1994), in the coastal belt, there is abundant evidence of recent tectonics and active uplift of a few blocks, which affected the pattern of several rivers draining the main escarpment. The eastward tilting of blocks of the Plio-Pleistocene Desset Formation north of Massawa is another indication of recent tectonics (see Chap. 2). The occurrence of still active tectonic systems was also reported by Angelucci et al. (1982) for the Dahlac Islands. Within this framework of active tectonics is not surprising to find the downstream increasing trend of SGI for the Falcat and the Medere rivers. Also the Quaternary uplift in the downstream reach of the Dandero (see Chap. 5) can account for high SGI values and its convex longitudinal profile, standing above the theoretical exponential curve considered by several authors (e.g. Knighton 1998) as the typical profile of equilibrium rivers. Drury et al. (1994) considered the Damas River as a good example of such recent tectonic activity. These authors, in fact, suppose that in the past the Damas River was flowing from the escarpment straight into the Red Sea. Nowadays, the river flows to NE in its upper reach, then turns to NNW near the village of Dengolo, flows in this direction for about 35 km, then turns to the east, crosses the coastal escarpment and enters the Red Sea north of Wekiro. A few kilometres upstream of Dengolo a small, dry channel departs from the Damas main channel and joins the Desset River (Fig. 4.30). In other words, in the recent past, according to Drury et al. (1994), the Damas and the Desset were part of the same river system, but recently,

Fig. 4.26 Bedrock shearing area crossed by the Mareb River, where a peak of SGI occurs

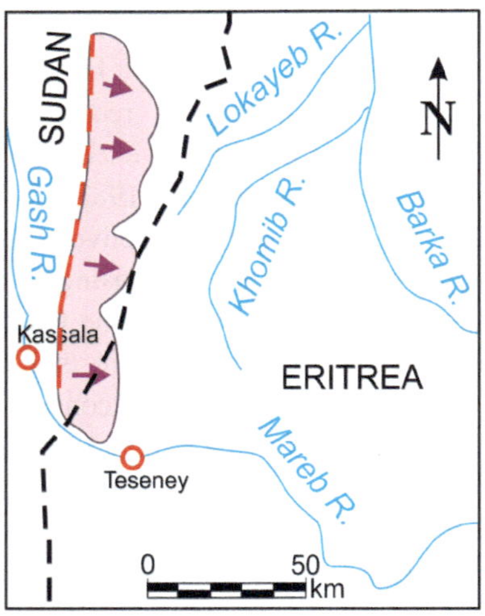

Fig. 4.27 Fault slope close to the Sudanese border reported by Merla et al (1979) (pink area). The red dashed line is the presumed fault line, and the purple arrows indicate the fault slope dip (modified from Merla et al. 1979)

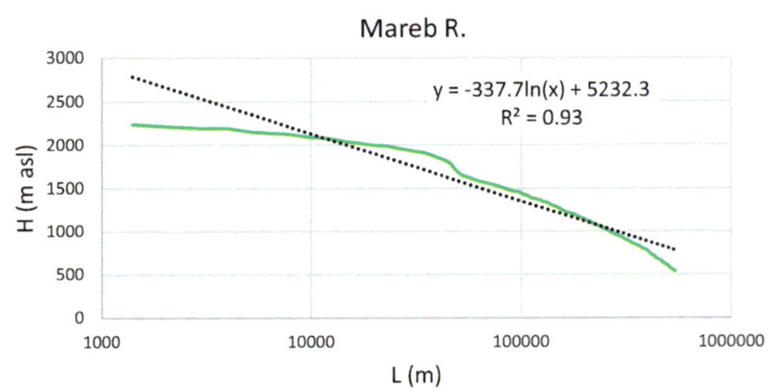

Fig. 4.28 Semi-log diagram of Mareb River longitudinal profile. H is the elevation above sea level, and L is the stream length form the upstream divide. Notice the knickpoint corresponding to the SGI peak

the Damas was disconnected from the Desset and was forced to flow northward. In support of this hypothesis, a few geomorphic data can be provided. Presently, the catchment size of the Damas and Desset rivers is 222.4 and 122.4 km^2, respectively, but the delta area is 21.56 km^2 for the Damas and 33.27 km^2 for the Desset. Thus, the larger river has presently a smaller delta, whereas the smaller river formed a larger delta. Since the two catchments are neighbouring and underlain by the same rock formations, it can be presumed that they should be subjected to the same climate and erosion conditions. The marked difference in the size of the deltas seems to confirm the hypothesis of the two rivers as part of a formerly singular river system, the lower reach of which is today taken up by the Desset. This result, however, is not reflected by the low SGI values less than 300 gradient*m measured in the downstream reach of the Damas and considered as indicative of no or low tectonic activity.

Examples of recent or ongoing river captures are present also in the Falcat and Medere catchments (Fig. 4.31). These captures and the case of an alluvial fan supplying two rivers flowing in opposite direction witness the recent tectonic activity of the southern part of the coastal belt and the tectonic systems Gulf of Zula-Danakil depression (Figs. 4.24 and 4.32).

Notwithstanding the evidence of recent tectonic activity that is also reflected by the SGI downstream increasing pattern of the Falcat, Medere and Dandero rivers, the SGI decreasing pattern of the Damas is opposite compared to the other coastal rivers (Figs. 4.29 and 4.33) and no clear explanation was found. In the Damas, the SGI is high (500–600 gradient*m) in the upstream reach, which virtually could reflect recent tectonics of the eastern part of the main escarpment, but it decreases to low values (100–200 gradient*m, characteristic of no or low tectonic activity—El Hamdouni et al. 2008; Mahmood and Gloaguen 2012) as it approaches the coastal plain where there is evidence of neo-tectonic and present-day tectonic activity. The semi-log diagram (Fig. 4.33) indicates a more mature profile of the Damas, with an incised middle to lower reach, which may be the results of a higher erosivity capacity exerted by former larger discharges when the Desset was a tributary of the Damas (Fig. 4.30).

4.6 Discussion

4.6.1 Channel Morphology

None of the Eritrean rivers has perennial flow; they are ephemeral or seasonal. In a region with a predominantly semi-arid climate and scarce precipitation, poor vegetation cover, steep slopes and, hence, high sediment supply, one would expect braided rivers as the most common channel morphology. Billi et al. (2018) have shown that ephemeral meandering rivers are very common in drylands and have highlighted the morphological differences with perennial rivers. Nevertheless, very little information is available for bedrock ephemeral meandering channels of mountainous areas.

The bedrock meandering channels with alluvium (BAC) considered in this study have an average streambed gradient of 0.0172 (range 0.0035–0.0323), which is rather steep for a meandering river and close to the threshold slope to braided streams (0.02—Rosgen 1994). The gradient of meandering channels within the confined alluvial plain (AMR) is one order of magnitude lower (0.0029) and typical of lowland alluvial meandering rivers. By contrast, BAC rivers sinuosity is higher (average 1.54, range 1.29–1.86) than in AMR (average 1.25, range 1.12–1.46). The very limited occurrence of AMR reaches (only five in Eritrea) may explain their unexpected low sinuosity. According to

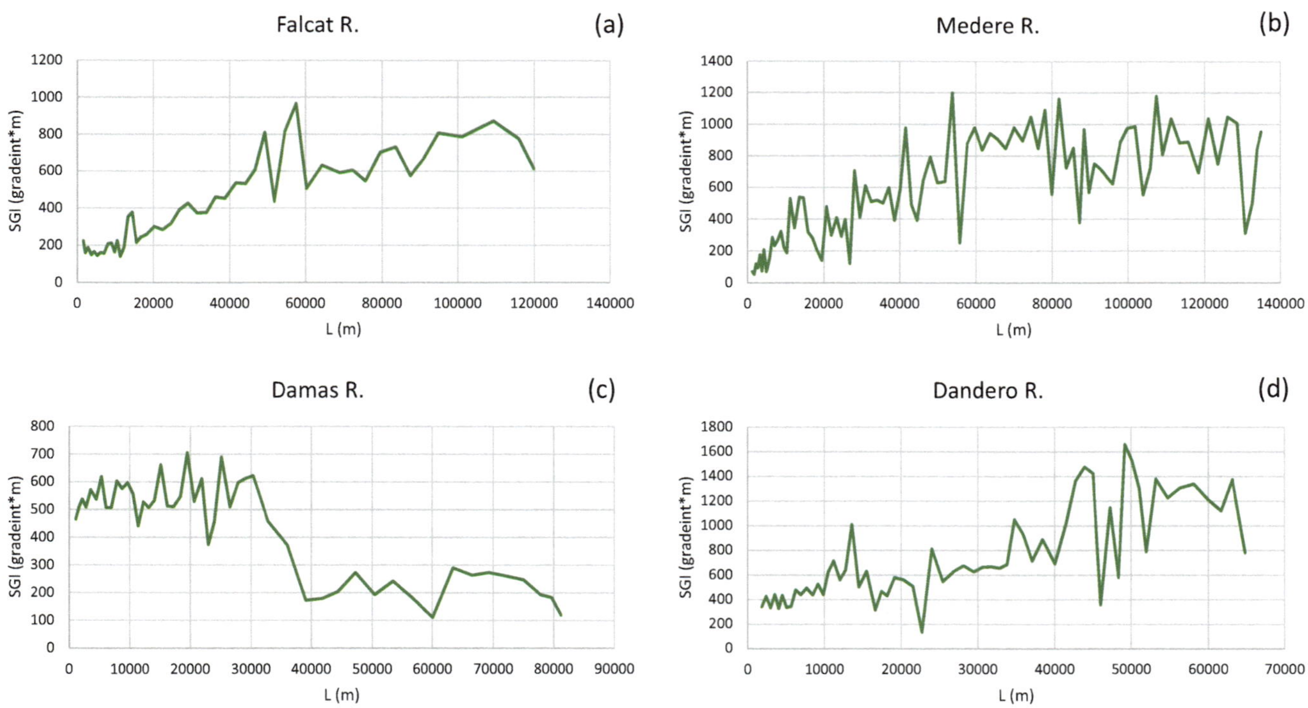

Fig. 4.29 Plot diagram of the stream gradient index (SGI) (Hack 1973) versus downstream distance from divide (L) for the three coastal rivers, **a** Falcat, **b** Medere and **c** Damas and the Danakil river, **d** Dandero

Fig. 4.30 Nameless channel connecting the Damas with the Desset River

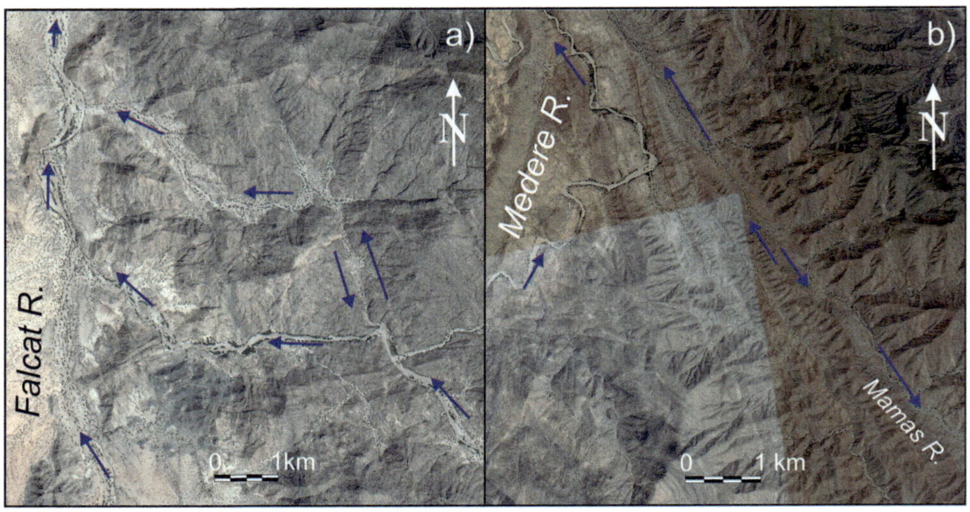

Fig. 4.31 River capture within the catchment of the Falcat River (**a**) and the Medere River (**b**). Both captures are ongoing or in their very beginning (Google Earth coordinates: capture **a**: 16° 56′ 13″ N—38° 29′ 08″ E; capture **b**: 17° 01′ 36″ N—38° 35′ 20″ E)

Eke and Wilcok (2015), confined meanders with persistent angular bends are associated with low transport rates of bed material and their radius of curvature tends to increase as the increased sediment supply from cutbanks is sufficient to foster the point bar development.

According to Stark et al. (2010) and Johnson and Finnegan (2015), though there are cases in which meandering have maintained an antecedent alluvial conditions and sinuosity through incision during uplift, not all cases can be explained with the old theory of Davis (1899) of entrenched meanders and the occurrence of slaking lithologies is predominant in controlling lateral erosion and bedrock meander formation rather than vertical incision. This study shows that BAC and AMR channels have the same curvature ratio, which would suggest that, though bank resistance to erosion is an important factor in controlling meander formation in mountain environments, other processes/factors should be considered as well to account for the strict similarity between bedrock meanders geometry and that of free, alluvial meandering channels.

In Eritrea, different kinds of granitic and metamorphic rocks crop out and the uplift of mainland went on uninterruptedly from the Oligocene to the Quaternary (Merla et al. 1979). In the western highlands, the occurrence of rivers, which inherited their meandering morphology while incising their alluvium and then bedrock, is plausible given the large-scale crustal doming. The marked influence of associated faults on river orientation and the common presence of bedrock meandering rivers also within the short and steep eastern escarpment suggest that processes other than simple vertical incision should be responsible for the formation and development of meanders cut into hard bedrock.

Fig. 4.32 Effects of recent tectonics on alluvial fan development. A wide alluvial fan in the mountains of the Gheden horst is supplying two rivers: one flowing to the north and entering the Red Sea a few kilometres south of Massawa and the other flowing to the south, to the Gulf of Zula

The average wavelength to channel width ratio of the study rivers is 7.9 (BAC 8.2; AMR 6.8), which is close to the values obtained by other authors (e.g. 13.3—Leopold and Wolman 1960; 6.3—Yalin 1971; 12—Williams 1986; 6—Garde and Ranga Raju 2000) for perennial rivers in more humid areas. Higher values (26 and 16) were obtained by Schumm (1968) for the seasonal Murrumbidgee River in Australia and by Billi et al (2018) from a very large dataset of ephemeral, free-meandering rivers in different drylands of the planet. From these results, it comes out that the geomorphic characteristics of bedrock meanders in Eritrean drylands are more similar to perennial rivers than to alluvial, free meanders of ephemeral or seasonal streams. In other words, though the channels and the alluvial, free-meandering rivers studied by Billi et al. (2018) share similar dry climate conditions, for comparable channel width the former develop a shorter wavelength than the latter. Constraints to larger wavelength in bedrock channels can include higher resistance to erosion of the boundary material, but also streambed gradient seems to play a role (Fig. 4.34). Steeper gradient may lead to higher unit stream power and shorter time of residence for the bed material in a reach (Lecce 1997). Under such circumstances, the formation of point bars and meander lateral migration are not favoured. However, lateral accumulation of scree material may force the channel towards the opposite bank and the downstream propagation of such flow perturbation may result in the generation of helical flow that is commonly observed in meandering rivers.

If the occurrence of slaking rocks, scree and colluvial deposits is a necessary condition to start the formation of bedrock meanders (Johnson and Finnegan 2015), their geomorphic characteristics indicate that hydraulic processes such as secondary currents (Shen 1983), macroturbulent eddies (Yalin 1971) or flow energy expenditure (Yang 1971) are prevalent also in controlling the morphology of meanders incised into bedrock or confined within a narrow alluvial plain bound by hillslopes of indurated old alluvium. The radius of curvature of meanders presented here, in fact, is well predicted by Eq. (4.2) of Langbein and Leopold (1966) (Fig. 4.19). This result confirms that a sine-generated curve best approximates geometry of bedrock meanders, which is not different from that of free-migrating alluvial rivers. This conclusion is supported also by the wavelength to channel width ratio, which is in the range 6–13, as also reported by several authors (e.g. Leopold and Wolman 1960; Yalin 1971; Richards 1982; Williams 1986; Garde and Ranga Raju 2000) for rivers in different environments.

The majority (80%) of the curvature ratios measured in the studied rivers is less than 2 and lies outside of the 2–3 range predicted by Leopold and Wolman (1960) and Bagnold (1960) from minimum flow resistance investigation in curved pipe experiments and field data. This result is, however, consistent with the findings of Billi et al. (2018) in their study on ephemeral, free-meandering streams in different drylands of the world. Bagnold (1960) argued that in natural rivers, channel cross section has not a regular semi-circular shape and turbulence distribution can be expected to be different from that in a circular pipe. On the base of reasoning of Leopold and Wolman (1960) and Leopold (1994), it follows that for bedrock streams resistance is increased by sharper bends, such as those in which the curvature ratio is less than 2. According to Leopold (1994), river channel morphology results from the tendency to achieve simultaneously two conditions of minimum work and uniform distribution of work. The gradient of bedrock meanders studied here (BAC) is higher than that typical of alluvial plain sinuous rivers, which should virtually result in instability condition. However, the tendency to minimize the total work in such steep channels is balanced by a more

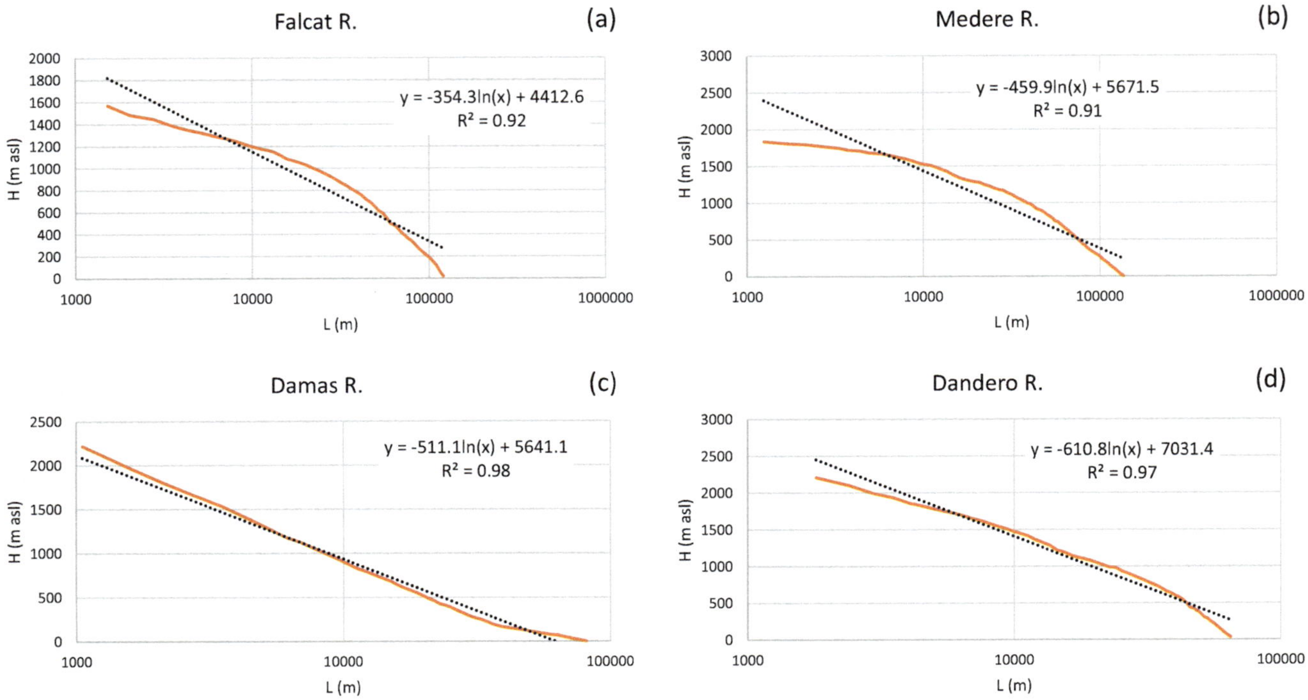

Fig. 4.33 Semi-log diagrams showing longitudinal profiles of coastal and Danakil rivers. H is the elevation above sea level, and L is the stream length from the upstream divide

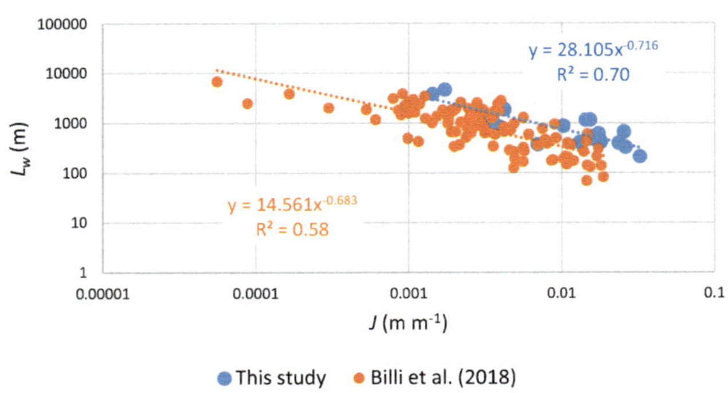

Fig. 4.34 Plot diagram of streambed gradient (J) versus meander wavelength (L_w). The bedrock meandering channels studied here and the ephemeral, free-meandering river of Billi et al. (2018) have similar interpolating equations, with the former characterized by higher gradients

uniform distribution of energy expenditure due to comparatively sharper bends.

The pattern geometry of bedrock meanders measured in this study is rather regular and is well approximated by a sine-generate curve (Eq. 4.2 and Fig. 4.19; $R^2 = 0.97$) which, according to Leopold (1994), is a minimum variance curve, i.e. the most probable condition. All these considerations may account for the very common occurrence of bedrock meanders, with bend geometry similar to their freely meandering alluvial counterparts, which are found in rugged highlands and steep escarpments such as those of Eritrea. The Eritrean main escarpment is very steep, crossed by many longitudinal faults, associated with the opening of the Red Sea, and drained by many short rivers, the pattern of which is substantially influenced by the progressive lower fault blocks degrading to the Red Sea. In such a tectonically active area, it is therefore difficult to imagine pre-existing, alluvial meandering rivers down-cutting the bedrock and maintaining their original sinuous morphology.

Though the development of bedrock meanders may be initiated in and affected by bedrock with poor mechanical properties (Johnson and Finnegan 2015), the occurrence of bedrock meandering channels also in areas underlain by hard and durable rocks, such as the Neoproterozoic granites and metamorphic rocks cropping out over large parts of Eritrea (Fig. 4.4), suggests that meanders can be formed also in hard bedrock under the control of the same hydraulic variables and physical processes, which typify alluvial, free-meandering rivers.

4.6.2 Channel Pattern and Profile

In Eritrea, three main different phases of very active tectonics can be recognized. The first and oldest one is associated with complex crustal deformations that affected the Precambrian and Paleozoic basement rocks (Drury and De Souza Filho 1998; Ghebreab and Talbot 2000), the second one is responsible for the Oligo-Miocene doming and rifting of the Arabian-Ethiopian region (Merla et al. 1979) and for the formation of the main escarpment (Fig. 4.1) and the last occurred during the Pliocene and Quaternary age and is ongoing (Drury et al. 1994), affecting mainly the eastern, coastal belt (Fig. 4.35) and the Danakil depression (Fig. 4.24). These three main structural events provided Eritrea with its present major landforms.

The drainage network reflects overlapping effects of such a complex sequence of folding, uplifting and faulting. The larger rivers are set on the main structural lineaments (Fig. 4.1), whereas the pattern of their tributaries and smaller rivers is largely influenced by faulting and marked lithological changes. The main and most recent fault lines are oriented roughly N-S, but also a secondary set of smaller, strike-slip faults almost perpendicular to the former is present (Drury et al. 1994) (Fig. 4.1). Therefore, the drainage network mainly shows a fragmented parallel to rectangular style (Fig. 4.36), and there are several examples of river capture (e.g. Figs. 4.30, 4.31) and uncertain flow direction (Fig. 4.32).

Within such a framework of active tectonics, the main Eritrean rivers appear as a good example to apply the SGI and to verify its use also in areas with such a complex history of crustal deformations. Several authors (e.g. Hack 1973; Wobus et al. 2006; El Hamdouni et al. 2008; Phillips and Lutz 2008; Troiani and Della Seta 2008; Demoulin et al. 2017) considered the longitudinal profile of a river as a useful diagnostic indicator of tectonic activity (mainly uplift or subsidence), variations in rock resistance, base-level changes and climate change in a region. The SGI is a simple index to be measured, capable to quantitatively express river response to such changes.

The rivers considered in this study are characterized by a downstream increase of SGI, with the exception of two rivers: the Mareb and the Damas (Figs. 4.22 and 4.29). All the rivers originating from the tableland of the Asmara plateau are characterized by very low SGI values near the source (which are expected given the almost flat morphology of the Asmara plateau—Figs. 4.3 and 4.23) and a very prominent SGI peak between 20 and 50 km from the source, followed by a mildly increasing trend. Also, two of the coastal rivers (Falcat and Medere) and the Danakil River (Dandero) show low SGI values (from 30 to 300 gradient*m) in their source areas, which are located in the eastern portion of the main escarpment. Here, much higher values of SGI would be expected given the large magnitude of crustal uplift during the Cenozoic, but the headwater channel run parallel to the main N-S normal faults (Drury et al. 1994) (Fig. 4.1) and the effect of this block faulting is just to host the main channel, the longitudinal profile of which in the upper reach lacks knickpoints. Moreover, these rivers flow parallel to the main fault lines, and as postulated by Keller and Pinter (2002) and El Hamdouni et al. (2008), the SGI values are expected to be low.

In the downstream reach, the Falcat and Medere turn to the east to reach the Red Sea, flow across the eastern margins of the faulted blocks and enter an area of recent uplift that is well recorded by the increasing values of the SGI. The

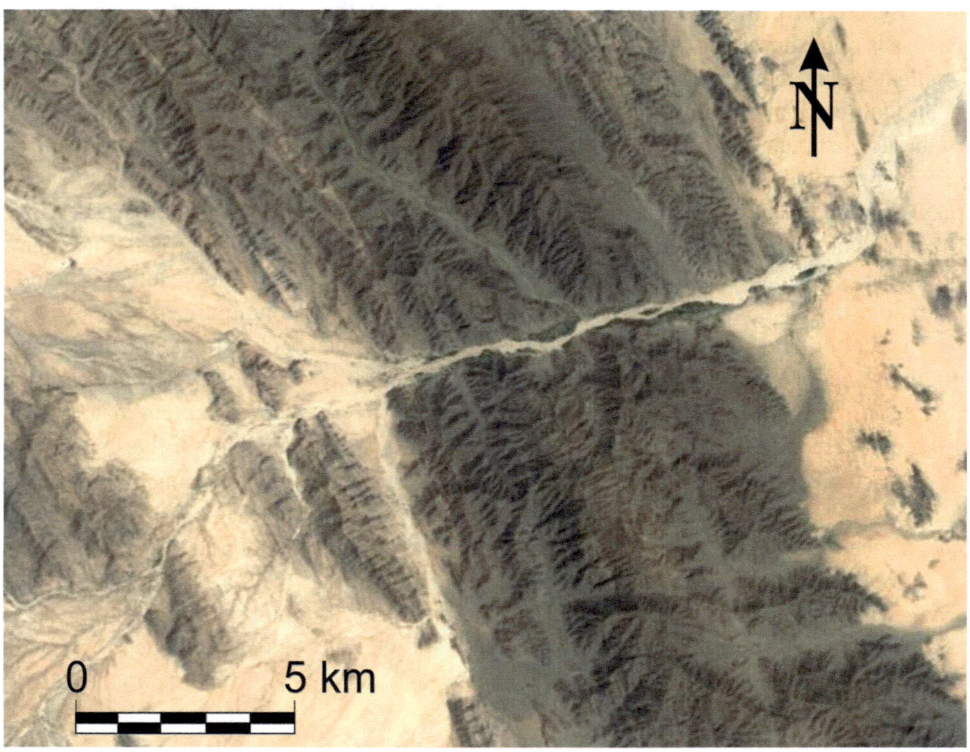

Fig. 4.35 Example of a strike-slip fault across a coastal tilted block. The fault is taken up by a river entering the alluvial coastal plain 35 km west of the Red Sea coastline

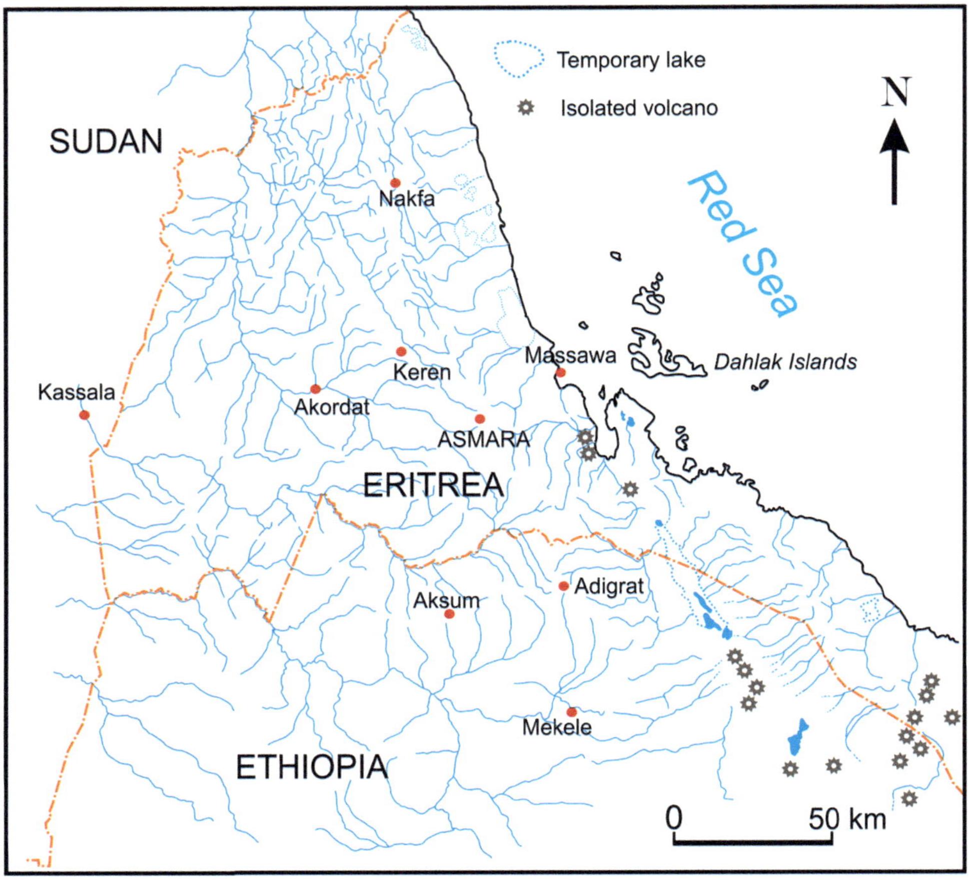

Fig. 4.36 Drainage network of Eritrea. The map includes the main rivers extracted from 1:500,000 topographic maps

Damas River catchment developed in a structural and lithological setting very similar (if not identical) to the Falcat and Medere, but, in spite of that, the SGI pattern of the Damas is opposite as it is characterized by high values (>500 gradient*m) in the upstream reach (where it flows along a main fault line) and by low values (around 200 gradient*m) in the downstream half, where this river crosses a wide belt (12 km) of tilted rocks before reaching the coastal alluvial plain. The Damas coastal reach is 160 km away from the Medere, but there is no strong evidence that the central portion of the coastal belt is subjected to a milder recent tectonic activity compared to the northern portion. Yet, lithological differences are not relevant. As illustrated in the previous section, the downstream reach of the Damas was formed or, more likely, captured from the Desset River system in recent times and the capturing river (now the downstream reach of the Damas) probably formed earlier (as it seems to be suggested by its much larger valley bottom) and reached the Red Sea along a strike-slip fault across the coastal blocks. According to this hypothesis, the pre-Damas River had larger discharges and enough time to develop a graded profile, favoured also by a fault across the coastal block, in which the river has formed a narrow and deep, 6 km long gorge.

In the upper reach of the Mareb, the SGI pattern is similar to that of the other large rivers draining the western slope (Barka and Anseba); i.e. low values occur on the Asmara plateau and a peak is recorded about 50 km downstream. Beyond this individual peak, the SGI pattern shows a markedly increasing trend. As the river enters the western lowlands and approaches the Sudanese border, the recent tectonic activity decreases substantially and even the effects of basement rock shearing show much less expression at the ground surface. Basement faulting in this area is rather old (600–800 Ma) and should no longer have much effect on the longitudinal profile of the river. Moreover, the bedrocks are not harder that those crossed by the river in its upstream reach. In the Mareb longitudinal profile, it is easy to identify the knickpoint that supports the peak of the SGI, but beyond this stretch the profile is concave upward and the classical interpolation with an exponential function has a high determination coefficient of 0.98, notwithstanding the knickpoint. The same knickpoint is visible in the river profile (Fig. 4.28), which is obviously based on the same data but plotted in a semi-log diagram for conformity with the analytical approach of Merritts and Vincent (1989).

Compared to the other rivers, in which tectonic activity is reflected by the downstream increase of the SGI, thus confirming the theory that recent crustal deformations are recorded in the river profile, the Mareb does not follow this general rule. In their investigation, carried out in different areas of the northern Apennines and the Calabrian Arc, Italy, on what a knickpoint in a river profile may represent, Cyr et al. (2014) concluded that "the interpretation of uplift rates or relative changes in lithologic resistance to channel incision from longitudinal profile morphology is ambiguous at best". This study seems to partially confirm this conclusion because, though the profiles of some rivers proved capable to reflect recent tectonic activity in some portions. In the rivers originating from the Asmara plateau (Barka, Anseba and Mareb), the abrupt change from flat morphology inherited from uplift and exhumation of an old planation surface underlain by harder rocks (a lateritic crust) to the sloping side in relatively more erosion-prone rocks resulted in a very prominent SGI peak. In this case, therefore, the contrast between the preserved, flat paleo-morphology and the rugged relief on metamorphic rocks further downstream was the main factor affecting the shape of the most upstream portion of river profiles, rather than or in combination with the remarkable Tertiary uplift.

The SGI proved to be sensitive to tectonic activity, but in an area subject to a long-lasting sequence of tectonic phases with different intensity and spatial influence, local conditions may be effective in forcing the river profiles to depart form the theoretical shape. Another factor that may be invoked to account for the unexpected pattern of the SGI is climate. Though no general information about climate variability in Eritrea during the Quaternary is available in the literature, with the exception of a few studies in the Eritrean Danakil (e.g. Albianelli and Napoleone 2004; Ghinassi et al. 2009; see also Chap. 3), studies of the eastern Sahara-Sahel region (e.g. see Hoelzmann et al. 2004, for a comprehensive review) suggest that from the early Pleistocene in the Danakil and from the Late Pleistocene in the eastern Sahel and Arabian Peninsula to present the climate of Eritrea was subjected mainly to arid and semi-arid conditions, interrupted by short wetter intervals of different duration.

Phillips and Lutz (2008) concluded that a long time (at least >1.3 Ma) is necessary for a river to develop a concave, equilibrium profile, devoid of knickpoints, i.e. a time that is normally longer than the typical timescale of environmental change during the Quaternary. According to Cyr et al. (2014), a profile without convexities is typically achieved by a river under conditions of uniform uplift rates, though high or low rates may result in a more or less steep profile, respectively. In Eritrea, all these conditions are not matched. The long dry intervals, which were probably more pronounced in the western and coastal lowlands, reduced the capacity of the rivers to incise their channels into bedrock. Combined with the Quaternary tectonic episodes overlapping the Tertiary deformations, this may have delayed or prevented some of the Eritrean rivers from developing the theoretical concave profile.

4.7 Concluding Remarks

The majority of Eritrean rivers are incised into bedrock or have the streambed covered by a thin, coarse-grained sediment layer. Both these bedrock channels are rather steep, but while the former are straight and steeper, the latter very commonly have a sinuous pattern. Bedrock meandering channels are typically associated with entrenched rivers that inherited their present morphology from antecedent alluvial conditions during down-cutting. In Eritrea, this classical hypothesis for bedrock meandering is only partially applicable since meandering rivers occur also in the very steep eastern escarpment generated by rifting associated with the opening of the Red Sea, whereas alluvial meandering rivers flowing within a narrow and confined flood plain are present only in the western lowlands, i.e. far from the most actively uplifted area. According to few authors (e.g. Stark et al. 2010; Johnson and Finnegan 2015), the formation of bedrock meanders is controlled by bank bedrock erosivity and lateral channel migration. These hypotheses, however, were not supported by this study based on simple geomorphic parameters such as channel gradient and width, meander wavelength and radius of curvature measured from Google Earth for 15 steep (average gradient 0.017) bedrock meanders with thin streambed alluvium and on five alluvial meandering reaches, the only ones present in Eritrea.

Data analysis revealed that meander geometry of both bedrock and alluvial channels is very similar to that of alluvial plain counterparts in more humid environments. Bedrock meanders, in fact, are very well approximated by a sine-generated curve as proposed by Langbein and Leopold (1966). The average wavelength to width ratio is 8.3, i.e. equivalent to that reported by several authors (e.g. Leopold and Wolman 1960; Yalin 1971; Williams 1986; Garde and Ranga Raju 2000) for lowland alluvial meanders. The average curvature ratio of the study rivers is 1.67, and the majority of the values is less than 2, that is outside of the 2–3 range predicted by Leopold and Wolman (1960) and Bagnold (1960). This discrepancy can be explained in terms of the tendency to minimize the total work in such steep channels, which is balanced by a more uniform distribution of energy expenditure due to comparatively sharper bends. Though the occurrence of slaked erosive rocks in the channel banks may be important in triggering the formation of channel curves, evident similarity with alluvial meanders suggests that their downstream propagation and development is mainly controlled by the same hydraulic factors governing the formation of alluvial free-meandering rivers. These results indicate that the hypotheses of Leopold (1994) of the most probable distribution of energy expenditure and that meanders are the most probable conditions in river channels is valid also for bedrock channels. More field data on bedrock meandering channels are, however, necessary to quantify the role of bank rock mechanical characteristics and hydraulic factors such as helicoidal flow and secondary currents in generating a sinuous channel in steep mountain environments.

The pattern of Eritrean rivers is substantially influenced by uplift and faulting. Many rivers are set on the main fault systems and a typical sequence of downstream variation in channel morphology can be identified. It consists of straight and very steep bedrock streams in the headwaters, straight or meandering channels with a thin streambed alluvium on the plateau and the main escarpment and braided rivers in the eastern coastal plains or meandering rivers within the confined alluvial belts of the western lowlands.

Tectonic activity is also recorded in the longitudinal profile of some rivers, representative of different areas of Eritrea. The analysis of the stream gradient index (SGI) data returned contradictory results. Although the profiles of some rivers are capable to reflect the recent tectonic activity in some portions of their length, in the case of the rivers originating from the Asmara plateau (Barka, Anseba and Mareb), the SGI diagrams are characterized by an initial very prominent peak followed by much lower SGI values in the Barka and Anseba and in a slightly increasing trend in the Mareb. The SGI peak is due to the abrupt transition from the flat morphology of the plateau (inherited from the exhumation of an old planation surface consisting of harder rock—a lateritic crust) to the sloping side incised in more erodible rocks. In this case, the contrast between the preserved ancient flat morphology and the rugged topography in metamorphic rocks further downstream is well witnessed by the SGI but it reflects a combination of rock hardness difference and land denudation factors rather than the remarkable Cenozoic uplift, though tectonic activity played an important role in uplift and uncovering the lateritic crust of the Asmara plateau.

Within the main escarpment, the SGI data are unexpectedly characterized by relatively low values. This seems to indicate that the effects of the Cenozoic uplift, associated with the doming and rifting of the area, were already cancelled from the river profile as they were "digested" by the protracted river erosion. In support of this hypothesis stands the increasing trend of the SGI in the coastal rivers. In these rivers, the longitudinal distribution of the SGI shows very low values (<500 gradient*m) in the headwaters, i.e. in the areas subjected to the highest uplift, and higher values as they approach the coastal plain, where recent (Quaternary) tectonics was locally active.

These results suggest that, in principle, the analysis of the longitudinal profile of a river by means of the SGI can be a useful, though complementary tool in investigating the uplift history of a region. For a better comprehension and evaluation of the robustness of the SGI in catching the main tectonic phases by the river profile, its main discontinuities have to be framed within the local physiographic and geomorphological characteristics. Additional data on a wider range of rivers in areas strongly affected by old and recent uplift are, however, needed for a better understanding of river response to active tectonics.

Acknowledgements Ernesto Abbate is greatly acknowledged for a critical and productive review of a preliminary version of the manuscript. This research was supported by funds from the International Platform for Dryland Research and Education (IPDRE), University of Tottori, Japan.

References

Abul-Haggag Y (1961) A contribution to the physiography of northern Ethiopia. The Athlone Press, London

Albianelli A, Npoleone G (2004) Magntostratigraphy of the *Homo*-bearing Pleistocene Dandiero Basin (Danakil depression, Eritrea). Riv Ital Paleontol Stratigr 110(supplement):35–44

Alemngus A, Amlesom S, Bovas JJL (2017) An overview of Eritrea's water resources. Int J Eng Res Dev 13(3):74–84

Angelucci A, Boni CF, Bono P, Carbone F, Ciancetti G, Civitelli G, D'alessandro L, Funiciello R, La Monica GB, Lombardi S, Palmieri EL, Mariotti G, Matteucci R, Toro B (1982) Il Ghubbet Entatu nell'arcipelago delle Isole Dahlak (Mar Rosso): un esempio di sedimentazione carbonatica. Boll Della Soc Paleontol Ital 21(2–3):189–200

Bagnold RA (1960) Some aspects of the shape of river meanders. U.S. Geological Survey Professional Paper 282-E, Washington, DC.

Bertarelli V (1929) Possedimenti e colonie. Isole Egee, Tripolitania, Cirenaica, Eritrea, Somalia. TCI, Milano

Billi P, Golla S, Tefferra D (2015) Ethiopian rivers. In: Billi P (ed) Landscapes and landforms of Ethiopia. Springer, Heidelberg, pp 89–116

Billi P, Demissie B, Nyssen J, Moges G, Fazzini M (2018) Meander hydromorphology of ephemeral streams: Similarities and differences with perennial rivers. Geomorphology 319:35–46

Bravard JP, Petit F (2009) Geomorphology of streams and rivers. In: Likens GE (ed) Encyclopedia of Inland waters. Elsevier, Amsterdam, pp 387–395

Buffington JM, Montgomery DR (2013) Geomorphic classification of rivers. In: Wohl E (ed) Treatise on geomorphology, vol. 9. Academic Press, San Diego, Fluvial Geomorphology, pp 730–767

Cyr AJ, Granger DE, Olivetti V, Molin P (2014) Distinguishing between tectonic and lithologic controls on bedrock channel longitudinal profiles using cosmogenic 10Be erosion rates and channel steepness index. Geomorphology 209:27–38

Davis WM (1899) The geographical cycle. Geogr J 14(5):481–504

Demoulin A, Mather A, Whittaker A (2017) Fluvial archives, a valuable record of vertical crustal deformation. Quat Sci Rev 166:10–37

Drury SA, Kelley SP, Berhe SM, Collier REL, Abraha M (1994) Structures related to Red Sea evolution in northern Eritrea. Tectonics 13(6):1371–1380

Drury SA, Souza Filho de CR (1998) Neoproterozoic terrane assemblages in Eritrea: review and prospects. J Afr Earth Sci 27:331–348

Eke EC, Wilcock PR (2015) Don't fence me in: free meanders in a confined river valley. American Geophysical Union, Fall Meeting 2015, abstract id. EP51A-0906

El Hamdouni R, Irigaray C, Fernández T, Chacón J, Keller KA (2008) Assessment of relative active tectonics, southwest border of the Sierra Nevada (southern Spain). Geomorphology 96:150–173
Fairbridge RW (1993) Incised meander. In: Geomorphology. Encyclopedia of Earth Science. Springer, Berlin
Finlayson DP, Montgomery DR (2003) Modeling large-scale fluvial erosion in geographic information systems. Geomorphology 53:147–164
Gaidzik K, Ramírez-Herrera MT (2017) Geomorphic indices and relative tectonic uplift in the Guerrero sector of the Mexican forearc. Geosci Front 8: 885e902. http://dx.doi.org/10.1016/j.gsf.2016.07.006
Garde RJ, Ranga Raju KG (2000) Mechanics of sediment transportation and alluvial stream problems. New Age International Publishers, New Delhi
Ghebreab W, Talbot CJ (2000) Red Sea extension influenced by Pan-African tectonic grain in eastern Eritrea. J Struct Geol 22:931–946
Ghebreab W, Talbot CJ, Page L (2005) Time constraints on exhumation of the East African Orogen from field observations and ^{40}Ar/^{39}Ar cooling ages of low-angle mylonites in Eritrea, NE Africa. Precambr Res 139:20–41
Ghebreab W, Greiling RO, Solomon S (2009) Structural setting of Neoproterozoic mineralization, Asmara district, Eritrea. J Afr Earth Sci 55:219–235
Ghinassi M, Libsekal Y, Papini M, Rook L (2009) Palaeoenvironments of the Buia *Homo* site: high-resolution facies analysis and non-marine sequence stratigraphy in the Alat formation (Pleistocene Dandiero Basin, Danakil depression, Eritrea). Palaeogeogr Palaeoclimatol Palaeoecol 280:415–431
Hack JT (1973) Stream-profile analysis and stream-gradient index. J Res U.S. Geol Surv 1(4):421–429
Hoelzmann P, Gasse F, Dupont LM, Salzmann U, Staubwasser M, Leuschner DC, Sirocko F (2004) Palaeoenvironmental changes in the arid and sub arid belt (Sahara-Sahel-Arabian Peninsula) from 150 kyr to present. In: Batterbee RW, Gasse F, Stickley CE (eds) Past climate variability through Europe and Africa. Springer, Dordrecht, pp 2019–2246
Hovius N, Stark C (2001) Actively meandering bedrock rivers. American Geophysical Union, Fall Meeting, abstract
Johnson KN, Finnegan NJ (2015) A lithologic control on active meandering in bedrock channels. Geologicl Soc Am Bull 127 (11/12):1766–1776
Keller EA, Pinter N (2002) Active tectonics: earthquakes, uplift, and landscape. Prentice Hall, Upper Saddle River, New Jersey
Kirby E, Whipple KX (2001) Quantifying differential rock-uplift rates via stream profile analysis. Geology 29(5):415–418
Knighton D (1998) Fluvial forms and processes: a new perspective. Arnold, London
Langbein W, Leopold LB (1966) River meanders – Theory of minimum variance. US Geol Surv. Professional Paper 422-H, Washington
Lecce S (1997) Nonlinear downstream changes in stream power on Wisconsin's Blue River. Ann Assoc Am Geogr 87(3):471–486
Leopold LB (1994) A view of the river. Harvard University Press, Cambridge
Leopold LB, Wolman MG (1960) River meanders. Geol Soc Am Bull 71(6):769–793
Leopold LB, Langbein WB (1962) The concept of entropy in landscape evolution. USGS Professional Paper 500-A, US Department of the Interior, Washington
Limaye ABS, Lamb MP (2014) Numerical simulation of bedrock valley evolution by meandering rivers with variable bank material. J Geophys Res Earth Surf 119:927–950
Mahmood SA, Gloaguen R (2012) Appraisal of active tectonics in Hindu Kush: Insights from DEM derived geomorphic indices and drainage analysis. Geosci Front 3(4):407–428
Merla G, Abbate E, Azzaroli A, Bruni P, Canuti P, Fazzuoli M, Sagri M, Tacconi P (1979) A geological map of Ethiopia and Somalia. CNR Italy, Firenze
Merritts D, Vincent K (1989) Geomorphic response of coastal streams to low, intermediate, and high rates of uplift, Mendocino triple junction region, northern California. Geol Soc Am Bull 101:1373–1388
Montgomery DR, Buffington JM (1997) Channel-reach morphology in mountain drainage basins. Bull Geol Soc Am 109(5):596–611
Montgomery DR, Gran KB (2001) Downstream variations in the width of bedrock channels. Water Resour Res 37(6):1841–1846
Morris PH, Williams DJ (1997) Exponential longitudinal profiles of streams. Earth Surf Proc Land 22:143–163
Phillips JD, Lutz JD (2008) Profile convexities in bedrock and alluvial streams. Geomorphology 102:554–566
Richards K (1982) Rivers: form and process in Alluvial channels. Methuen, New York
Roberts GG, White N (2010) Estimating uplift rate histories from river profiles using African examples. J Geophys Res 115:B02406
Rosgen DL (1994) A classification of natural rivers. CATENA 22:169–199
Sembroni A, Molin P (2018) Long-term drainage system evolution in the Wabe Shebele River basin (SE Ethiopia—SW Somalia). Geomorphology 320:45–63
Schumm SA (1968) River adjustment to altered hydrologic regimen-Murrumbidgee river and paleochannels, Australia. U.S. Geological Survey Prof. Pap. 598, Washington, DC
Shen HW (1983) Examination of present knowledge of river meandering. In: Elliot CM (ed) River meandering, Proceedings of the conference Rivers-83. American Society of Civil Engineers, New York, pp 1008–1012
Singh VP (2014) Entropy theory in hydraulic engineering: an introduction. ASCE, Reston, Virginia
Snow RS, Slingerland RL (1987) Mathematical modeling of graded river profiles. J Geol 95:15–33
Stark CP, Barbour JR, Hayakawa YS, HattanjiT HN, Chen H, Lin CW, Xu KQ, Fukahata Y (2010) The climatic signature of incised river meanders. Science 327:1–5
Sutfin NA, Shaw J, Wohl EE, Cooper D (2014) A geomorphic classification of ephemeral channels in a mountainous, arid region, southwestern Arizona, USA. Geomorphology 221:164–175
Troiani F, Della Seta M (2008) The use of the stream length-gradient index in morphotectonic analysis of small catchments: a case study from Central Italy. Geomorphology 102:159–168
Williams GP (1986) River meanders and channel size. J Hydrol 88:147–164
Wobus C, Whipple KX, Kirby E, Nyder N, Johnson J, Spyropolou K, Crosby B, Sheehan D (2006) Tectonics from topography: procedures, promise, and pitfalls. In: Willett SD, Hovius N, Brandon MT, Fisher DM (eds) Tectonics, climate, and landscape evolution: geological society of America special paper 398, penrose conference series, pp 55–74
Wohl E (2013) Mountain rivers revisited. American Geophysical Union, Water Resources Monograph Book 19, Washington D.C.
Yalin MS (1971) On the formation of dunes and meanders. In: Proceeding of the 14th world congress international association of hydraulic research, vol. 3, pp. 101–108
Yang CT (1971) On river meanders. J Hydrol 13(3):231–253

5

Pleistocene Landscape Variability Recorded in the Homo-Bearing Dandiero Basin (Eritrea)

Massimiliano Ghinassi, Mauro Papini, Lorenzo Rook, Oriol Oms, and Federico Sani

Abstract

Plio-Pleistocene fluvio-lacustrine continental successions can provide unique insights into link transformations of terrestrial ecosystems with related landscapes in East Africa. Continental deposits of the Dandiero Basin (Eritrean Rift margin) form a 450–500 m sedimentary succession recording paleoecological and landscape transformations that occurred at the Early-Middle Pleistocene transition in the northernmost sector of the African Rift Valley. The Dandiero Basin is bounded by two main, roughly NNW-SSE trending faults, which were superimposed on the N-S trending faults that delimited the basin at early stage of development. The basin-fill succession consists of alternating fluvial and lacustrine deposits. The quality of magnetostratigraphic data and its comparison with radiometric dating and vertebrate paleontology, constrains deposition between ~ 1.2 up to 0.75 Ma. Integrated sedimentological, paleopedological and isotopic data suggest that although sedimentation occurred under the dominant control of tectonics repeated shifts from fluvial to lacustrine deposition were also triggered by climate forcings.

Keywords

Pleistocene • Climate • Fluvio-lacustrine deposition • East Africa • Eritrea

M. Ghinassi (✉)
Dipartimento di Geoscienze, Università di Padova, Padova, Italy
e-mail: massimiliano.ghinassi@unipd.it

M. Papini · L. Rook · F. Sani
Dipartimento di Scienze della Terra, Università di Firenze, Firenze, Italy

O. Oms
Department of Geology, Universitat Autònoma de Barcelona, Bellaterra, Spain

5.1 Introduction

Plio-Pleistocene environmental dynamics played a crucial role in shaping the modern landscapes and in distributing different biota worldwide. In East Africa, ecological and landscape changes, associated with Plio-Pleistocene environmental instability (e.g., de Menocal 2004; Maslin et al. 2014), had considerable impact on mammal dispersal and diversification (Fernández and Vrba 2006; Patterson et al. 2014) and on hominid evolution as well (Potts 2013; Antón et al. 2014). Plio-Pleistocene marine and lacustrine deposits provide continuous, high-resolution proxies, which are widely investigated to understand time and mode of paleoecological transformations. However, fluvio-lacustrine continental successions are no less important and can provide unique insights to link transformations of terrestrial ecosystems with the landscape change (Wolde Gabriel et al. 2005; Aronson et al. 2008). In East Africa, Early to Middle Pleistocene paleoecological and landscape changes are essentially documented in the continental succession of the Olduvai Gorge (Domínguez-Rodrigo et al. 2013) in Tanzania, the Daka Member of the Bouri Formation (Gilbert and Asfaw 2008) in Ethiopia, and the Olorgesailie Formation (Owen et al. 2009) in Kenya.

The Early-Middle Pleistocene continental deposits of the Dandiero Basin (Eritrean Rift margin) form a unique sedimentary succession recording paleoecological and landscape transformation during the Early-Middle Pleistocene transition (Abbate et al. 2004) in the northernmost sector of the Red Sea through of the African Rift Valley system. The fluvio-lacustrine deposits filling the basin contain abundant remnants of fossil mammals and artifacts and, over the past 25 years, provided also several human remains, including a complete cranium. This paper summarizes a number of previous multidisciplinary works (Abbate et al. 2004; Delfino et al. 2004, 2018; Ghinassi et al. 2009, 2013, 2015; Papini et al. 2013; Rook et al. 2010, 2013; Sani et al. 2017; Scarciglia et al. 2018) and aims at outlining the major

P. Billi (ed.), *Landscapes and Landforms of the Horn of Africa*, World Geomorphological Landscapes,
https://doi.org/10.1007/978-3-031-05487-7_5

landscape transformations that have occurred in the Dandiero Basin since its origin during the Middle Pleistocene.

5.2 Geological Settings

The Dandiero Basin is located 35 km south of the Gulf of Zula (Fig. 5.1), along the N-S trending western marginal graben system of the Danakil depression (Abbate et al. 2004). The Danakil depression extends for ca. 300 km in a NNW-SSE direction, and it is bounded by the Eritrean–Ethiopian plateau and by the Danakil block to the west and east, respectively. The alignment of the Quaternary Mahalta-Afrera volcanic centers marks the southern termination of the Danakil depression. The sedimentary infill of the Danakil depression is mainly made of Plio-Pleistocene marine evaporitic deposits, which onlap Palaeozoic-Mesozoic sedimentary units, the Precambrian basement of the Eritrean–Ethiopian plateau and the Danakil horst. Moreover, huge volumes of flood basalts, mainly referred to as the Stratoid Series, were outpoured since the Middle Miocene.

The Dandiero Basin is filled with Early-Middle Pleistocene fluvial and lacustrine deposits (Fig. 5.2), which were mainly fed from the south (Abbate et al. 2004; Ghinassi et al. 2013, 2015). Two major NNW-SSE trending and east-dipping normal faults roughly bound the basin. The westernmost fault delimits the basin from the Eritrean Plateau, whereas the easternmost one marks the boundary between the basin-fill deposits and the Samoti Plain, where fluvio-aeolian sedimentation currently occurs (Fig. 5.2). The age of the basin-fill deposits has been constrained to the Early-Middle Pleistocene transition by means of integrated magnetostratigraphic, geochronological and paleontological data (Abbate et al. 2004; Ghinassi et al. 2015). The sedimentary succession filling the Dandiero Basin (Fig. 5.2) has been referred as the Dandiero Group (Abbate et al. 2004; Ghinassi et al. 2009; Papini et al. 2013), and it has been partitioned into three unconformity-bounded stratigraphic units (Synthems): Maebele Synthem (Early to Middle Pleistocene), Curbelu Synthem (Late Pleistocene) and Samoti Synthem (Late Pleistocene to Holocene). The Maebele Synthem is the focus of the present study and consists of six formations (Abbate et al. 2004; Papini et al. 2013): fluvial Bukra Sand and Gravel (the oldest), fluvio-deltaic Aalat Formation, fluvial Wara Sand and Gravel, lacustrine Goreya Formation, fluvio-deltaic Aro Sand and alluvial Addai Fanglomerate (the youngest).

5.3 The Basin-Fill Succession

5.3.1 Sedimentology and Magnetostratigraphy

The Dandiero Basin infill succession is ca. 450–500 m thick and has been investigated in four main sections: Dandiero River, Aro Mt., Aalat well and Aalat Mt. (Fig. 5.3). The Dandiero river section is located to the south, in the upstream reaches of the basin, whereas the Aalat Mt section is in the downstream portion of the basin (Fig. 5.3). These sections include the same stratigraphic units and have been physically correlated through integration of remote sensing analyses and a detailed ground checking. Correlations across these sections allowed one to depict a basin-scale stratigraphy, along with detecting downstream transformations of depositional environments and landscapes. The Bukra Sand and Gravel Fm. is the lowermost stratigraphic unit of the succession and consists entirely of fluvial deposits, which onlap the metamorphic bedrock. This unit is at least 120 m thick and includes 2–6 m thick channelized sandy and gravelly bodies with overbank mud (Fig. 5.4a). Its uppermost part consists of dominant overbank mud that commonly lacks clear evidence of pedogenic processes. Fluvial deposits of the Bukra Sand and Gravel Fm. are abruptly covered by lacustrine sediments, which contain a major calcareous layer bearing peculiar fossil traces (Fig. 5.4b, c; Abbate et al. 2012). This layer originated in shallow water setting, as pointed out by the local occurrence of desiccation cracks and represents a basin-scale key bed ("Donut key bed" in Abbate et al. 2012; Fig. 5.3) which, in the southern part of the basin, directly overlies the metamorphic bedrock. The donut key bed marks the base of the Aalat Fm., which is ca. 70 m thick, although at the Aalat well site (Fig. 5.3) it thins to 35 m. The Aalat Fm. consists of two major fluvial intervals embedded within deltaic and lacustrine deposits (Fig. 5.4d, e). The fluvial deposits occur in the lower and upper part of the Aalat Fm. and mainly consist of amalgamated, channel-like bodies ranging in thickness between 1 and 3 m, whereas deltaic sediments were mainly accumulated by sandy shoal water deltaic systems, which formed coarsening-upward, compensationally stacked lobate units up to a few meters thick. The uppermost part of these deltaic deposits, that commonly host fluvial distributary channels, locally contains centimeter to decimeter-thick, laminated sandy layers cemented with secondary gypsum. Gilbert-type delta deposits occur only in the central part of the Aalat Fm. and consist of clinostratified sands. The Gilbert-type delta system was sourced from the south and reached a maximum

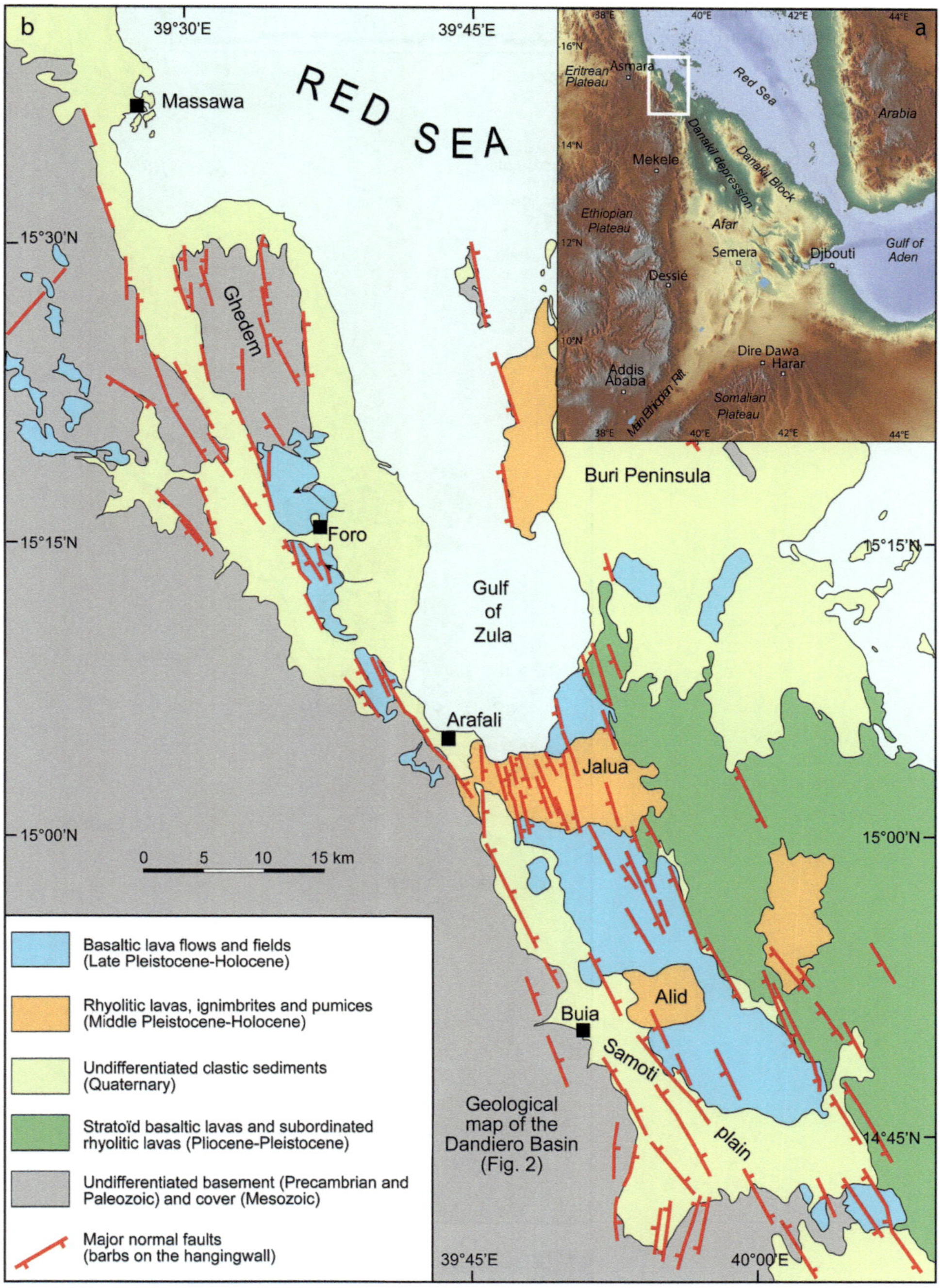

Fig. 5.1 Geographic location of the Dandiero Basin (see Fig. 1 for location)

thickness of ca. 10 m, pointing to a 10 m-deep lake. The lacustrine flooding event, which created accommodation space for the emplacement of this Gilbert-type delta, allowed for the development of a clear stratigraphic surface (i.e., flooding surface) that is the second basin-scale key bed (Fig. 5.3). Lacustrine deposits mainly consist of massive to laminated mud. The major fossiliferous horizon of the basin-fill succession is sited in the fluvial deposits occurring in the upper part of the Aalat Fm. This horizon provided abundant fossil mammal remains and artifacts, along with several human bones, which were mainly accumulated at the base of channelized sand. Fluvial beds in the upper part of the Aalat Fm. show also paleopedological features, consisting of iron-oxide/hydroxide mottles and coatings approximately conformable to bedding and laterally continuous hardpans up to 50 cm thick.

The overlying Wara Sand and Gravel Fm. is entirely made of fluvial deposits but, unlike the Bukra Sand and Gravel Fm., it mainly consists of isolated fluvial channels embedded within alluvial plain deposits (Fig. 5.5a). The thickness of the Wara Sand and Gravel Fm. is ca. 90 m at the Aalat Mt. section, but it increases to 120 m upstream, in the

Fig. 5.2 Geological map of the Dandiero basin (modified after Sani et al. 2017)

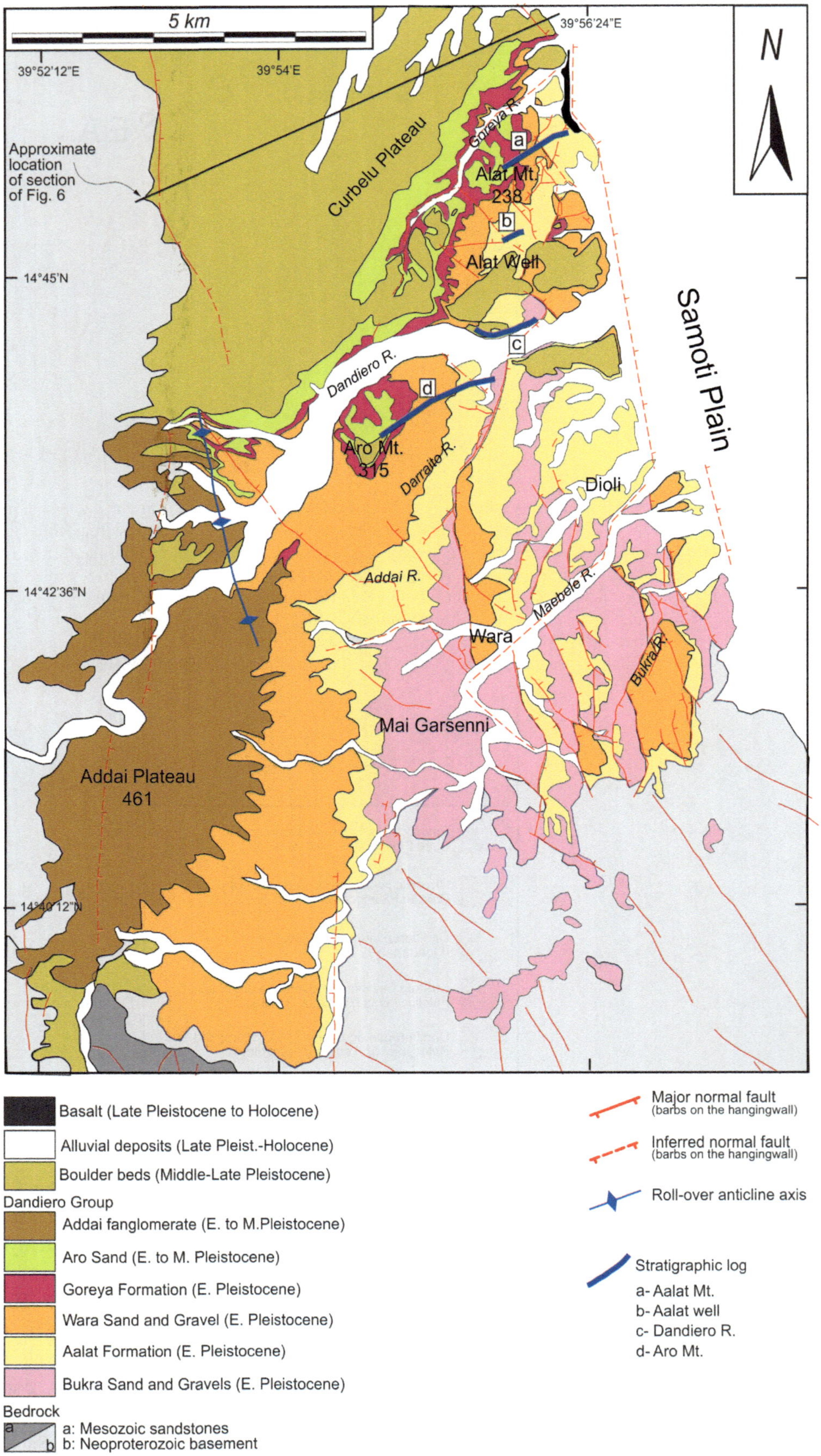

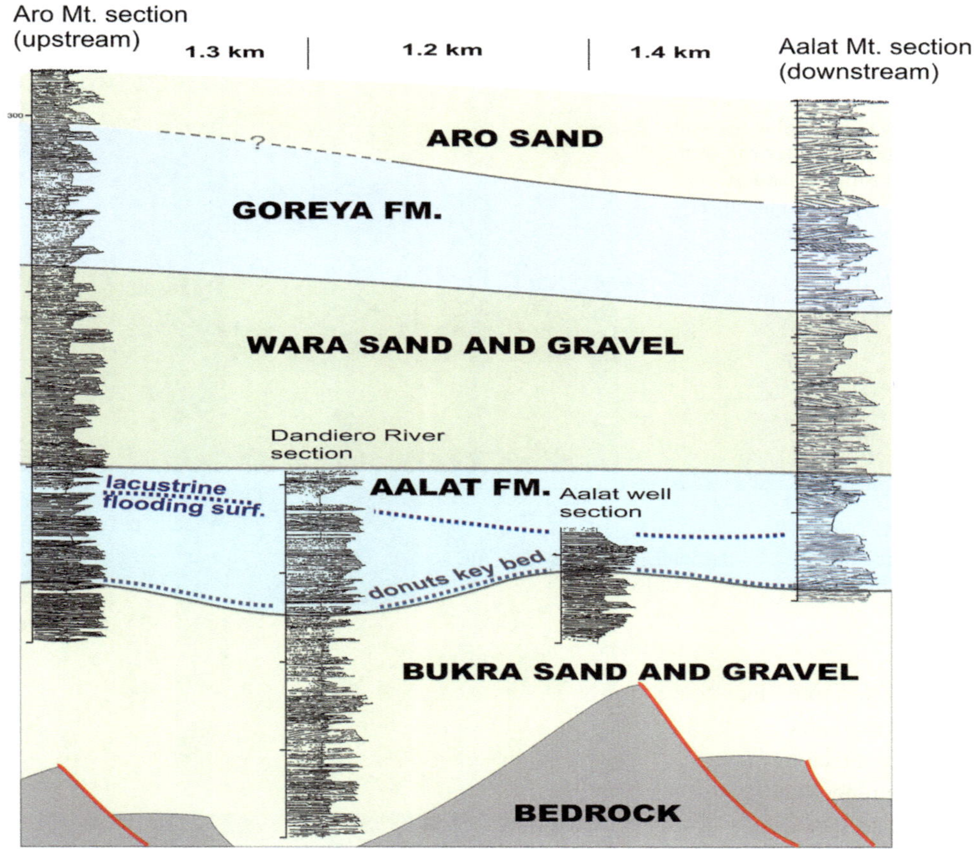

Fig. 5.3 Downdip variability of different formations composing the Dandiero basin-fill succession

Aro Mt. section (Fig. 5.3). Channelized deposits are essentially sandy and tend to increase in thickness from 2–3 m at the base to 7–8 m in the top of the formation. Overbank deposits are mainly made of crevasse sand with subordinate overbank mud, the latter showing local pink to pale brown and reddish yellow iron-oxyhydroxide mottles and sub-vertical to sub-horizontal carbonate-filled veins.

The overlying Goreya Fm. is about 65 m thick and consists of fluvial and lacustrine deposits, though fluvial sediment is dominant in the upstream reaches of the system (Aro Mt. section). In the distal part of the basin, two major lacustrine intervals are separated by a fluvial unit (Fig. 5.5a). The lacustrine deposits are essentially muddy and are characterized by a high content of carbonates and the presence of freshwater shells (Fig. 5.5b). The fluvial sediments are sandy and form amalgamated channelized bodies up to 7 m thick, which, in their upper part, show yellowish iron-oxyhydroxide mottles.

The Goreya Fm. is covered by the fluvial Aro sand (Fig. 5.5c), which thins northward from 80 to 60 m (Fig. 5.3). The Aro sand consists of channelized sand and overbank sandy mud occurring in a similar proportion. Channelized sand deposits form 6–8 m thick bodies, whereas overbank deposits are made of crevasse sand with subordinate overbank mud and levee materials. Overbank mud also includes secondary carbonate coatings on aggregate surfaces and small nodules.

In the Aalat Mt. area, paleoflow measurements document an overall provenance from the south, whereas close to the Aro Mt. zone flows were mainly sourced from the west (Fig. 5.6), following the gradual transition of the deposits embedded between the Bukra Sand and Gravel Fm. and the Aro Sand into the conglomerates of the Addai Fm. These conglomerates are a few hundred meters thick and consist of poorly organized, channelized gravel emplaced in an alluvial fan setting.

The Dandiero basin-fill succession is erosively overlain by alluvial fan gravels, which contain clasts of ignimbrites emitted by the Alid volcano (Duffield et al. 1997) in their uppermost part.

Detailed magnetostratigraphic analyses through the basin-fill succession were carried out (Ghinassi et al. 2015; Sani et al. 2017), and the results were correlated with the polarity standard scale (Fig. 5.7), observing the constraints provided by fission-track radiometric ages (Bigazzi et al. 2004) and the general age attribution of mammal biochronology (Delfino et al. 2004, 2018; Ferretti et al. 2003; Martínez-Navarro et al. 2004, 2010; Medin et al. 2015; Rook et al. 2010, 2013). Stratigraphic key horizons of the Aalat Fm., represented by the lacustrine flooding surface

Fig. 5.4 Basin-fill deposits—the lower part of the succession. **a** Fluvial facies of the Bukra Sand and Gravel Fm. **b**, **c** Top surface of the donut key bed at the Aalat Oasi section. **d** Lower part of the fluvio-lacustrine Aalat Fm. **e** Fluvial deposits in the upper part of the Aalat Fm

Fig. 5.5 Basin-fill deposits—the upper part of the succession. **a** Fluvial deposits of the Wara Sand and Gravel Fm. covered by fluvio-lacustrine deposits of the Goreya Fm. **b** Shell-rich lacustrine carbonates of the Goreya Fm. **c** Uppermost part of the Goreya Fm and the overlying fluvial Aro Sand

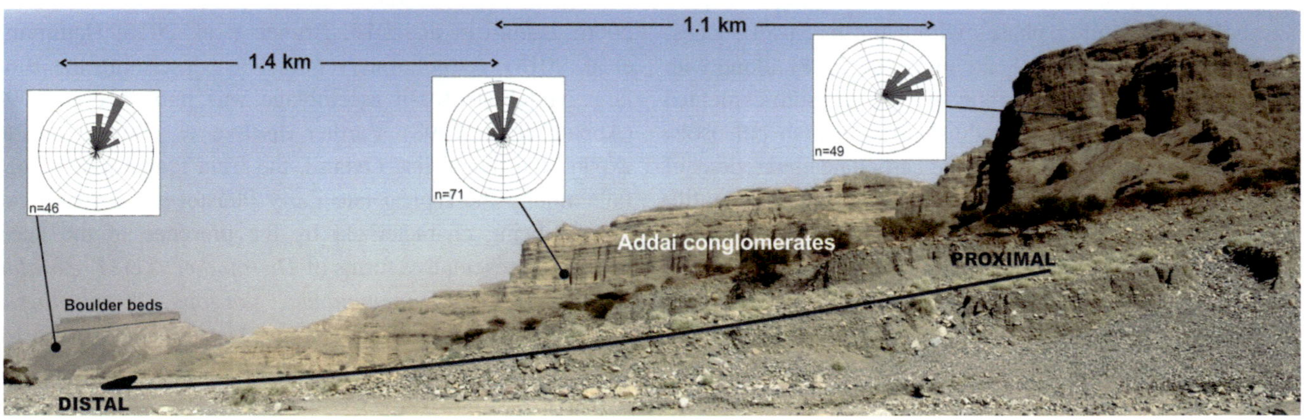

Fig. 5.6 Addai Conglomerates forming the upstream part of the fluvio-lacustrine succession. Note progressive northward shift of flow moving from the proximal to the distal part of the fan

associated with the Gilbert-type delta deposits and the "donut key bed" provide a fully consistent correlation scheme throughout the whole basin. Magnetostratigraphic analyses place the Mathuyama–Brunhes transition at the base of the Aro Sand, and the Jaramillo subchron in correspondence with the Aalat Fm., with its upper boundary (i.e., 0.988 Ma) sited at the stratigraphic level that provided hominin and artifact remains. These results allow one also to establish that the Aro and Aalat sections were characterized by sedimentation rates of ca 0.95 and 0.70 mm/yr,

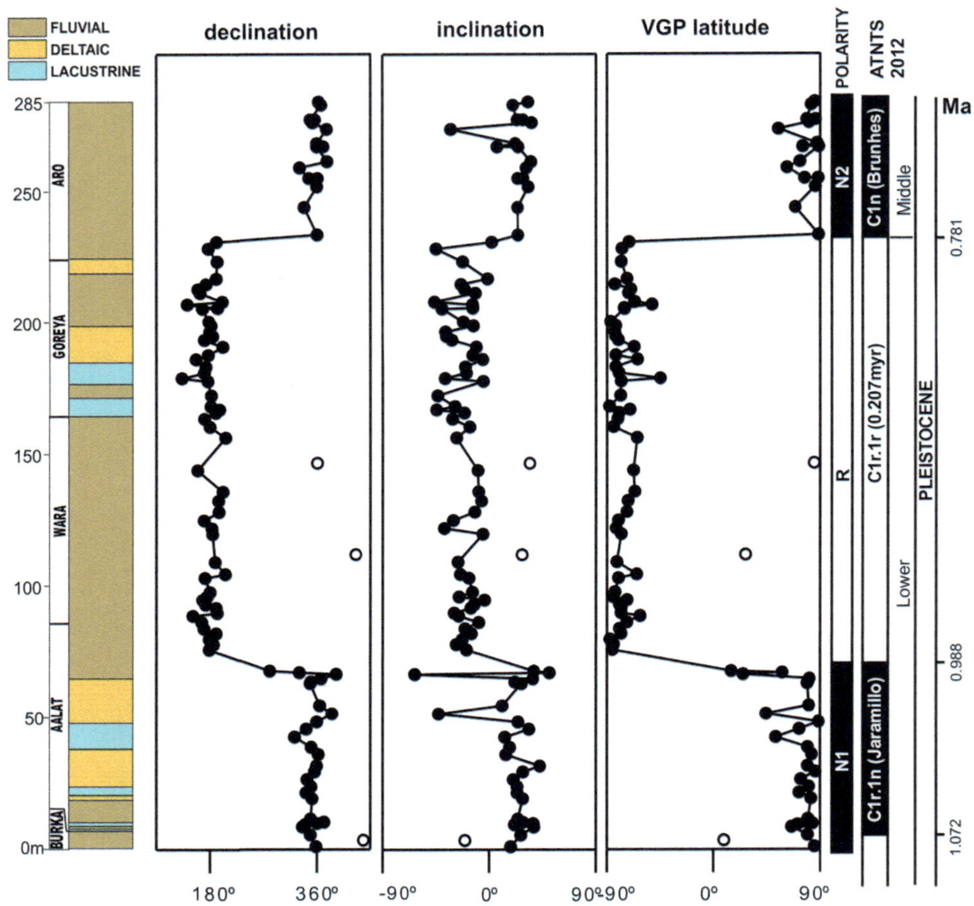

Fig. 5.7 Schematic log of the Aalat section and results of paleomagnetic analyses. N and R indicate normal and reverse polarity zones, respectively. Chrons 1r.1n and 1r.1r lasted 0.084 and 0.207 Myr, respectively

respectively. Specifically, fluvio-lacustrine deposits of the Aalat Fm. probably represent the most expanded continental sedimentary record of the Jaramillo subchron up to date.

Paleopedological investigations also highlighted geomorphological stability phases within the basin-fill succession (Ghinassi et al. 2015; Scarciglia et al. 2018), along with cyclical alternation of carbonate- or gypsum-cemented horizons with iron-stained, reddish to yellowish paleosols. Stable isotope data from pedogenic carbonate concretions of the upper Aalat Fm. and Wara Sand and Gravel Fm. identify four main negative shifts of D18O (Scarciglia et al. 2018), which occurred where wetter conditions are documented by sedimentary facies (e.g., increase in depth of fluvial channels; onset of lacustrine conditions).

5.3.2 Mammal Remains

A number of paleontological studies have documented the occurrence of a faunal assemblage in the Aalat Fm. along the Dandiero succession, including a variety of reptiles and large mammals and, to a lesser extent, of fish, birds and small mammals (Ferretti et al. 2003; Delfino et al. 2004, 2018; Martínez-Navarro et al. 2004, 2010; Rook et al. 2010, 2013; Medin et al. 2015). The assemblage in known in the literature as the Buia faunal assemblage, and it is world renown for the occurrence of well-preserved *Homo erectus* remains (Abbate et al. 1998; Macchiarelli et al. 2004; Bondioli et al. 2006; Zanolli et al. 2014; Bruner et al. 2016; Hammond et al. 2018). A preliminary faunal list describing the Buia large vertebrate fossil assemblage was published in 1998 (Abbate et al. 1998). Further discoveries allowed one to expand the faunal list. Overall, the Buia faunal assemblage falls among the typical late Early Pleistocene East African faunas, being characterized by the presence of the latest derived representative forms of *Theropithecus* cf. *T. oswaldi*, *Elephas recki*, *Hippopotamus gorgops*, *Kolpochoerus olduvaiensis*, *Kolpochoerus majus*, *Metridiochoerus* aff. M. *modestus* and *Bos buiaensis*, in association with taxa still living today, such as *Ceratotherium simum*, *Equus* cf. *E. grevyi*, *Tragelaphus* cf. *T. spekei* and *Kobus* cf. *K. ellipsiprymnus*. The indications obtained through mammal biochronology point to the final phases of the Early Pleistocene and such an evidence is in agreement with the geological and sedimentological framework, paleomagnetism and fission-track thermochrnology (Ghinassi et al. 2015). Taxa characterized by strong water dependence, such as hippopotamus, waterbuck, sitatunga, Kolpochoerus, a

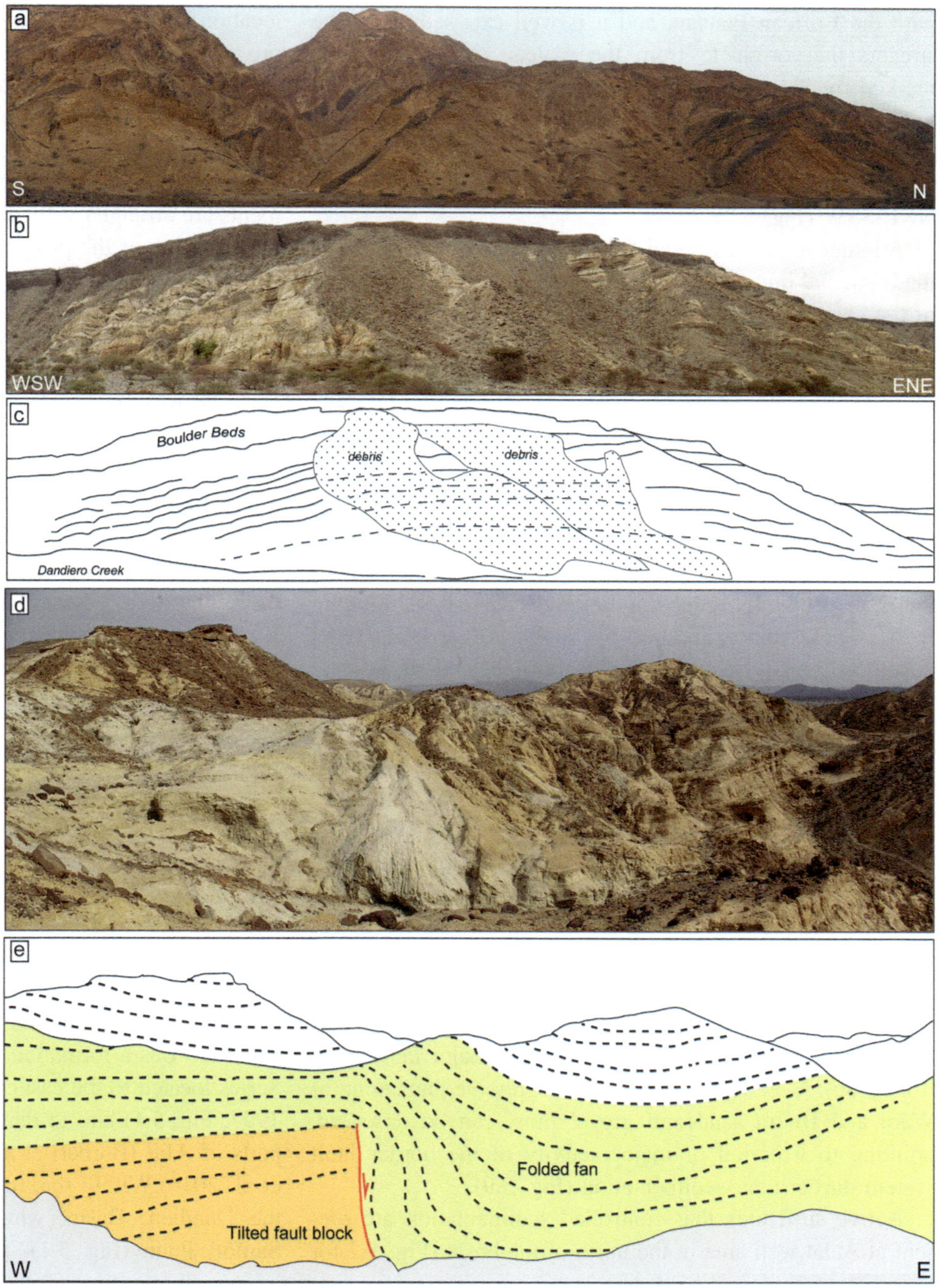

Fig. 5.8 Tectonic evidence along the basin margin and within the infill succession. **a** Basaltic dikes cutting bedrock strata and striking parallel to major faults bounding the western margin of the basin. **b, c** Roll-over anticline in the Addai Conglomerates along the western margin of the basin. **d, e** Panoramic view of Aalat Well area showing the progressive tilting of a faulted block that generates progressive unconformities and growth folding

Thryonomyid rodent, crocodile, pelomedusid chelonian, Nile monitor lizard and African rock python predominate in this vertebrate assemblage (Martínez-Navarro et al. 2004; Ghinassi et al. 2009). The characteristics derived from the faunal assemblage allow one to reconstruct a paleoenvironment characterized by moist, grassy habitats adjacent to persistent water bodies. The fossil bird taxa identified at Buia (including rare specimens of kori bustard, *Ardeotis kori*, the largest living flying bird native to Africa) suggest the occurrence of both open water with reedbeds and dry areas nearby, with local or distant presence of well-developed savannah with stable grass and scattered trees (Rook et al. 2013; Delfino et al. 2018), i.e., a much more humid environment compared to the arid desert of today, devoid of almost any kind of vegetation but scattered weeds and bush in the dry beds of ephemeral streams.

5.4 Structural Geology

Sedimentation in the Dandiero Basin has been mostly controlled and conditioned by two major NNW-SSE trending, east-dipping normal faults that roughly delimit the basin. The westernmost major fault system marks the boundary with the Eritrean Plateau, and it is well exposed along the streams that originate from the plateau. Moreover, it is marked by the occurrence of many basaltic dykes aligned with the major fault planes, evidencing the depth of these structures (Fig. 5.5a). This system is composed of various fault segments, whose trends vary from NNW-SSE to NNE-SSW (Fig. 5.2).

Another regional normal fault marks the eastern limit of the basin and links the basin sediments with the Samoti Plain of the Danakil Depression (Fig. 5.2). This latter fault is not exposed, but the abrupt and straight contact between the sediments of the Dandiero Basin and the alluvial deposits of the Samoti Plain, as well as the occurrence of basaltic dykes aligned for more than 1 km along this margin suggest the presence of an important structure at depth.

Many other minor faults, generally oriented NNE–SSW to NNW-SSE, affect the basin infill (Fig. 5.2). They dip either to the east or to the west and are therefore synthetic or antithetic, respectively, in relation to the two major east-dipping faults. These minor, mesoscopic faults are generally well developed, easily recognizable in the field and show clear kinematic indicators that allowed Sani et al. (2017) to make a complete structural analysis. This intense tectonic activity, expressed by widespread faulting that affected the basin fill, mainly occurred during sedimentation. This is evidenced by throws of faults that reduce up-section, faults sealed by younger sediments, and soft sediment deformations like disaggregation bands (Fossen et al. 2007; Fossen 2010).

However, the best example of interaction between tectonics and sedimentation is the major fault system located along the western margin of the basin. These normal faults show roll-over anticlines with thick wedges of coarse sediments (mainly conglomerates) toward the major fault plane (Fig. 5.8b). Moreover, a westward progressive thickening of Wara and Bukra sand and gravel Fms. can be observed, pointing to a period of major activity of the border fault system during their sedimentation (Fig. 5.9).

Active structures that controlled sedimentation are present at Aalat well area in the north of the basin (Fig. 5.2 for location). Here, an east-dipping normal fault delimiting a progressively tilted block was active during sedimentation creating the accommodation space for a folded fan of fine sediments in the lowered block (Fig. 5.8c). Above the tilted block, the thickness of the succession is strongly reduced in comparison with the adjacent areas. The oldest tilted sediments are arranged vertically, and they flatten upwards, as a consequence of the progressive deactivation of the fault.

5.5 Discussion: Evolution of the Basin and Forcings on Landscape Transformations

Integration between different proxies, along with the definition of an accurate chronostratigraphic frame, provides insights to discuss the factors controlling the shifts between depositional environments and to make comparisons with the evolution of adjacent areas (Fig. 5.10).

The onset of sedimentation in the Dandiero Basin cannot be established with accuracy, although sedimentation rates estimated for the Aro and Aalat sections can be used to extrapolate back the age of the base of the 130 m thick Bukra Sand and Gravel Fm. at ca 1.2 Ma (Sani et al. 2017). Accumulation of the Bukra Sand and Gravel deposits indicate a well-established fluvial drainage that flowed to the north along the axis of a N-S trending graben, the eastern margin of which is still visible. This basin probably represents the remnant of a marginal graben, which is roughly aligned with other similar basins formed to the south along the Eritrean and Ethiopian margins of the plateau (Zwaan et al. 2020). The inferred 1.2 Ma for the beginning of the Dandiero Basin formation is close to the age of the Stratoid Series located to the east of the Alid volcano (Duffield et al. 1997, Fig. 5.1) and to that of other Stratoid Series in other parts of Afar (Barberi et al. 1972, Zumbo et al. 1995; Stab et al. 2016).Within this framework, the eastern shoulder of the Dandiero Basin, which recently subsided below the Samoti Plain (Fig. 5.1), probably prevented basaltic lava

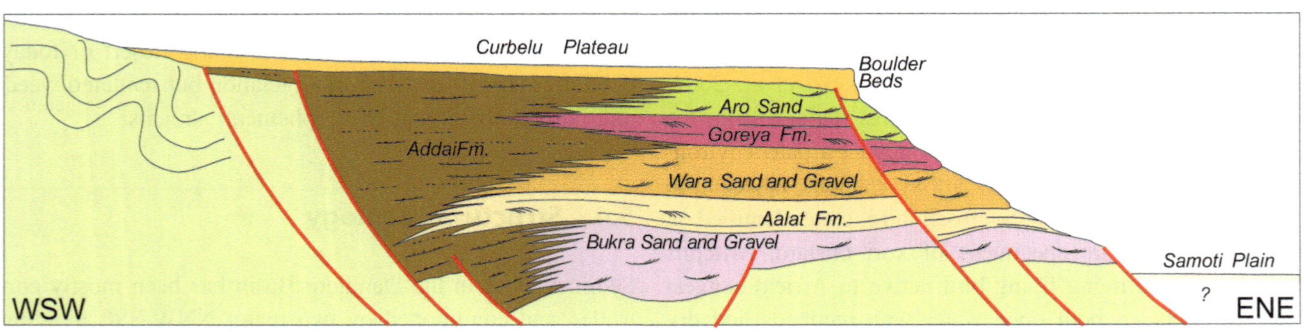

Fig. 5.9 Schematic cross-section through the Dandiero Basin (not to scale). Length of the section is ca 8 km. Thickness of the basin infill is ca. 450–500 m

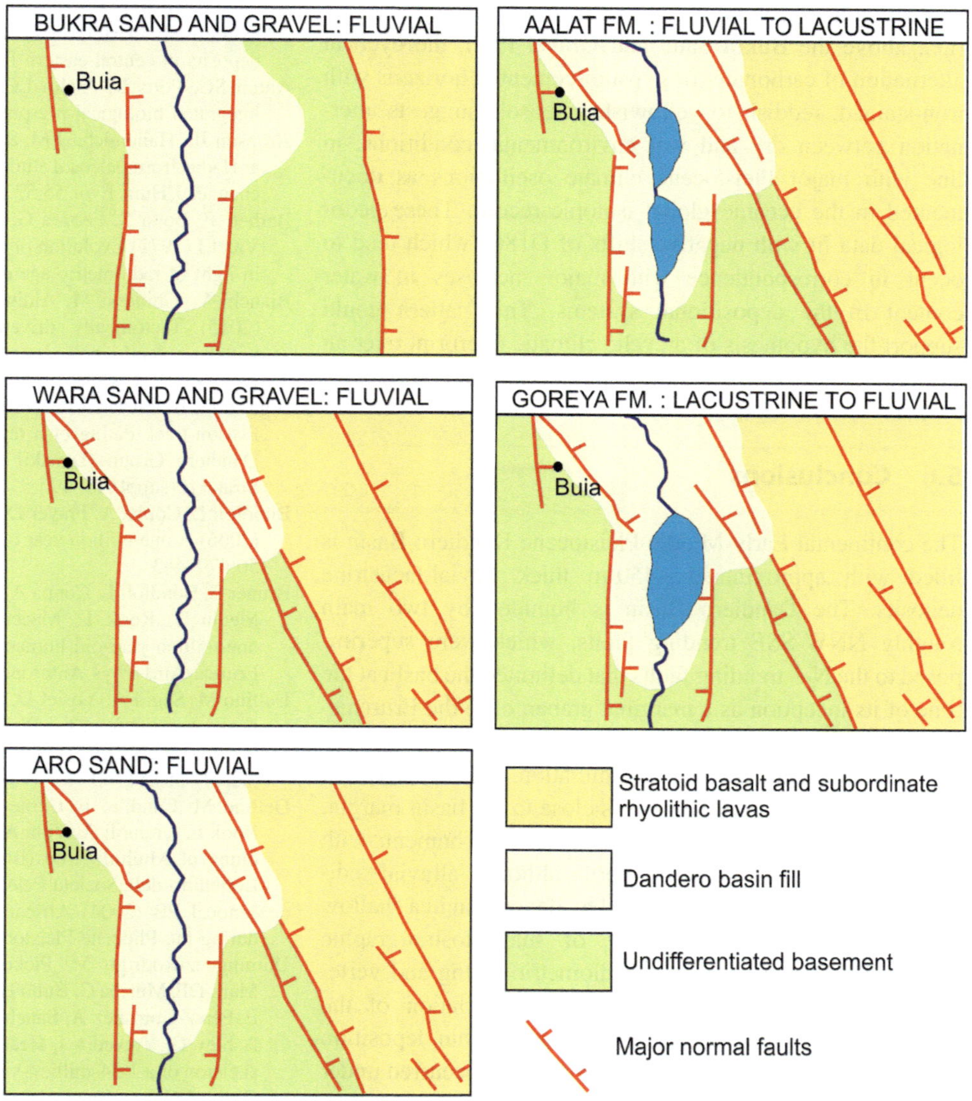

Fig. 5.10 Chronological scheme of the evolution of the Dandiero Basin within two main extensional tectonic phases. The early phase occurred during deposition of the Burka Sand and Gravel, whereas the second extensional phase started at ca. 1 ma with accumulation of the Aalat Fm

flows to enter the basin, where fluvial deposits of the Bukra Sand and Gravel accumulated. The onset of Aalat Fm. deposition marked a new step in the basin evolution and corresponds to the nucleation and growth of NNW-SSE trending normal faults, which controlled sedimentation of all the units overlaying the Bukra Sand and Gravel. These NNW-SSE faults are consistent with the northern propagation of the Afar branch of the triple junction and are superimposed on the previously formed marginal N-S trending graben, which was dissected and lowered below the modern sediments of the Samoti Plain. Activation of these faults is dated at ca. 1 Ma (i.e., base of Jaramillo subchron), when sediments were sourced from the escarpment to form the Addai alluvial fan and deflected toward the north, along the axis of the basin. This basin configuration persisted until ca. 0.7–0.6 Ma (i.e., lower part of the Brunhes Chron), when, after deposition of the fluvial Aro sand, the Dandiero Basin was affected by widespread erosion due to lowering of the Samoti Plain and uplift of the fault block corresponding to the basin.

Although a definite structural basin configuration persisted until the onset of this erosional phase, a marked tectonic control on sedimentation is clearly documented during deposition of the Aalat Fm. by the common occurrence of syn-sedimentary faults and growth structures. These localized deformations caused considerable changes in thickness along the basin axis (Fig. 5.3) by developing localized fluvial base levels (cf. Bianchi et al. 2015), but also induced a remarkable downstream variability of depositional landscapes, which varied from alluvial to lacustrine in poorly and highly subsiding zones, respectively. The lack of relevant deformations in deposits accumulated above the Aalat Fm. demonstrates that no significant localized tectonic activities influenced deposition. In the deposits accumulated after the nucleation and growth of NNW-SSE trending normal faults (i.e., above the Bukra Sand and Gravel Fm.), the cyclical alternation of carbonate- or gypsum-cemented horizons with iron-stained, reddish to yellowish paleosols suggests alternation between dry and wet environmental conditions, in line with major Pleistocene climate oscillations as documented in the oceanic global isotopic record. These pedological data fit with negative shifts of D18O, which tend to occur in correspondence with major increases in water content in the depositional systems. This pattern could support the hypothesis of a cyclic climatic overprint over an overall tectonic-driven sedimentation (Ghinassi et al. 2016).

5.6 Conclusions

The continental Early-Middle Pleistocene Dandiero Basin is filled with approximately 450 m thick fluvial-lacustrine deposits. The Dandiero Basin is bounded by two main roughly NNW-SSE trending faults, which were superimposed to the N-S trending faults that delimited the basin at the time of its inception as a marginal graben of to the Eritrean–Ethiopian plateau. Sedimentation occurred mainly along the basin axis and allowed for accumulation of sand and mud deposits, with subordinate gravels close to the basin margin. Deposition mainly occurred in a fluvial environment, with paleoflow directed toward the north, although alluvial sedimentation was repeatedly replaced by deposition in a shallow lacustrine setting. The quality of magnetostratigraphic data, and its comparison with radiometric dating and vertebrate paleontology, provides a robust correlation of the recorded paleomagnetism and allows to constrain deposition between $\sim$1.2 up to 0.75 Ma. Sedimentation occurred under the dominant control of tectonics, although integrated sedimentological, paleopedological and isotopic data suggest climate forcing on repeated shifts from fluvial to lacustrine deposition. The basin-filling facies association, the faunal assemblage and the paleosol characteristics indicate climatic oscillations between dry and wet environmental conditions, similarly to the major Pleistocene climate alternations documented in the oceanic global isotopic record.

References

Abbate E, Albianelli A, Azzaroli A, Benvenuti M, Tesfamariam B, Bruni P, Cipriani N, Clarke RJ, Ficcarelli G, Macchiarelli R, Napoleone G, Papini M, Rook L, Sagri M, Medhin Tecle T, Torre D, Villa I (1998) A one-million-year-old Homo cranium from the Danakil (Afar) depression of Eritrea. Nature 393:458–460

Abbate E, Beraky W, Bruni P, Falorni P, Papini M, Sagri M, Simret G, Tewolde MT (2004) Geology of the homo-bearing Pleistocene Dandiero Basin (Buia Region, Eritrean Danakil Depression). Rivista Italiana di Paleontologia e Stratigrafia 110:5–34 (supplement)

Abbate E, Bruni P, Landucci F, Pellicanò G (2012) Unusual ichnofossils in Homo erectus-bearing beds of the Pleistocene lake deposits in central-eastern Eritrea, East Africa. Palaios 27:97–104

Antón SC, Potts R, Aiello LC (2014) Evolution of early Homo: an integrated biological perspective. Science 345:1236828–1236913

Aronson JL, Hailemichael M, Savin SM (2008) Hominid environments at Hadar from paleosol studies in a framework of Ethiopian climate change. J Hum Evol 55:532–550

Barberi F, Borsi S, Ferrara G, Marinelli G, Santacroce R, Tazieff H, Varet J (1972) Evolution of the Danakil Depression (Afar, Ethiopia) in light of radiometric age determinations. J Geol 80:720–729

Bianchi V, Ghinassi M, Aldinucci M, Boaga J, Brogi A, Deiana R (2015) Tectonically driven deposition andlandscape evolution within upland incised valleys: Ambra Valley fill, Pliocene–Pleistocene, Tuscany, Italy.Sedimentology 62:897-927

Bigazzi G, Balestrieri ML, Norelli P, Oddone M, Tewolde MT (2004) Fission-track dating of a tephra layer in the Alat formation of the Dandiero Group (Danakil Depression, Eritrea). Riv Ital Paleont Strat 110(suppl):45–49

Bondioli L, Coppa A, Frayer DW, Libsekal Y, Rook L, Macchiarelli R (2006) A one million year old human pubic symphysis. J Hum Evol 50:479–483

Bruner E, Bondioli L, Coppa A, Frayer DW, Holloway RL, Libsekal Y, Medin T, Rook L, Macchiarelli R (2016)The endocast of the one-million-year-old human cranium from Buia (UA 31), Danakil Eritrea. AmJ Phys Anthropol 160:458–468

Delfino M, Segid A, Yosief D, Shoshani J, Rook L, Libsekal Y (2004) Fossil reptiles from the Pleistocene Homo-bearing locality of Buia (Eritrea, Northern Danakil Depression). Riv It Paleont Strat 110 (suppl) 110:51–60

Delfino M, Candilio F, Carnevale G, Coppa A, Medin T, Pavia M, Rook L, Urciuoli A, Villa A (2018) The earlyPleistocene vertebrate fauna of Mulhuli-Amo (Buia area, Danakil Depression, Eritrea). Bollettino dellaSocietà Paleontologica Italiana 57:27–44

deMenocal PB (2004) African climate change and faunal evolution during the Pliocene-Pleistocene. Earth Plan Sci Letters 220:3–24

Domínguez-Rodrigo M, Pickering TR, Baquedano E, Mabulla A, Mark DF, Musiba C, Bunn HT, Uribelarrea D, Smith V, Diez-Martin F, Pérez-Gonzalez A, Sanchez P, Santonja M, Barboni D, Gidna A, Ashley G, Yravedra J, Heaton JL, Arriaza MC (2013) First partial skeleton of a 1.34-million-year-old Paranthropus boisei from Bed II, Olduvai Gorge, Tanzania. PloS One 8(12):e80347

Duffield WA, Bullen TD, Clynne MA, Fournier RO, Janik, Kahsai G, Mariam K, Tesfai T (1997) Geothermal Potential of the Alid Volcanic Center, Danakil Depression, Eritrea. US Geological Survey Open File Report 97–291, 62pp

Fernández MH, Vrba HS (2006) Plio-Pleistocene climatic change in the Turkana Basin (East Africa): evidence from large mammal faunas. J Hum Evol 50:595–626

Ferretti MP, Ficcarelli G, Libsekal Y, Tecle TM, Rook L (2003) Fossil elephants from Buia (northern Afar Depression, Eritrea) with remarks on the systematics of Elephas recki (Proboscidea, Elephantidae). J Vert Paleont 23:244–257

Fossen H (2010) Deformation bands formed during soft-sediment deformation: observations from SE Utah.Mar Pet Geol 27:215–222

Fossen H, Schulz RA, Shipton ZK, Mair K (2007) Deformation bands in sandstone—a review. Geol Soc Lond 164:755–769

Ghinassi G, Libsekal Y, Papini M, Rook L (2009) Palaeoenvironments of the Buia Homo site: high-resolution facies analysis and non-marine sequence stratigraphy in the Alat formation (Pleistocene Dandiero Basin, Danakil depression, Eritrea). Paleogeogr Paleoclimatol Paleoecol 280:415–431

Ghinassi M, Ielpi A, Aldinucci M, Fustic M (2016) Downstream-migrating fluvial point bars in the rockrecord. Sedim Geol 334: 66–96

Ghinassi M, Billi P, Libsekal Y, Papini M, Rook L (2013) Inferring fluvial morphodynamics and overbank flow control from 3D outcrop sections of a Pleistocene point bar, Dandiero Basin, Eritrea. J Sediment Res 83:1065–1083

Ghinassi M, Oms O, Papini M, Scarciglia F, Carnevale G, Sani F, Rook L, Delfino M, Pavia M, Libsekal Y, Bondioli L, Coppa A, Frayer DW, Macchiarelli R (2015) An integrated study of the Homo-bearing Aalat stratigraphic section (Eritrea): an expanded continental record at the early—Middle Pleistocene transition. J Afr Earth Sci 112:163–185

Gilbert WH, Asfaw B (2008) Homo erectus. Pleistocene evidence from the Middle Awash, Ethiopia. University of California Press, London

Hammond AS, Almécija S, Libsekal Y, Rook L, Macchiarelli R (2018) A partial Homo pelvis from the EarlyPleistocene of Eritrea. J Human Evol 123:109–128

Macchiarelli R, Bondioli L, Chech M, Coppa A, Fiore I, Russom R, Vecchi F, Libsekal Y, Rook L (2004) The late early Pleistocene human remains from Buia, Danakil Depression, Eritrea. Riv It Paleont Strat 110(suppl):133–144

Martínez-Navarro B, Rook L, Segid A, Yosief D, Ferretti MP, Shoshani J, Tecle TM, Libsekal Y (2004) The large fossil mammals from Buia (Eritrea): systematics, biochronology and paleoenvironments. Riv It Paleont Strat 110(suppl):61–88

Martínez-Navarro B, Rook L, Papini M, Libsekal Y (2010) A new species of bull from the Early Pleistocene paleoanthropological site of Buia (Eritrea): parallelism on the dispersal of the genus Bos and the Acheulian culture. Quat Int 212:169–175

Maslin MA, Brierley CM, Milner AM, Shultz S, Trauth MH, Wilson KE (2014) East African climate pulses and early human evolution. Quat Sci Rev 101:1–17

Medin T, Martínez-Navarro BM, Rivals F, Libsekal Y, Rook L (2015) The late early Pleistocene suid remains from the paleoanthropological site of Buia (Eritrea): systematics, biochronology and eco-geographical context. Paleogeogr Paleoclimatol Paleoecol

Owen RB, Renaut RW, Scott JJ, Potts R, Behrensmeyer AK (2009) Wetland sedimentation and associateddiatoms in the Pleistocene Olorgesailie Basin, southern Kenya Rift Valley. Sedim Geol 222:124–137

Papini M, Ghinassi M, Libsekal Y, Rook L (2013) Facies associations of the northern Dandiero Basin (Danakil depression, Eritrea, including the Pleistocene Buia Homo site). J Maps 10:126–135

Patterson DB, Faith JT, Bobe R, Wood B (2014) Regional diversity patterns in African bovids, hyaenids, andfelids during the past 3 million years: the role of taphonomic bias and implications for the evolution ofParanthropus. Quat Sci Rev 96:9–22

Potts R (2013) Hominin evolution in settings of strong environmental variability. Quat Sci Rev 73:1–13

Rook L, Ghinassi M, Libsekal Y, Martínez-Navarro B, Medin T, Papini M (2010) Stratigraphic context and taxonomic assessment of the large cercopithecoid (Primates, Mammalia) from the late Early Pleistocene palaeoanthropological site of Buia (Eritrea). Hum Evol 59:692–697

Rook L, Ghinassi M, Carnevale G, Delfino M, Pavia M, Bondioli L, Candilio F, Coppa A, Martínez-Navarro B, Medin T, Papini M, Zanolli C, Libsekal Y (2013) Stratigraphic context and paleoenvironmental significance of minor taxa (Pisces, Reptilia, Aves, Rodentia) from the late Early Pleistocene palaeoanthropological site of Buia (Eritrea). J Hum Evol 64:83–92

Sani F, Ghinassi M, Papini M, Oms O, Finotello A (2017) Evolution of the northern tip of Afar triangle:inferences from the Quaternary succession of the Dandiero-Massawa area (Eritrea). Tectonophysics 717:339–357

Scarciglia F, Mercatante G, Fondevilla V, Anadón P, Oms O, Donato P, Agnini C, Papini M, Rook L, GhinassiM (2018) Pleistocene paleosol development and paleoenvironmental dynamics in East Africa: amultiproxy record from the Homo-bearing Aalat pedostratigraphic succession, Dandiero basin (Eritrea).Quat Sci Rev 191:275–298

Stab M, Bellahsen N, Pik R, Quidelleur X, Ayalew D, Leroy S (2016) Modes of rifting in magma-rich settings: tectono-magmatic evolution of Central Afar. Tectonics 35:2–38. https://doi.org/10.1002/2015TC003893

WoldeGabriel G, Hart WK, Katoh S, Beyene Y, Suwa G (2005) Correlation of Plio-Pleistocene Tephra in Ethiopian and Kenyan rift basins: temporal calibration of geological features and hominid fossil records. J Volc Geotherm Res 147:81–108

Zanolli C, Bondioli L, Coppa A, Dean CM, Bayle P, Candilio F, Capuani S, Dreossi D, Fiore I, Frayer DW, Libsekal Y, Mancini L, Rook L, Medin TT, Tuniz C, Macchiarelli R (2014) The late early Pleistocene human dental remains from Uadi Aalad and Mulhuli-Amo (Buia), Eritrean Danakil: macromorphology and microstructure. J Hum Evol 74:96–113

Zumbo V, Féraud G, Vellutini P, Piguet P, Vincent J (1995) First 40Ar/39Ar dating on early Pliocene to Plio-Pleistocene magmatic events of the Afar—Republic of Djibouti. J Volcanol Geotherm Res 65:281–295

Zwaan F, Corti G, Sani F, Keir D, Muluneh AA, Illsley-Kemp F, Papini M (2020) Structural analysis of theWestern Afar Margin, East Africa: evidence for multiphase rotational rifting. Tectonics 39 (7):e2019TC006043

Landscape and Landforms of the Samoti Plain, Eritrean Danakil

6

Paolo Billi

Abstract

The Samoti plain is a structural basin and it is located in the northern part of the Danakil depression. The environment is that typical of a hot and dry desert. The landscape has been shaped by the recent tectonic activity and by hydromorphological and aeolian processes. The rivers are ephemeral and have a dry bed for the most of the time. Data on channel morphology and bed material grain size were collected in the field. An abrupt change from a boulder to a prevailing sandy bed was found to occur within a very short distance (about 100 m), and an explanation based on the abrupt decline of shear stress for water infiltration is presented. The prevailing type of wind dunes is the transverse dunes, originated from the coalescing of barchan dunes. The wavelength and the wavelength-to-height ratio of the Samoti plain dunes are smaller and higher, respectively, compared to the dunes of other larger deserts in the world. An explanation based on the limited sediment supply in the windward side of the Samoti plain is discussed. Several elements observed in the basin infilling sediment indicate that the Samoti plain was subjected to a wetter climate around 7–5000 years BP in coincidence with the African Humid Period.

Keywords

Structural basin • Ephemeral streams • Distributary systems • Wind dunes • Dryland

P. Billi (✉)
International Platform for Dryland Research and Education, Tottori University, Tottori, Japan
e-mail: bli@unife.it

6.1 Introduction

The Danakil is a long and narrow depression of structural origin. It is about 300 km long and less than 42 km wide south of the Gulf od Zula, to become wider and wider southward as far as the Afar region. The Eritrean portion of the Danakil is only about 100 km long, but similarly to the southern portion, it is characterized by impressive landscapes and landforms witnessing their origin from erosion and deposition processes that have accompanied the complex tectonic and volcanic events, which have affected this area since the Neogene (Abbate et al. 2004). Though the Eritrean Danakil is separated from the Red Sea by a low mountain range and lines of recent volcanoes, not exceeding 645 and 371 m asl, respectively, the ground in the central part of this structural depression is commonly below sea level (−100 m in the southern portion and −20 m in the Samoti plain, in the northern part). The Eritrean Danakil is one of the hottest and driest inhabited places in the world. In the Samoti plain, the maximum temperatures range from 44 °C, in winter, to over 50 °C, in summer, and annual precipitation is probably less than 150 mm. This chapter reports about the geomorphological landscape of the Samoti plain, and in particular, it deals with the modern fluvial and eolian landforms and processes.

6.2 Study Area

The Samoti plain takes up the bottom of a depositional, structural basin oriented parallel to the Red Sea. It is located within the Eritrean Danakil depression, about 36 km southeast of the Gulf of Zula. The Samoti plain is famous for the finding of one million years old *Homo* erectus bones (Abbate et al. 1998) (Fig. 6.1). The basin is actually divided into two portions by the recent (Pleistocene to Holocene) emplacement of a large extensions of ryholitic and basaltic lava emitted from the volcanoes of the Alid group when a

P. Billi (ed.), *Landscapes and Landforms of the Horn of Africa*, World Geomorphological Landscapes,
https://doi.org/10.1007/978-3-031-05487-7_6

much larger structural basin was already formed. These volcanoes are placed on the eastern master fault that likely originated the larger structural basin, which virtually includes also the Gulf of Zula, stretching for about a length of 100 km. The larger basin is now fragmented in 3–4 portions by transverse faults and by the basalt flows that have reached the center of the basin and in places have reached the Pleistocene deposits on the other side of the basin, of which the latter are an older filling. Here, the recent (probably Holocene—Sani et al. 2017) lava flows of the Alid group have split the Samoti plain into a northern and a southern portion (Figs. 6.1 and 6.2).

A simplified sketch of the geology and stratigraphy of the northern portion of the Danakil depression area is reported in Fig. 6.2; a focus on the main geological characteristics of the Samoti plain is outlined in Figs. 6.3 and 6.4.

Unfortunately, no meteostation is available in the study area. The nearest ones are on the Red Sea coast: Mersa Fatma, 38 km ENE of the study area, Massawa, about 115 km to the NNW (Fig. 6.2) and Assab, which is rather far south (340 km), but it is the only meteostation with some wind data. In a similar inland position, there is Dallol, in Ethiopia (Fig. 6.2), but its climatic data are very old (Pedgley 1967). Nevertheless, also the data of Dallol were considered in order to give a more comprehensive information about the climate of the study area.

The temperatures in the Samoti plain are very high during any season (Fig. 6.5). The hottest temperatures are recorded in summer with mean monthly maximum values close to 40 °C in Massawa and Mersa Fatma e over 45 °C in Dallol, where peaks of 58 °C have been recorded. In the Samoti plain, the author of this paper has measured maximum daily temperatures of 42–44 °C in November 2007. Such hot temperatures are due to the inland position and the occurrence of a coastal range that prevent this area from the refreshing action of the sea, as it is instead observed in the coastal towns of Massawa, Mersa Fatma, and Assab. Also, the mean monthly lower temperatures are relatively high with a lower value around 25 °C in January and the highest values around 31–32 °C in the summer at Dallol. In the other meteostations, the minimum temperatures are on average 5–7 °C lower than in Dallol. In November 2007, in the Samoti plain, temperatures around 30 °C in the night and 25–26 °C at dawn were measured by the author of this paper.

Rainfall is very unusual in Dallol (a few millimeters per year) and in the Samoti plain as well, though the closer vicinity with the basin western shoulder mountains, with peak between 1800 and 2000 m asl, may favor the development of orographic rains reaching also the Samoti plain. The annual precipitation, however, is expected to be less than 100 mm. In Massawa annual rainfall is 186 mm, in Mersa Fatma 238 mm, and in Assab 71 mm. The higher monthly rain occurs at the end of autumn and in winter (Fig. 6.6). In all the weather stations considered in this study, with the exception of Dallol, the average monthly rain is typically less than 10 mm.

The only wind data available are from the meteostation of Assab. These data cover only a few years and are the only information on the wind regime based on field measurements for the study area. Nevertheless, the wind data available are useful to give even a rough an idea of the

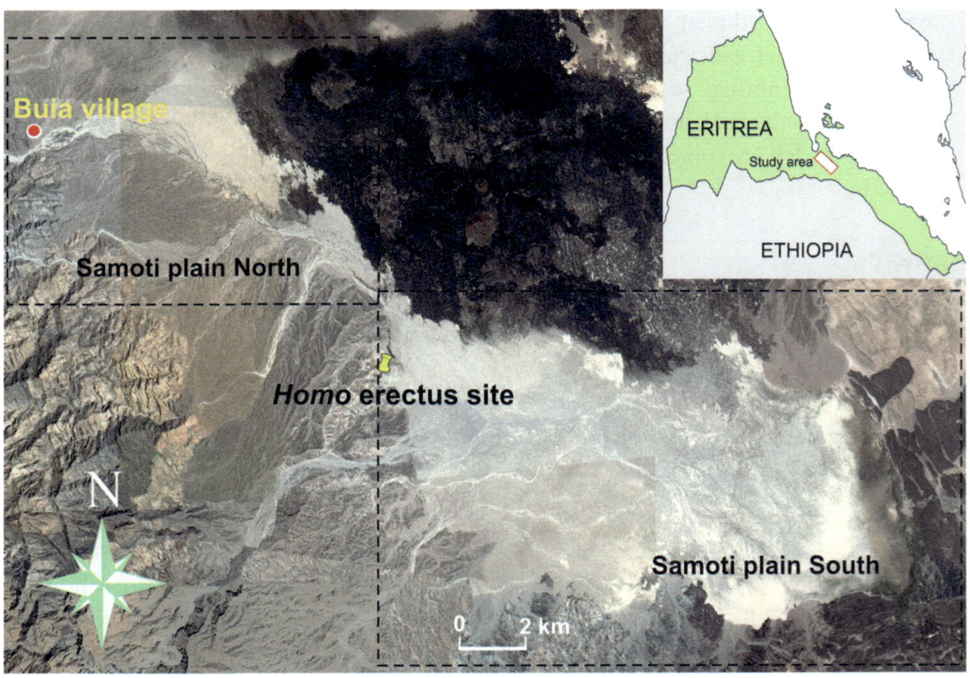

Fig. 6.1 Location map of the Samoti plain. For a more detailed geomorphological map of the northern and southern portions, see Figs. 6.3 and 6.4, respectively. The yellow pin indicates the *Homo* erectus site

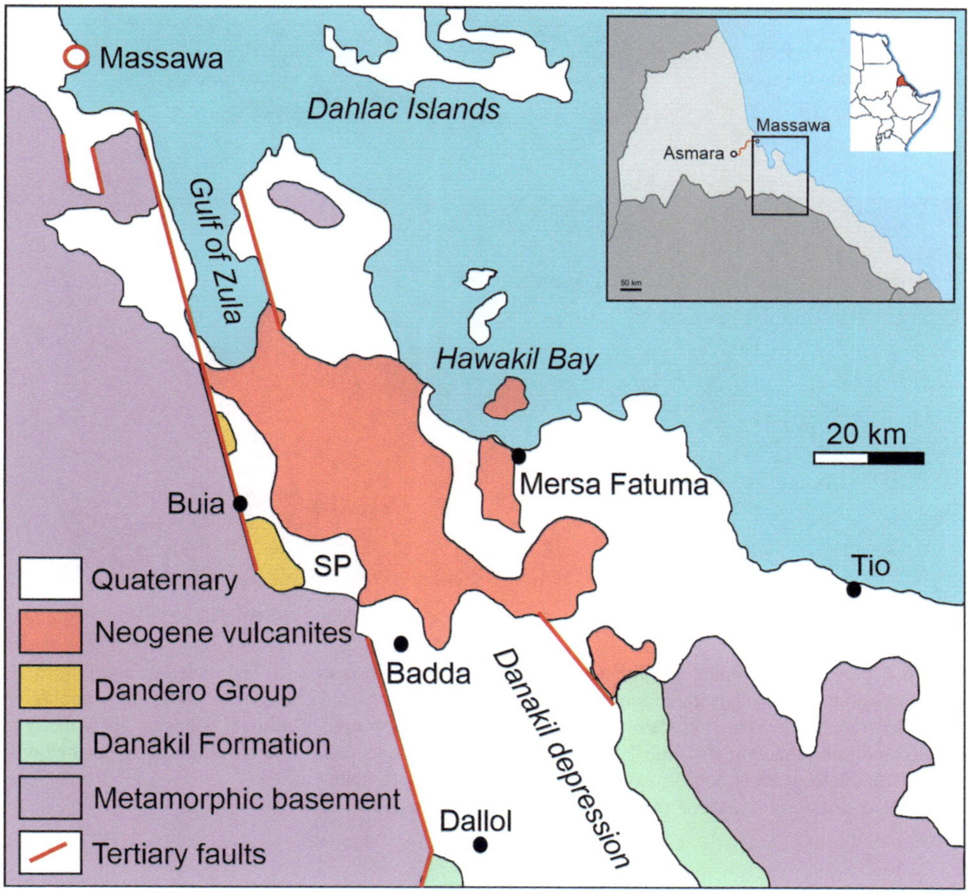

Fig. 6.2 Sketch geological and stratigraphic map of the northern portion of the Danakil depression. SP stands for Samoti plain. The basement rocks include Neoproterozoic and Paleozoic slates, marbles, metadolostones, chloritoschists, sandstones, and limestones; the Danakil formation includes Miocene to Pliocene sandstones and shales; the Dandero group consists of Early to Middle Pleistocene fluvial, fluvio-deltaic and lacustrine to palustrine sand and gravel deposits; the Neogene vulcanites are mainly rhyolitc lavas, ignimbrites, pumices, and basaltic lava flows and fields [simplified and redrawn from Abbate et al. (2004), Ghinassi et al. (2009), and Sani et al. (2017)]

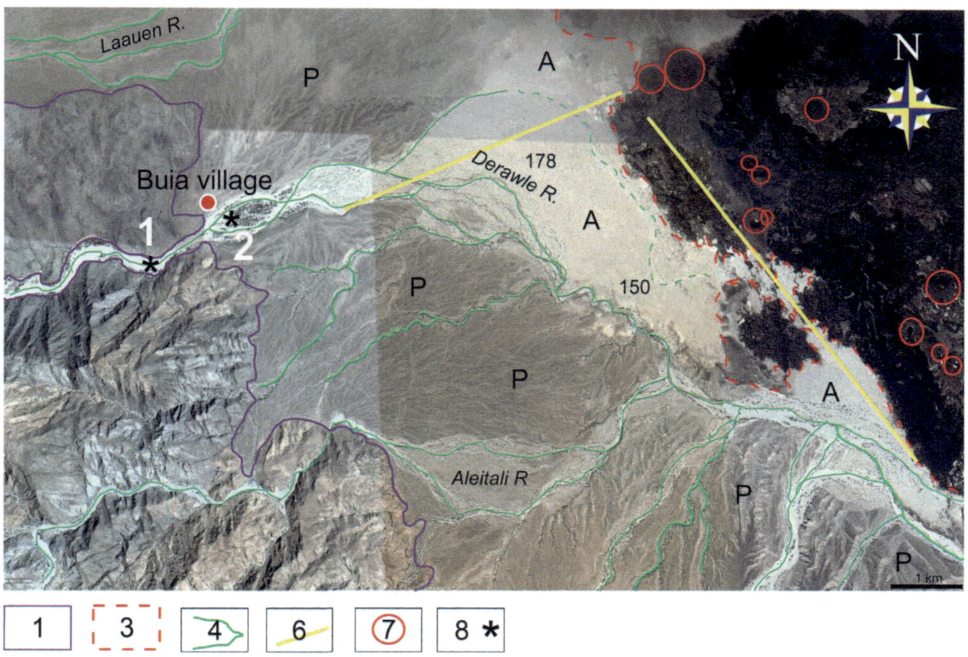

Fig. 6.3 Map of the main geo-lithological and geomorphological elements of the northern portion of the Samoti plain: (1) Neoproterozoic and Paleozoic metamorphic basement; (3) Basaltic lava flows and field (Late Pleistocene-Holocene); (4) ephemeral streams main channels; (6) main faults; (7) volcanic center; (8) bed material sampling site (*) and sample identification number (in white); P = Pleistocene deposits; A = recent alluvial deposits; black numbers indicate the elevation above sea level (modified from Sani et al. 2017)

prevailing wind directions and velocity, which can be considered representative also for the study area. The prevailing winds blow from SE from October to May and in February–March, whereas in November–December, the highest average velocities of 8.5–9.5 ms^{-1} (Fig. 6.7) are recorded. Rosen et al. (1999) compiled a larger number of wind speed

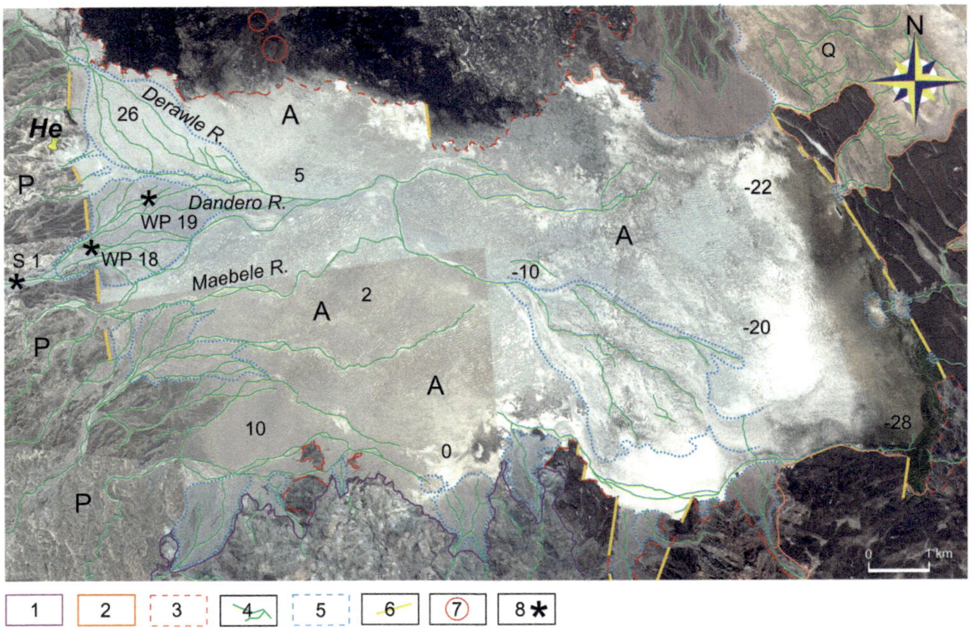

Fig. 6.4 Map of the main geo-lithological and geomorphological elements of the southern portion of the Samoti plain: (1) Neoproterozoic and Paleozoic metamorphic basement; (2) Pliocene–Pleistocene stratoid basalts and subordinately rhyolitic lavas; (3) Basaltic lava flows and field (Late Pleistocene-Holocene); (4) ephemeral streams main channels; (5) distributary systems; (6) main faults; (7) volcanic center; (8) bed material sampling site (*) and identification number; P = Pleistocene deposits; Q = undifferentiated Quaternary deposits; A = recent alluvial deposits; He = the *Homo* erectus site; black numbers indicate the elevation above or below (−) sea level (modified from Sani et al. 2017)

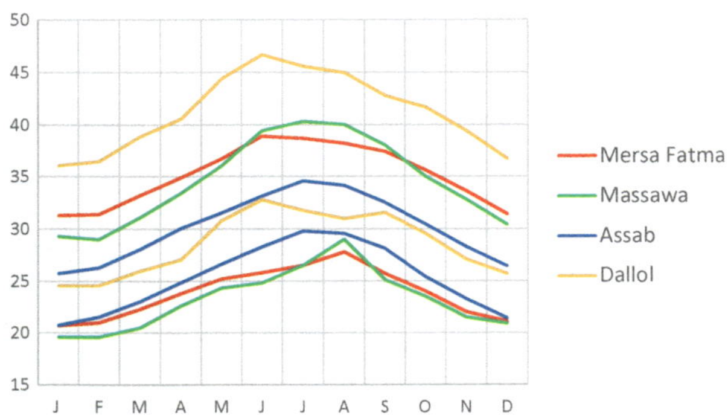

Fig. 6.5 Mean monthly minimum and maximum temperatures measured at selected meteostations near the study area

observation recorded every three hours at the Assab airport and archived by the United States Air Force Environmental Technical Application Center. Their frequency distribution diagram indicates that wind velocities higher than 10 ms^{-1} were recorded but their frequency is less than 5% (Fig. 6.8).

The prevailing wind direction from SE, associated with the highest velocities, is consistent with the Khamsin wind, which is not a seasonal nor or a cyclic wind blowing for intermittent long period across the Sahara. Khamsin is an Arab word that means 50, i.e., the number of consecutive days that, according to the tradition, this wind blows with a certain regularity. Though no wind data in available for the Samoti plain, the local people confirm the occurrence of the

Fig. 6.6 Mean monthly precipitation measured at selected meteostations near the study area

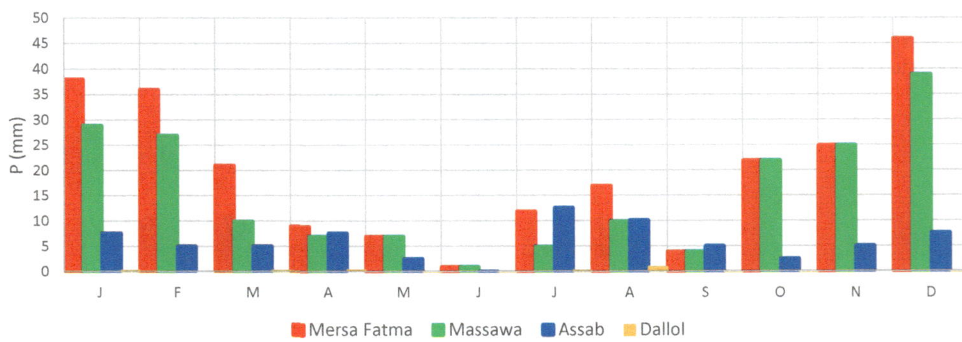

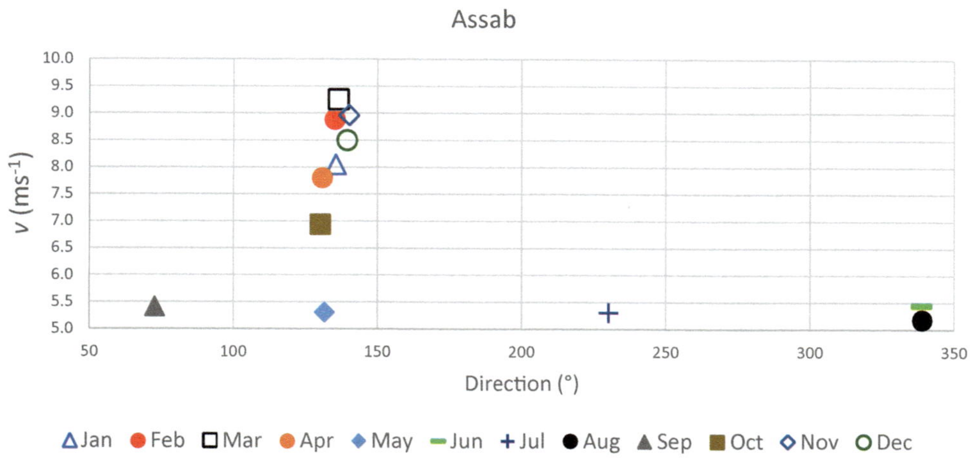

Fig. 6.7 Mean monthly wind velocity measured at Assab

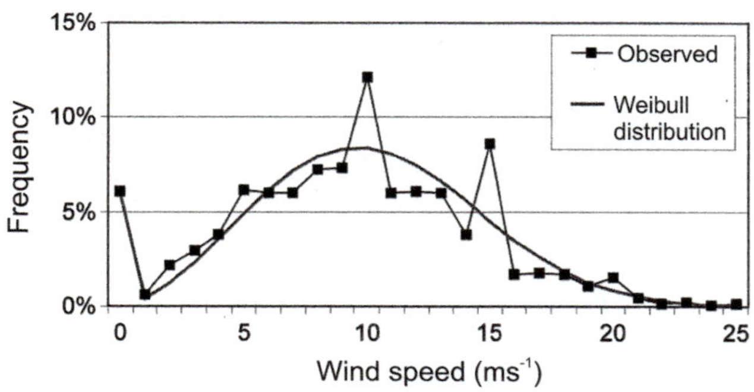

Fig. 6.8 Diagram of wind velocity frequency distribution. Weibull distribution is actually a three-parameter Weibull probability density function (modified from Rosen et al. 1999)

Khamsin, but its blowing duration seems to be shorter, from days to a couple of weeks. This wind is capable to move the fine sediment of the Samoti plain and to form aeolian dunes as it will be illustrated farther on.

Under such hot and dry conditions in any season, a desert environment is not unexpected. The Samoti plain area, in fact, is devoid of vegetation with the exception of sparse bush trees and grass tufts in the dry river beds or in their distributary systems. However, this poor vegetation and the fresh water emerging to the surface as small puddles, especially in areas below sea level, seem to be sufficient for a few desert gazelles (Dorcas) to survive. They are probably the largest wild mammals living today in the Samoti plain.

6.3 Geomorphological Landscapes

The landscape characteristics of the study area are strongly influence mainly by the recent geological events. At larger scale, the Samoti plain is located within the northern portion of the Danakil depression which terminates into the Gulf of Zula graben (Fig. 6.2). This structural basin is separated from the Badda basin and the main portion of the Danakil depression by earlier (Pliocene–Pleistocene) volcanic emissions and since then the sedimentary and tectonic evolution of these two grabens followed distinct patterns. The Gulf of Zula graben is markedly asymmetric, with the western shoulder, coinciding with the main escarpment of the Red Sea rift branch, characterized by high elevations close 2000 m asl, from which most of the modern river systems originate. By contrast, the elevations of the eastern horst of the Zula graben are much lower, commonly less than 600 m asl, and only very few small rivers originate from there, impeded in their course development by the Alid volcanoes and their extensive lava emissions. (Fig. 6.9).

In the Zula graben, the river courses run perpendicular to the basin axis in the mountainous escarpment and then, as they reach the garben bottom, they turn into two different axial directions: The northern rivers flow to the north, reaching the Red Sea; the southern rivers flow to the south and end up in the Samoti plain, which is a closed basin (Fig. 6.9). The collapse of the Alid caldera (36–15 ka—Sani et al. 2017) and the following basalt emplacement acted as a sort of divide, thus forcing the main rivers of the Samoti area (Derawle, Aleitali, Dandero, Meeble—Figs. 6.3 and 6.4) to flow southward, in that also favored by the progressive lowering of the southern portion of the Samoti plain bottom, the southwestern part of which has now elevations below the sea level (Fig. 6.4).

The studies of Ghinassi et al. (2009) and of Sani et al. (2017) on the Pleistocene to Holocene deposits filling the Samoti graben depict a complex structural, environmental, and paleogeographic evolution of this area (see Chap. 5 of this publication for more details). The boulder beds topping the filling sequence still preserve their original gradient dipping to NNE (Figs. 6.9 and 6.10), which roughly corresponds to the flow direction of the rivers that formed the entire depositional sequence (about 80 m thick) and that was maintained from the fluvio-lacustrine deposits resting on the basin Neoproterozoic metamorphic basement upward.

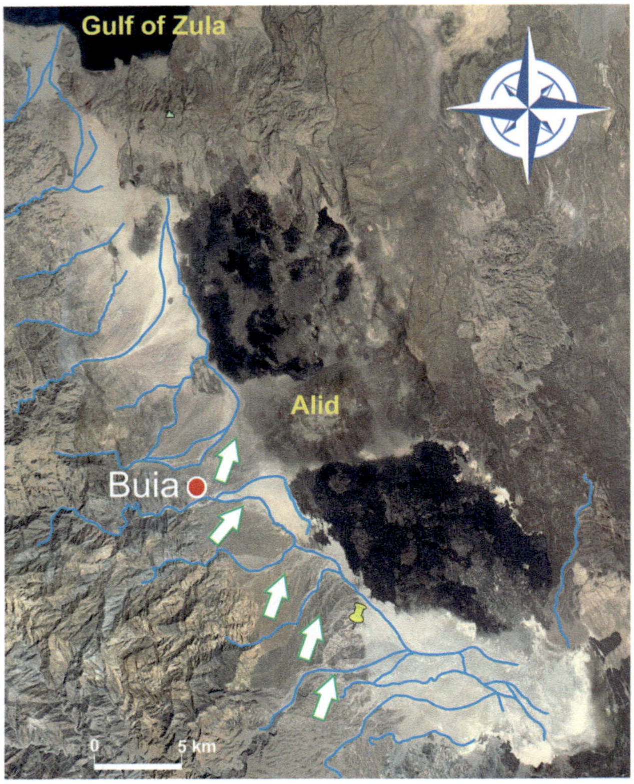

Fig. 6.9 Southern portion of the Gulf of Zula tectonic depression including also the Samoti plain. The recent lava flow of the Alid volcanoes has forced the rivers draining the western margin of the basin in two opposite directions. The white arrows indicate the dip of the Pleistocene fluvial deposits upper surface which correspond to the original depositional gradient (Sani et al. 2017, their Fig. 6) and roughly to the paleocurrents. The yellow pin indicates the *Homo* erectus site

Fig. 6.10 Samoti plain with the flat-topped Pleistocene deposits in the background. The boulder beds on top of the sequence still preserve their original gradient dipping to NNE

As the recent erosion phase started, the Pleistocene deposits were deeply incised by rivers that are still oriented to NNE in the Zula graben north of the Buia, whereas those south of Buia were deflected to the south. This change of flow direction seems to have occurred very recently since one branch of the Derawle river is still flowing to the northeast against the Alid volcano (Fig. 6.11), where it is then deflected to the south. Other channels of the distributary system, formed by the Derawle beyond the narrow valley incised into the Pleistocene deposits, follow a similar pattern.

The rivers entering the Samoti plain form very flat fan-shaped bodies. The alluvial fans that punctuate the plain borders margin (Fig. 6.4) have low gradients, in the 0.005–0.05 range, which are from one to two orders of magnitude lower than dryland fans gradients reported in the literature (e.g., Harvey 2011). The Samoti plain flat fans have morphological characteristics more similar to the distributive systems of ephemeral streams observed in other drylands by Billi (2007). The lack of alluvial fans is probably due to the differential action of the border faults which are probably more active on the eastern margin, resulting in the eastward tilting of the basin bottom and, hence, in the lower elevation of the eastern portion of the plain. Another explanation can be found in the high sedimentation rate and the decreasing accommodation space due to the progressive expansion of the Alid volcanoes toward the center of the basin. In the southern part of the Alid volcanoes, the progradation of the basaltic lavas in the early Holocene (Duffield et al. 1997) reached the opposite margin of the basin, de facto splitting the Samoti plain into a northern and a southern portion (Fig. 6.1). The Derawle river has bypassed the basalt obstruction by cutting a narrow gorge at the contact between the volcanic rocks and the distal outcrops of the Pleistocene deposits (Fig. 6.12).

Throughout the whole study area, desert weathering processes are rather common. Soils are very rare and poorly developed. In places, soil patches may be found on top of alluvial plain deposits, where some overbank flow fine sediment may accumulate (Fig. 6.12). The hill slopes and the flat surface of the Pleistocene deposits are devoid of any kind of soil. They are covered with a rocky weathered mantle forming a typical stone pavement (Fig. 6.13).

Fig. 6.11 Derawle river distributary system prograding toward the Alid mountains. The northern branch runs to NE along the incision in Pleistocene deposits and reaches out to the Alid mountains, where it is deflected to the south to rejoin the main river channels sharply turned to the south into the Samoti plain

Fig. 6.12 Derawle river downstream of the gorge cut at the contact between the basaltic lava flow prograding to the left (coming from the Alid volcanoes) and the Pleistocene deposits on the right (Google Earth coordinates: 14°47′06″ N–39°56′05″ E)

The accumulation of coarse particle on the surface of the Pleistocene deposits plateau is predominantly of primary origin as the coarse particles are the main component of the underlying Boulder bed formation. Here, deflation of fine particles is the prevailing process for the concentration of coarse particle. On the hillslopes underlain by the metamorphic basement rocks or other consolidated sedimentary rocks, weathering processes play a more relevant role in bedrock disintegration (Fig. 6.14), boulder splitting, rock pitting, and alveoles formation (Fig. 6.15). The large availability of sharp-edged rock fragments was an endless supply of raw material for *Homo* erectus to make his tools (Fig. 6.16).

6.3.1 Fluvial Geomorphology and Processes

The rivers entering the Samoti plain are all ephemeral. Water flows only in response to heavy downpours in the headwaters, close to the continental divide between the Red Sea and the Mediterranean drainage systems. Here, the monthly precipitation in June and July can be around 150 mm. On the base of data on dryland rivers in different parts of the world, Kempf et al. (2018) defined an empirical relation between catchment area and event precipitation according to which in the most of the study area rivers 25 mm of rain are necessary to generate some flow. The same authors indicate that stream

Fig. 6.13 Stone pavement on the Pleistocene deposits paleo depositional surface. Notice the absence of soil, the exfoliation on the boulder in the foreground, and the typical black and red desert varnish coating the most of surface stones

flows 0–5 and 3–11 times per year in hyper-arid and semi-arid environments, respectively.

On the coast, the three wettest months are in winter (December-February), but the average monthly precipitation ranges between 25 and 45 mm, and, seldom, they can generate some stream flow in the Samoti plain rivers. From these considerations, it is evident why the study area rivers are dry for most of the time, and though there is no instrumental data of river flow, floods are expected to be rather infrequent.

In the basin western shoulder, rivers have cut deep and narrow valleys and show the typical features of mountain streams; that is, no or a very small alluvial plain is present, the streambed takes up the whole valley bottom, coarse sediment is supplied directly to the stream channel from the hillslopes, bed material is very coarse, and gradients are steep (Fig. 6.17), typically in the 0.03–0.02 range.

As the rivers exit the basin shoulder narrow valley, they enter the Pleistocene deposits in which they have cut wider channels (for instance, the Derawle average streambed width is 561 m, which is more than three times the width of the river in the upstream mountain reach), given the weaker resistance to erosion of these loose sandy and gravel sediments. In these middle reaches, the streambed gradient decreases to values ranging from 0.014 to 0.018 and the rivers assume the typical braided channel morphology (Fig. 6.18). However, bed material is still coarse.

As the rivers proceed beyond the Pleistocene deposits reaches, they enter the basin bottom and form large distributary systems. The distributary channels gradient tends to decrease further, compared to the upstream middle reaches, but the change is less marked as it ranges from 0.015 to 0.009 (see also Table 6.1). Notwithstanding such a relatively small change in gradient, bed material grains size decreases remarkably and the transition from coarse cobbles dominated to sandy bed occurs within a very short distance (for example, no more than 100 m in the Dandero river) (Figs. 6.19 and 6.20). In the Derawle and Dandero rivers, bed material D_{50} and D_{90} do not show any significant change between the mountain reach and the lower gradient, middle reaches cut into the Pleistocene deposits (Table 6.1).

In the Dandero river, though the change in streambed gradient from site WP18–WP19 is almost negligible (Table 6.1), all the characteristic diameters show a very

Fig. 6.14 Mudstone rock weathering and blocky boulders formation

Fig. 6.15 Hard rock weathering processes: **a** boulder splitting and exfoliation; **b** rock pitting; **c** alveoles

Fig. 6.16 *Homo* erectus artifacts on a Pleistocene preserved, depositional paleosurface. Hand ax, chopper, and parent stone are all present in one site

marked decrease of about an order of magnitude, whereas the sand content increases to a value (about 82%) typical of sand bed rivers.

Though in textbooks, it is commonly reported that in rivers grain, size declines with distance downstream according to an exponential low (e.g., Robert 2003), field observations (e.g., Sambrook Smith and Ferguson 1995), and flume experiments (Paola et al. 1992; Sambrook Smith and Ferguson 1996; Venditti et al. 2015) demonstrated that the abrupt gravel-sand transition is rather common in rivers. All these authors provided different explanations, including selective deposition, local base level control on excess of sand supply, abrasion/breakdown of fine gravel, change in slope, and a decrease in shear stress. With the exception of the change in slope, that is, not relevant in the case of the study area rivers, the other factors may play some role in the abrupt gravel-sand transition. However, in dryland ephemeral streams, flood water transmission loss, due to infiltration in the downstream reaches, is a common process (e.g., Costa et al. 2012). A marked decrease of discharge, accompanied by a decrease in shear stress, may result is a sudden decline in sediment transport capacity and deposition of the coarser particle, whereas the sand is selectively transported farther downstream.

In a few cases, one or two river channels may continue beyond the distributary systems, some distance beyond which they form another a much less developed and active distributary system or simply terminate as floodouts (Tooth 2000).

Big floods are not common in the study area. Extreme, high intensity rainfall, however, may generate very large floods that occasionally reach the southeastern margin of the Samoti plain. The large floods are characterized by high sediment transport rates that are capable to transport large boulders as far as the basin center. In fact, individual boulders, 100–200 mm in mean diameter, are found on the stream bed and in the basin filling deposits (Fig. 6.21), very far from the upstream reaches and from the gravel-sand transition (Fig. 6.20).

6.3.2 Aeolian Geomorphology

Aeolian processes are very active in the study area, and their most outstanding results are large areas covered by wind-blown dunes in the southern portion of the Samoti plain (Fig. 6.22). These dunes are basically of the crescentic type that coalesced to form transverse dunes (Fig. 6.23) (Lancaster 1995 and 2011). These dunes have a sinuous ridge, perpendicular to the main wind direction and typically form, where wind direction is constant for long intervals as it is the case of the Samoti plain, where strong and weeks lasting wind is predominantly from southeast (locally known as Khamsin) (Fig. 6.7).

In the study area, the transverse dunes have an average wavelength of 36 m (range 19–61 m) and an average height of 1.5 m (range 0.7–3.0 m). These dunes are much smaller

Fig. 6.17 Mountainous reach of the Derawle river incised into the metamorphic basement (Google Earth coordinates: 10°49′16″ N–39°50′22″ E). The blue arrows indicate the flow direction

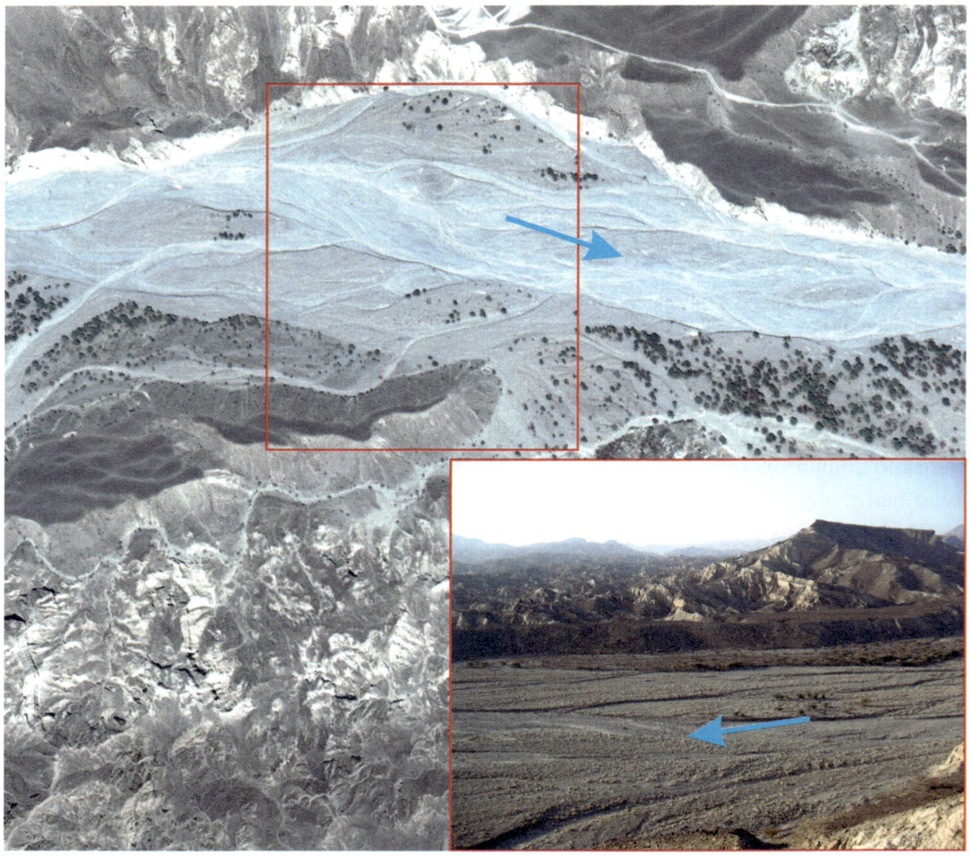

Fig. 6.18 In their middle reaches, cutting through the Pleistocene deposits, the stream bed of the main rivers becomes very large and assumes a braided stream morphology. The braided channel morphology of the Dandero river in the reach is cut into the Pleistocene deposits (Google Earth coordinates: 14°44′32″ N–39°55′56″ E). The blue arrows indicate the flow direction

than those observed in larger deserts, whose wavelength and height vary from 100 to 1000 m and 2–50 m (Fig. 6.24), respectively. The Samoti plain dunes have a larger dune wavelength/height ratio of 24 compared to that of about 5 reported for similar dunes in other deserts of the world (Lancaster 1995). That is, probably due to the limited sediment supply since only very small rivers enter the plain from S-SE-E. The Samoti plain transverse dunes, however, seem

Table 6.1 Main gran size characteristics of bed material in the Derawle and Dandero rivers

River	Site	Gradient	D_{10} (mm)	D_{50} (mm)	D_{90} (mm)	Sand %
Derawle	1	0.026	0.40	1.48	235.3	60.0
	2	0.026	0.31	1.12	181.0	53.6
Dandero	S1	-	0.29	6.35	125.8	44.9
	WP18	0.0012	0.21	7.13	117.1	43.5
	WP19	0.0010	0.08	0.30	22.63	81.7

For sampling site location, refer to Figs. 6.3 and 6.4

Fig. 6.19 Abrupt transition from cobble dominated to sandy streambed in the Dandero river. Flow is toward the reader

to be a smaller scale end member in the great variability of dune size confirming that wind-blown dunes are a classic example of self-organization in a geomorphic system (Hallet 1990). The interdune distance is about 18 m, which corresponds to an interdune area cover of about 50%, a value very close to that of 60% reported for larger dune fields in the world (Lancaster 1995). The interdune area is of deflationary (or non-depositional) dry type (Ahlbrandt and Fryberger 1980; Lancaster 1995). The dunes, in fact, rest and move on a few centimeters thick layer of consolidated muddy sediment (the white parts in Fig. 6.23) which is found all across the most of the Samoti plain southern portion on top of the filling sedimentary sequence.

The dune migration rate was calculated by taking a few shrubs as a fixed reference and comparing Google Earth images of 2011 and 2013. Though the method is not accurate, because many variables could not have been considered, an approximate migration rate of 6 m yr^{-1} was obtained. This value is smaller if compared to larger desert data (Fig. 6.25) (Lancaster 1995), but it is within the range of the global data.

Wind ripples are ubiquitous in any desert and in the Samoti plain as well. Unfortunately, no geometry data for wind ripples are available for the study area, whereas fluvial ripples wavelength and height were measured in the Dandero river. Plotting these data within a wavelength/height bivariate diagram including also wind ripple data from different desert areas in the world (Lancaster 1994) (Fig. 6.26), it is evident that river current ripples cannot be distinguished from wind ripples simply on the base of their geometry, unless for the orientation of the former perpendicular to flow when the direction of the latter is different from that of the prevailing wind.

6.4 Environmental Change

The Samoti plain filling deposits are well exposed for a depth of 2–4 m along the main river channels cut banks. These sediments are predominantly of fluvial origin (Fig. 6.27) and consist mainly of massive, horizontal planar laminate, and cross-bedded sand with silt and clay, thin layers occurring as weathered channel clay plugs (Fig. 6.27) or as a few centimeters (5–30) thick horizontal beds that are mainly found in the upper part of the sedimentary sequence, which is topped by the thickest one. This layer of consolidated clay forms the base on which the wind dune rest and move. No clear evidence of old aeolian deposits was found in the exposed filling sequence.

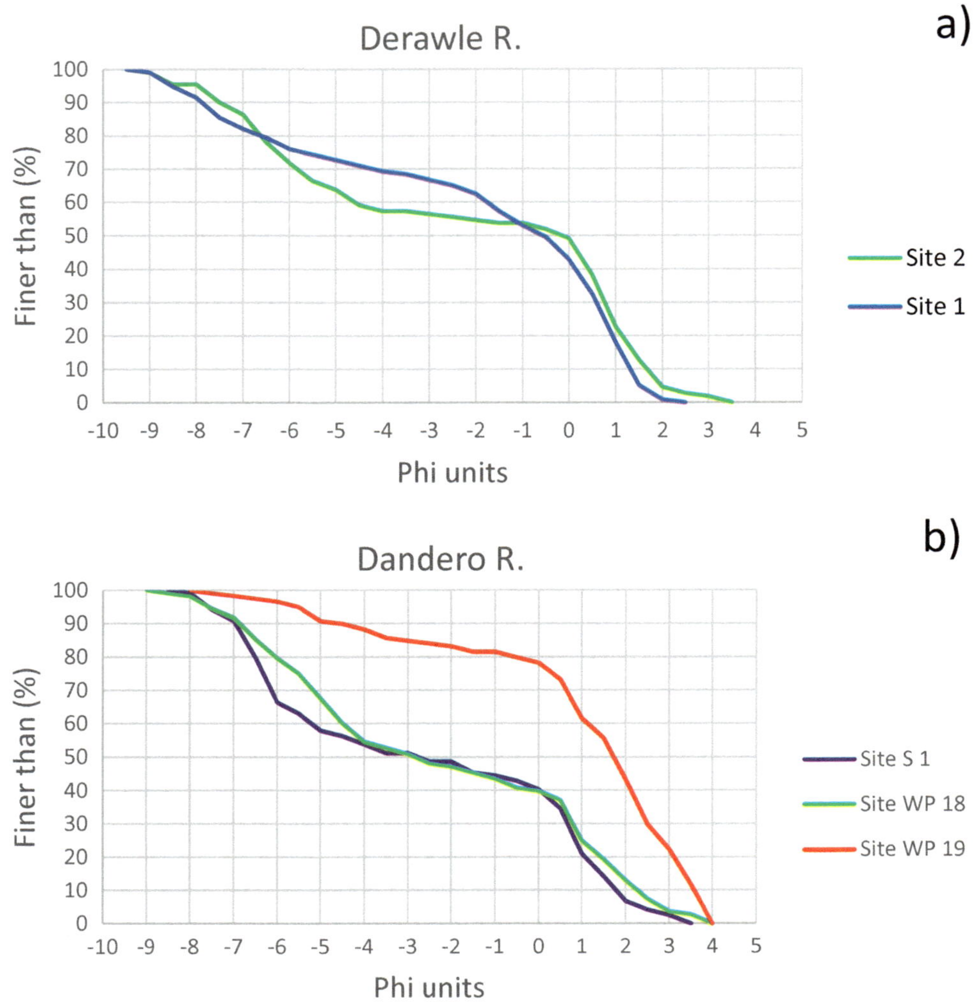

Fig. 6.20 Grain size frequency distribution of the Derawle (**a**) and Dandero (**b**) streambed in different sites (see Figs. 6.3 and 6.4 for the sample location)

Paleochannels range in width from a few meters (Fig. 6.27 a) to about 30 m (Fig. 6.27 b), which is approximately the same width of modern rivers in the study area. The thickness of the weathered clay plug in the largest paleochannel indicates that flow depth was at least 1 m and the redoximorphic features of the channel filling indicate periods of water saturation thus reflecting climate conditions more humid than today. In the modern channels, in fact, fine sediment was observed only as a very thin (1–3 mm thick) drape on the streambed, whereas no soil development on top of the river alluvial deposits where fine sediment deposition from overbank flow should be expected. Moreover, on the upper part of the sequence and on the surface of the top thick muddy layer, there is evidence of saturated sediment deformation (Fig. 6.28a, b) and desiccation mud cracks (Fig. 6.28c). Soft sediment deformation implies water saturation, and two main processes are involved: compaction and/or liquefaction. Compaction is due to overburden, and according to Alsop et al. (2017), also a limited amount of overburden may be sufficient to generate compaction fabrics in unconsolidated sediment. In horizontal beds, however, soft sediment deformation can be considered the result of earthquake-induced liquefaction (Allen 1986).

Some footprints of a big bovidae were found on the top muddy layer (Fig. 6.29). It is not easy to identify the animal that made them but, given the large size of the footprints, for sure they were not made by the gazelles that occasionally can be seen in the Samoti plain. Gazelle expert (Ito, personal communication) suggests that the footprint of Fig. 6.29 could be of a big antelope, the size of a greater kudu, which is a scrub and bush woodland antelope, nowadays found in the Ethiopia and Eritrea highlands, whereas no similar big antelope seems to be present in the study area. In the past, more humid conditions may have favored a denser vegetation cover, likely enough to feed also such a big antelope.

All these elements point at past wetter conditions, probably coinciding with the African Humid Period. During such humid interval, rivers discharge was intermittent but, likely, with longer than today base flows and temporary swamps may have formed in the alluvial plain. As the climate

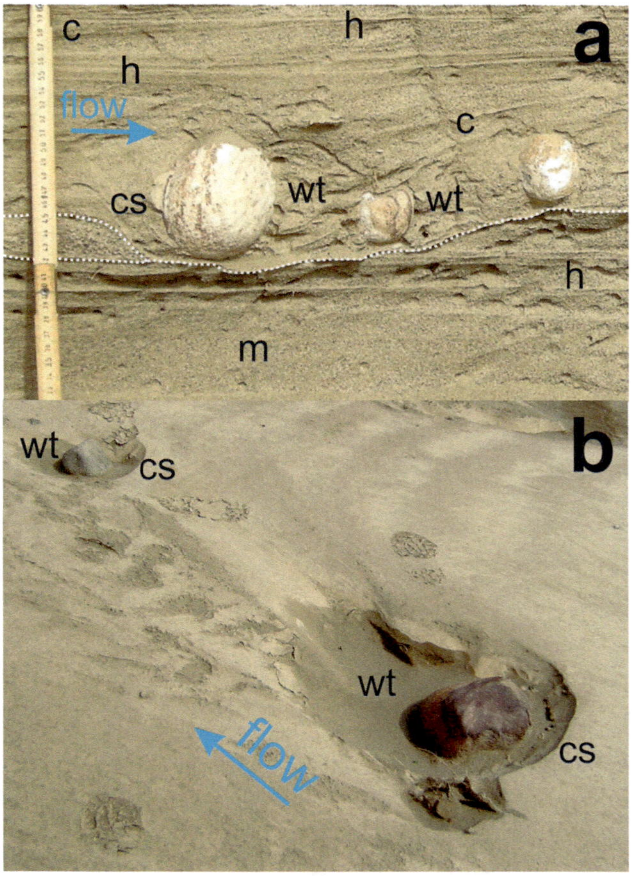

Fig. 6.21 Isolated cobbles within the basin filling sandy deposits (**a**) and on the modern streambed (**b**) in the center of the Samoti plain; cs = crescent scour upstream of the individual boulder; wt = wake turbulence area and disturbed sediment; h = horizontal lamination; c = cross-bedded or inclined lamination; m = massive sand; white dotted line = erosion surface in the massive sand layer under the cobbles

changed, becoming very dry within a relatively short time, conditions were set for the development of the Khamsin wind and the formation of Aeolian dunes. The soft sediment deformation, associated with water saturation and earthquakes, may witness the reactivations of faulting along the Samoti plain with a progressive lowering of its southeastern portion, thus triggering the incision of the modern rivers into the basin filling deposits.

Unfortunately, no dating of this deposits is available, but the most recent of them should be at least Holocene. A humid period is documented for the Sahara (de Menocal et al. 2000; de Menocal and Tierney 2012). It occurred between 10 and 5 yr. ago, and it is witnessed by higher discharges and delta sedimentation rates of the Nile (Sun et al. 2019), higher than present water level in many lakes across Central and Northern Africa (de Menocal and Tierney 2012), including lakes in Ethiopia (e.g., Benvenuti et al. 2002) and by the humidity index (Tjallingii 2008), which show two peaks around 7–6 yr. BP.

According to de Menocal et al. (2000), the early Holocene humidity period ended up abruptly (within a few hundred years—Holmes and Hoelzmann 2017) 5 yr. ago with the onset of very dry conditions. On the base of soil stable isotopic and elemental analyses, radiocarbon dating and historical data, Terwilliger et al. (2011) concluded that in Northern Ethiopia and Eritrea, the interval between 4400 and 1200 kry BP was characterized by increasing aridity interrupted by relatively short wetter periods, though in a context of general rainfall decrease.

Though no dating is available for the Samoti plain filling, it seems reasonable to associate the large river channels with weathered clay plug and the top thick clayey layer with the

Fig. 6.22 Main dune fields of the Samoti plain (areas within the dashed lines). The thick arrows indicate the main directions of potential sediment supply; the thin lines indicate the wind direction and the numbers the angle to the north of their provenance

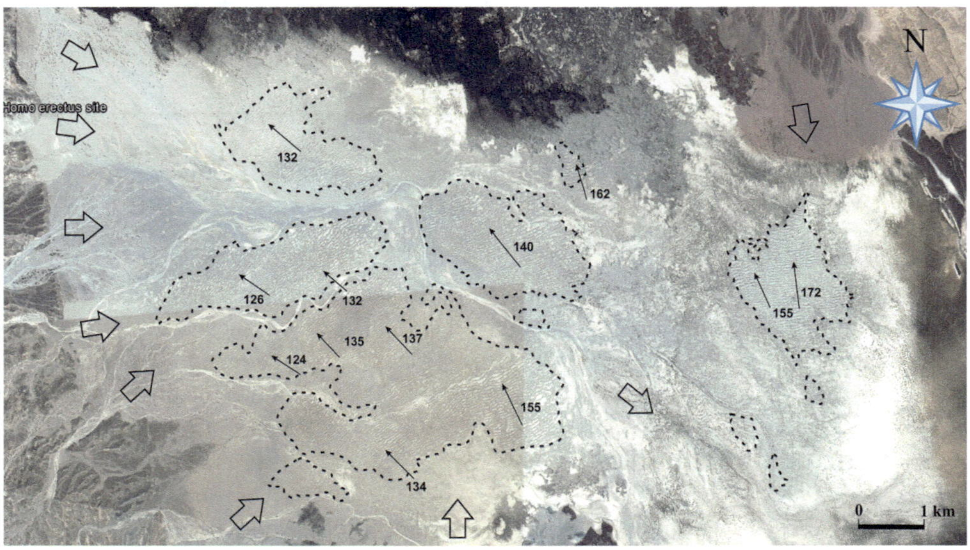

Fig. 6.23 Satellite and ground view of the crescentic dunes merged into transverse dunes

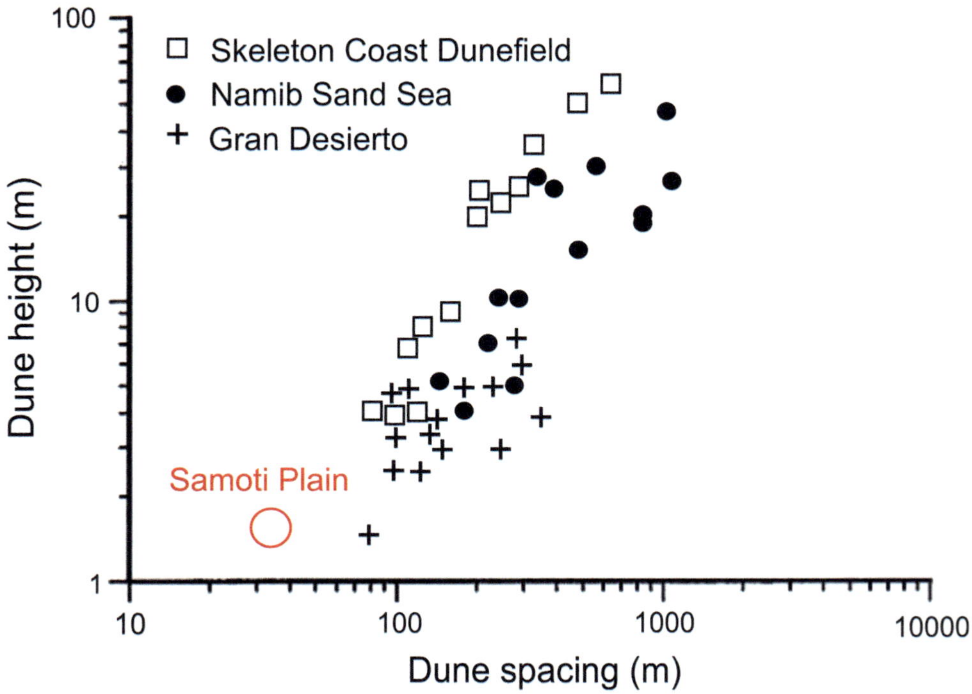

Fig. 6.24 Dune spacing vs. dune height. The Samoti plain dunes are smaller than those in other deserts (modified from Lancaster 1995)

African Humid Period and the formation of the aeolian dunes with the onset of dryer conditions that initiated 5000 years ago.

6.5 Concluding Remarks

The Samoti plain is the floor of a structural basin in the northern part of the Danakil depression. In spite of the sediment filling, the southeastern part of the plain is about 20 m below sea level. The main geomorphological characteristics of the study area reflect the combination of the recent geo-structural events with weathering, erosive, and depositional fluvial and aeolian processes. The climate is hyper-arid. Rainfalls occur mainly on the main Eritrean escarpment, on which the headwaters of the main rivers are located. On the Samoti plain, rains are very sporadic and temperatures are very hot providing the region with the typical characteristics of a desert environment.

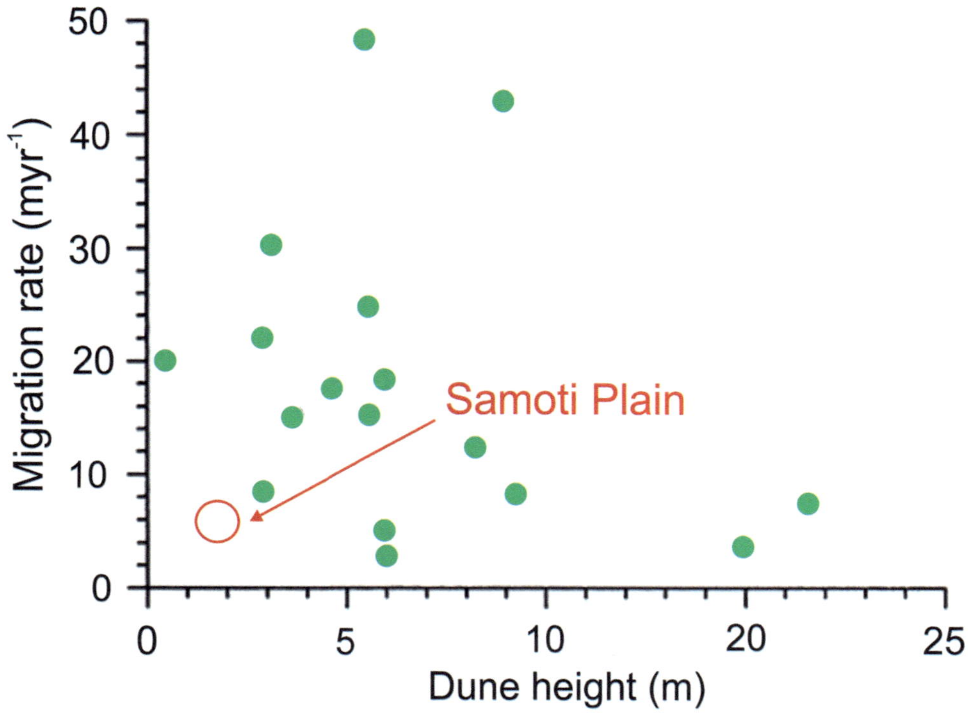

Fig. 6.25 Dune height vs. migration rates. The Samoti plain dunes are smaller and migrate at slower rate compared to dunes of other deserts in the world (modified from Lancaster 1995)

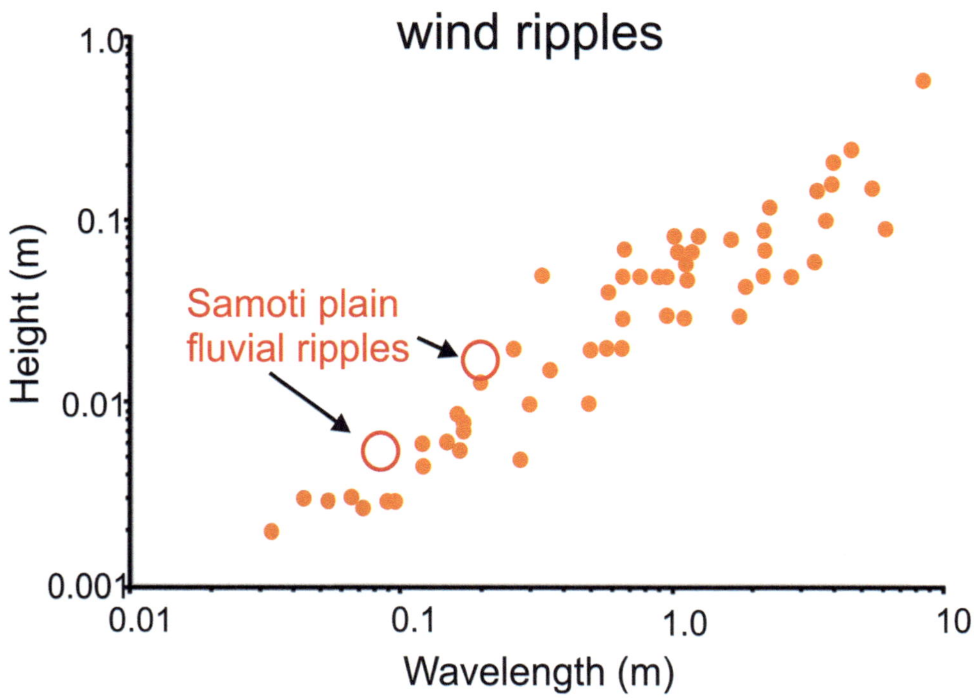

Fig. 6.26 Ripples wavelength vs. height. The fluvial ripples of the Samoti plain rivers cannot be distinguished from wind ripples of different deserts in the world (modified from Lancaster 2009)

The rivers entering the closed basin of the Samoti plain are ephemeral with a dry bed for most of the time. Water flows are resumed only in response to very intense rainstorms, which, occasionally, my turn these dry streambeds into furious rivers. In the mountain valleys and in the middle reaches, the streambed consists of large boulders, but the transition from a very coarse to a prevailing sandy bed material occurs within a very short distance (around 100 m) without a marked change in the streambed gradient. Such a fast transition from coarse gravel to sand has been observed also in rivers of more humid environments, but no specific explanation is shared within the scientific community. In dryland ephemeral streams, however, the high rate of flood water infiltration, especially in the downstream reaches, results in a sharp decrease in discharge. The consequent decrease in shear stress implies a decline of sediment transport capacity and the

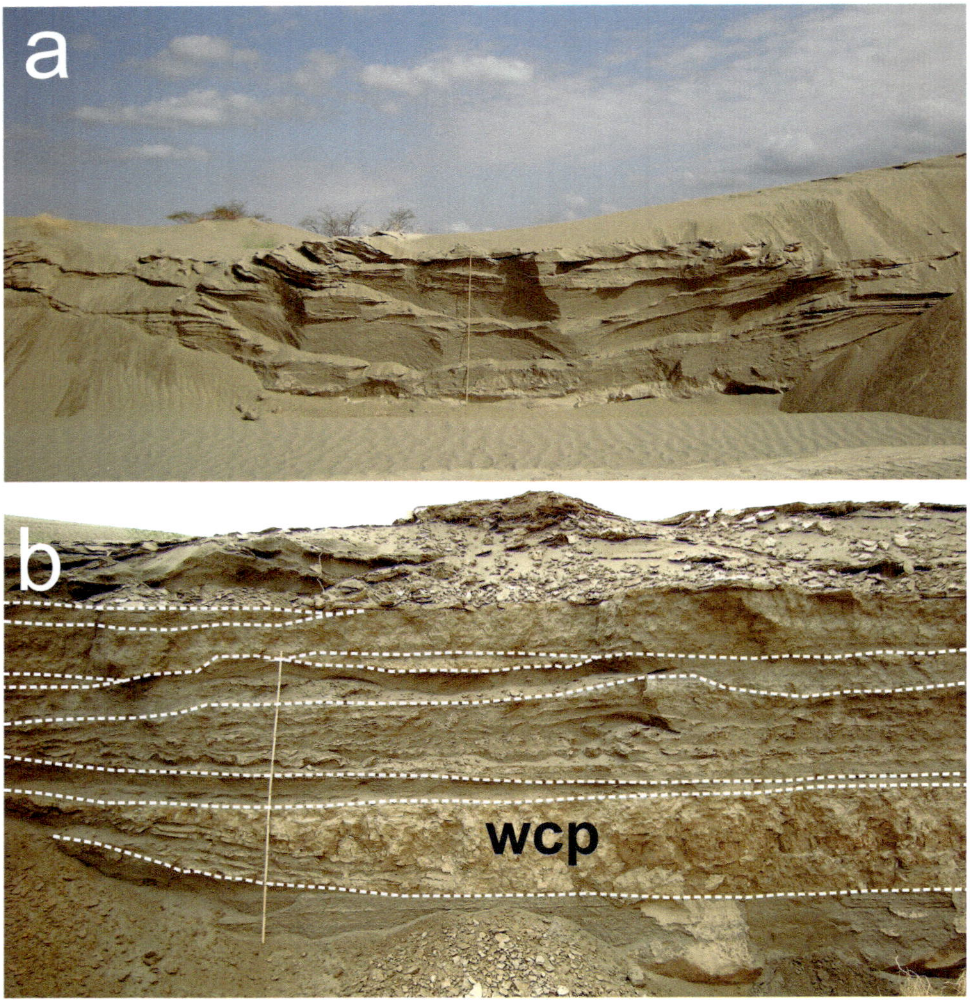

Fig. 6.27 Paleochannels in the Samoti plain filling deposit: **a** small channel (6–7 m) crossbedding; **b** large channel, about 30 m wide, with stacked old channels. wcp = weathered clay plug. The stick ruler is 2 m long

Fig. 6.28 Example of: **a** overload deformation in the infilling sedimentary sequence; **b** saturated sediment deformation of unknown origin (probably caused by overburden and/or earthquake) in the thick clay layer on top of the infilling sedimentary sequence; c) mud cracks in the thick clay layer on top of the infilling sedimentary sequence

Fig. 6.29 Foot print of a large antelope, probably a greater kudu, in the thick clay layer on top of the infilling sedimentary sequence. This animal is no longer living in the study area.

deposition of the coarser particle, whereas the sand is selectively transported farther downstream.

In the Samoti plain, wind processes are very active and a large proportion of the plain is covered with sand blown dunes. Transverse dunes, formed by the coalescing of barchan dunes, are the most common type. They are shaped by the Khamsin wind blowing from southeast. These dunes are much smaller than those of larger deserts but have a larger dune wavelength/height ratio. This may be the result of the scarce availability of sediment in the southern quadrant of the basin, which is a closed one, since the larger rivers with the higher sediment contributions enter the plain from the northeastern quadrant, whereas only very small rivers enter the plain from the southern quadrant, thus providing the Khamsin wind with supply limited conditions.

A few facts indicate that in the recent past (7–5000 years BP), the Samoti plain was subject to a wetter climate. They include:

(a) The lack of any evidence of old aeolian deposits in the exposed filling sequence
(b) The occurrence in the plain filling deposits of paleochannels as wide as the modern ones, but topped by a thick clay plug with evidence of persistent water submergence and indicating a flow depth of about 1 m. In modern rivers, instead, only a very thin (1–3 mm thick) drape may cover small parts of the streambed.
(c) Examples of saturated sediment deformations due to overload and, probably, to seismic activity are rather common in the river banks cut into the plain infilling deposits.
(d) Foot prints of a big bovidae, probably a greater kudu, were found on the thick mud layer at the top of the basin infilling sequence. This top layer is also characterized by liquefactions features and dissection cracks. Nowadays, the greater kudo lives in scrub and bush woodlands of East Africa, but it seems it is no longer present in the study area.

All these facts suggest that 7–5000 years ago, more humid conditions (African Humid Period) prevailed in the Samoti plain. The transition to a drier climate seems to have been very fast and synchronous with the structural lowering of the basin floor which caused the basin infilling incision by the northwestern rivers, while sand dunes started to form on the thick mud layer on the top of the basin infilling sequence as the climate became drier.

References

Abbate E, Albianelli A, Azzaroli A, Benvenuti M, Tesfamariam B, Bruni P, Cipriani N, Clarke RJ, Ficcarelli G, Macchiarelli R, Napoleone G, Papini M, Rook L, Sagri M, Tecle TM, Torre D, Villa I (1998) A one-million-year-old Homo cranium from the Danakil (Afar) Depression of Eritrea. Nature 393:458–460

Abbate E, Woldehaimanot B, Bruni P, Falorni P, Papini M, Sagri M, Girmay S, Tecle TM (2004) Geology of the homo-bearing Pleistocene Dandiero basin (Buia region, Eritrean Danakil depression). Riv It Paleont Strat 110(supplement):5–34

Ahlbrandt TS, Fryberger SG (1980) Sedimentary features and significance of interdune deposits. In: Ethridge FG, Flores RM (eds) Recent and ancient nonmarine depositional environments: models for exploration. Society of economic palaeontologists and mineralogists. Tulsa, Oklahoma, pp 293–314

Allen JRL (1986) Earthquake magnitude-frequency, epicentral distance, and soft-sediment deformation in sedimentary basins. Sed Geol 46:67–75

Alsop GI, Weinberger R, Levi MT (2017) Identifying soft-sediment deformation in rocks. J Struct Geol 125:248–255. https://doi.org/10.1016/j.jsg.2017.09.001

Benvenuti M, Carnicelli S, Belluomini G, Dainelli N, Di Grazia S, Ferrari GA, Iasio C, Sagri M, Ventra D, Atnafu B, Kebede S (2002) The Ziway-Shala lake basin (Main Ethiopian Rift, Ethiopia): a revision of basin evolution with special reference to the Late Quaternary. J Afr Earth Sc 35:247–269

Billi P (2007) Morphology and sediment dynamics of ephemeral streams terminal reaches in the Kobo basin (northern Welo, Ethiopia). Geomorphology 85:98–113

Costa AC, Bronstert A, de Araujo JC (2012) A channel transmission losses model for different dryland rivers. Hydrol Earth Syst Sci 16:1111–1135

de Menocal P, Ortiz J, Guilderson T, Adkins J, Sarnthein M, Baker L, Yarusinsky M (2000) Abrupt onset and termination of the African Humid Period: rapid climate responses to gradual insolation forcing. Quatern Sci Rev 19:347–361

de Menocal PB, Tierney JE (2012) Green Sahara: African humid periods paced by Earth's orbital changes. Nat Educ Knowl 3(10):12

Duffield WA, Bullen TD, Clynne MA, Fournier RO, Janik, CJ, Lanphere MA, Lowenstern J, Smith JG, Wolde Giorgis L, Kahsai G, Wolde Mariam K, Tesfai T (1997). Geothermal potential of the Alid Volcanic Center, Danakil Depression, Eritrea. US Geological Survey Open File Report 97–291, Reston

Ghinassi M, Libsekal Y, Papini M, Rook L (2009) Palaeoenvironments of the Buia Homo site: high-resolution facies analysis and

non-marine sequence stratigraphy in the Alat formation (Pleistocene Dandiero Basin, Danakil depression, Eritrea). Palaeogeogr Palaeoclimatol Palaeoecol 280:415–431
Hallet B (1990) Spatial self-organization in geomorphology: from periodic bedforms and patterned ground to scale-invariant topography. Earth-Sci Rev 29:57–76
Harvey A (2011) Daryland alluvial fans. In: Thomas DSG (ed) Arid zone geomorphology: process, form and change in drylands. Wiley, New York, pp 333–370
Holmes J, Hoelzmann P (2017)The Late Pleistocene-Holocene African Humid Period as Evident in lakes. Oxford Res Encyclop Clim Sci 1. https://doi.org/10.1093/acrefore/9780190228620.013.531
Kampf SK, Faulconer J, Shaw JR, Lefsky M, Wagenbrenner JW, Cooper DJ (2018) Rainfall thresholds for flow generation in desert ephemeral streams. Water Resour Res. https://doi.org/10.1029/2018WR023714
Lancaster N (1995) Geomorphology of desert dunes. Routledge, New York
Lancaster N (2009) Dune morphology and dynamics. In: Parsons AJ, Abrahams AD (eds) Geomorphology of desert environments, 2nd edn. Springer, Heidelberg, pp 557–596
Lancaster N (2011) Desert dune processes and dynamics. In: Thomas DSG (ed) Arid zone geomorphology: process, form and change in drylands, 3rd edn. Wiley, New York, pp 487–515
Paola C, Parker G, Seal R, Sinha SK, Southard JB, Wilcock PR (1992) Downstream fining by selective deposition in a laboratory flume. Science 258:1757–1760
Pedgley DE (1967) Air Temperature at Dallol, Ethiopia. Meteorol Mag 96:265–271
Robert A (2003) River processes. An introduction to fluvial dynamics. Arnold, London
Rosen K, Van Buskirk R, Gabresi K (1999) Wind energy potential of coastal Eritrea: an analysis of sparse wind data. Sol Energy 66 (3):201–213
Sambrook Smith GH, Ferguson RI (1995) The gravel-sand transition along river channels. J Sediment Res 65(2a):423–430
Sambrook Smith GH, Ferguson RI (1996) The gravel-sand transition: flume study of channel response to reduced slope. Geomorphology 16:147–159
Sani F, Ghinassi M, Papini M, Oms O, Finotello A (2017) Evolution of the northern tip of Afar triangle: inferences from the Quaternary succession of the Dandiero—Massawa area (Eritrea). Tectonophysics 717:339–357
Sun Q, Liu Y, Salem A, Marks L, Welc F, Ma F, Zhang W, Chen J, Jiang J, Chen Z (2019) Climate-induced discharge variations of the Nile during the Holocene: Evidence from the sediment provenance of Faiyum Basin, north Egypt. Global Planet Change 172:200–210
Tjallingii R, Claussen M, Stuut J-BW, Fohlmeister J, Jahn A, Bickert T, Lamy F, Rohl U (2008) Coherent high-and low-latitude control of the northwest African hydrological balance. Nat Geosci 1:670–675
Tooth S (2000) Process, form and change in dryland rivers: a review of recent research. Earth-Sci Rev 51:67–107
Terwilliger VJ, Eshetu Z, Huang Y, Alexandre M, Umer M, Gebru T (2011) Local variation in climate and land use during the time of the major kingdoms of the Tigray Plateau in Ethiopia and Eritrea. CATENA 85:130–143
Venditti JG, Domarad N, Church M, Rennie CD (2015) The gravel-sand transition: sediment dynamics in a diffuse extension. J Geophys Res Earth Surf 120:943–963

Forest Landscape Evolution in Eritrea Throughout the Last Century—A Review

7

Lorenzo Orioli and Ermias Lulekal Molla

Abstract

The purpose of this review is to present the changes in the forest landscape in Eritrea over a period of approximately 120 years. Data on Eritrean forest resources with their status, distribution and challenges were gathered and compiled after assessing all available published documents pertinent to forests in Eritrea. The review has identified that a number of earlier Italian botanists played a crucial role in identifying vegetation types of Eritrea and reporting its changes over time. Moreover, referring to the reviewed documents, the present work provides a synthesis of the principal phyto-geographical descriptions of Eritrean vegetation and the historical cumulative environmental drivers that contributed to the degradation of existing Eritrean forests. Very credible reviewed documents also depicted that Eritrean woodland cover has decreased significantly in the past century despite the efforts from the Eritrean government to cover the country, especially the highlands with dense forests. Generally, the output of this review can serve as a basis for further overall and in-depth floristic composition investigation of Eritrean forests. Readers are referred to original articles on Eritrean forests for detailed analytical methods and interpretation of results; all resources used for this review are also duly cited.

Keywords

Forest landscape • Environmental history • Bioclimatic zoning • Eritrea

L. Orioli (✉)
DAGRI Department, School of Agriculture, University of Florence, Florence, Italy
e-mail: lorenzo.orioli@unifi.it

E. L. Molla
Department of Plant Biology and Biodiversity Management, College of Natural and Computational Science, Addis Ababa University, Addis Ababa, Ethiopia
e-mail: ermias.lulekalm@aau.edu.et

7.1 Introduction

A *National Biodiversity Strategy and Action Plan* (NBSAP) was prepared by Eritrea in 2000. As part of a quinquennial plan (2011–2015) for the *Great Green Wall Initiative* (GGWI), the government of Eritrea developed a *Sustainable Natural Resources Management* (SNRM) plan aimed at increasing vegetation cover by 35% and soil fertility improving by 10% with respect to situation described by the Action Plan of the Ministry of Land, Water and Environment in 2012. The stabilization of soil erosion (gully erosion) was projected through hillside terracing and tree planting on degraded lands. The objective was to cover the land with one million tree seedlings and 20,000 mangrove seedlings to stabilize 400 ha of coastal dunes. Moreover, 1000 ha of marginal lands were afforested with *Acacia seyal* and *Acacia senegal,* considered as profitable trees. Recently, the *Revised National Biodiversity Strategy and Action Plan* (2014–2020) was launched again in line with the *Strategic Plan for Biodiversity* (2011–2020). A complementary aim was also to increase the awareness of the Eritrean people about the value of biodiversity, its conservation and sustainable use. The planners' intention was to halve deforestation or possibly reduce it to zero by 2020 in order to restore at least 15% of the degraded ecosystems. For this purpose, the development of five protected areas, covering a total area of 392,214 ha, was scheduled (MLWE 2012).

The "environmental needs" that have accumulated over the Eritrean territory for years are an indirect consequences of decades of war. The occurrence of impoverished regions in East Africa is both structural and incidental. An historical approach towards understanding the main environmental drivers leading to deforestation and land degradation is proposed in the present chapter. The prevailing vegetation cover existing in the past is known from data collected particularly during the Italian colonial period (1882–1941). Moreover, changes over time up to the recent years are documented. Additional data were produced across a 120-

P. Billi (ed.), *Landscapes and Landforms of the Horn of Africa*, World Geomorphological Landscapes,
https://doi.org/10.1007/978-3-031-05487-7_7

year interval by the former Istituto Agronomico per l'Oltremare (IAO) (Overseas Agronomic Institute) of Firenze in Italy within its institutional monitoring activities. The documents of the IAO, archives including articles, monographs, maps and photos, provided a unique set of information that was used to describe the forest landscape changes in Eritrea over the last century. It is worth noticing that the current dramatic environmental constraints have been recurrent in the recent past. Cycles of deforestation and reforestation have characterized the forest history in Eritrea in response to changing and contrasting socio-economic interests.

7.2 Geobotanical Features of Eritrea: An Overview

Eritrea has a surface area of 117,600 km^2 (World Bank Data 2022). It is located in the northern part of the Horn of Africa, between 12°22′ and 18°02′ N latitude and 36°26′ and 43° 13′ E longitude. Its extension includes various islands and a coastline spanning some 1900 km (MLWE 2012). The country shares borders with Sudan to the north and west, Ethiopia to the south, Djibouti to the south-east and the Red Sea to the east, respectively. In Eritrea, there are approximately 390 islands, among which the Dahlak Archipelago is the most prominent (MLWE 2014). The country is divided into six administrative regions, namely Maekel, Debub, Anseba, Gash-Barka, Northern Red Sea and Southern Red Sea (MLWE 2012).

The country exhibits a varied topography and climate, with elevations ranging from 120 m below sea level to over 3000 m asl (Ghebrezgabher et al. 2016). Mean annual temperature ranges from 15 °C in the moist and arid highlands to 32 °C in the desert areas. Annual precipitation varies from less than 200 mm in the deserts to 1100 mm in the subhumid areas in the south-eastern part of the country. The rainfall pattern is extremely variable, both within and between years, and with marked variations over very short distances. The south-west monsoon winds are responsible for both the major and minor summer rains, whereas the northern and north-eastern continental air streams are responsible for winter rains along the coast and in the southern part of the central highland escarpments (MLWE 2014). The northern and north-eastern winds are dry in nature, but acquire moisture while crossing the Red Sea.

7.2.1 Botanical Explorations in Eritrea

Before the colonial era, there was very scarce information on the classification of forests and other types of vegetation. Various Italian botanists played a crucial role in identifying vegetation and reporting its changes over time.

A systematic field survey of both woodlands and woody plants of Eritrea was carried out in 1909 by the Italian botanist, Adriano Fiori. The work of this author made partial reference to the initial observations of Georg August Schweinfurth (1901), regarding useful plants from Eritrea. Fiori provided the description of vegetation of Eritrea, designating eight vegetation zones (Maritime, Lowlands, Inland Danakil, Wooded Areas on the Eastern Slopes, Marginal Areas of the Highlands, Main Highlands, Valley Areas of Western Slope Zones and the Sudanese Plain Zone). These eight vegetation zones corresponded to eight climatic zones that were classified on the basis of average annual temperature, daily thermal variation, relative humidity and atmospheric precipitation. Today, this classification scheme is considered as an *ante litteram* bioclimatic classification because the vegetation cover and the edaphic properties influence the climatic zoning. Fiori defined the zones on the basis of the different vegetal landscapes he encountered during his journey in Eritrea in January–April 1909. This author also included a particular zone, classified as the Sudanese Plain Zone, which was in turn subdivided into three further zones and five sub-zones based on the traditional Abyssinian land use. The additional three zones included the *Quolla* zone (0–1700 m a.s.l), situated on the western side of the Eritrean Abyssinian Plateau, the *Woina-Degà* zone (1700–2400 m a. s.l), characterized by a Mediterranean climate, and the *Degà* zone (2400–3700 m a.s.l) (Fiori 1937). In the first version of this classification, Fiori included an additional zone named the Samhar and Sudanese lowland zone ranging from 0 to 600 m a.s.l. In turn, this zone was subdivided into four subzones. The *Quolla* zone was the richest in terms of flora, with 311 botanical species and twenty endemisms (Fiori 1912). Along an increasing altitudinal range, the *Woina-Degà* zone vegetal landscape was characterized by the following key species: *Olea europaea* subsp. *cuspidata* (former *Olea chrysophylla* or wild olive*), Acacia etbaica,Acacia abyssinica* and *Juniperus procera,* dominant between 2200 and 3000 m a.s.l (Fiori 1937), which also included a part of the *Degà* zone. According to Fiori (1912), on the top of the Eritrean Mountains, only a few wooded species were present. In general, in all of the above-mentioned bioclimatic zones, evergreen species with persistent leaves were reported to prevail. They included 60, 53 and 61 species in the *Quolla* zone, *Quolla-Woina-Degà* intermediate zones and *Woina-Degà* zone, respectively (Fiori 1912).

Fiori identified a total of 430 wooded species (including suffrutices and succulents with woody stems) belonging to 76 botanical families and 226 genera. He was essentially focussed on the causes of deforestation and on government measures for the protection of woodlands in the Eritrean Colony. He was concerned with providing a solution to rebuild forests, arboreta and homestead tree plantings (Fiori 1912).

By contrast, the objective of the Italian agronomist Baldrati (1907) was to provide ecological classification of the Eritrean territory, with the purpose of extending coffee plantations. On the basis of rainfall data and temperature gradient (about 0.7 °C per 100 m of altitude), Baldrati identified seven floristic and climatic-agricultural zones or regions, i.e. Coastal Region, Eastern Lowland, Eastern Escarpment, Main Highland, Western Escarpment, Western Lowland and Danakil Region (Baldrati 1928). This classification scheme was used by Nastasi (1993) for a synthetic research on floristic and climatic areas in Eritrea (Table 7.1).

For Fiori and Baldrati, Eritrea represented a transitional region among Abyssinia, Sudan and Arabia, as regards both climatic and vegetal aspects. However, they did not provide quantitative data on the actual extension of the vegetal zones. Another Italian botanist, Negri (1940), further elaborated on an existing phyto-geographical map of the Horn of Africa (Eritrea, Ethiopia and Somalia). Previous authors, such as Schweinfurth (1868), Dove (1890), who mapped the northern part of Abyssinia and Ethiopia, were all cited by Negri (1940). The main classification scheme developed by Negri (1940) consisted of vegetation units defined on the basis of their physiognomy, which represents "a characteristic reaction of the vegetal life to the environmental condition". Even if the forest landscapes proposed by Negri (1940) were not exclusive for the Eritrean territory, they nonetheless represent a valid general description.

Another, more comprehensive vegetation map was produced about twenty years later by the Italian botanist, Pichi-Sermolli (1957), who described the vegetation landscape of the Horn of Africa (including Eritrea, Ethiopia and Somalia). Through an analysis of the photos archived in the former IAO, he adopted a physiognomy criterion to describe the vegetation cover. Moreover, he studied more than 900 papers published by the first explorers, travel and technical reports and unpublished official documents dealing with East Africa. Regarding this work, Friis et al. (2011) were quoted as saying. "In spite of Pichi-Sermolli's decision to map the vegetation as it appeared at the time of the mapping, this author did not indicate the widespread and extensively cultivated areas in the highlands, but classified farmlands by the surrounding vegetation, a highly man-imposed landscape". This is a valid assumption for the Eritrean territory as a whole, because human impact (agriculture and pasture) remains embodied in the vegetal landscape. This characteristic was observed and documented by all the above-mentioned naturalist explorers. Pichi-Sermolli produced a vegetation map (1:5,000,000) without providing corresponding vegetation types in the English language (a translation of the original Italian denomination was recently compiled by Friis et al. 2011). In total, Pichi-Sermolli included twenty-four vegetation types. Pichi-Sermolli reported that in "aiming at listing the various types of vegetation, I took into consideration first the climatic formations and then if they are typical and edaphic". With these last words, he explained synthetically how the vegetation formations were grouped.

According to Fiori (1912), the geographic characteristics would provide Eritrea with the capacity to potentially sustain a full forest cover. In the past, the natural rate of forest regeneration was maintained below the rate of deforestation. This is attributed to a predatory farming system whereby extensive land use was increased around the villages. Fiori (1912) reported that "The herbaceous plants are devoured

Table 7.1 Comparison between the bioclimatic zones of Fiori (1912) and Baldrati (1928) and Nastasi (1993)

Fiori's (1912) bioclimatic zones	Agro-climatic classification of Baldrati (1928), Nastasi (1993)	Main tree and small tree species
Maritime Zone	Coastal Region	*Avicennia* spp. *Commiphora* spp.
Eritrean Lowland	Eastern Lowland Region	*Acacia orfota; Acacia spirocarpa; Acacia mellifera; Dobera glabra; Salvadora persica; Balanites aegyptica; Tamarix nilotica; Cassia angustifolia* Vahl
Inland Danakil	Danakil Region	*Acacia spirocarpa; Acacia orfota; Hyphaene dankaliensis; Avicennia* spp.
Wooded Areas on Eastern Slopes	Eastern Escarpment Region	*Acacia* spp.; *Ficus* spp.; *Combretum* spp. *Loefl; Teclea nobilis; Trichialia emetica; Mimusops kummel; Stereospermum kunthianum; Celtis kraussiana; Juniperus procera*
Marginal Highland	Main Highland Region	*Ficus vasta; Ficus palmat; Ficus* spp.; *Acacia albida; Acacia abyssinica; Acacia spirocarpa; Croton macrostachys; Erica arborea; Cordia abyssinica; Eucalyptus* spp.;
Main Highland		
Valley and Western Slope Zone	Western Escarpment Region	*Acacia* spp.; *Adansonia digitata; Erytrina abyssinica; Bauhinia reticulata; Ximenia americana; Ficus sycomorus; Tamarindus indica*
Sudanese Plain	Western Lowland Region	*Acacia seyal; Acacia Senegal; Balanites aegyptica*

before seed production; trees and shrubs are bowed and twisted, whereas rain waters wash away the organic matter, thereby sterilizing the soil. [...] When the herbaceous vegetation is arid and exhausted, the shepherds cut down the young trees, especially Acacia species, feeding livestock with the tender leaves and branches".

7.2.2 Geobotanical Studies and Maps

In Fig. 7.1, a vegetation transect starting from Massawa at the Red Sea level, on the east side, to the Sudanese Plains (1000 m asl) on the west side is shown (Abul-Haggag 1961). Similarly to Fiori (1912) (Table 7.1), eight main vegetation types are recognized. The Samhar and Sudanese Plains (lowlands) are situated on the eastern and western sides of the highlands, respectively. The eastern slope ranging from 1000 up to 2350 m asl (Asmara) is characterized by deciduous, non-thorny woods and, then, by thorny bushes and acacias, euphorbias and evergreen woods with olive trees and dense underground at the highest altitude. On the less inclined western slopes, there is a vegetation gradient characterized by euphorbia, thorny woods and steppes, acacias and sycamores. Baobab trees (*Adansonia digitata*) and doum palms (*Hyphaene nodularia*) are found in the valleys. On the plateau, there are instead thickets of the highly deforested surface with rare occurrence of junipers and olive trees.

In Fig. 7.2, a map of the forest cover of the Eritrean territory is shown. This map was elaborated in 1998 by the FAO National Food Information System (NFIS) (1998) for Eritrea. In this map, part of the Central Highland Zone—Northern Midland subzone with the North-Western Zone are dominated by Grassland, Wooded Grassland, Bushland and Shrubland. The Coastal Plain Zone is represented by areas of barren soil, Bushland and Shrubland, Grassland and Wooded Grassland. In the Green Belt Zone only in part Closed and Medium Closed Forest are present. The most densely forested area in Eritrea, commonly referred to as the green belt, is found between Asmara and Ghinda and extends to the north as far as Merara on the Eastern Escarpment, about 43 km north of Asmara (MEM 2004) (Fig. 7.3). The map of the woody areas of Eritrea was prepared by Di Gregorio and Weepener in 2002 (Di Gregorio and Weepener 2002), and it is a thematic map of the FAO Land Cover Map of Eritrea. In the years 2000–2001, the IAO produced a land cover map of Eritrea at 1:100,000 scale, based on photointerpretation of Landsat satellite images (Sarfatti 2005), within an agreement with FAO regarding the AFRICOVER project. In this map, the vegetation classes are a generalization of the original 62 *Africover* classes for Eritrea and the vegetation is classified into: Closed Trees, consisting of both evergreen broadleaved and needle leaved trees, Open Trees, Closed Woody Vegetation distributed in thickets, Areas with Sparse Trees, Open and Closed Shrubs and Closed Mangrove.

Bein et al. (1996) classified the Eritrean territory into six agro-climatic zones: (1) the Coastal Plains; (2) the Eastern Escarpment; (3) the Central Highlands; (4) the Western Escarpment; (5) the South-Western Lowlands and (6) the North-Western Lowlands (Fig. 7.4). These zones are mainly based on the FAO (1994) document on Eritrea. Bein et al. (1996) were quoted saying "This zone (The Eastern Escarpment) is a unique area where annual rainfall exceeds 1000 mm. It encompasses numerous micro-ecological zones determined by the interrelationship of altitude, rainfall, exposure and soils [...]. The relief is steep and requires terracing for successful farming". This zone also includes the so-called green belt because of vast annual and permanent tree crops, such as coffee, and part of the 53,000 ha of

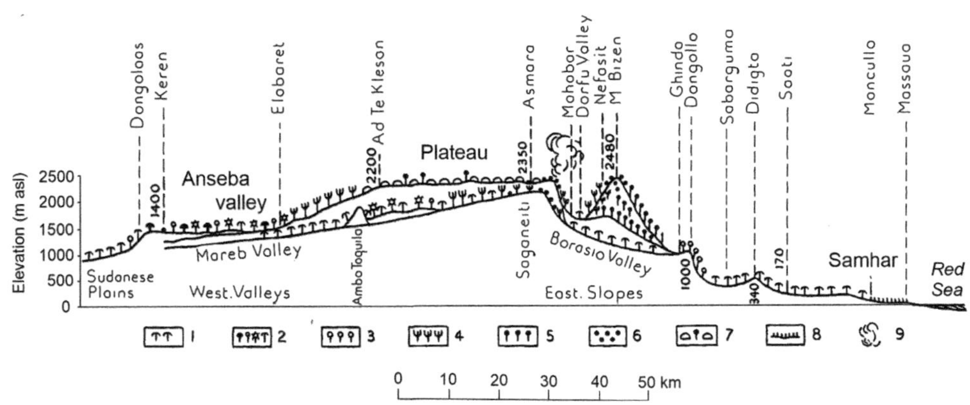

Fig. 7.1 A latitudinal west–east vegetation transect of Eritrea. The altitudes range from below sea level (Red Sea) up to 2480 m asl of Mount Bize. The vegetation units are represented by the following types: (1) thorny bushes and acacias, and doum palm; (2) thorny wood and steppe with acacias, sycamores and baobabs in the west (Anseba Valley); (3) deciduous, non-thorny woods; (4) euphorbias, acacias and other trees species; (5) evergreen woods and wild olive trees; (6) mist bushes and mist-trees in the eastern slopes, especially junipers; (7) thickets of highly deforested summit surface of the plateau; and (8) semi-desert of the coastal zone (Samhar), western limit of winter mists and mist vegetation (modified from Abul-Haggag 1961)

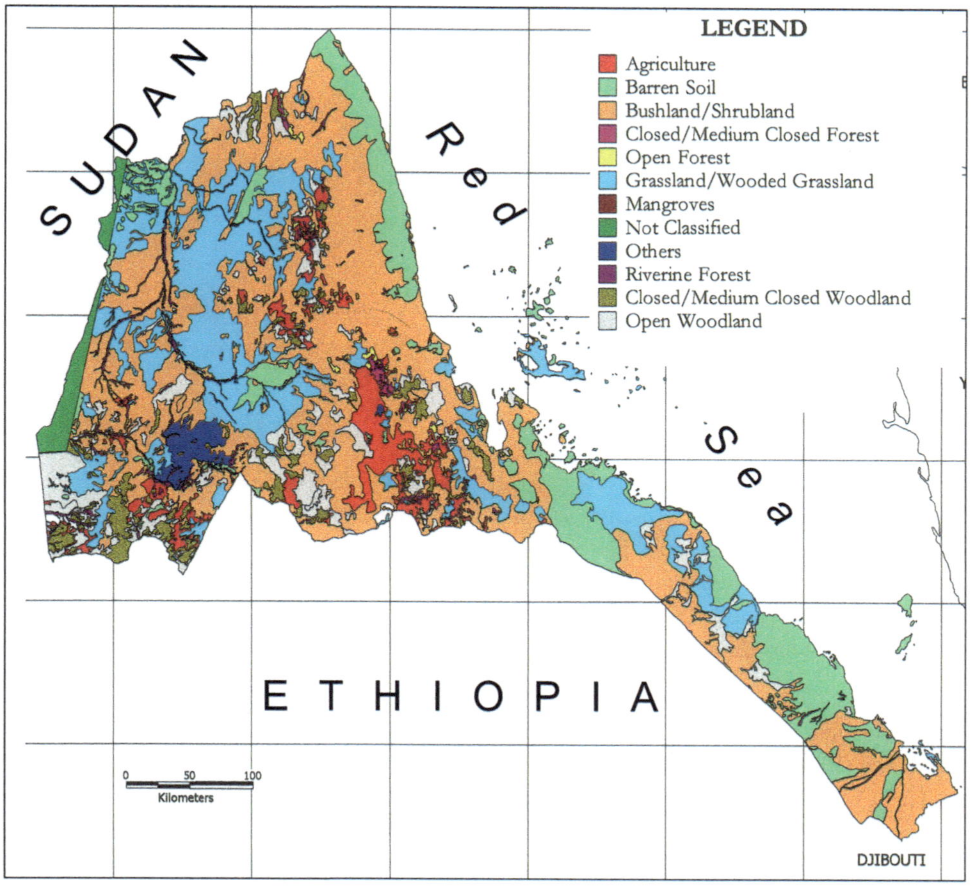

Fig. 7.2 Forest cover of Eritrea. Modified from FAO-NFIS (1998) Map 1:4,000,000 (Archive of the former Istituto Agronomico per l'Oltremare 1998)

Fig. 7.3 A forest area in Merara, on the Eastern Escarpment, about 43 km north of Asmara (Photographic Archive of the former Istituto Agronomico per l'Oltremare 1928)

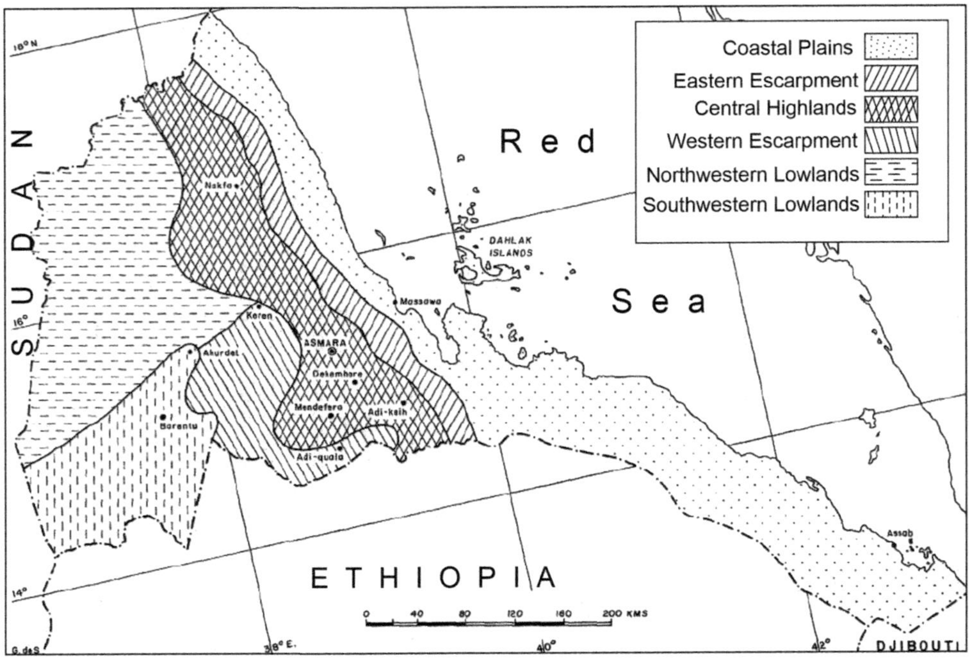

Fig. 7.4 Main agro-climatic zones of Eritrea (modified from Bein et al. 1996)

coniferous forest that once covered a large proportion of the highlands (Bein et al. 1996).

The Coastal Plain zone records less than 200 mm of annual precipitation, and, similarly to the highlands, the annual evapotranspiration may exceed 1800–2000 mm. The Coastal Plain zone is characterized by agro-pastoralism as the most important type of farming system. Agro-pastoralism practices include the seasonal nomadic movements towards the uplands (in mid-April) and the return to the wadis (lowland ephemeral streams) (in mid-September). In the Highlands, rainfed cereal/pulse-based farming systems and irrigated horticultural systems prevail. According to Bein et al. (1996) "The Western Escarpment lies at an altitude of 600–1500 m asl and has a warm-to-hot semi-arid climate. It is a transition zone between the highlands and the western lowlands […]. The dominant production system is agro-pastoralism". The South-Western Lowlands, at an altitude of 600–750 m a.s.l., are generally flat, with hot semi-arid conditions, and an annual rainfall of 400–600 mm. In this region, the farming systems include nomadic pastoralism, semi-sedentary agro-pastoralism, mixed crop and livestock production. The North-Western Lowlands lie between 400 and 1500 m asl, and the climate is hot and arid, with an average annual rainfall of 300 mm. Evapotranspiration is between 1500 and 2000 mm. The principal type of farming system is nomadic pastoralism, very similar to the one practised in the South-Western Lowlands (Bein et al. 1996).

For comparative purposes, the FAO National Food Information System (NFIS) (1998) map of the agro-ecological zones of Eritrea is reported in Fig. 7.5. This classification is more detailed and includes eight zones. The Central Highland Zone was subsequently split into two sub-zones: the Northern Midland and the Southern Midland, which are less elevated (1500–2000 m asl) than the main Central Highland Zone (over 2000 m asl). These two sub-zones are distinguished by differences in rainfall, as the Northern Midland receives an annual precipitation of about 700 mm, whereas the Southern Midland is essentially arid, with less than 400 mm. In Fig. 7.5, the Green Belt Zone, localized in between the Central Highland Zone and the Coastal Plain Zone, is evident.

Eritrea is also rich in flora. In the Central Highlands, the most widespread tree species in the scattered forest patches include *Juniperus procera* and *Olea europaea* subsp. *cuspidata* community mixed with deciduous trees like *Acacia abyssinica*, *Rhus abyssinica*, *Carissa spinarum*, *Terminalia brownii*, *Mimusops kummel* and *Combretum molle*. According to Bartolommei Gioli (1903), the flora found between 900 and 2000 m asl along the Eastern Escarpment was characterized by *Olea europaea* subsp. *cuspidata* (former *Olea crysophylla*). After first attempts in 1892, in 1908 some trees of *Olea europaea* were imported from Central Italy (Pescia) and introduced to Addiché, a few kilometres to the south-west of Asmara. The acclimatization of these trees failed because of the strong winds and felling activities by the local people (Pajella 1934). In 1919, in Adi Caieh, a town located 80 km to the south-east of Asmara, olive trees imported from Sicily (Catania) were planted (Pajella 1947). Once again, this endeavour resulted in failure due to incorrect selection of planting sites, traditional land tenure regulations, lack of knowledge on the biology of *Olea europaea* subsp. *cuspidata* (former *Olea crysophylla*) and a lack of

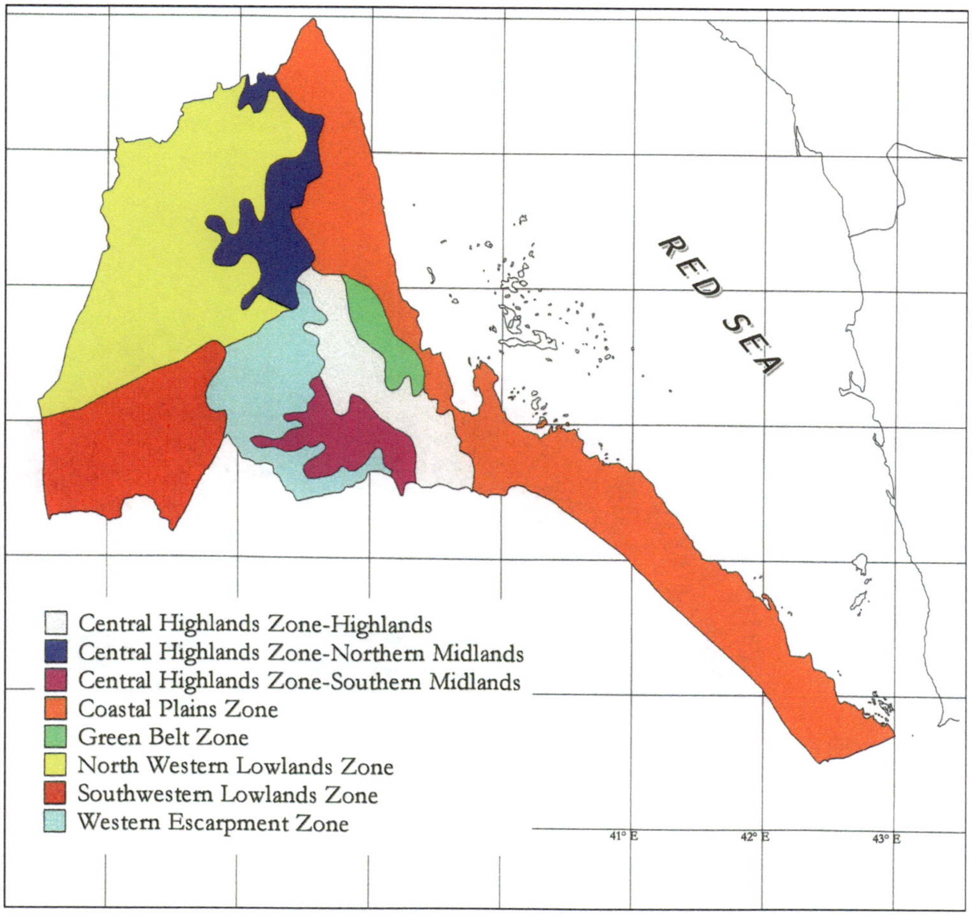

Fig. 7.5 Main agro-ecological zones of Eritrea. Modified from FAO-NFIS Map 1:4,000,000 (Archive of the former Istituto Agronomico per l'Oltremare 1998)

cultivation skills of the local farmers (Pajella 1947). In Fig. 7.6, a rocky slope in Anagulè (a few kilometres north of Asmara) with wild olive trees is portrayed. Towards the edge of the Eastern Escarpment, however, groves of olive and juniper are found, which tend to be particularly abundant in areas with higher rainfall. Trees are often grown in household compounds, while many villages also have small plantations of eucalyptus. Larger plantations of eucalyptus still exist on lands once occupied by former Italian concessionaires and in the vicinity of towns, where they were originally planted by the Italian colonial government (Boerma 2005b). Figure 7.7 shows a village surrounded by a plantation of eucalyptus trees.

7.2.3 The Mangrove Formations

The vegetation of the Coastal Plains is diverse. It includes swamps along the seasonal rivers, acacia woodlands, semi-desert acacia woodlands, bushland and thickets, desert scrubs and mangrove swamps (Haile et al. 1998). Along the Eritrean Red Sea coast, three mangrove species prevail, namely *Avicennia marina* (Forsk.) Vierh., *Rhizophora mucronata* Poir. and *Ceriops tagal*. The first two species are indigenous and concentrated along the muddy coasts, whereas the third one is limited to the northern coast with a few sparse individuals. In 1994, the mangrove forest covered an area of approximately 8900 ha, which declined by 7.7% in 2014 (Hailemichael 2015). Large continuous mangroves are also found in the Asseb Bay. The major threats to the mangroves are represented by water quality deterioration, rapid and short floods, land-filling for development and changes in coastal dynamics. High sedimentation rates, as observed in Zula, tree felling, dying mangrove trees and the increasing frequency of droughts are noticeable threats along Eritrea's coast. Grazed or cut mangrove forest is characterized by short, dwarf stunted trees and scattered old mangrove (Hailemichael 2015). Natural mortalities are caused by sediment movement, cut-off of fresh water flow due to sand deposition or bank erosion or dieback diseases (Hailemichael 2015).

Fig. 7.6 Wild olive trees in Anagulè, a few kilometres north of Asmara. Photo by Maugini 1933 (Photographic Archive of the former Istituto Agronomico per l'Oltremare)

Fig. 7.7 Village protected by a eucalyptus plantation (photo by Billi 2011)

Tree felling, severe grazing of camels, sheep, goats and donkey and the impact of large settlements are additional degradation factors. Recently, pollution was observed to affect the mangrove areas in the form of polythene bags and bottles, plastic and metal can waste, which are disposed in small quantities by the coastal towns and villages. This may have a serious physical impact by covering the young seedlings and pneumatophores, blocking the tidal channels and causing disturbance to the mangrove associated fauna. Oil pollution is due to a few oil spills from shipping routes and commercial fishing activities, especially in the area of Massawa (MLWE 2014).

In 2003, a mangrove plantation was conducted by the Manzanar Project situated on an area near the port of Massawa. Around 700,000 mangrove seedlings, chiefly *Avicennia marina*, were grown on the treeless mud flats (Sato et al.

2005). Since Eritrea's coast is characterized by gradually sloping beaches, shallow bays and low-energy waves, mangrove forests could be established on at least 10,000 ha of the intertidal zone. It has been proposed to plant 10 million trees in this area to ensure thick canopies at maturity. Since the tidal zone extends inland for about 500 m, the forests could eventually cover 50,000 ha. Harvesting these trees for lumber in a sustainable way would bring annual revenues of US$200 million and significantly boost the Eritrean economy (Opec Fund Newsletters 2003).

7.3 Forest Cover and Soil Erosion

Unfortunately, little information exists on the Eritrean landscape before the Italian colonization in 1890. Available information suggests that forest resources were already scarce before the Italian colonization (Boerma 2005b). While the vegetation of the Eastern Escarpment of the Central Highlands was rather rich, richer than today, the plateau hosted very little woodlands of any kind (Boerma 2005b). In 1941, Eritrea went under the stewardship of the British Military Administration (BMA) and the British, like the Italians, were greatly concerned about paucity of wood resources in the country. In an initial estimate of tree cover in 1945, it was calculated that roughly 5% of the entire country was wooded. However, this estimate was revised two years later, after a more comprehensive survey, which indicated that 10% of the three Central Highland provinces included wooded areas (Boerma 2005b). This estimate was similar to that inferred by Fiori in 1912. On the base of archival material, oral testimonies and photos, it was possible to confirm that, in the past, woodlands and forests covered not more than 7.5% of Eritrea land (7575 km^2). In the 1970s, approximately 2966 km^2 of Eritrea was covered by forests. This area declined to 1401 km^2 in 2014 (Ghebrezgabher et al. 2016). Over a period of 44 years, Eritrea lost approximately 1565 km^2 of forestland, thereby attaining a maximum rate of deforestation during the 1980–2014 period, with an annual rate of 28.30 km^2 per year (Ghebrezgabher et al. 2016). This phenomenon may appear recent, but, on the contrary, it is the result of very ancient exploitation of forested lands. The United Nations Food and Agriculture Organization (FAO 1997) reported that the total forest area including woodland forests was approximately 15,276 km^2. In 2014, the forest and the woodland were 15,078 km^2 (Ghebrezgabher et al. 2016). These data highlight an important contradiction between the documental evidence, which indicates that already by the first decade of the twentieth century forests were scarce and limited to the Eastern Escarpment and some valleys, and the population belief that in the recent past large forests existed. This convincement is also fostered by the Eritrean government which, in the National Environmental Management Plan of 1995, states that at the beginning of the twentieth century 30% of Eritrea was covered with forests, but then this percentage was reduced to the 1% of today (Boerma 2005b). Nyssen et al. (2004) confirmed that there is no evidence of a supposed 30–40% of forest cover in Eritrea in 1900. Instead, the relationship between deforestation and erosion remains a valid assumption.

In 1995, a team of agronomists and pedologists of the former IAO carried out a phyto-sociological survey in the upper Mareb river basin, located 20 km south of Asmara and with a catchment area of 535 km^2 (15°00′ and 15°15′ N–38° 40′ and 38°55′ E). The upper Mareb basin, upstream of the town of Debarwa, is located in the southern region, administrative subzone of Debarwa. The drainage area is about 200 km^2 with elevation varying between 2550 and 1905 m asl (Gehbrehiwot et al. 2019). The main aim of this work was to evaluate the possibility of estimating the *Acacia etbanica* woody biomass available as fuel wood. The calculated productivity of this tree species (0.2 m^3/ha/year of wood) could sustain 10,000 people (Viti et al. 2001).

When comparing the estimates of the Agricultural Ministry of Eritrea for 48,000 people living in the area, a marked disequilibrium between offer and demand emerged (Viti et al. 2001). About 70% of the basin is cultivated, while the remaining part is used for extensive grazing (Colombo and Sarfatti 2001). These data are confirmed by the historical map of the Mareb basin (Fig. 7.8).

Two sub-catchments of the same upper Mareb river basin (the Shiketi, located at 15°10′ N and 38°51′ E, and the Emni Tselim, located at 15°02′ N and 38°44′ E l) were selected in order to calculate the rates of soil erosion by the Universal Soil Loss Equation (Colombo et al. 2001). The area of the Shiketi sub-basin is 585 ha, and denudation processes are very active, with hill slopes affected by mass movements, gully and ravine erosion. The lateritic formation gives rise to structural terraces due to differential erosion which uncovers the surface of the old buried peneplain. This flat surface represents the local base level for the deposition of colluvial sediments and talus slope coming from the weathering of above-lying volcanic rocks. The valley bottom is covered by ancient alluvial sediments and by recent colluvial and fan debris (Colombo et al. 2001). The Emni Tselim sub-basin is 1172 ha. Here, volcanic rock outcrops prevail and slope erosion processes produce large accumulation landforms (glacis) that are periodically eroded by deep, linear gullies and progressively destroyed by lateral fluvial erosion. In the Tselim plain, portions of ancient terraces, buried by a colluvial layer variable in depth, are also found (Colombo et al. 2001). This layer is also eroded, and screen deposits accumulate at the footslopes.

The rates of soil erosion are controlled by three principle factors: geomorphological setting, soil type and land use.

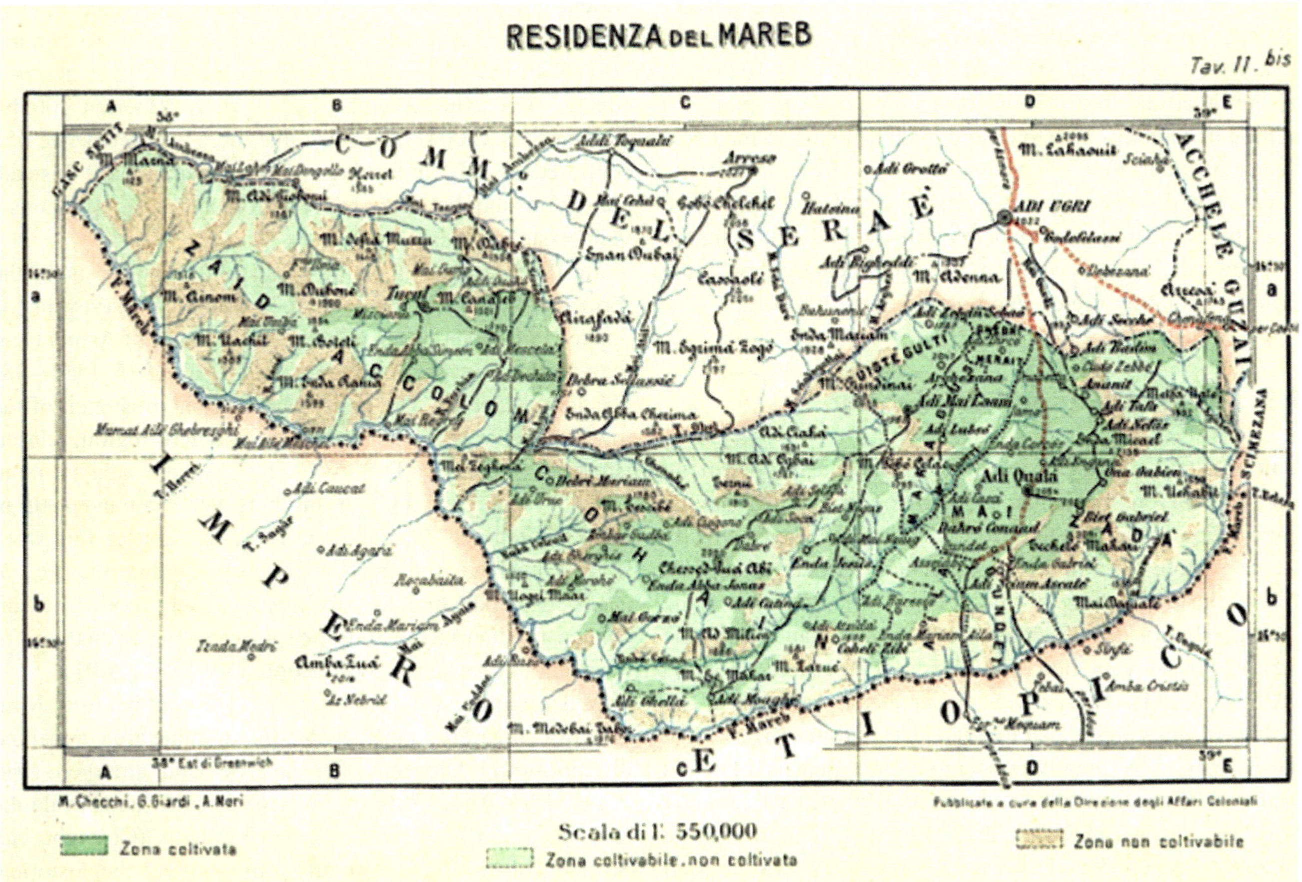

Fig. 7.8 The Mareb basin in an Italian historical map 1:550,000. Cultivated areas in dark green, non-cultivated areas in light green and no-cultivable areas in ochre colour (Checchi et al. 1907) (Photographic Archive of the former Istituto Agronomico per l'Oltremare)

The geomorphological units include alluvial plains, alluvial terraces, recent upper and lower accumulation glacis, plateaus with sheet flood cover, concave-shaped valleys, a volcanic fault escarpment, residual volcanic hills, volcanic structural terraces and recent lower buried glacis. The FAO soil types are eutric cambisol, eutric vertisol, ass. eutric fluvisol–calcaric cambisol–regosol, ass. eutric vertisol–eutric cambisol, gleyic cambisol, ass. eutric leptosol–eutric regosol and ass. eutric regosol–eutric cambisol. In the same river basin, Micconi (1998a, b) classified six land units: seepage slope (Unit 2), convex creep slope (Unit 3), fall face (Unit 4), transportation mild slope (Unit 5), colluvial footslope (Unit 6) and alluvial toeslope (Unit 7) (Fig. 7.9). Land use consists of controlled grazing, no grazing, improved agricultural techniques and degraded land.

In both sub-basins, the largest soil loss, calculated with the USLE, was found to occur in degraded lands with 255 and 129 t ha^{-1} yr^{-1} in the Shiketi and the Emni Tselim, respectively. In the degraded land of the Shiketi, conservation practices such as terracing of hillslopes underlain by volcanic deposits and cultivated colluvial deposits could lead to a 1/3 reduction of erosion and to a 50% reduction for the whole basin (Colombo et al. 2001). Soil loss resulted to be lower in non-grazing land than in controlled grazing areas, 60 and 15% less in the Shikeli and Emni Tselim, respectively. A SCS-CN method was also applied to the two sub-basins to calculate the annual runoff which resulted in 95 mm yr^{-1}in the Shikeli and 84 mm yr^{-1} in the Emni Tselim. In Eritrea, it is estimated that $35–70\times10^6$ t of fertile soil is eroded annually from an area of 2240,000 ha (MLWE 2012). Figure 7.10 shows the soil map of Eritrea and indicates the prevalence of leptosols (FAO-UNESCO soil classification1998) in the Central Highlands.

Yet in 1995, a more specific research on gully erosion was carried out in Halhale Experimental Station of the Eritrean Ministry of Agriculture, near Debarwa. This study was carried out in a small (about 260 ha) sub-basin of the upper Mareb river, ranging in elevation between 1910 and 2020 m asl (Rodolfi et al. 1998). The entire Halhale watershed is included in the region defined as "basaltic plateau" by Abul-Haggag (1961). The study site was subdivided in four land units (Table 7.2).

This study pointed out susceptibility to gully erosion by applying De Ploey's Es Model (Rodolfi et al. 1998). The

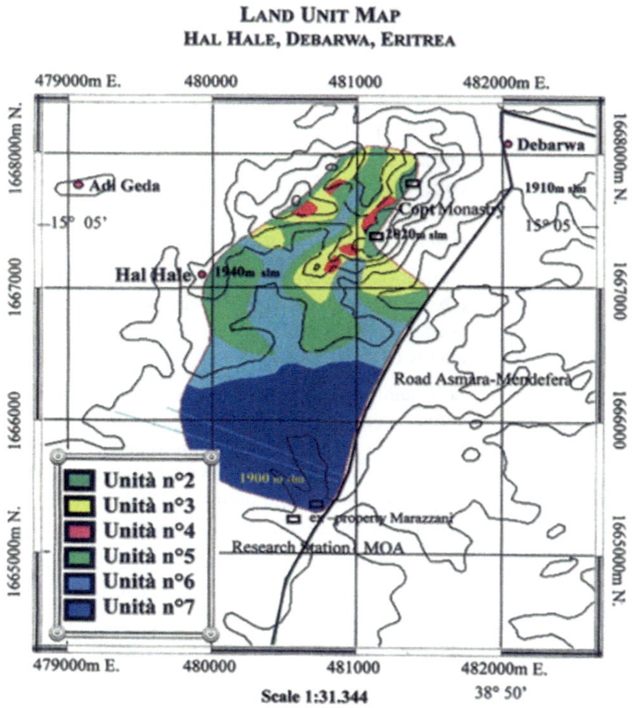

Fig. 7.9 Land Unit Map in the upper Mareb river basin in the area between Debarwa and Halhale localities. In this map is also indicated as an important site the Coptic Monastery Modified from Micconi (1998a, b)

amount of eroded soil volumes during the great rainy season of 1995 (260.2 mm rainfall) was calculated for two gullies (G1 with a drainage area of 145,854 m^2 and G2 with a drainage area of 157,018 m^2) as 42.05 m^3 (G1) and 13.61 m^3 (G2). This level of susceptibility is worrying. The highest values of soil loss are reached at the beginning of the rainy season, while they tend to decrease with the progress of the rainy season. This is probably due to the increase in the protective effect of vegetation cover, as it keeps on growing during the rainy season. Careful observations of the two representative gullies allowed one to obtain objective, though approximate, information regarding the magnitude of soil and gully erosion within a very short time interval (Rodolfi et al. 1998). Figures 7.11 and 7.12 show two typical aspects of Eritrean landscape linked to water erosion.

7.4 Discussion

7.4.1 The Consequences of Colonialism on the Forest Land

Although during the Italian colonial period much work was made on the classification of vegetation, ironically, this period posed a threat to the forests. The main threats were

Fig. 7.10 Soil map of Eritrea 1:4,000,000 based on the FAO-UNESCO soil classification (Photographic Archive of the former Istituto Agronomico per l'Oltremare Archive 1998)

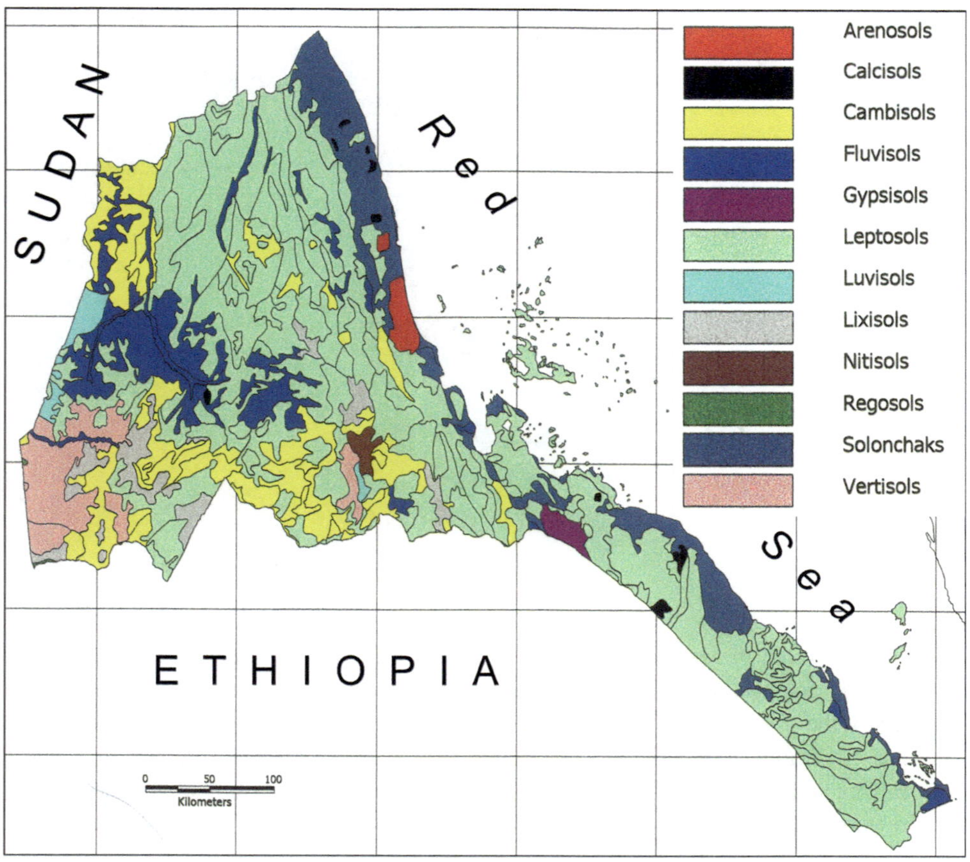

Table 7.2 The Land Unit subdivision in the Halhale experimental catchment (modified from Rodolfi et al. 1998)

Land unit	Soil type (FAO-UNESCO)	(ha)	Main description
Convex interfluves	Rock outcrops	70.1	Vegetation cover is an open shrub, dominated by *Acacia etbaica*. This unit is considered as "marginal land" by the local communities (villages) and utilized as common grassland and for wood collection. Vegetation cover never exceeds 60%, even during the rainy season; water erosion has no obstacle at all
Seepage and creep slopes	Mollic-eutric leptosols	125.5	
Transportational midslope	Mollic-eutric leptosols, eutric cambisols	62.5	Utilization pressure is heavier, and this causes clear erosional features: many rills and sporadic but deeper channels (gullies) can be observed near clear effects of sheet erosion
Colluvial footslope	Eutric vertisols	25.6	Soils can present drainage problems and a high shrink coefficient; however, they are deep and fertile. The main land utilization type is the cultivation of cereals and other annual rainfed crops. In some places, cultivation is hindered because of the development of deep gullies
Alluvial toeslope	Fluvic-eutric cambisols	38.8	These areas can become prone to sheet erosion hazards, where cultivation is extensive and mechanized: the use of a disc plough can lead to the crushing of the soil structure and the formation of surface crusts during successive rains, so that the erosion hazard is increased. Along the riverbanks, basal undercutting can be very active, especially where banks exceed 3 m in height

Fig. 7.11 Eritrean landscape: contour bounds on bare slopes (photo by Rodolfi 1995)

economic, through the expansion of coffee plantations, and the resultant impact on the local population that, in turn, threatened the forests. The main causes of deforestation included traditional and colonial agricultural practices, overgrazing and tree felling for new pastures, the use of trees for firewood, house construction (the traditional Abyssinian *hedmò)* and railway construction (Massawa-Biscia in 1930s). Senni (1915) reported that in 1905 the livestock carrying capacity of Eritrean pastureland was 0.03 big head per hectare and, as a whole, the animal population amounted to 1,063,665 heads, the larger part of which was represented by goats and sheep (69.2%). Furthermore, an important

Fig. 7.12 Eritrean landscape: typical gully erosion within a flat plain (photo by Rodolfi 1995)

factor contributing to the loss of forested land was the use of wood for military purposes in the form of mobile garrisons. According to Senni (1915), during the 1898–1901 interval, a company of indigenous soldiers consumed about 14,000 trunks of acacia species and about 38,000 poles of junipers and other tree species for the construction of traditional dwellings. The industrial activity consumed more than 1800 m^3 of firewood per year. Vegetation was destroyed in periods of both war and peace, with demographic growth as the principal reason for deforestation. Yet, the colonial government favoured deforestation for agricultural settlements by the concessional land system. In 1933, comprehensive legislation was introduced, which laid down which type of wood could be cut by the Italian and native population, respectively, in which areas and under which circumstances, and specified the conditions of allocation of concessions (Lätt 2004). Fifty years of Italian colonial dominion in Eritrea resulted in the systematic degradation of its forest assets. In this regard, it is interesting to report here a short description of the vegetal landscape by Guidotti (1934) before the massive forest degradation: "Along the border of the Setit river there are incense plants (Fig. 7.13) and *Acacia gummifera* trees, which are intercalated with small ebony plants, covering sets of ten square kilometres of land. Along the rivers of the Sudanian lowland, it is possible to encounter palms, tamarind, mahogany and sycamore trees, and on the top hills, there are spiny bushes without any importance. From Akordat to Cheren, tamarisk, baobab (Fig. 7.14), poinciana flame trees and many types of Acacia provide the landscape with a particular physiognomy. On the slopes and highlands, wild olive, *Euphorbia candelabrum, Combretum* plants, terminalia trees, *Anogeissus and Junipers* trees such as *Juniperus procera* (Fig. 7.15) are prevailing on different zones as a consequence of soil characteristics and water content, rainfall distribution and general climatic conditions".

A typical example of deforestation in Eritrea was provided by the Metaten Forest case. This area is in the Woina-Degà zone on a high mountain (2727 m asl), where evergreen rain forest with coniferous plants prevailed. In the 1920s, this forest consisted of 100,000 adult trees with a diameter exceeding 25 cm and approximately 100,000 m^3 of wood were calculated to be present in the region. In that time, however, about 30% of the trees were either already dead or dying. A further 30% were deformed trunks, with the remaining 40% being represented by trees still standing (Mangano 1920). Thus, the Metaten Forest seemed like an ecological island surrounded by a territory with scarce vegetation cover. This forest was the most expansive juniper woodland in Eritrea but, like the other forests, it was rapidly deteriorating. The disproportionately high percentage of dead trees and trees with rotted inner wood compared to the low percentage of young trees required for the forest renewal was impressive. The causes of the forest fatal decline are manifold, including the lack of a rational forest management system (Mangano 1920). At that time, there was no efficient forest service at the national level. Incorrect utilization of adult trees, the fires set by indigenous people to extend their small cereal crops, local practices of felling medium-size trees and parasitic phyto-patologies were among important causes leading to the degradation of the Metaten Forest. The colonial administration of Eritrea attempted to reduce the loss of forest cover. In Fig. 7.16, the forest zones existing during the Italian occupation of Eritrea are shown in green

Fig. 7.13 *Boswellia papyrifera* trees along the Tekezé-Setit River. This is a small tree well known for its frankincense. It is a multipurpose deciduous species (photo by Maugini 1933) (Photographic Archive of the former Istituto Agronomico per l'Oltremare)

Fig. 7.14 Western lowlands of Eritrea are rich in baobab trees (Photographic Archive of the former Istituto Agronomico per l'Oltremare 1937)

on the map drawn by the *Milizia Forestale* (Italian forest rangers) in 1930s.

In 1890, the first governmental ban prohibited fires next to woodlots and the felling of large trees, particularly tamarind and ebony trees. Other bans followed in 1898, 1907, 1927 and 1932–1933. In the last period, the Forest Rule was set up and woods were classified into three types: open access woods, reserved woods and reserved woods where grazing was forbidden. In the 1910s, assignments were carried out by Bartolommei Gioli (1903), Senni (1915), Baldrati (1907) and Fiori (1908–1909), with the predominant objective to solve the problem of incipient deforestation and soil degradation. Proposals of reforestation and re-afforestation measures were reiterated by these authors. Fiori (1912) reported that some tree species were already experimented in situ giving good results in terms of forestry. These tree species were *Cupressus sempervirens* and *Cupressus macrocarpa* in the highlands (Fig. 7.19). In 1911, the first tree plantation site for *Schinus molle*, eucalyptus and acacia species, along with species of pines, *Angophera sub-velutina*

Fig. 7.15 *Juniperus procera* trees (Photographic Archive of the former Istituto Agronomico per l'Oltremare Archive 1927)

and casuarina trees was setup in the Asmara Region (Fig. 7.18).

Between 1922 and 1931, new tree nurseries were setup by the colonial forest service, involving participation of both indigenous people and colonialist farmers. As a result, over a period of twelve years, 1,998,700 trees were planted (Guidotti 1934). This venture consisted of 22 indigenous tree nurseries, distributed over three different altitudinal levels (lowlands, highlands and intermediate altitude plains), in sites most suitable for cultivation (Fig. 7.17). Exotic seeds, derived from America (Tecoma Spp, *Ipomea arborea, Acacia picnanta, Acacia tabehnia, Acacia leucaena*), Ethiopia and Kenya (*Acacia australiana*) and Italy (many coniferous, eucalyptus and robinia tree species), were introduced in 1932–1933 for experimentation. Moreover, seeds of the spontaneous Eritrean flora (*Eritrina* spp., *Euphorbia* spp., *Eugenia* spp. and *Juniperus procera*) were also used (Guidotti 1934).

Though traditional agriculture and goat, sheep, cattle and camel breeding were in part banned in order to protect natural forestlands, the colonial administration not only agreed but also favoured the expansion of cotton and coffee plantations. This fact represented an anomaly. The first attempt to introduce the Arabica coffee on the highlands (>1400 m asl) was done directly by Yemenite growers invited by the colonial administration. The Italian forestry inspector Senni, who periodically visited the Eritrean Colony, identified the northern and eastern mountain slopes (from 1100 to 1900 m asl) as the preferred sites for the extension of coffee plantations, though Brizioli (1930) reported: "In the recent past, this sufficiently wide region was covered by a massive dense tree vegetation and in some sites, it was similar to the forest in the moist tropical regions". Nevertheless, after preliminary forest cleaning and terracing, coffee was cultivated over an area of about 25,000 ha, under both the former natural forest canopy and in artificial plantations of *Poinciana regia* and *Melia azedarach.* In 1923, the coffee farming system was formally initiated, and seven years later, more than 350,000 young coffee plants were planted, approximately 233,000 of which were already producing (Senni 1930). Exotic coffee species such as *Coffea excelsa*, *Coffea robusta* and *Coffea canephora* were also introduced. In 1923–1934, the coffee yield increased from 6.5 to 120 t (Brizioli 1936). In order to protect coffee plantations at higher altitudes, the Italian foresters proposed to plant tree lines of eucalyptus species, namely *Eucalyptus globulus* and *E. corynocalix.* These species were planted along with *E. mollissima*, a well-adapted species able to grow on the slopes of the Eastern Escarpment, and *Leucaena glauca,* a useful tree able to grow on the *bahari* land (Brizioli1930). Other eucalyptus species such as *E. resinifera, E. rostrata, E. robusta, E. citriodora, E. cornuta, E. amygdalina, E. dives, E. piperita* and *E. macarthuri* were introduced for phyto-therapeutic purposes and for the oil extraction industry (Rovesti 1928). Larger eucalyptus plantations still exist on the land of former Italian concessionaires and in the vicinity of towns (Boerma 2005b) (Fig. 7.20).

Undoubtedly, eucalyptus represented a valid alternative in response to the domestic firewood demand, which was increasing by both demographic and industrial growth. In recent years, the wood collecting continues to be a traditional women task (Fig. 7.21).

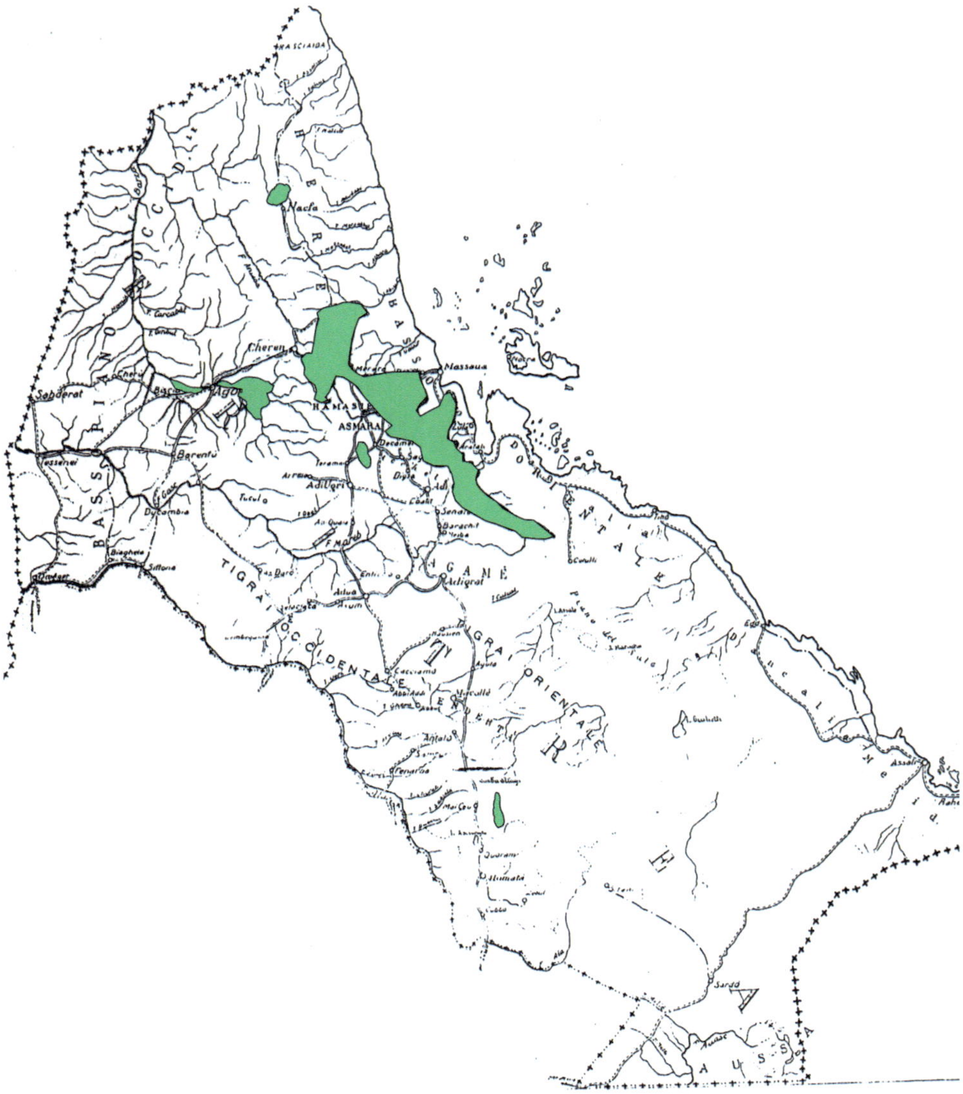

Fig. 7.16 Forested areas (green colour) in Eritrea Italian Colony Map 1:2,000,000 (unpublished, Documentation Centre of the former Istituto Agronomico per l'Oltremare, undated)

7.4.2 Reforestation and Landscape Protection

The rapid demographic increase that occurred on the highlands after the Italian occupation not only created the conditions for the displacement of local people towards zones characterized by a double rainfall regime in search of better crop lands with higher productivity, but also left the forestland subject to plundering and grabbing (Brizioli 1930). According to Boerma (2005b), on the eastern side of the Eritrean highland, "trees are often grown in household compounds, while many villages also have small eucalyptus plantations". During the Eritrean Civil Wars, many former, large Italian plantations and local community plantations were plundered, leading the vegetation landscape back to the situation before the 1920s and 1930s (Boerma 2005b). A large proportion of the old eucalyptus plantations, completed before the liberation, was reduced to degraded lots because of expansion of agricultural fields and soil erosion (Micconi 1998b). According to recent statistics (FAO 2010), from 1990 to 2010, there was a 240% increase in planted forests in Eritrea to attain 34,000 ha by the year 2010. Nevertheless, deforestation is still identified as one of the most serious threats to the indigenous forests of Eritrea. According to MLWE (2014), most of the Eritrean land, originally covered by *Juniperus* forest, has been extensively harvested, mostly for fuel wood and traditional housing construction.

In areas suitable for cultivation, forestlands were replaced by croplands (FAO 1997). Since 80% of the Eritrean population is dependent on farming, the negative impact of subsistence cultivation and overgrazing on forest cover is significant. The dependence on firewood and traditional housing are also factors contributing to excessive deforestation of the country (Ghebrezgabher et al. 2016). Approximately 63 km^2 of forestland and woodland were lost annually in the 1970–2014 interval (Ghebrezgabher et al.

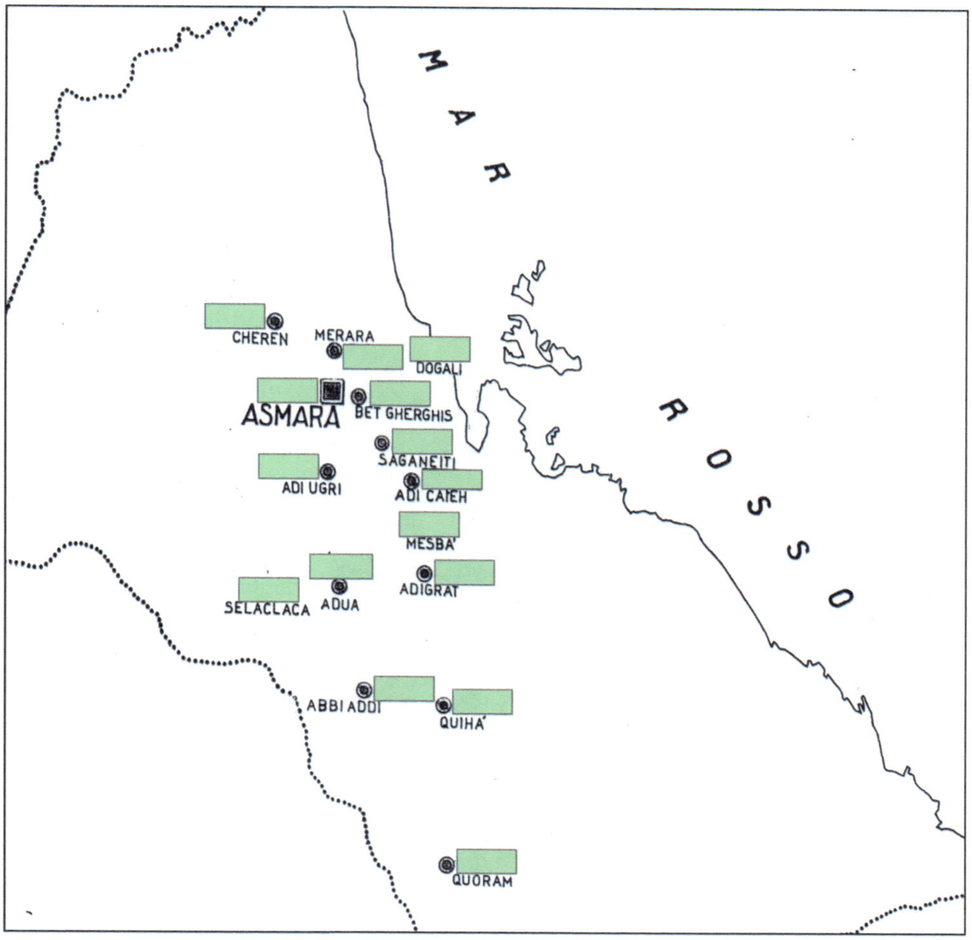

Fig. 7.17 Forest nurseries location in Eritrea in the 1930s on a 1:2,000,000 map (unpublished, Documentation Centre of the former Istituto Agronomico per l'Oltremare, undated)

Fig. 7.18 Reforestation area near Asmara (Photographic Archive of the former Istituto Agronomico per l'Oltremare 1937)

2016), and it has been estimated a devastation of about 70% of woodland during the conflict between Eritrea and Ethiopia (1961–1991). Forests, in fact, were not considered by armies as a mere livelihood resource, but they also were used as shelters, and for this reason, they had to be completely destroyed (usually using fires) (Micconi 1998b).

During warfare periods, in the Eritrean Highlands, the ratio between the number of people living within an area and

Fig. 7.19 Reforestation area with *Cupressus* spp. in Bet-Ghirghis at 2400 m asl (Photographic Archive of the former Istituto Agronomico per l'Oltremare 1937)

the number of people that actually could be sustained based on the available wood resources was calculated to be 4:1, which implied wood consumption four times larger than what the environment could normally sustain (Micconi 1998b). In recent years, invasive alien species (such as *Opuntia ficus indica* in the highland forests and *Prosopis juliflora* in the riverine forests), excessive pollarding of multipurpose trees (such as *Balanites aegyptiaca, Faidherbia albida* and for dry season fodder *Terminalia brownii)*, desert locusts and the excessive use of pesticide sprays are reported to affect forest resources (MLWE 2012, 2014). Grazing pressure is also common throughout the country, and it is particularly severe during the dry season. There are no systematic data on the effect of livestock population size on the forest cover. However, diffuse degradation and the lack of regeneration of many tree species due to overgrazing are ubiquitously evident throughout the country. In the highlands, the growth rate of herbaceous species is extremely low; therefore, the foliage of shrubby and arboreous species (mainly *Acacia* spp.) constitutes the primary food source for pasture. This implies that pastures represent the main cause of the decrease in the number of healthy trees and the decreased opportunity of both regrowth and new seedling birth. All that leads to a vicious circle, in which, as the pasture resources decrease, the potential of rearing resources is also destined to decrease, thereby obliging the farmers to increase the number of animals to satisfy their needs with unpredictable consequences for the environment (Micconi 1998b). Where natural woods were felled and the land degraded, relatively aggressive pioneer species (e.g.

Fig. 7.20 Reforestation plan by eucalyptus trees and anti-erosion techniques on a mountain slope at the Asmara—Nefarit road (photo by Pollera 1961) (Photographic Archive of the former Istituto Agronomico per l'Oltremare)

Fig. 7.21 In Eritrea, wood collection is yet a traditional woman task (photo by Rodolfi 1995)

Opuntia vulgaris, Dodonea viscosa, Carissa spinarum, Euclea schimperi and *Nicotiana* spp,) started to invade the most compromised territories in the upland and at intermediate altitudes (Micconi 1998a), giving way to an ecological succession. In 1994, FAO proposed a reforestation programme to combat the incipient deforestation. It was proposed to create closed areas, inaccessible to both people and farm animals for at least 7–8 years to permit a sort of rotation among many wood lots within the territory. The success of such measures would inevitably depend on the extent to which the local community is involved (Micconi 1998b). The Eritrean government made many efforts and deployed many initiatives for environmental and soil conservation over a 10-year period in the 2000s. In that period, 206,000 ha of permanent forest were closed, more than 90 million tree seedlings were planted, more than 300 dams and millions of kilometres of hillside terracing were constructed, and, since 1995, five sites were proposed as protected areas (MLWE 2012). Moreover, it is interesting to note that the areas surrounding the local Copt Monastery, found for example in the upper Mareb basin (Fig. 7.8), represent an environmental exception because of the presence of plants, including, *Olea* spp, *Euphorbia candelabrum*, and *Prunus* spp. growing on mollisols, which constitute a kind of secondary forest, still very disturbed, but potentially extensible (Micconi 1998a). This secondary forest, if not disturbed, was envisaged to lead to the return of the original vegetation (Micconi 1998b). Figure 7.22 shows a Copt Monastery surrounded by tree vegetation such as an acacia woodlot that creates an ecological island on the Mandefera Plain. Knowledge about forested areas developed around churches and monasteries is yet scarce in Eritrea. During the Italian occupation, the colonial government confiscated the land of one of the most famous monastery of Eritrea, the Dabra Bizen, located in Hamasièn, about 25 km east of Asmara. The surrounding forest was destroyed by the colonialists (Breton 2009). Traditionally, the land tenure system in Eritrea was distinguished according to highlands and lowlands, respectively. On the plateau, the Eritrean Orthodox Tewahido Church was the main owner of a considerable amount of land (Houtart 1980).

In the neighbouring Ethiopia, researches on the church-forests are more advanced. The only areas where one can yet observe natural forests, in northern highlands of Ethiopia, for example, are concentrated in the surroundings of churches. These patches of natural forest have survived as a result of traditional conservation effort of the Ethiopian Orthodox Tewahido Churches (EOTC) (Mekonen et al. 2019). Pilot studies suggest that church-forests might be relics of ancient and largely lost forest ecosystems. They are hotspots of biodiversity for indigenous species, and therefore, they might serve as priority areas for in situ conservation (Wessie Eshete 2007). The effectiveness of these church-forests to provide ecosystem services for the landscape and serve as "stepping stones" for restoration will depend on their long-term sustainability (Wessie Eshete 2007). *Juniperus procera, Podocarpus falcatus, Olea europaea* L., *Ficus vasta* Forssk, *Ficus sur* Forssk, *Hagenia abyssinica* and *Dovyalis abyssinica* are common native trees grown in church-forests (Endalew et al. 2020). In general, sacred groves are the ideal sites for species afforestation programmes in their specific localities (in situ) and could serve as models of sustainable forest management and biodiversity conservation (Mekonen et al. 2019). The benefits of church-forests are categorized under non-market goods because the market has no bid to estimate the monetary value of these goods and services (Endalew et al. 2020).

Fig. 7.22 Coptic Monastery localized in the Mandefera Plain. The woodlot developed around the church is like an ecological island (photo by Billi 2011)

In Eritrea, the indirect economic values of Eritrean forests include catchment protection (57%), carbon sequestration (41%) and erosion control (2%). Emerton and Asrat (1998) estimated these indirect benefits to be nearly 29.8 million euro at the current value (2020) in total. Most recently, a study carried out on the change of ecosystem service monetary values in response to the shift in Eritrea's forest cover during the period 1970–2020 has estimated these losses: "from US$ 5.05 billion in 1970 to US$ 4.75 billion, US$ 3.74 billion and US$ 3.57 billion in 1980, 2014 and 2020, respectively. The overall ecosystem service value in the study period decreased by US$ 1.48 billion due to a decline in woodland and forest land cover 1341 and 1785 km^2, respectively, resulting from deforestation and desertification" (Gebremariam et al. 2021).

The economic loss due to the soil erosion was estimated, in terms of crop losses, as about 46,500 euro per year at the current value (2020). The cost for water supply micro-dams maintenance, in the presence of on-farm soil erosion, was calculated to be around 516,000 euros per year at current value (2020). This amount is equivalent to the value of natural vegetation in mitigating local soil erosion (Emerton and Asrat 1998). The economic value of carbon sequestration by forests, woodlands and grasslands as a whole was estimated to be about 11.8 million euro per year at the current value (2020). The cost to the Eritrean economy of replacing the catchment protection services provided naturally by plant resources would require expenditures of nearly 17.89 million euro per year at the current value (2020).

The direct economic value of biological resources expressed in term of forest products utilization has been calculated to be about 46.6 million euro per year at the current value (2020). This budget does not include local non-timber forest products utilization (fruits, fibres, roots, bark, leaves, seeds, fodder) and the potential income from terrestrial protected areas (Emerton and Asrat 1998).

7.5 Conclusion

In 1998, Micconi stated that "significant data on forest resources in Eritrea are nearly non-existent especially because of the former occupation of Ethiopian armed forces, the consequent situation of compromised safety and finally due to the remarkable lack of interest by the Ethiopian Regime during the last decades". It is worth recalling that in that same year, in June, a new conflict between Eritrea and Ethiopia exploded. This event could explain the existence of two particularities characterizing the environmental history of Eritrea. The first is that the actual knowledge on forest patrimony is linked to the military occupation during the Italian colonial period. The second is that an endogenous process of scientific improvement was never carried out by national institutions. Sometimes, the knowledge about the Highlands, for example, is deduced from parallel studies carried out in an analogous environmental context in Ethiopia. In this *excursus* on the woodlands in Eritrea, we have through necessity followed the *fil rouge* of history because of the scarcity of detailed data and information available today. As historical environmental background, this article referred to the so-called narrative of reforestation that is an expression coined by McCann (1997) and recalled by Lätt (2004), who stated: "There is a widespread belief in Eritrea, that the woodland cover has decreased significantly during last century and that the Highlands were covered with dense forests before or even during Italian colonization" (Lätt 2004). As such, we have outlined the contradictory nature of recent data sources and the difficulty in attaining a comprehensive description of the state of art of the Eritrean forests. We recall here the fact that the prevailing belief about 30% of wood cover in Eritrea just before the Italian colonization would go back to Andrea Branca, leader of the Forest and Wildlife department under the British Administration in Eritrea, who published this figure in a report in 1947 (Lätt 2004).

Surely, Eritrean forests necessitate serious conservation attention. Documented findings from previous studies, presented in this article, have shown that the country has been losing its floristic wealth over a long period. Anthropogenic activities are noted as the principal factors resulting in deforestation. Forest resources are utilized for subsistence cultivation, firewood, timber, urbanization and road construction. Overgrazing, population growth and climate changes are also noted as additional factors aggravating deforestation in the country, and deforestation in turn can exacerbate the effects of climate change; in fact, a quite recent study on the spatio-temporal dynamics of vegetation cover in Eritrea in response to climate (precipitation and temperature) and drought from 2000 to 2017 confirms that "low precipitation was mainly attributed to the slowly declining vegetation trends and increased drought conditions in the semi-arid region of the country" (Measho et al 2019).

Besides boosting the existing reforestation effort by the government, conservation attention should be currently given to the few remaining forest patches in the country. According to the FAO's country programme in Eritrea (2017–2021), the management of natural resources, including soil conservation measures, water harvesting and afforestation, represents "areas that the government of Eritrea wishes to explore in the next phase with FAO" (FAO 2021).

References

Abul-Haggag Y (1961) A contribution to the physiography of northern Ethiopia. University of London. The Athlone press

Baldrati I (1907) L'Eritrea sotto l'aspetto agricolo. Memoria letta alla Società Agraria della Provincia di Bologna 10 febbraio 1907. Tipografia di Paolo Cuppini, Bologna

Baldrati I (1928) Note ecologiche sulla Colonia Eritrea. Tipografia Editrice Pacini Mariotti, Pisa

Bartolommei Gioli G (1903) L'Agricoltura nell'Eritrea. Tipografia della Camera dei Deputati, Roma

Bein E, Hatbe B, Jaber A, Birnie A, Tengnas B (1996) Useful Trees and Shrubs in Eritrea. RSCU/SIDA, Nairobi

Boerma P (2005b) Assessing forest cover change in Eritrea. A historical prospective. Mountain Res Dev 26(1):41–47. https://doi.org/10.1659/02764741(2006)026[0041:AFCCIE]2.0.CO;2

Breton JF (2009) Monastères d'Erythrèe. Edition «Culture et Patrimoines d'Erythrée» Alliance Française d'AsmaraVol. 3, Asmara

Brizioli F (1930) Piante da rimboschimento, copertura dei terreni e sovescio per la regione delle Pendici Orientali dell'Eritrea. L'agricoltura Coloniale 3:114–126

Brizioli F (1936) La coltivazione del caffè sulle Pendici Orientali dell'Eritrea. L'agricoltura Coloniale 3:63–65

Checchi M, Giardi G, Mori A (1907) Colonia Eritrea, Volume II, Carte Speciali. Studio Cartografico G. Giardi, Firenze

Colombo R, Sarfatti P (2001) Hydrological analysis of two subcatchment of the Mareb River (Eritrea)) In: Delli G, Abraha M, Teclai R (eds) Land resources and Land evaluation of the Aini Mereb and Halhale area (Eritrea), Main Report. 17th Training Course on Remote Sensing and Natural Resources Evaluation, IAO, Firenze, pp 57–66

Colombo R, Martucci A, Rodolfi G, Sarfatti P, Tadesse K, Afewerky Y (2001) Soil erosion estimation in the upper Mareb basin (Eritrea) In: Delli G, Abraha M, Teclai R (ed) Land resources and Land evaluation of the Aini Mereb and Halhale area (Eritrea), Main Report. 17th Training Course on Remote Sensing and Natural Resources Evaluation, IAO, Firenze, pp 52–56

Di Gregorio A, Weepener HL (2002) Eritrea Land Cove. Map 1:100,000 FAO Africover Project, Italy

Dove K (1890) Kulturzonen von Nord-Abessinien. Ergänzungsheft No. 97. zu Petermanns Mitteilungen

Emerton L, Asrat A (1998) Eritrea biodiversity: economic assessment. IUCN/MLWE, Asmara

Endalew B, Wondimagegnhu BA, Tassie K (2020) Willingness to pay for church forest conservation: a case study in northwestern Ethiopia. J Forest Sci 66:105–116. https://doi.org/10.17221/154/2019-JFS

FAO (1994) Eritrea. Agriculture sector review and project identification report: Annex 3. Rome

FAO (1997) Support to forestry and Wildlife sub-sector: preinvestment study TCP/ERI/6712(F). FAO Investment Centre, Rome

FAO-NFIS (1998) Forest Cover of the Eritrean territory Map 1:400,000. FAO/GCPS/ERI/002/ITA for Ministry of Agriculture, Istituto Agronomico per l'Oltremare, Firenze

FAO (2010) Country report eritrea, global forest resources assessment (FRA). Food and agriculture organization of the United Nations, Rome

FAO (2021) Review of FAO's country programme in Eritrea 2017–2021. Country Programme Evaluation Series 12, Rome

Fiori A (1912) Boschi e piante legnose dell'Eritrea. Edizioni dell'Istituto Agricolo Coloniale Italiano, Firenze

Fiori A (1937) La vegetazione dell'Impero, le zone forestali dell'Eritrea. L'Alpe: 404–408

Friis I, Sebsebe D, Van Breugel P (2011) Atlas of the Potential Vegetation of Ethiopia. University Press & Shama Books, Addis Ababa

Gebremariam TM, Ping F (2021) Assessment of deforestation and desertification induced impacts on ecosystem services in Eritrea. Int J Sci Res Publ 11(3):18–29. https://doi.org/10.29322/ijsrp.11.03.2021.p11104

Gehbrehiwot A, Kozlov D (2019) GIUH-Nash based runoff prediction for Debarwa catchment in Eritrea. E3S Web of Conferences 97 05001. In: XXII international scientific conference "construction the

formation of living environment" (FORM-2019). https://doi.org/10.1051/e3sconf/20199705001
Ghebrezgabher MG, Yang T, Yang X, Wang X, Khan M (2016) Extracting and analysing forest and woodland cover change in Eritrea based on landsat data using supervised classification. Egyptian J Rem Sens Space Sci 19:37–47. https://doi.org/10.1016/j.ejrs.2015.09.002
Guidotti R (1934) Boschi e servizio forestale in Eritrea. L'agricoltura Coloniale 3:113–124
Haile A, Gebretatios I, Ogbazghi W, Omer MK, Araia W, Gebremariam T, Gebreselassie G (1998) Rehabilitation of Degraded lands in Eritrea, 2nd edn. University of Asmara, Asmara
Hailemichael TM (2015) Mangrove forest extent and status along the Eritrean Red Sea coast. Master Thesis in Fisheries Biology and Management, Department of Biology, University of Bergen, Norway.https://allafrica.com/stories/201806280746.html. Accessed 10 June 2020
Houtart F (1980) The Social Revolution in Eritrea In: Davidson B, Cliffe L, Selassie BH, Behindthe War in Eritrea. Spokesman, Nottingham
Lätt L (2004) Eritrea re-photographed: landscape changes in the Eritrean highlands 1890–2004, An environmental-historical study based on the reconstruction of historical photographs. Master Thesis. Faculty of Natural Science, University of Berne, Switzerland. https://web.archive.org/web/20060304115009/http://www.cde.unibe.ch/University/pdf/TTD/Laett_Eritrea.pdf. Accessed 13 May 2020
Mangano G (1920) I boschi di ginepro della regione Mataten nell'Acchelè Guzai. L'agricoltura Coloniale 11:454–456
Measho S, Chen B, Trisurat Y, Pellikka P, Guo L, Arunyawat S, Tuankrua V, Ogbazghi W, Yemane T (2019) Spatio-temporal analysis of vegetation dynamics as a response to climate variability and drought patterns in the semiarid region, Eritrea. Rem Sens 11 (6):724. https://doi.org/10.3390/rs11060724
McCann JC (1997) The plow and the forest: narratives of deforestation in Ethiopia, 1840–1992. Environ Hist 2(2):138–159
MEM (2004) Asmara power distribution and rural electrification project. ESA Report Including ESMMP, Asmara. http://documents.worldbank.org/curated/en/427141468770937983/pdf/E87210PAPER.pdf. Accessed 13 May 2020
Mekonen AB, Gebreegziabher BG, Wassie WA, Tsegay BA (2019) Review: Church forests—the green spots of Ethiopian highlands. Asian J Forestry 3(2):45–53. https://doi.org/10.13057/asianjfor/r030201
Micconi A (1998a) Environment survey carried out to evaluate land degradation in a semi-arid area of the Highlands of Eritrea (First Part). J Agric Environ Int Dev 92(2/3):155–190
Micconi A (1998b) Environment survey carried out to evaluate land degradation in a semi-arid area of the Highlands of Eritrea (Second Part). J Agric Environ Int Dev 92(2/3):263–287
MLWE (2012) Eritrea's five years action plan. For the great green wall initiative (GGWI) Draft (2011–2015), Asmara
MLWE (2014) The 5th National report on the implementation of the UNCBD. Department of Environment, Asmara
Mereb and Halhale Area (Eritrea), Main Report. 17th Training Course on Remote Sensing and Natural Resources Evaluation, IAO, Firenze, pp 46–52
Nastasi P (1993) Cenni sulle regioni climatiche e floristiche dell'Eritrea. Eritrea: Documento inedito, Cooperazione Italiana-GSO, Asmara
Negri G (1940) Per uno schema cartografico della vegetazione dell'Africa Orientale Italiana. Rivista Geografica Italiana 47:2–16
Nyssen J, Poesen J, Moeyersons J, Deckers J, Haile M, Lang A (2004) Human impact on the environment in the Ethiopian and Eritrean highlands a state of the art. Earth Sci Rev 64:273–320. https://doi.org/10.1016/S0012-8252(03)00078-3
OPEC fund (2003) Mangroves: salt-resistant allies in the fight against hunger and poverty. Newsletter, vol 11, no 3. http://themanzanarproject.com/media.html. Accessed 8 May 2022
Pajella A (1934) Descrizione dell'*Olea Chrysophylla* Lamkc. Istituto Agricolo Coloniale Italiano, Firenze
Pajella A (1947) L'Olivo nell'Eritrea e nell'Etiopia. Olearia 2–3:96–102
Pichi-Sermolli REG (1957) Una carta geobotanica dell'Africa Orientale (Eritrea, Etiopia, Somalia). Webbia XIII 1:15–129
Rodolfi G, Vigiak O, Ongaro L (1998) An application of the De Ploey's Es Model for a quick appraisal of the gully erosion activity in a small watershed in the Eritrean Highlands (Halhale, Debarwa). Geogr Fis Din Quat 21:255–266
Rovesti P (1928) Le piantagioni d'Eucalipto in Eritrea e la loro industrializzazione per la produzione delle essenze. L'agricoltura Coloniale 1:54–66
Sarfatti P (2005) Activities of the Istituto Agronomico per l'Oltremare (IAO) with Eritrea from 1994 to 2001. J Agric Environ Int Dev 99 (1–2):117–123
Sato G, Fisseha A, Gebrekiros S, Karim H, Negassi A, Fischer S et al (2005) A novel approach to growing mangroves on the coastal mud flats of Eritrea with the potential for relieving regional poverty and hunger. Wetlands 25(3):776–779. https://doi.org/10.1672/02775212(2005)025[0776:anatgm]2.0.co;2
Schweinfurth GA (1868) Pflanzengeographische Skizze des gesammten Nil-Gebiets und der Uferländer des Rothen Meeres. Peterm Geogr Mitteil 113-129, 155–169, 244–248
Schweinfurth GA (1901) Le piante utili dell'Eritrea. Boll Soc Africana d'Italia X:129
Senni L (1915) Note sulla legislazione forestale in Eritrea. L'agricoltura Coloniale 2:65–166
Senni L (1930) La coltivazione del caffè e i boschi dell'Eritrea. L' Agricoltura Coloniale 11:563–569
The Manzanar Project (2003) Mangroves: salt-resistant allies in the fight against hunger and poverty. Open Fund Newsl 3(XI):24–26. file:///F:/articolo%20eritrea/eritrea%20revisione%20giugno%202020/NL11_3.pdf. Accessed 10 June 2020
Viti ML, Delli G, Ongaro L (2001) Using SPOT data for woody biomass in upper Mareb catchment(Eritrea). In: Delli G, Abraha M, Teclai R (eds) Land resources and Land evaluation of the Aini
Wassie Eshete A (2007) Ethiopian Church Forests: opportunities and challenges for restoration. PhD thesis, Wageningen Univeristy, Wageningen, The Netherlands. file:///F:/articolo%20eritrea/eritrea%20revisione%20giugno%202020/ethiopian_church_forests_opportunities_and_challe-wageningen_university_and_research_27911.pdf. Accessed 20 June 2020
World Bank (2022) Eritrea Country Profile. https://databank.worldbank.org/views/reports/reportwidget.aspx?Report_Name=CountryProfile&Ib450fd57&tbar=y&dd=y&inf=n&zm=n&country=ERI. Accessed 6 Jan 2022

8 Landscape Change, Land Degradation, and Sustainable Land Management in the Central Highlands of Eritrea

Thomas Kohler and Hans Hurni

Abstract

This study deals with environmental challenges in the central highlands of Eritrea, a semi-arid region and the most populous part of the country. We use three thematic lenses to focus on landscape change, land degradation, and land rehabilitation. First, we look at deforestation and afforestation since the beginning of the colonial era in 1890; second, we examine soil erosion and soil and water conservation; and third, we consider dam and reservoir construction for the provision of water for domestic use and irrigation. The paper recommends an integrated watershed planning approach to promote regional development and sustainable land management.

Keywords

Land degradation • Landscape change • Soil and water conservation • Sustainable land management • Eritrea highlands

8.1 Introduction

Eritrea, located in the northern part of the Horn of Africa, faces increasing sustainable land management challenges. The country lies along the Red Sea and is bordered by Sudan to the north and west, Djibouti to the southeast, and Ethiopia to the south. Its backbone are its highlands, which extend over the country's central North–South axis at altitudes ranging from 1500 to 3000 m above sea level (Fig. 8.1).

Geomorphologically, the highlands of Eritrea are as diverse as their geology, shaped by tectonic vertical movements and river downcutting leading to the evolution of landforms, the basis on which the more recent development of vegetation and soils, as well as the history of human land use takes place. Much of the Eritrean highlands lies above 2000 m of altitude and comprise extensive areas of structurally horizontal land. This is an expression of the essentially peneplained Precambrian basement rocks covered in part by Tertiary flood basalts (Mohr 1971; Daniel et al. 2009). On a smaller scale, denudational, depositional, and volcanic processes have been significant in shaping the surface of the highlands, while the rivers have incised valleys into the highlands and helped to form relict landscapes (Henricksen et al. 1984).

The highlands include geologic formations of nearly all geologic eras (cf. Kazmin 1973): beginning with Proterozoic deeply weathered and folded metamorphic rocks, Palaeozoic granites and quartz diorites, Mesozoic sediments, Cenozoic plateau basalts, and post-tectonic granites. Tectonic uplifting during the great swell of the Horn of Africa in the second half of the Tertiary intensified fluvial processes due to increased relief energy. Today, the Barka River flows northwards to the Sudanese lowlands and towards the Red Sea, the Mereb River flows westwards into the Sudan lowlands, and a number of rivers along the eastern escarpment flow towards the coastal lowlands, and eventually, the Red Sea, and to the Danakil depression. During the Pleistocene, a series of climate changes, from warm to cold and from wet to dry, propelled soil formation and provided the highlands with their present woody vegetation. In the Holocene, human land users gradually removed the forests, particularly in the highlands, and replaced them with cultivated and grazing land, which resulted in widespread degradation of biodiversity, soil and water.

Covering only one-seventh or 16,750 km^2 of Eritrea's surface, the highlands are home to two-thirds of the country's people—and the population is growing. Most of the

T. Kohler (✉)
Centre for Development of Environment (CDE), University of Bern, Bern, Switzerland
e-mail: thomas.kohler@cde.unibe.ch

H. Hurni
Institute of Geography and Centre for Development and Environment (CDE), University of Bern, Bern, Switzerland

P. Billi (ed.), *Landscapes and Landforms of the Horn of Africa*, World Geomorphological Landscapes,
https://doi.org/10.1007/978-3-031-05487-7_8

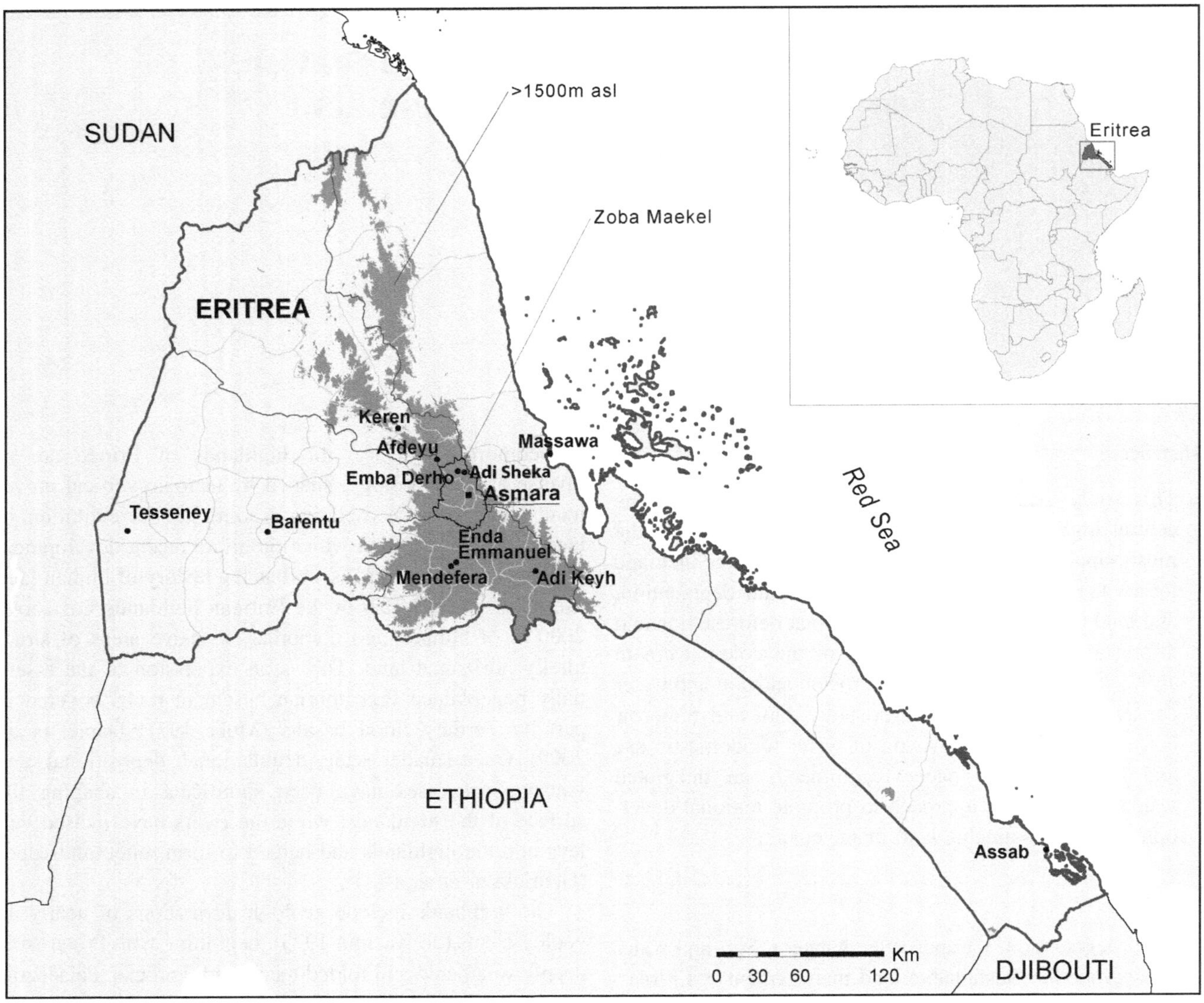

Fig. 8.1 *Map of Eritrea. Highland areas above 1500 m.a.s.l. in dark grey. Map prepared by J. Krauer, CDE University of Bern*

people live in rural areas and derive their livelihoods from small-scale, mixed, rain-fed farming, on farms with about one hectare of cropland per household, combining cereal production—mostly wheat and barley—with livestock keeping (Fig. 8.2). However, annual rainfall is critically low for supporting rain-fed farming. Rainfall records at Asmara meteorological station show an annual average of 509 mm for the period 1914–2015. The main rainy season (June to August) accounts for 76% of the rainfall on average, inter-annual variability is high, and the number of consecutive days without rainfall has been increasing over the last 100 years (Fessehaye et al. 2019). This has led to ever more precarious farming conditions for crop production and to recurrent regional food deficits (FAO 2005). Moreover, with no perennial rivers in the area, the potential for irrigation is limited. And, there is the challenge of soil erosion in what is an area of uninterrupted land use over many generations—probably over many centuries. The highlands are home to the majority of Eritrea's farmers and most of the growing urban population. They include the capital city, Asmara, which is by far the largest urban area in Eritrea and rapidly expanding (Araya and Hergarten 2008). The highlands are also the location of important industrial sites, such as mining enterprises.

The present contribution investigates these challenges with a focus on landscape change, land degradation, and land rehabilitation through three thematic lenses: first, by looking at deforestation and afforestation; second, by examining soil erosion and soil and water conservation; and third, by considering dam/reservoir construction for the provision of water. The geographical focus of this study is Zoba Maekel, the central administrative zone of the highlands (which includes Asmara)—a region also simply known as "the central highlands".

Fig. 8.2 Farmers threshing cereals with oxen, central highlands of Eritrea (photo by Roden 2006)

8.2 Landscape Change: The History of Forest Cover and Current Afforestation Efforts in the Central Highlands of Eritrea

Government policy on land management has been shaped by the belief that forested lands in Eritrea decreased from 30% in 1890 (around the start of Italian colonization) to below 1% in the early 1990s. The figure of 30% has been widely cited in scientific publications, development reports, Eritrean textbooks, and travel guides, laying the foundation for the country's marked efforts in tree planting and afforestation.

The reasons given for this century-long decline in forest cover are the combined effect of destructive land use practices by the Italian colonizers, the war of liberation against Ethiopia, and the unsound land and resource management by the peasantry, i.e. Eritrea's small-scale farmers, including extensive use of wood and timber for cooking and for construction of traditional housing (*Hidmos*). Deforestation is held responsible for a variety of Eritrea's problems, including widespread soil erosion, desertification, lack of surface water, and persistent food insecurity (Liebi 1993; Laett 2004; Boerma 2006). This perception is in line with the narrative of rapid and widespread deforestation in Africa, including Ethiopia and the Greater Horn Area, during the course of a century (McCann 1997; FAO 2001; Boerma 2006). In Eritrea, the image of destroyed forests has served as a metaphor for the oppression and suffering the country had to endure for 100 years starting with colonization. Similarly, it provides a powerful tool for general mobilization for the purpose of national reconstruction. This includes large afforestation campaigns using rural and urban communities and young people during their national service. Forests, thus, symbolize the hope of a green future for all, based on patriotism and nation building. They are seen as a cornerstone to overcoming the country's main environmental problems (Laett 2004; Boerma 2006).

However, research carried out in recent decades casts doubt on the belief that deforestation in Africa is a phenomenon of the recent past—i.e. of the last century. A growing body of evidence suggests that deforestation rates for this period have been largely overestimated, also in the Horn of Africa, and especially in Ethiopia (Wolde-Mariam 1991; McCann 1997; Ritler 2003; Nyssen et al. 2009). There is less research on the history of forest cover in Eritrea, especially relating to the whole country. But for the central highland plateau, there is increasing evidence that forest cover, and hence, forest loss over the last 100 years, has been overestimated (Liebi 1993; Laett 2004; Boerma 2006; Fig. 8.3).

Travel accounts and reports of expeditions in the nineteenth century and early twentieth century all indicate that the Eritrean highland plateau contained very little woodland already then, in contrast especially to the eastern escarpment, which was likely richer in vegetation than today (Matteucci 1880; Munzinger 1890; in Boerma 2006). The Italian colonial administration soon restricted the cutting of trees, restrictions that became increasingly tighter over the years, and very early on began to import timber to meet local

Fig. 8.3 Forest cover and land use at the village of Enda Emmanuel near Mendefera, Eritrean highlands, 1920 (left) and 2004 (right). Both photos show a landscape largely devoid of trees. The 2004 photo shows less open bush on the slope, but more trees in the village, which consists mainly of brick instead of wooden housing (Photos: left: photographer unknown, photo taken around 1920, Istituto Agronomico per l'Oltremare, Firenze. Right: photo by L. Laett, CDE University of Bern, 2004)

construction demands (Martini 1900). A study undertaken by an Italian forestry expert in 1905–07 puts the country's forest cover at about 7% (Senni 1915, in Boerma 2006). A few years later, a second and more detailed study highlighted the stark contrast between the highlands and the eastern escarpment: "While the escarpment is in most places generally wooded the highland plateau instead is completely bare or almost so". The study did not quantify forest cover (Fiori 1913, in Boerma 2006). Recent research based on repeat photography, which compares historical with recent photos, confirms the impression of the highland plateau as an area mostly devoid of trees, suggesting that whatever forest cover there was must have largely disappeared before the Italians arrived around 1890 (Boerma 1999, 2006; Laett 2004). In contrast, on hillsides and steeper slopes, wooded species were often more widespread than today, and olives (*olea africana*) and junipers *(juniperus procera)* have been replaced with acacias in some areas, a change indicating a qualitative deterioration of the woods. To increase tree cover and give the highlands a more hospitable shape, the colonial administration, Italian settlers, and local Eritrean communities engaged in tree planting in the highlands, mostly relying on eucalyptus and sisal. The colonial administration also established eucalyptus plantations in the immediate surroundings of larger settlements and towns, for example in and around Asmara, in Segeneiti and Mendefera (Boerma 1999). On hillsides and steeper slopes, shrub was replaced by *beles (opuntia ficus indica)*. Catholic missionaries reportedly introduced this cactus plant in the early nineteenth century on the eastern escarpment, where it is widespread today. It is also prominent across the highlands, where it forms part of the communal land use system mainly for its fruits (Laett 2004).

From the period of the Ethiopian rule (1952–1991), statistical data on forests and forest cover are largely lacking. Studies carried out by the Ethiopian government in 1974 and later by international agencies (FAO 1984) were hampered by logistical and technical problems at a time of escalating warfare, and their results relating to forest cover appear not to be reliable (Boerma 2006). An extensive reforestation and soil conservation programme was established by the Ethiopian administration, for example along the Massawa-Asmara road on the eastern escarpment. Local communities were incited or coerced to participate in food-for-work programmes, which continued later under the DERG regime after the government of Haile Selassie was overthrown in 1974. By that time, the liberation war was in full rage, but both sides, the Ethiopian and the Eritrean, continued to engage in afforestation and soil and water conservation in the regions under their control. In retrospect, the impact of this 30-year war (1960–1991) on forests is difficult to appraise in quantitative terms. On the one hand, many trees were pillaged and plantations destroyed by the Ethiopian army. Trees were cut to sell in Asmara and other towns and for building trenches by both sides, which consumed a lot of wood. On the other hand, increased inaccessibility of certain areas during the war may have had a positive impact on wood cover in the country. Military activity and mines kept animals and herders away from these often more remote areas. The fall in livestock numbers during the war may also have relieved pressure on vegetation cover, including woods (Boerma 2006).

After liberation in 1991, the new Eritrean government strongly promoted tree planting and other ways of landscape restoration, as it had done already in the years preceding independence. The administration launched massive campaigns to implement terracing combined with afforestation, woodland closures, catchment protection, and dam construction, in an act of faith in future of the country. The high level of resources devoted to afforestation and tree planting has resulted in a remarkable regeneration of much of the highlands in terms of forest cover since the early 1990s (Laett 2004; Boerma 2006). According to the Eritrean Land Use Survey of 2014, tree plantations covered over 16,000 hectares or close to 9% of Zoba Maekel, mostly eucalyptus and some olives (*Olea africana*). The region ran nine nurseries from which trees were distributed to local communities for planting. Some of these nurseries had been established under Ethiopian rule between 1982 and 1987 (Selamawit and Mihreteab 2016). The region also had 7346 ha (7%) of temporary enclosures, i.e. areas closed for grazing. Such enclosures, based on local traditional practices, have been widely implemented by authorities across the whole highlands to facilitate rehabilitation of degraded overgrazed areas, mainly hillsides, by natural regeneration. Enclosures were often combined with hillside terracing and eucalyptus plantations. Such combined efforts have been used, especially in dam catchment areas, in order to reduce siltation of reservoirs, with a focus on those reservoirs that secure water supply for Asmara (Selamawit and Mihreteab 2016). Farmers' access to the trees they planted is restricted, and grazing in plantations is banned altogether.

The efforts to restore vegetation cover have been complemented over the last two decades by policies to relieve pressure on wood resources. The most prominent example is the promotion of improved cooking stoves (locally called *Adhanet*), mostly through village campaigns. By 2015, over 30,000 stoves, which cut firewood consumption by half, are reported to have been established in Zoba Maekel alone, which is very close to full coverage of all rural households in the Zoba (Selamawit and Mihreteab 2016).

To summarize, there is little written evidence on the extent of forest cover throughout the country in the early days of Italian colonization. For the central highland plateau, neither archival evidence and old travel accounts, nor the results of repeat photography, support the thesis of extensive forest cover in this region around 1890, and an overall, rapid, and general decrease of forest cover since then. If there were ever extensive forests there, they must have disappeared earlier. The story of the highland plateau may well be comparable to that of the landscapes around the Mediterranean Sea. As with the Mediterranean landscapes, the highlands represent an old cultural landscape whose largely sloping land has been tilled and ploughed for centuries, by a dense population largely reliant on wood resources. The loss of forest cover in the highlands is thus very likely to go much further back in history.

8.3 Land Degradation and the Promotion of Soil and Water Conservation

8.3.1 Soil and Water Conservation in the Highlands of Eritrea

Land degradation has negative effects on large tracts of land and renewable natural resources across the world. Drylands are particularly vulnerable, due to unsustainable land management and soil erosion (Schwilch et al. 2012; UNCCD 2014). The highlands of Eritrea present a case in point: erratic, often high-intensity rainfall, scarce natural vegetation, sloping land, and increasing pressure on land resources by a growing population, all contribute to land degradation. Soil erosion has been a key development issue in Ethiopia and Eritrea, well documented since the 1970s and especially since the 1980s (Hurni 1978, 1983, 1987, 1993; Stillhardt et al. 2002; Hurni et al. 2016). In those years, measures against soil erosion were taken by the Ethiopian administration (at that time Eritrea was a province of Ethiopia). After independence in 1993, the Eritrean authorities took up the fight against soil erosion with renewed vigour and widespread action across the country. The 1995 National Environmental Management Plan identified soil erosion as the most important issue relating to natural resource use in the country (Government of Eritrea 1995).

Soil and water are the basis for production in a country where about 80% of the population depend on small-scale, rain-fed farming for their livelihoods. The situation is particularly critical in the highlands. Home to 65% of the country's population, they cover only 16% of the country's territory, and farming is suitable or marginally suitable on only an estimated 30% of the land (Selamawit and Mihreteab 2016). Moreover, mounting pressure on grazing land and wood resources has reduced vegetation cover, especially grass cover, to critical levels in many places. Soil erosion is affecting agricultural production, causing crop yields to decline at an estimated rate of 0.5% per annum (Hurni 1993; Juel 2002). Set against the background of the national production, which covers only about half of the food requirements in a year of adequate rainfall, this is a worrying decline in the long run. Soil erosion is, thus, a threat not only to individual farmers' livelihoods, but also to national food security.

The extent of soil and water conservation measures across the highlands is considerable. While official statistics are hard to come by, satellite imagery can help present an estimate. Automatic classification from highest-resolution satellite imagery (geo-eye), done on a test area of 2330 ha in the central highlands (Eckert et al. 2017), identified different types of overlapping indigenous and introduced conservation structures, typical of the highland areas (Negassi

Fig. 8.4 Extensive soil and water conservation works (terracing) on cropland during a government-led campaign in 1999–2000 near Afdeyu, central highlands of Eritrea (photo by Gurtner 2004)

Fig. 8.5 Hillside terracing at Afdeyu in the foreground, with some newly planted eucalyptus on the terraced hills in the background. Previously used for grazing, the hill area was closed off once the terraces were built and the trees planted (photo by Gurtner 2004)

et al. 2002; Ogbazghi et al. 2011). For the entire test area, the analysis found over 1500 km (1,538,370 m) of terraces and bunds on cropland, or about 940 m per ha if 70% of the test area was cultivated, a figure confirmed by ground truthing. According to the Ministry of Agriculture, if we assume that one person built about 8–10 m of terraces or bunds per day, this many conservation structures involved 150,000–190,000 person-days. These figures relate to cropland conservation only (Fig. 8.4), without accounting for the construction of the narrow-based hillside terraces and the tree planting on these (Fig. 8.5). The costs of a comprehensive conservation campaign, including cropland and hillsides, are difficult to establish, but estimates made in 2001 by Denmark's development cooperation agency, Danida, present a figure of about USD 450 per ha (Juel 2002), based on their extensive conservation work with the Ministry of Agriculture across the country. Taking this figure, conserving the cropland of our study area of 2330 ha would have incurred costs of just over USD 1 million, and this for what is but a small fraction of the central highlands (1,042,000 ha). Based on such a small area, this figure might be taken as anecdotal evidence, but it gives an idea of the size of investment needed to protect the soil resources in the area from being washed away.

When soil and water conservation (SWC) activities in the highlands of the Horn of Africa, i.e. in Ethiopia and Eritrea, gained momentum in the late 1970s and in the 1980s, there were no data on erosion rates or on the importance of the factors that defined these rates. Practical measures such as establishing terraces, for example, had to rely on guesswork, or at best, on experiences from other parts of Africa or elsewhere in the world—such as the USA. However, soil erosion is site-specific; while the factors that trigger it can be generalized, their scope and importance vary from one region to the next. Erosion models and conservation work must therefore consider these regional specificities.

To collect the regionalized field-based data required, the Soil Conservation Research Programme (SCRP) was set up in Ethiopia in 1981 (Hurni 1982). Funded mainly by the Swiss Agency for Development and Cooperation, the programme was jointly run by the Ethiopian Ministry of Agriculture and the Institute of Geography, and later, the Centre for Development and Environment, both at the University of Bern, Switzerland. The main objective was to support implementers of soil and water conservation by monitoring soil erosion, testing alternative conservation measures, building local capacity, and informing policy and decision-makers on key aspects of conservation approaches and technologies (Fig. 8.6). The project ran seven field stations, one of which was in the central highlands of Eritrea at Afdeyu, 20 km north of Asmara, at an altitude of 2300 m. Established in 1984, Afdeyu research station remained staffed with records kept through the time of the liberation war. It was handed over to the newly created National Agricultural Research Institute (NARI) of Eritrea in 2006.

Within its research design, Afdeyu had four experimental plots: three for testing different conservation measures and a fourth as the control plot (see Box). The plots were identical in size, slope (31%), and soil properties and represented an average slope angle of the farm (crop) land in the region (Stillhardt et al. 2002). Plot erosion measurements were supplemented with mapping of rill and gully erosion in the catchment area of the station. The station also collected hydrometeorological data, such as temperature, precipitation, evaporation, runoff, and sediment yield of the small watercourse draining the 200 ha-catchment. Key results relating to rainfall, soil erosion, and conservation are presented in Table 8.1 for the years 1989 to 2001. Land use, soil, and conservation mapping in the catchment complemented the experimental work.

The first thing to notice in Table 8.1 is the great inter-annual variability of rainfall (in Asmara, the variation

a b c d

Fig. 8.6 Selection of conservation structures on cropland in the central highlands, from left to right: **a** stone terraces, **b** level bunds, **c** level bunds with tied micro-basins to retain water during rainfall and make it available to crops; **d** effect of SWC on crop growth, here above a stone bund; see text on conservation and crop yields (photos **a, b, d,** by Gurtner 2004; photo **c** by Burtscher 2002)

Table 8.1 Soil loss and runoff for different conservation measures on experimental plots, Afdeyu research station 1989–2001 (Tesfay et al. 2009) *Fanya juu* is a Swahili term meaning "to throw upwards" (i.e. the soil from the trench); the measure was developed in Eastern Africa

Year	Annual rainfall (mm)	Annual soil loss (tonnes per hectare)			
		Control plot	Level bund	Level *fanya juu*	Level double ditch
Mean 1989–2001	480	49.2	11.6	4.6	2.8
Wettest year 1995	658	87.6	21.9	5.5	7.0
Driest year 1990	244	10.4	8.6	1.5	0.4
Year	Annual rainfall (mm)	Annual runoff (mm)			
		Control plot	Level bund	Level *fanya juu*	Level double ditch
Mean 1989–2001	480	222.3	111.4	69.1	57.4
Wettest year 1995	658	248.4	148.4	106.4	108.8
Driest year 1990	244	65.7	90.7	15.7	8.1

coefficient of annual precipitation is 0.32, Billi personal communication). The wettest year received 2.7 times more rainfall than the driest one. Moreover, the annual sequence of precipitation follows no pattern that would allow predictability from one year to the next (Tesfay et al. 2009). Annual totals are low, with three humid months (June–August), and the rest arid (Burtscher 2003, following the classification of Walter et al. 1975). This makes overall conditions for crop production difficult and unpredictable. The data also show that soil erosion was heavy without conservation. On the control plot, on average 49 tonnes of soil/ha were lost per year over the whole period. Soil loss increased to 87 tonnes in the wettest year. The different conservation measures reduced soil loss considerably: for the whole period (1989–2001), annual average soil loss varied between 11 tonnes/ha (level bunds) and as little as 2.8 tonnes/ha (level double ditches). These values correspond to 23 and 6%, respectively, of the loss on the control plot. Assuming a local soil formation rate of 1–4 tonnes/ha per year (Hurni 1983), which would be the maximal tolerable soil loss, two of the three measures tested were thus within these limits and hence appear to be ecologically sustainable. One measure—level bunds—led to soil losses above tolerable levels on average and in the wettest year, but in drier years, also level bunds are close to being sustainable, i.e. in terms of soil formation. Level bunds were the most widespread local conservation practice at Afdeyu and in most parts of the highlands (Fig. 8.6). Farmers prefer them because they require less labour and land compared to other measures.

The soil erosion records at Afdeyu station revealed that a small number of heavy rainfall events (generally storms with over 50 mm in less than one hour) accounted for disproportionally high soil loss. Over the whole recording period (1984–2007), such storms accounted for 15% of all rainfall events and totalled 27% of annual rainfall—but they caused 70% of soil loss (Tesfay et al. 2009).

Experimental plots for testing Soil and Water Conservation (SWC) measures

The plots were 30 m long, 6 m wide, with a slope of 31% and identical soil properties. The following measures were implemented on these plots:

- Plot 1: *Control plot* (untreated)
- Plot 2: *Level bund:* a level embankment 0.3–0.5 m high, made of stones or earth depending on availability of material. Width of bund is 0.5 m. Level means following the contour.
- Plot 3: *Level fanya juu:* a trench dug by throwing the soil upwards to form a level embankment. Over time, the embankment develops into a terrace. *Fanya juu* is a Swahili term meaning "to throw upwards" (i.e. the soil from the trench); the measure was developed in Eastern Africa. Width of trench is 1 m.
- Plot 4: *Level double ditch*: a level embankment with a ditch on the lower and upper side. Ditch height is about 0.5 m, width of each ditch 1.5 m. Level means following the contour.

8.3.2 Soil Conservation and Water Availability

Table 8.1 shows that soil conservation reduces runoff very considerably, particularly during dry years. Without conservation, runoff amounted to 46% of rainfall on average over the whole period. Conservation reduced runoff to 23% (level bunds) and even 12% (double ditches) of total rainfall. In 1990, the driest year in the period, runoff was negligible in *fanya juu* (see box for explanation of this term) and double ditches plots: with over 90% of rainfall retained on the plots with such measures (the control plot retained 74%).

Fig. 8.7 Annual rainfall and river runoff at Afdeyu before and after the 1999–2000 SWC campaign. The Figure shows event (storm)-based rainfall and river discharge (Burtscher 2003)

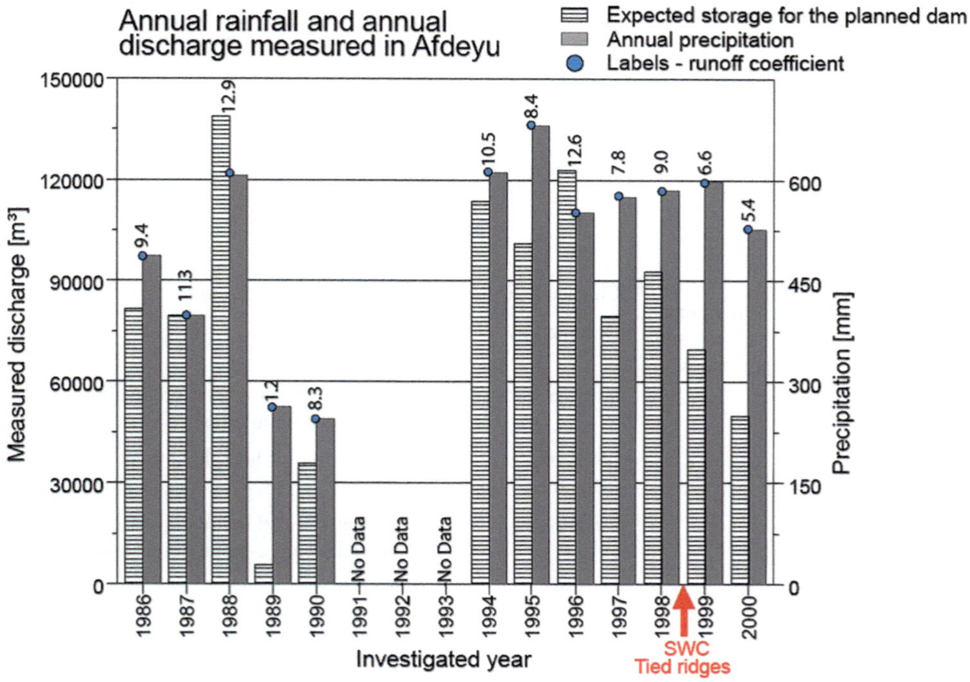

Conservation measures thus store moisture and make it available for crop production in dry years when it is needed most. This has a positive effect on crop yields, as shown below.

Soil and water conservation had already been carried out in Mayketin Catchment prior to the establishment of Afdeyu research station in 1984. The effectiveness of soil and water conservation was evidenced after a second SWC campaign in 1999 and 2000. The campaign, proposed by the Ministry of Agriculture, was carried out through a cash-for-work programme involving the whole community. It covered the catchment in which the Afdeyu research station was located. Funding came from a programme run by the University of Bern and financed by the Syngenta Foundation for Sustainable Agriculture. The whole catchment (200 ha) was treated at a total cost of USD 60,000—or USD 300 per ha—substantially lower than the Danida figure mentioned above. Yet conservation measures were extensive and included bunds with tied ridging on most of the land to prevent siltation of a small dam planned downstream. After the campaign, runoff from the catchment decreased markedly. The runoff coefficient (percentage of rainfall that flowed off as runoff) was 9–13% before the campaign, dropping to 6.6% after the 1999 campaign, and to 5.4% after completion of the work over the whole catchment in 2000 (Burtscher 2003) (Fig. 8.7). These figures represent the first field-based accounts of the effect of soil conservation on runoff and catchment water yields in Eritrea. They thus provide key data for locating and dimensioning rural dams and reservoirs across the highlands. Such dams are the main source of water for domestic use and irrigation in the central highlands of Eritrea, where river flow is very scarce, ephemeral, or at best seasonal (see Chap. 3 of this book for more details).

The campaign at Afdeyu also led to a higher water table and perennial base flow in the main valley. Prior to the campaign, the flow had dried up in the dry season. The additional groundwater enabled farmers to extend garden irrigation along the valley bottom. The campaign was thus very effective in controlling erosion and improving year-round groundwater availability. Crop yields in the lower section of the plots were higher than in the rest of the plot, most likely due to soil, water, and nutrient accumulation caused by the conservation structures (Stillhardt et al. 2002) (Table 8.2). Improved storage of water in the soil is particularly important, especially if climate trends observed over the last 100 years (no change in total annual rainfall, but higher rainfall intensity and increasing number of consecutive dry days) remain in future. However, farmers were only partly happy because of the loss of arable land due to the relatively space-intensive conservation measures that included micro-basins between the tied ridges. In the years after the campaign, many farmers ploughed these basins over to reclaim land for cultivation.

8.3.3 Local Traditional or External International? Merits and Demerits of Different Conservation Approaches

Farmers are well aware of the threat of soil erosion and of the importance of soil and water conservation. A study in a central highland community documented over 20 traditional

Table 8.2 Average crop yields (medians, in g/m^2) on terraced land (soil and stone bunds) for the period 1984–1998 (Stillhardt et al. 2002)

Position of yield samples	Barley	Wheat	Potato	Onion	Linseed	Maize
Above lower bunds	206	148	3913	933	37	738
Between bunds	198	112	2638	1571	33	368
Below upper bunds	183	124	2601	1735	30	375
Number of yield samples	44	31	11	8	11	3

Table 8.3 General characteristics of local traditional and external international soil and water conservation approaches, Eritrea

Characteristics	Local traditional SWC approach	External international SWC approach
Designed by	Farmers (local)	Engineers, development planners (national, international)
Purpose	Maintain and enhance farm production, especially relating to crops	Maintain regional resource base (soil and water) and ecosystem services
Level of integration	Farm operations at household level	Catchment, watershed
Design	Modifiable; site and household-specific. Structures take up a minimum of land	Standardized design. Structures may take up much land
Area of application	Cropland (plots) of individual household	Whole catchment: uncultivated hillsides, communal grazing land, and cropland
Approach	Individual household initiative, farmer-to-farmer exchange; no external support	External initiative; campaign, often by FFW/CFW, external technical support
Implementation	Incremental and individual, fitting into household labour availability	One-time, collective campaign
Labour requirement	Variable and preferably low	High to very high
Inputs / costs	Low, if any, external material input	Generally high; involves external input
Returns	Benefits go to individual farm	Benefits go to local community and to downstream and external users
Documentation	Practices and effects poorly documented, but handed down locally	Substantial body of research and implementation experience

FFW Food-for-Work, *CFW* Cash-for-Work (Gurtner et al. 2006, modified by authors)

conservation measures still in local common practice (Gurtner et al. 2006). Such practices typically focus on securing food production and hence on cropland (Table 8.3). To keep labour investment low, conservation work is done when household labour is available, and integrated into other fieldwork as a side activity. Land ownership in the highlands is vested in the village community and reallocated among its households at regular intervals, generally spanning seven-year cycles. Farmers were clear that this tenure regime does not provide sufficient incentive for heavy investment in conservation. Given that land holdings are generally around one hectare per household, conservation measures are chosen to occupy as little of the scarce cropland as possible.

In comparison, external approaches, following the international mainstream, often require much more land and labour. If the labour is paid, for example through cash-for-work or food-for-work, such approaches are also extremely expensive. Conserving one square kilometre of land by standard measures such as those applied in government campaigns in Eritrea would then cost about USD 45,000, as mentioned earlier (Juel 2002). However, external approaches have their merits, too. Normally, they are regionally inclusive as they treat entire watersheds or catchments, thus addressing erosion comprehensively in terms of space. After all, cropland is only one source of erosion and not necessarily the most important one. Especially in Eritrea: in the central highlands, for example, grazing lands including steep hillsides accounts for over 60% of the land (Selamawit and Mihreteab 2016). In contrast to traditional measures, their focus in safeguarding critical ecosystem services extends beyond the local scale and thus supports regional development. One example is securing the provision of water for urban development by preventing siltation of local reservoirs.

Future efforts in soil and water conservation should be integrated into collaborative watershed management and regional planning, particularly in the central highlands of Eritrea with their increasing demand for land and water for

Fig. 8.8 Dam with reservoir and gauging station, central highlands of Eritrea. In the background, the village of Emba Derho (photo by S. Tesfay)

Table 8.4 Dams and reservoirs in the central highlands of Eritrea (Zoba Maekel) 1890–2007 (Daniel et al 2009)

Year of construction	Total number of dams	Total design capacity (m^3)	Mean design capacity (m^3)	Main water use	Number of dams with local irrigation	Irrigation (ha per dam)
1890–1951	8	7,110,000	888,800	Urban	4	4–5 ha
1952–1991	32	36,043,000	1,126,300	Urban and rural	19	2–33 ha
1992–2007*	23	21,220,000	922,600	Urban and rural	9	2–29 ha
2008–2015**	18	No data	No data	Urban and rural	No data	No data
Total	81	No data	No data	Urban and rural	No data	No data

According to Eritrean classification, dams have a designed reservoir capacity of $\geq 70{,}000$ m^3
* 1992–2007: *Source* Daniel et al. (2009)
** 2008–2015: *Source* Selamawit and Mihreteab (2016)
Note Main water use: *Urban*: domestic, services, and industry, mainly for the capital, Asmara. *Rural:* livestock, watering and small-scale, village-based irrigation

farming, industry, services, and domestic use. Collaborative management will need flexible approaches, in order to better reconcile local farmers' needs with regional development agendas (Table 8.3).

8.3.4 Dam Construction and Sustainable Land Management

The central highlands have no perennial watercourses and ground water availability is limited, owing to the rocky underground of the area (mainly a lateritic crust covering the planation surface of the basement rocks in the Asmara plateau—see Chap. 1 of this book for more details). The region's water supply thus critically hinges on dam construction and rainwater stored in their reservoirs (Fig. 8.8). The central highland zone (Zoba Maekel, 1042 km^2) had 125 artificial reservoirs in 2015, thereof 81 dams, i.e. reservoirs with a storage capacity of 70,000 m^3 or above in accordance with the classification applied in Eritrea; see Selamawit and Mihreteab (2016), and (Table 8.4). Storage capacity varied greatly; while most dams had between 70,000 and several hundred thousand m^3, the few largest ones were designed to store several million m^3 and the

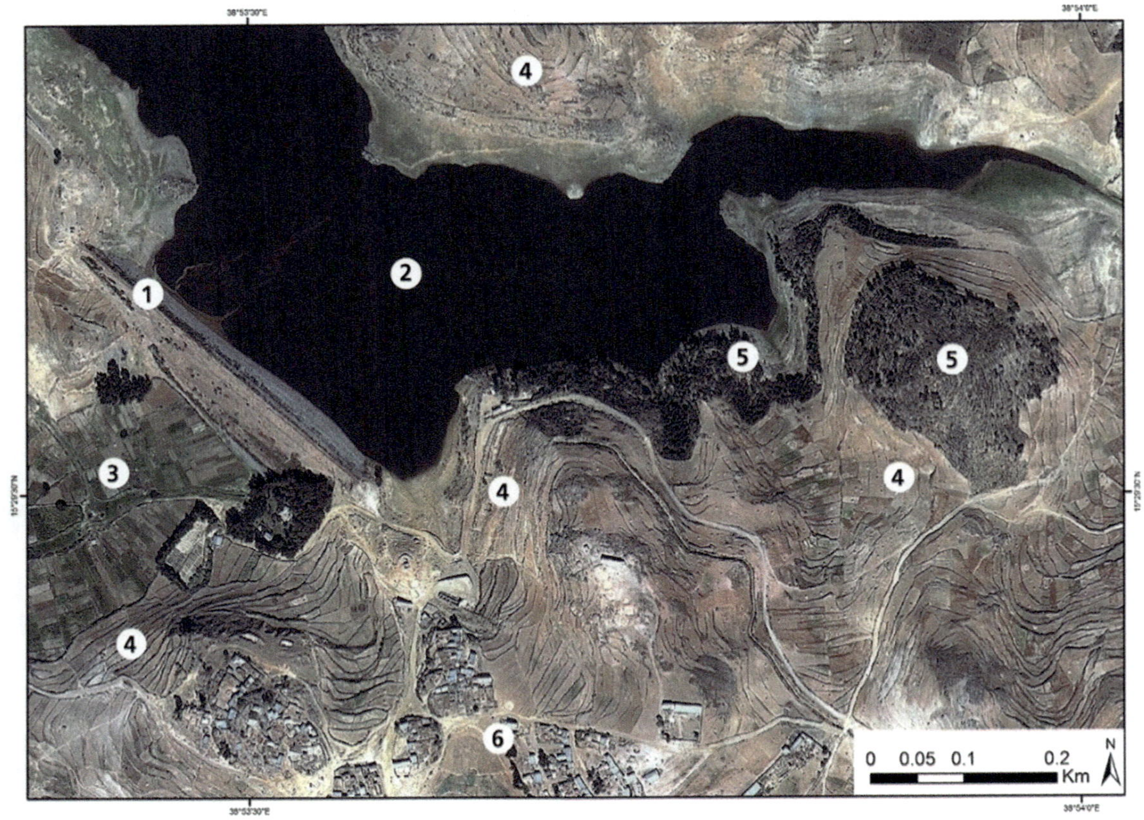

Fig. 8.9 This satellite picture shows the key elements of landscape change in the central highlands of Eritrea as described in this paper. Following the numbers on the picture, we can identify a dam **(1)** and reservoir **(2)** (in this case the dam of Adi Sheka, constructed before 1930 by the Italian colonial administration). The dam, its water body, and garden irrigation **(3)** are new features of the regional landscape and new to Eritrean small-scale farming. Terracing on cropland **(4)** and on hillsides together with afforestation **(5)** aim to prevent soil erosion and siltation of the reservoir. Village areas are indicated by **(6)** (*Source* Geoeye-1 subset acquired in December 2013 showing a true colour RGB image. Image © 2013 DigitalGlobe, Inc. All Rights Reserved. Numbers (1–6) on image inserted by S. Eckert and S. Kummer, CDE University of Bern, 2020)

largest reservoir, built to supply water to Asmara, was designed to hold 26 million m^3 (Daniel et al. 2009).

The first few dams were constructed in colonial times by the Italian administration, mainly to secure the water supply of Asmara, and are located near the town. Dam construction continued during Ethiopian times between 1952 and 1991, especially in the 1980s, when Ethiopia attempted a leap forward in Eritrea in its failed Red Star military campaign. Dams in this decade were funded by the international donor and aid community. This support came in response to the devastating famine in Northern Ethiopia 1983–85, which also affected Eritrea. Generally, activities followed a sustainable land management approach, including soil conservation, afforestation, and area closures in the catchment area of the dams.

The main aim of dam construction in this period was to supply water to rural areas, including small-scale irrigation for village communities, to improve food production and local livelihoods. Construction was done in food-for-work campaigns (Hurni 1985). In Zoba Maekel, the irrigation areas are mostly located in the valley bottoms below the reservoirs. They are small, rarely exceeding 10 ha (Table 8.4), but could be more than doubled if managed more effectively. This would mean reducing present conveyance losses due to unlined furrows, and improving irrigation schedules, i.e. abandoning untimely or excessive irrigation (Daniel et al. 2009).

Dam construction intensified after independence, not only in Zoba Maekel with more reservoirs created on an annual basis than before (Table 8.4), but also elsewhere in the highlands and especially in the western lowlands, in a bid to increase national food security. Across history, the majority of the dams (more than 40 out of the total of 81) were built by government bodies, mostly the Ministry of Agriculture, nine by the Italian colonial administration, and the rest by various civil society organizations including institutions close to the Catholic and Protestant Churches (Daniel et al. 2009). Estimates show that all 81 dams are designed to store as much as 70% of the runoff from their catchment. As rainfall is given, and in the perspective of a decreasing annual precipitation trend (see Chap. 2 of this book for more details), the potential for construction of additional reservoirs in Zoba Maekel is limited, especially if soil conservation in the catchments is intensified and hence runoff reduced. The figure of 70% storage refers to design capacity, meaning that reservoir siltation is not considered. Siltation is a problem, but reliable data on its magnitude are not available. Estimates from a rudimentary bathymetric survey in nine reservoirs established an average cumulative siltation rate of 23%, reporting two dams constructed in the 1930s and 1940s as completely silted up and one dam as broken (Daniel et al. 2009). Siltation of the 81 dams in service in the study area may vary greatly from one reservoir to the other, depending on reservoir age, nature of catchment, and amount and effectiveness of conservation work done therein.

8.4 Concluding Remarks

The Eritrean highlands—especially the central highlands, with their rapidly growing rural and urban population—are key to the future development of Eritrea. But natural limitations are significant, with a critically low level of annual rainfall for rain-fed farming and high inter-annual variations. The effects of climate change on farming are palpable, with an increase in the number of consecutive dry days as well as in rainfall intensity, over the last 100 years. With no perennial rivers or substantial groundwater resources in the area, most of the water supply is provided by rainfall stored in reservoirs. Soils have been worked for many generations and over many centuries, and soil degradation is widespread. Wood resources for fuel and construction are scarce.

However, not all is doom and gloom. Forest cover has increased considerably over the last decades, thanks to successive afforestation campaigns. What needs rethinking is the exclusive use of eucalyptus trees. Considering mixed stands including indigenous trees would reduce the threats associated with monocultures and increase biodiversity. In terms of combatting soil erosion, much has been achieved in soil and water conservation since the 1980s. Overall, afforestation, soil conservation works, and dams with their reservoirs are new physical features of the central highlands; they have led to a remarkable change in the regional landscape over the last 100 years and have increased its diversity (Fig. 8.9).

For future works, a watershed approach could help reconcile the needs of rural with those of urban areas and balance the interests of upstream with downstream users. Structures should be optimized to protect soils while occupying as little of the scarce cropland as possible and allow enough runoff to enter streams to fill the reservoirs that secure regional water supply. Irrigation, a relatively new form of land use for most farming communities, could be substantially expanded by water-saving conveyance and more effective irrigation schedules—a task for agricultural extension. These are but a few examples that could help lead the way to a sustainable future of land management. Urban areas and small central places have an important function too, as they provide services and employment for a growing rural population running short of land. A regional masterplan for sustainable land and resource management could provide the necessary guidance on how to coordinate such initiatives and others.

References

Araya YH, Hergarten C (2008) A comparison of pixel and object-based land cover classification: a case study of the Asmara region, Eritrea. In: Mander U, Brebbia CA, Martin-Duque JF (eds) Geo-environment and landscape evolution III, vol 100. WIT Press, Southampton, pp 233–243. https://doi.org/10.2495/GEO080231

Boerma P (1999) Seeing the wood for the trees: deforestation in the central highlands of Eritrea since 1890. PhD Dissertation, University of Oxford

Boerma P (2006) Assessing forest cover change in Eritrea—a historical perspective. Mt Res Dev 20(1):41–47. https://doi.org/10.1659/0276-4741(2006)026[0041:AFCCIE]2.0.CO;2

Burtscher R (2003) Analysis of rainfall-runoff data of Afdeyu and its implication on planning of small dams. SLM Eritrea Papers 1, Centre for Development and Environment, University of Bern

Daniel A, Tesfaslassie F, Tesfay S (2009) An appraisal of the current status and the potential of surface water in Upper Anseba Catchment, Eritrea. AEAS/ESAPP/SLM Eritrea Joint Report, Centre for Development and Environment, University of Bern

Eckert S, Tesfay S, Hurni H, Kohler T (2017) Identification and classification of structural soil conservation measures based on very high resolution stereo satellite data. J Environ Manage 193 (2017):592–606. https://doi.org/10.1016/j.jenvman.2017.02.061

FAO (1984) Assistance to land use planning Ethiopia. Report, Food and Agriculture Organization, Rome

FAO (2001) Global Forest Resources Assessment 2000. Main Report, Food and Agriculture Organization, Rome

FAO (2005) Crop and food supply assessment mission to Eritrea. Global information and early warning system on food and agriculture. Special report, Food and Agriculture Organization, Rome

Fessehaye M, Brugnara Y, Savage MJ, Brönnimann S (2019) A note on air temperature and precipitation variability and extremes over Asmara: 1914–2015. Int J Climatol 39(14):1–13. https://doi.org/10.1002/joc.6134

Fiori A (1913) Boschi ed ordinamento forestale nell'Eritrea. In: Martini F (ed) L'Eritrea economica: Prima serie di conferenze tenute in Firenze sotto gli auspici della Societa di Studi Geografici e Coloniali. De Agostini, Novara, pp 353–373

Government of Eritrea (1995) National Environmental Management Plan for Eritrea (NEMP-E), Asmara

Gurtner M, Zewengel G, Iyassu H, Zeraj T, Hadgu Y, Stillhardt B, Roden P (2006) Land management in the central highlands of Eritrea: a participatory appraisal of conservation measures and soils in Afdeyu and its vicinity. SLM Report 6, Geographica Bernensia, Bern

Henricksen BL, Ross S, Tilimo S, Wijntje-Bruggeman HY (1984) Ethiopia—geomorphology and soils. LUPRD Field Document 3, DP/ETH/78/003, FAO, Rome, 425 pp. http://www.fao.org/3/ar860e/ar860e.pdf

Hurni H (1978) Soil erosion forms in the Simen Mountains-Ethiopia (with map 1:25 000). Geographica Bernensia, Bern

Hurni H (1982) Soil Conservation Research Project. Inception Report, Vol. 1, University of Bern and United Nations University

Hurni H (1983) Soil formation rates in Ethiopia (with scale 1:1 000 000). Ethiopian Highlands Reclamation Study, Working Paper 2, Food and Agriculture Organization, Rome

Hurni H (1985) The design and construction of small-scale earth microdams. A field manual for assistant technicians working under the supervision of agricultural or irrigation engineers. Ministry of Agriculture, Addis Abeba

Hurni H (1987) Erosion–productivity–conservation systems in Ethiopia. In Pla Sentis I (ed) Proceedings of the IV international soil conservation conference, soil conservation and productivity, Maracay, November 1985, pp 654–674

Hurni H (1993) Land degradation, famine, and land resource scenarios in Ethiopia. In: Pimentel D (ed) World soil erosion and conservation. Cambridge Studies in Applied Ecology and Resource Management, Cambridge, pp 27–61

Hurni H, Berhe WA, Chadhokar P, Daniel D, Gete Z, Grunder M, Kassaye G (2016) Guidelines for development agents on soil and water conservation in Ethiopia, second revised edition. Centre for Development and Environment, University of Bern, with Bern Open Publishing

Juel M (ed) (2002) Watershed development: economic and institutional sustainability of integrated watershed development at community level. Evidence from case studies. Proceedings of the International DANIDA Workshop on Watershed Development, November 2001, Asmara

Kazmin V (1973) Geological map of Ethiopia 1:2,000,000. Geological Survey of Ethiopia, Ministry of Mines, Addis Abeba (includes Eritrea)

Laett L (2004) Eritrea re-photographed: Landscape changes in the Eritrean Highlands 1890–2004. MSc thesis, University of Bern

Liebi F (1993) Landnutzungsstruktur und Landschaftsentwicklung im Hochland Eritreas 1800–1952. MSc thesis, University of Bern

Martini F (1900) Relazionesulla Colonia Eritrea 1899–1900. Rome

Matteucci P (1880) In Abissinia. Treves, Milan

McCann J (1997) The plow and the forest: narratives of deforestation in Ethiopia 1840–1992. Enviro Hist Durh N C 2(2):138–159. https://doi.org/10.2307/3985505

Mohr PA (1971) The geology of Ethiopia. University College of Addis Ababa Press, 268 pp. (includes Eritrea)

Munzinger W (1890) Studi sull' Africa Orientale. Voghera, Rome

Negassi A, Bein E, Ghebru K, Tengnäs B (2002) Soil and water conservation manual for Eritrea. Regional Land Management Unit (RELMA), Swedish International Development Cooperation Agency, Stockholm and Asmara

Nyssen J, Haile M, Naudts J, Munro N, Poesen J, Moeyersons J, Frankl A, Deckers J, Pankhurst R (2009) Desertification? Northern Ethiopia re-photographed after 140 years. Elsevier Sci Total Environ 407(8):2749–2755. https://doi.org/10.1016/j.scitotenv.2008.12.016

Ogbazghi W, Stillhardt B, Herweg K (2011) Sustainable Land Management. A textbook with a focus on Eritrea. Geographica Bernensia and Hamelmalo Agricultural College, Bern and Keren

Ritler A (2003) Forests, land use and landscape in the Central and Northern Ethiopian Highlands 1865–1930. Geographica Bernensia, Bern

Schwilch G, Hessel R, Verzandvoort S (2012) Desire for Greener Land. Options for sustainable land management in drylands. World Overview of Conservation Approaches and Technologies, Bern and Wageningen

Selamawit T, Mihreteab Y (2016) Atlas of Zoba Maekel. Compilation of spatial database and atlas map of Maekel Zone. Eastern and Southern Africa Partnership Programme Report. Department of Agriculture and Land, Maekel Administrative Region, Asmara, and Centre for Development and Environment, University of Bern

Senni L (1915) Note sulla legislazione forestale Eritrea. L'Agricoltura Coloniale 9(2):65–86/148–166

Stillhardt B, Herweg K, Hurni H (2002) Long-term monitoring of soil erosion and soil and water conservation in Afdeyu, Eritrea (1984–1998). Soil erosion and soil conservation data base. Centre for Development and Environment, University of Bern; Ministry of Agriculture, Asmara; Syngenta Foundation for Sustainable Agriculture, Basel

Tesfay S, Beyene S, Zewengel G, Iyassu H, Okbagabriel R (2009) Afdeyu Research Sub-Station. Hydrometeorological Data Analysis 1984–2007. SLM Eritrea Report 10. Centre for Development and Environment, University of Bern; Syngenta Foundation for Sustainable Agriculture, Basel; National Agricultural Research Institute, Asmara

UNCCD (2014) The land in numbers. Livelihoods at a tipping point. Secretariat of the United Nations Convention to Combat Desertification, Bonn

Walter H, Harnickell E, Mueller-Dombois D (1975) Klimadiagramm-Karten der einzelnen Kontinente und die ökologische Klimagliederung der Erde: Eine Ergänzung zu den Vegetationsmonographien. Fischer, Stuttgart

Wolde-Mariam M (1991) Suffering under God's environment. A vertical study of the predicament of peasants in North-Central Ethiopia. African Mountains Association, Geographica Bernensia, Bern

9 Land Degradation in Eritrea and Djibouti

Jasper Knight, Mohamed A. M. Abd Elbasit, and Elhadi Adam

Abstract

Land degradation is a significant issue in many areas of sub-Saharan Africa, including in Eritrea and Djibouti. This study presents and interrogates different environmental data sets, including rainfall, dust flux and land use/land cover change, over the period 2001–2017, that can help evaluate land degradation in this region. From these data, patterns of water and wind erosivity can be calculated. Results show that spatial variations in erosivity reflect climate forcing upon a topographically diverse landscape. The potential for addressing land degradation risk is also influenced by issues such as development of conservation agriculture techniques, training and institutional support for farmers and land tenure. Successful adaptation will most likely be determined by these social and political factors rather than any future changes in climate, which, however, remains a key factor in land degradation projections.

Keywords

Agricultural systems • Drought • Erosivity • Land degradation • MODIS • NDVI • Sustainability • TRMM

9.1 Introduction

Land degradation is a significant issue affecting both the physical landscape and local communities throughout sub-Saharan Africa (Kassahun et al. 2008). This situation arises as a combination of factors in both the physical and human environments and, once initiated, is difficult to arrest because of the strong coupling and positive feedback between physical and human environmental systems (Kiage 2013). In Africa, land degradation is also set in the wider context of the impacts of ongoing climate change on vulnerable farming communities. The Intergovernmental Panel on Climate Change (IPCC) states that 'Africa is one of the most vulnerable continents to climate change and climate variability, a situation aggravated by the interaction of multiple stresses' (Boko et al. 2007, p.435). Here, multiple stresses describe different elements of physical and human environmental systems that interact to contribute to land degradation (Fig. 9.1). There is no single definition of the meaning of land degradation. A commonly used definition is that of the Land Degradation Assessment in Drylands (LADA) project, run by the United Nations Food and Agriculture Organization (FAO), which states that land degradation is 'the reduction in the capacity of the land to provide ecosystem goods and services, over a period of time, for its beneficiaries' (Biancalani et al. 2013, p. 7). This definition therefore focuses on land degradation as the outcome of a set of processes, but does not specify what types of processes may be responsible for land degradation, or their causes. Land degradation as a concept emerged mainly in the 1970s as a shift towards greater aridity in many already-marginal areas globally was recognized, in particular from new remote sensing data sets, and in combination with rapid population growth and increased stock density (Chisholm and Dumsday 1987; Kiage 2013). Key driving factors contributing to land degradation are listed in Table 9.1.

The causes of land degradation represent an important area of research in sub-Saharan Africa because of the vulnerability of local populations to changes in agricultural productivity and thus food security (Webb et al. 2017). Over large spatial scales, land degradation can potentially result in significant drought disaster and famine risk (Baudoin and

J. Knight (✉) · M. A. M. Abd Elbasit · E. Adam
School of Geography, Archaeology and Environmental Studies, University of the Witwatersrand, Johannesburg, 2050, South Africa
e-mail: jasper.knight@wits.ac.za

M. A. M. Abd Elbasit
School of Natural and Applied Sciences, Sol Plaatje University, Kimberley, 8301, South Africa

P. Billi (ed.), *Landscapes and Landforms of the Horn of Africa*, World Geomorphological Landscapes,
https://doi.org/10.1007/978-3-031-05487-7_9

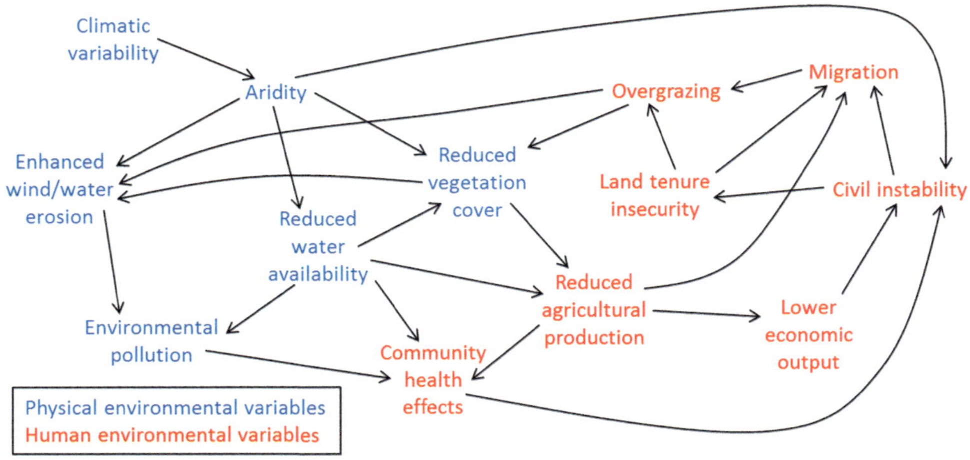

Fig. 9.1 Flow diagram illustrating how different physical and human environmental stresses can both contribute to and arise from land degradation factors. Note that there is no 'start point' on this flow diagram

Table 9.1 List of key elements of the physical and human environments that contribute to land degradation (Chisholm and Dumsday 1987)

Key elements of the physical environment	Key elements of the human environment
Degradation of vegetation cover	Agronomy and tillage practices
Dispersive soils	Deforestation
Erosion by wind and water	Development of mining
Geology and topography	Farmers' knowledge and expertise
Increased aridity	Fire
Salinization of soils	Institutional policies
Spread of unpalatable invasive species	Irrigation and shallowing water table
	Land ownership and tenure patterns
	Overgrazing/overstocking

Wolde-Georgis 2015; Sandstrom and Johola 2017). The Horn of Africa region is particularly sensitive to climatic variability and shifts in the timing and magnitude of the long rains (March–May) and short rains (October–December). Since the 1980s, the long rains experienced in the wider region have become dryer and the short rains wetter (Liebmann et al. 2014). This sets the scene for enhanced land degradation over the last decades (Kassahun et al. 2008; Solomon et al. 2018), discussed in this paper.

In detail, this study analyses the susceptibility of the Eritrea and Djibouti region to land degradation, based on different remote sensing data sets. Remote sensing is known to be a useful tool for evaluating different aspects of land degradation (Vågen et al. 2016; Dubovyk 2017) and is particularly useful in large and difficult to access regions with low cloud cover. This study presents spatial data on vegetation and rainfall for the period 2001–2017 and then calculates wind and water erosivity values, which can be used as a measure of land degradation potential.

Study Area and Methods

The region of Eritrea and Djibouti (Fig. 9.2) is located along the coastal fringe of the southwest Red Sea, Strait of Hormuz and western Gulf of Aden. Geologically, Djibouti lies at the northern continental end of the Great Rift Valley triple junction system, and this has greatly influenced both rock types and topography. Rift Valley faulting results in horsts and basins being formed including the Danakil horst and depression. Basalts of the Danakil Formation were extruded from the Miocene to Pliocene (Schlüter 2008). In Eritrea, Precambrian metamorphic rocks reflect the African basement and have been affected by Neogene rifting. This tectonic activity has shaped the geomorphology of the margins of the Ethiopian Highlands through Djibouti and southern Eritrea, which are characterized by steep and erosive rocky slopes with alluvial fans at footslope locations formed in particular during the late Cenozoic (Neogene and Quaternary). Neogene and Quaternary evaporites are common along the Red Sea coast of Eritrea in particular, and Cenozoic evaporites, colluvial and alluvial deposits, coastal and inland sand dunes are found across the region. These recent deposits attest to switches between arid and humid phases during the late Quaternary and Holocene which affected the stability of the land surface and gave rise to pulses of geomorphic change (Lanckriet et al. 2017). Arid phases tend to be associated with a dry and unstable land surface that is susceptible to wind deflation and erosion by flash floods. Today, there is strong dust export from this region eastwards across the Red Sea, in particular from the Ethiopian Highlands. Unlike other areas of the Horn of Africa, the Eritrea and Djibouti region has high to very high vulnerability to

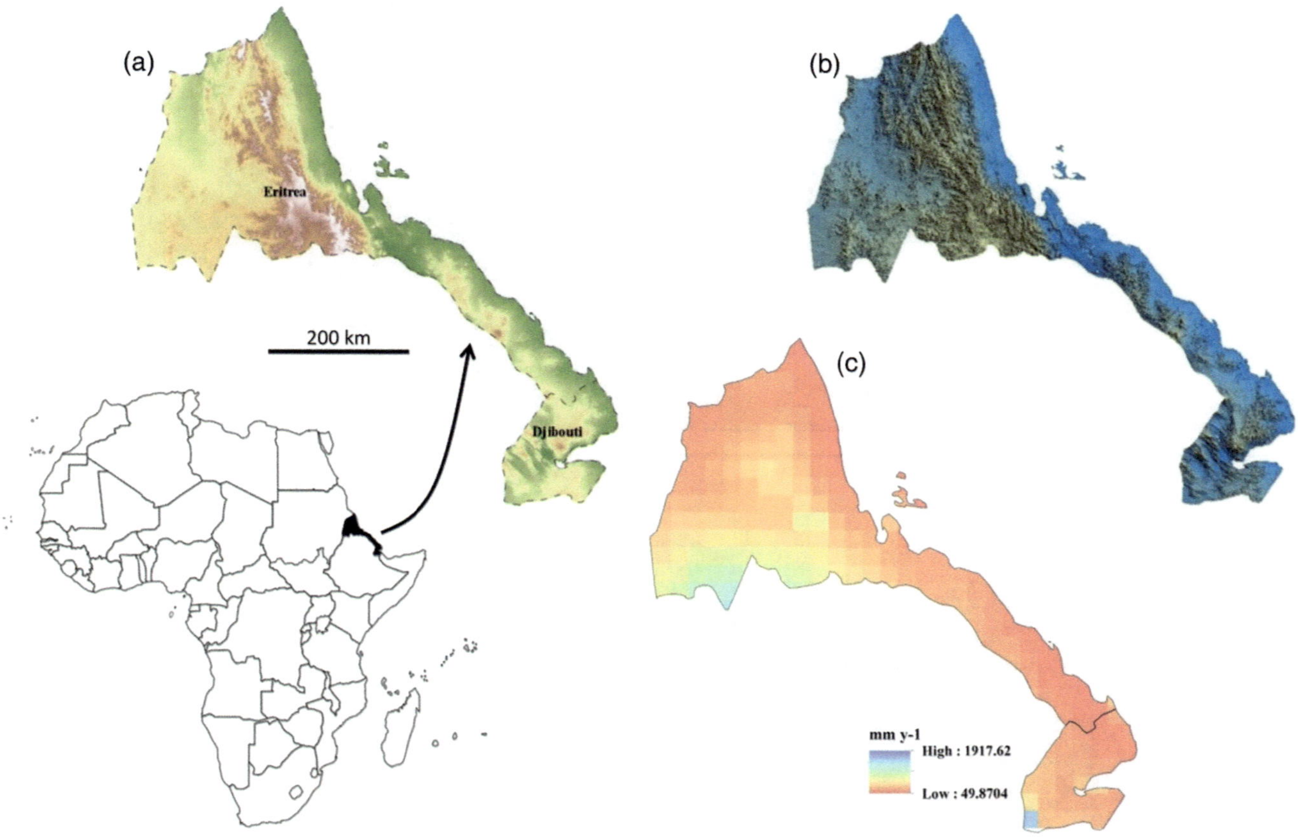

Fig. 9.2 **a** Location of Eritrea and Djibouti in the Horn of Africa region; **b** digital elevation model of the region; **c** annual rainfall (mm) derived from TRMM data

desertification, based on soil and climate data (see Chap. 2 of this book). This makes it a good location to examine land degradation processes and dynamics, due to its sensitivity to changes in hydro-climatic regimes controlled by equatorial Atlantic and Indian Ocean sea surface temperatures. Field observations from this region show evidence for land degradation in the form of enhanced soil erosion operating in both sloping and flat terrain settings (Fig. 9.3). This sets the scene for consideration of specific land degradation forcing factors affecting this region over the past decades.

Climatic and environmental conditions in the region were analysed based on remote sensing data sets. Monthly rainfall time series for Eritrea and Djibouti (1998–2017) were extracted from the Tropical Rainfall Measuring Mission (TRMM) platform, which provides rainfall estimates on grid cells between 10 and 25 km in size (available at http://trmm.gsfc.nasa.gov). TRMM data have been previously successfully validated against rain gauge measurements (e.g. Abd Elbasit et al. 2017). The product 3B43 monthly rainfall with resolution 0.25 was used to extract the spatio-temporal distribution of rainfall over the study area. The product represents a fusion between satellite rainfall data estimated by TRMM and other sources such as ground measurements. In turn, these data were used to calculate mean rainfall erosivity at a grid cell size of 10 km, applying the equation developed by Renard and Fremund (1994) for erosivity estimation using coarse resolution rainfall data (Eq. 9.1):

$$R - \text{factor} = 0.04830P^{1.510} \tag{9.1}$$

where R-factor is the rainfall/run-off erosivity factor (expressed as MJ mm ha^{-1} h^{-1} y^{-1}) in the Universal Soil Loss Equation and P is the mean annual rainfall (mm). Soil erosion in general is a function of climatic erosive agents and soil susceptibility. Soil erosivity by wind and water has been calculated using soil primary properties extracted from the harmonized world soil database v1.2 (Fischer et al. 2008). The soil water erosivity was calculated using the simplified relationship of Wischmeier and Smith (1978) (Eq. 9.2):

$$K \cong 2.1 \times 10^{-6} \times M^{1.14} \times (12 - \text{SOM}) \tag{9.2}$$

where

$$M = (\text{Si\%} + \text{Sa\%})(100 - \text{Cl\%}) \tag{9.3}$$

where K is the soil erodibility factor (t ha h ha^{-1} MJ^{-1} mm^{-1}); SOM is the percentage of soil organic matter; and Si

Fig. 9.3 Field evidence for land degradation in Eritrea. **a** Steep bedrock slopes with accumulation of rocky debris at footslope locations, and alongside an active river channel; **b** bare rock slopes revealed as a consequence of soil erosion, with patches of soil remaining; **c** upland plateau surface with an armour of bedrock-derived boulders, exposed as a consequence of soil erosion; **d** deflated and water-eroded land surface, with residual boulders present. Note the absence of vegetation indicating unsuitable conditions for agriculture

%, Sa% and Cl% are silt, sand and clay particle percentages, respectively. Soil erosivity by wind was calculated using Woodruff and Siddoway's (1965) erodibility factor (Eq. 9.4):

$$I = 26.69 + 0.031 \cdot (Sa\%) + 0.17 \cdot (Si\%) - 0.33 \cdot (Sa\%/Cl\%) - 4.66 \cdot (OC) - 0.95 \cdot (CaCO_3) \quad (9.4)$$

where I is soil wind erodibility (tons $ha^{-1}\ y^{-1}$); OC is soil organic matter content; and $CaCO_3$ is soil calcium carbonate content.

Land use/land cover (LULC) change was assessed using Moderate Resolution Imaging Spectroradiometer (MODIS) Land Cover Type (MCD12Q1) for three periods 2001, 2010 and 2017. The land cover data resolution was 500 m. The MODIS Global Land Cover Type (MLCT) consists of five land cover classification legends (Friedl et al. 2010). Among these classifications, the Plant Functional Type (PFT) scheme was used as land degradation indicator. The PFT consists of 13 classes, namely water, evergreen needleleaf trees, evergreen broadleaf trees, deciduous broadleaf trees, shrubs, grass, cereal crops, broadleaf crops, urban and built-up areas and barren or sparse vegetation. In addition, there are two classes that do not exist in the study area, which are deciduous broadleaf trees and snow and ice. The MLCT was developed using MODIS bands 1–7 in addition to enhanced vegetation index (EVI) and auxiliary layers such as land surface temperature and MODIS albedo products. These inputs were processed using pixel-based supervised classification algorithm to generate class probability for each class at each pixel (Friedl et al. 2010). The dust emission time series was retrieved from the second Modern-Era Retrospective Analysis for Research and Applications (MERRA-2) aerosol products. The dust products are a result of reanalysis of historical meteorological and atmospheric chemistry data. The study site time series was extracted from the global database for dust mass in the period December 1997 to December 2017.

To support the data on LULC change, a set of 16-day images of normalized difference vegetation index (NDVI) values from the new MODIS products available was downloaded from the USGS Earth Explorer website (https://earthexplorer.usgs.gov/) at a pixel resolution of 250 m. The images extracted cover the years 2008, 2010, 2012, 2015 and 2017 for Eritrea and Djibouti. The images were atmospherically corrected to minimize cirrus cloud and aerosols effects.

9.2 Results

9.2.1 Climate Data

Time series of key climatic and environmental data from Eritrea, applicable to the study area as a whole, are shown in Fig. 9.4. Rainfall patterns are variable, between 296 and 500 mm annually, and show a bimodal pattern with seasonal peaks in May and August/September (Fig. 9.4a). In the Eritrean highlands and the western lowlands, there is only one rainy season (June to August). In the eastern lowlands, along the coast and in Djibouti, the rainy season is later (October to March). Despite this variability, the relative proportion of summertime precipitation is fairly constant as 53–67% of the annual total. Over the time period of analysis, there is an averaged long-term decrease in mean monthly rainfall by ~10 mm/decade and an increase in dust flux by a third across the same time period ($R^2 = 0.565$). These trends are genetically related, in which decreased rainfall and surface moisture increase the likelihood of wind erosion (Maurer et al. 2009). Long-term decreases in rainfall, and interannual and seasonal changes in rainfall variability, have been noted in some other studies from the Horn of Africa region (e.g. Wang et al. 2017).

9.2.2 Vegetation Data

Based on LULC classification from MODIS data, barren/sparse vegetation covers over half the area (Fig. 9.5), especially in lowlands and coastal locations. This confirms what is seen in the field (e.g. Fig. 9.3). Vegetation of different types is present in inland locations and found in particular in higher elevation areas in the Eritrean Highlands. Across the time period of LULC analysis (2001–2017), there are no significant changes for most classes, but some small but persistent changes are identified for some vegetation classes. Shrubs, typical of degraded land, show a consistent increase, whereas grass, cereal crops and broadleaf crops all decrease (Table 9.2). Similar vegetation changes have been previously identified, related in particular to shifts in agricultural systems (Brink and Eva 2011).

The NDVI results show relatively consistent spatial patterns over the time period examined (2008–2017) (Fig. 9.6). There is a clear arid coastal fringe, located seaward of the mountains; and thus this is an ecological as well as a topographic and climatic barrier. Landward, and across the Eritrean Highlands region as a whole, there is a close correspondence between LULC and NDVI values. In detail, however, the categories of LULC (Fig. 9.5) do not capture the nuances of NDVI which reflects ecosystem productivity. The distribution of high NDVI values most closely corresponds to grassland, and it is also notable that drainage systems are well imaged through high NDVI values. NDVI values across Djibouti are more subdued, likely reflecting lower precipitation and thus drier soil conditions.

The most likely trigger for these LULC and NDVI changes in combination is precipitation over the Ethiopian Highlands, controlled by the strength of El Niño–Southern Oscillation (ENSO), seasonal Indian Ocean monsoon and the Indian Ocean Dipole (IOD) (Williams et al. 2012; Viste and Sorteberg 2013). There is no significant relationship between annual precipitation and IOD (Fig. 9.7); other combinations of monthly/seasonal precipitation and different forcing factors show no relationship.

9.2.3 Land Surface Erosivity

The calculated values of erosivity are consistent with results from regional-scale analyses (e.g. Elagib 2011). The very similar spatial patterns and magnitude of water and wind erosivity indices (Fig. 9.8b, c) reflect the role of climate forcing on the land surface. The lowest erosivity values are located in western Eritrea located within the Atbara subcatchment of the River Nile and which is fed by the seasonal rains over the Ethiopian Highlands (Viste and Sorteberg 2013). Highest erosivity values are found on Ethiopian Highlands slopes (e.g. Nyssen et al. 2005) and in Red Sea coastal areas of eastern Eritrea seaward of the coastal mountains where the lowest precipitation values are found and where sand dunes dominate.

Controls on rainfall erosivity in the region are orographic-driven rainfall and large droplet size as a result of unstable atmospheric circulation (Nyssen et al. 2005). Erosivity may also be exacerbated by variations in rainfall seasonality and event intensity (Elagib 2011), not shown in Eq. 9.1. Rain and wind erosion are also closely coupled because both correspond to instability of the land surface

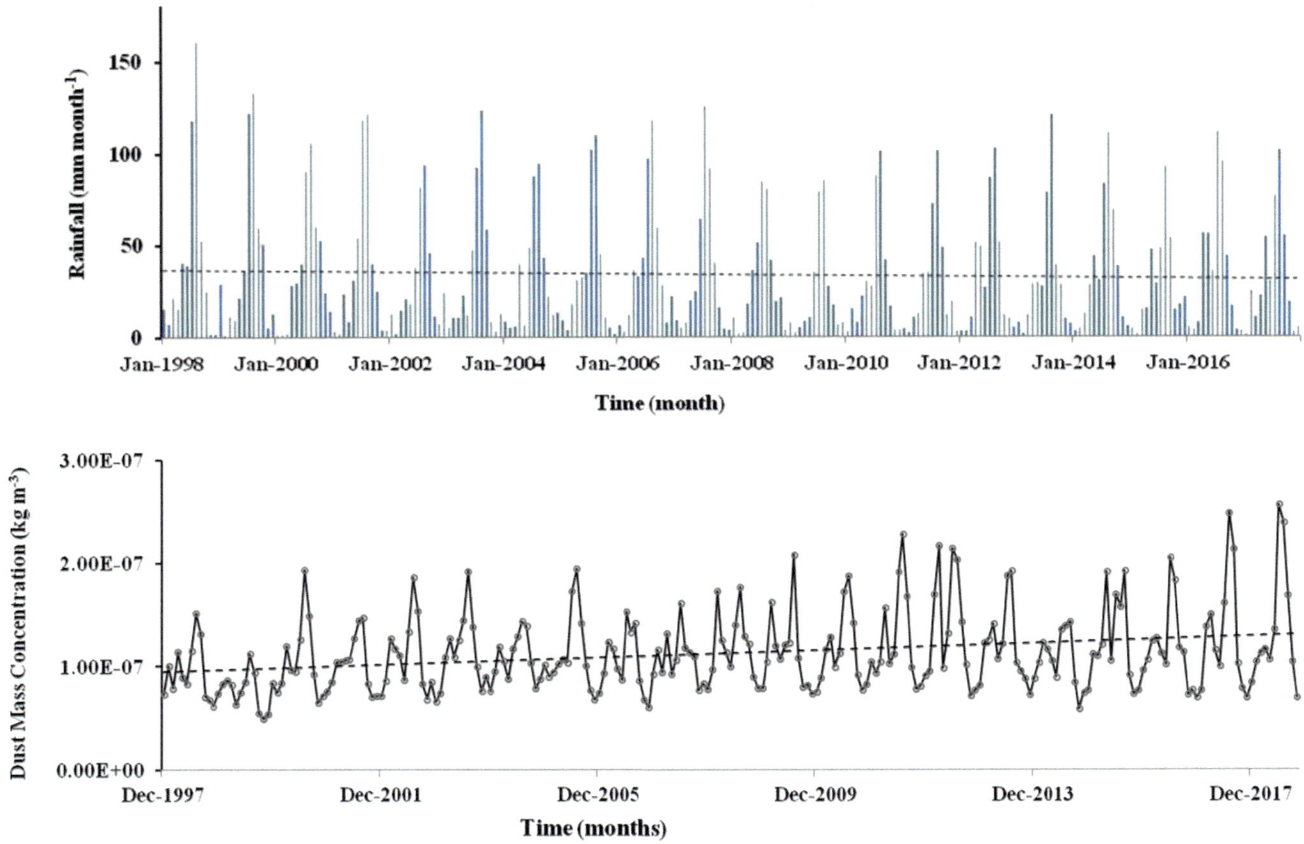

Fig. 9.4 Time series of environmental variables from Eritrea and Djibouti showing averaged trend lines. **a** Monthly rainfall time series (TRMM and GPM data); **b** dust mass concentration retrieved from MERRA-2 data

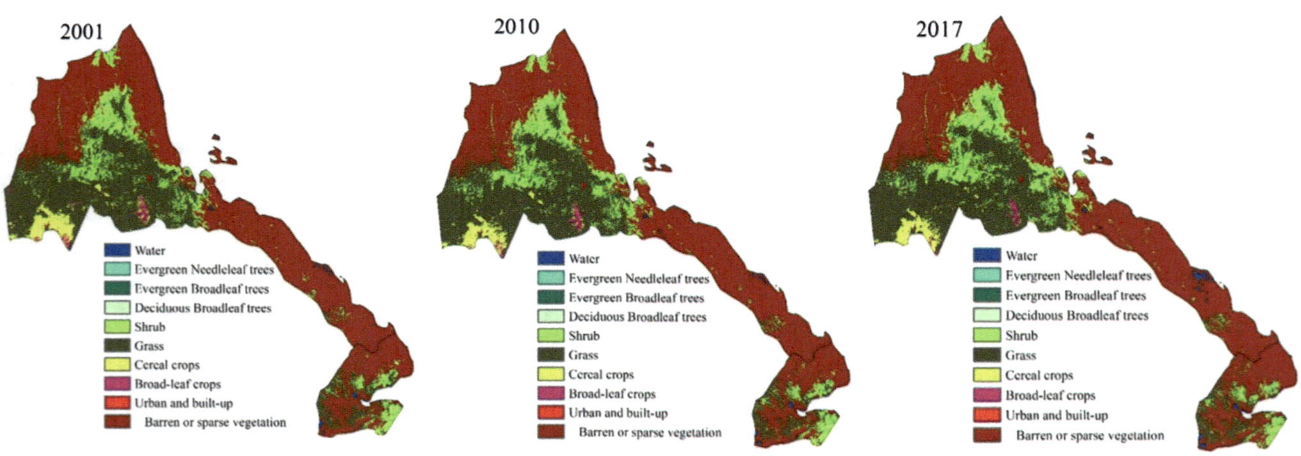

Fig. 9.5 LULC patterns in Eritrea and Djibouti, based on MODIS data, for 2001, 2010 and 2017

and can lead to enhanced erosion. Models based on field data from the Sahel show that wind erosion can change on an order of magnitude under different agricultural practices (Pierre et al. 2018). This highlights the fact that human activity is intimately linked with land surface destabilization and thus erosivity and land degradation potential (Lanckriet et al. 2015; Webb et al. 2017; Solomon et al. 2018).

9.3 Discussion

Results of this study show that there are spatial differences across the region in the degree of potential land degradation, as measured using the proxies of water and wind erosivity (Fig. 9.8). This arises as a function of topography (Fig. 9.2

b), rainfall patterns (Fig. 9.2c) and vegetation change (Fig. 9.5). Different remote sensing methods have been used to evaluate vegetation changes in the Horn of Africa region, commonly based on NDVI values (e.g. Brink and Eva 2011). Further, remote sensing methods including analysis of NDVI data and using MODIS imagery have been used in order to evaluate and then monitor land degradation and desertification (Vågen et al. 2016; Dubovyk 2017). Results of this study show that subtle changes in vegetation patterns took place in the period 2001–2017 (Figs. 9.5, 9.6; Table 9.2). A decrease in crops may be indicative of aridity and/or greater rainfall variability which have been noted from this region for this time period (Williams et al. 2012; Liebmann et al. 2014), and an increase in shrubs may reflect land abandonment as a result of rural depopulation. This rainfall variability is also associated with higher dust flux (Fig. 9.4b).

Although such physical factors have a first-order control on land surface erosivity and thus land degradation potential, factors in the human environment are also of significance (Table 9.1). The linked relationship between climate variations (in particular of rainfall) and agricultural success is a strong determinant of the stability of both human populations and the integrity of the land surface (Kiage 2013; Pierre et al. 2018). Thus, in the Horn of Africa region, land degradation has been closely linked to agricultural practices of cattle grazing, and the conventional viewpoint sees land degradation as an inevitable outcome of grazing (Kassahun et al. 2008). This is amplified by rainfall variability as a consequence of climate change (Sandstrom and Juhola 2017). However, other studies consider farmers as having an important role in reducing land degradation risk by undertaking more sustainable and adaptive land management and agricultural practices (Webb et al. 2017; Gebremeskel et al. 2018). Thus, farmers are neither passive agents of land degradation nor inactive in responding to such environmental challenges (Kiage 2013).

Several studies have highlighted other constraints to adaptation. A key issue is land tenure and administrative management structures, from community to national scales, that influence where farmers may farm, their security of tenure and their decisions regarding investment, soil sustainability and conservation measures (Lanckriet et al. 2015). These issues are particularly pertinent in Eritrea which has a low-density population and where agriculture is only found in a small part of the country and is limited by its topography and water scarcity. Through the twentieth century to present, linguistic, religious and ethnic differences have been used as guiding principles for establishing political and administrative regions and boundaries throughout the Horn of Africa (Schlee 2003). Within these regions, there are implications for cultural identity, the role of traditional authorities and national/international political agencies. War between Eritrea and Ethiopia in 1998–2000 contributed to regional political and social turmoil (Lyons 2009), leading to large-scale population migration (refugees), rural depopulation, disrupted trade and agricultural systems and greater food insecurity in a region already under environmental stress (White 2005; Solomon et al. 2018).

In order to address land degradation, maintaining the physical integrity of the land surface is a key first step. Studies have shown that subtle and low-cost but sustained changes to agricultural practices, including better water management, can result in significant community and socio-economic benefits (Webb et al. 2017; Gebremeskel et al. 2018). In areas of the Ethiopian Highlands, woodland density has also increased in areas of high population density, contrary to common belief (e.g. Boerma 2006). Thus, the viewpoint that land degradation is an inevitable consequence of human activity, amplified by climate change, is increasingly being contested.

Table 9.2 Percentage areal cover of different LULC classes in Eritrea and Djibouti at the three different time slots examined in this study (Fig. 9.5)

Land use/land cover class	2001 (%)	2010 (%)	2017 (%)
Water	0.86	0.88	1.16
Evergreen needleleaf trees	0.01	0.01	0.01
Evergreen broadleaf trees	0.04	0.03	0.04
Deciduous broadleaf trees	0	0	0
Shrubs	10.29	11.55	12.38
Grass	33.94	33.35	32.40
Cereal crops	2.43	1.92	1.62
Broadleaf crops	0.55	0.47	0.32
Urban and built-up areas	0.06	0.06	0.07
Barren or sparse vegetation	51.83	51.73	52.01
Total (%)	100	100	100

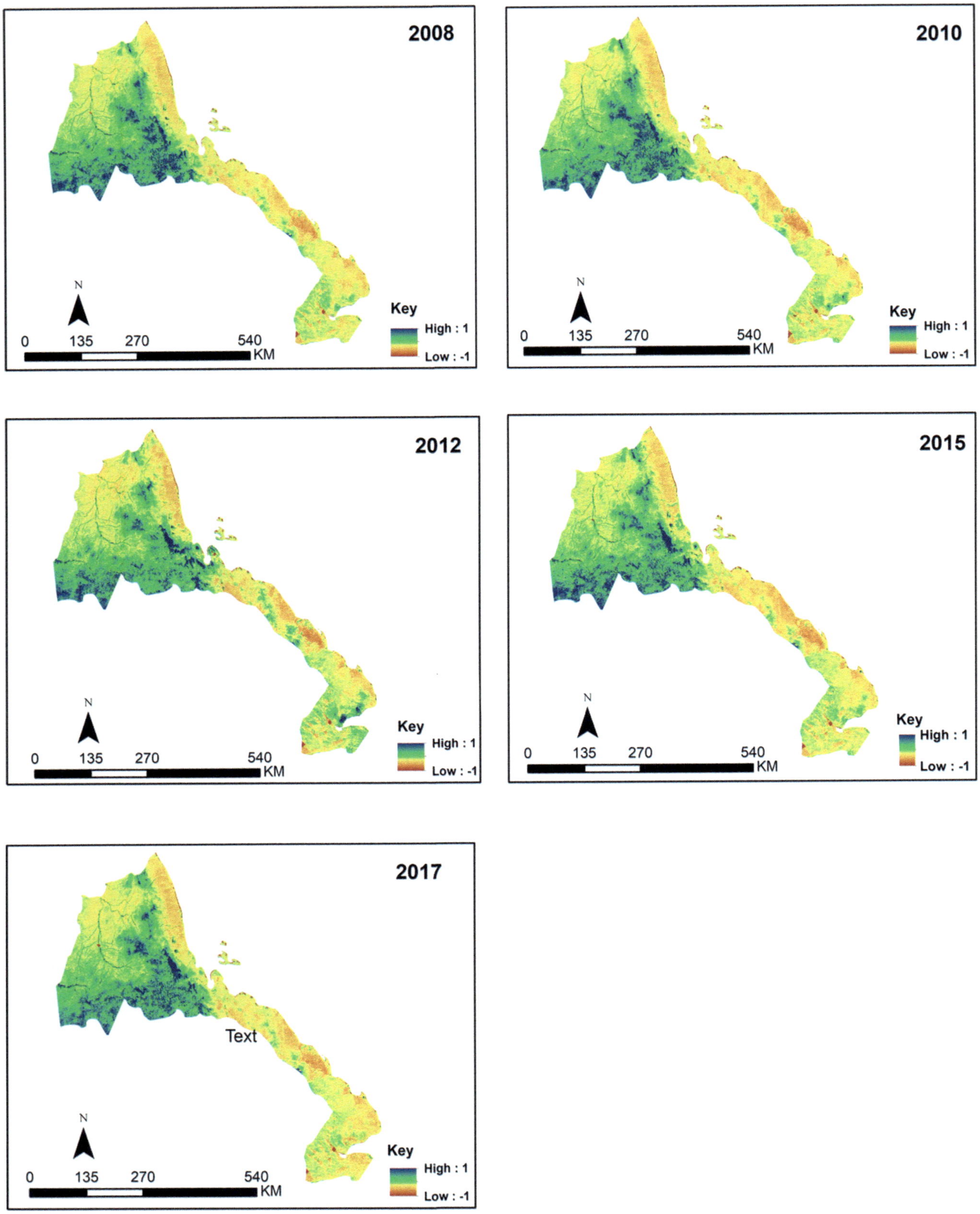

Fig. 9.6 Spatial distribution of NDVI values calculated from MODIS (250 m resolution) for the years 2008, 2010, 2012, 2015 and 2017

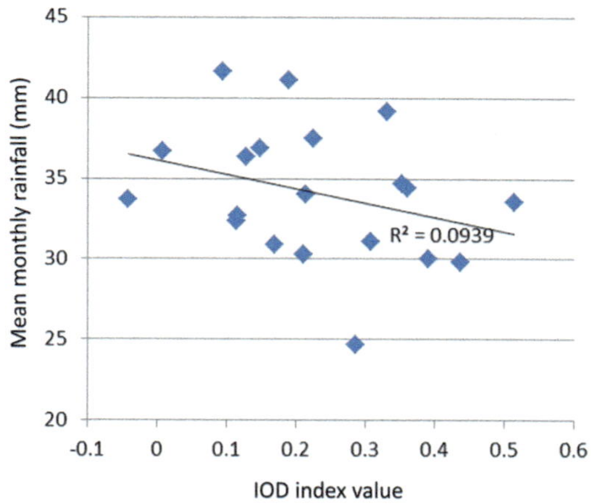

Fig. 9.7 Graph of mean monthly rainfall (mm) and IOD index values, from the study area, based on TRMM data for 1998–2017

9.4 Conclusions

Land degradation is a significant issue throughout the Horn of Africa region, including Eritrea and Djibouti. This study shows the extent and dynamics of land degradation in this region, based on proxies of climate, dust flux and land surface erosivity. The remote sensing approach adopted in this study shows its value in monitoring long-term changes in land degradation potential across the region. The results of this study also suggest the close relation between agricultural practices and land degradation and, conversely, the potential for sustainable agricultural practices as a key tool to reduce land degradation severity. Thus, conservation agriculture is the most desirable management practice in this region. However, this cannot be done without appropriate strategic

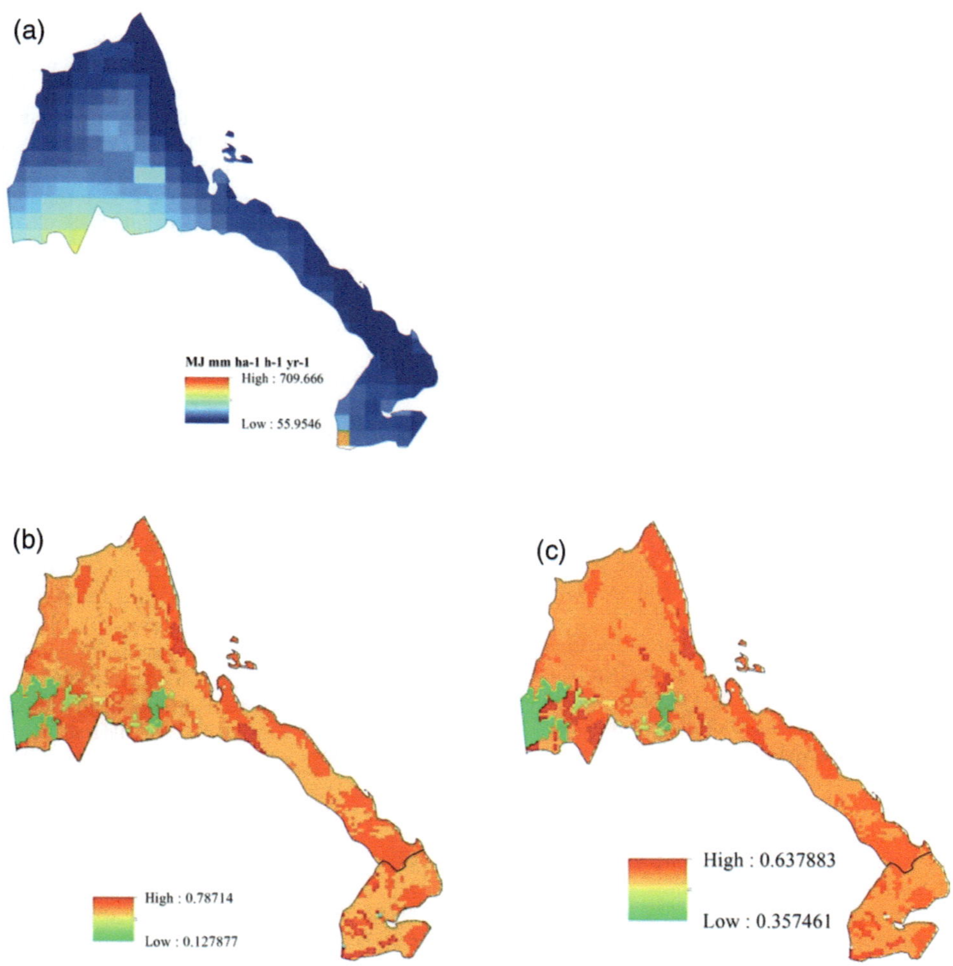

Fig. 9.8 Model outputs of land degradation measures from the study area. **a** Rainfall erosivity, **b** soil water erosivity (t ha MJ^{-1} mm^{-1}), **c** soil wind erosivity (t ha^{-1} yr^{-1})

support focused on agricultural extension and training (Baudoin and Wolde-Georgis 2015; Webb et al. 2017). It also requires an understanding of ecohydrological sensitivity to climate forcing (Sandstrom and Juhola 2017), as well as how human activity can variously amplify or suppress both forcing and response. Recent and continuing drought in the Horn of Africa region (Nicholson 2014), alongside ongoing and future climate change, highlights the increasing need for environmental sustainability and management, of which addressing land degradation is a major part.

References

Abd Elbasit MAM, Adam EO, Abu-Taleb K, Ahmed F, Yasuda H, Ojha CSP (2017) Space-borne rainfall measurement over arid regions. In: Ojha CSP, Surampalli RY, Bárdossy A, Zhang TC, Kao C-M (eds) Sustainable water resources management. American Society of Civil Engineers, pp 209–227

Baudoin M-A, Wolde-Georgis T (2015) Disaster risk reduction efforts in the Greater Horn of Africa. Int J Disaster Risk Sci 6:49–61

Biancalani R, Nachtergaele F, Petri M, Bunning S (2013) Land degradation assessment in drylands, methodology and results. FAO, Rome, p 56

Boerma P (2006) Assessing forest cover change in Eritrea—a historical perspective. Mt Res Dev 26:41–47

Boko M, Niang I, Nyong A, Vogel C, Githeko A, Medany M, Osman-Elasha B, Tabo R, Yanda P (2007) Africa. In: Parry ML, Canziani OF, Palutikof JP, van der Linden PJ, Hanson CE (eds) Climate change 2007: impacts, adaptation and vulnerability. Contribution of Working Group II to the Fourth Assessment Report of the Intergovernmental Panel on Climate Change. Cambridge University Press, Cambridge, pp 433–467

Brink AB, Eva HD (2011) The potential use of high-resolution Landsat satellite data for detecting land-cover change in the Greater Horn of Africa. Int J Remote Sens 32:5981–5995

Chisholm A, Dumsday R (eds) (1987) Land Degradation, problems and policies. Cambridge University Press, Cambridge, p 404

Dubovyk O (2017) The role of Remote Sensing in land degradation assessments: opportunities and challenges. Eur J Remote Sens 50:601–613

Elagib NA (2011) Changing rainfall, seasonality and erosivity in the hyper-arid zone of Sudan. Land Degrad Dev 22:505–512

Fischer G, Nachtergaele F, Prieler S, van Velthuizen HT, Verelst L, Wiberg D (2008) Global agro-ecological zones assessment for agriculture (GAEZ 2008). IIASA, Laxenburg, Austria/FAO, Rome, Italy

Friedl MA, Sulla-Menashe D, Tan B, Schneider A, Ramankutty N, Sibley A, Huang X (2010) MODIS Collection 5 global land cover: algorithm refinements and characterization of new datasets. Remote Sens Environ 114:168–182

Gebremeskel G, Gebremicael TG, Girmay A (2018) Economic and environmental rehabilitation through soil and water conservation, the case of Tigray in northern Ethiopia. J Arid Environ 151:113–124

Kassahun A, Snyman HA, Smit GN (2008) Impact of rangeland degradation on the pastoral production systems, livelihoods and perceptions of the Somali pastoralists in Eastern Ethiopia. J Arid Environ 72:1265–1281

Kiage LM (2013) Perspectives on the assumed causes of land degradation in the rangelands of Sub-Saharan Africa. Prog Phys Geogr 37:664–684

Lanckriet S, Derudder B, Naudts J, Bauer H, Deckers J, Haile M, Nyssen J (2015) A political ecology perspective of land degradation in the north Ethiopian Highlands. Land Degrad Dev 26:521–530

Lanckriet S, Rangan H, Nyssen J, Frankl A (2017) Late quaternary changes in climate and land cover in the Northern Horn of Africa and adjacent areas. Palaeogeogr Palaeoclimatol Palaeoecol 482:103–113

Liebmann B, Hoerling MP, Funk C, Bladé I, Dole RM, Allured D, Quan X, Pegion P, Eischeid JK (2014) Understanding recent Eastern Horn of Africa rainfall variability and change. J Clim 27:8630–8645

Lyons T (2009) The Ethiopia-Eritrea conflict and the search for peace in the Horn of Africa. Rev Afr Polit Econ 120:167–180

Maurer T, Herrmann L, Stahr K (2009) The effect of surface variability factors on wind-erosion susceptibility: a field study in SW Niger. J Plant Nutr Soil Sci 172:798–807

Nicholson SE (2014) A detailed look at the recent drought situation in the Greater Horn of Africa. J Arid Environ 103:71–79

Nyssen J, Vandenreyken H, Poesen J, Moeyersons J, Deckers J, Haile M, Salles C, Govers G (2005) Rainfall erosivity and variability in the Northern Ethiopian Highlands. J Hydrol 311:172–187

Pierre C, Kergoat L, Hiernaux P, Baron C, Bergametti G, Rajot J-L, Abdourhamane Toure A, Okin GS, Marticorena B (2018) Impact of agropastoral management on wind erosion in Sahelian croplands. Land Degrad Dev 29:800–811

Renard KG, Freimund JR (1994) Using monthly precipitation data to estimate the R-factor in the revised USLE. J Hydrol 157:287–306

Sandstrom S, Juhola S (2017) Continue to blame it on the rain? Conceptualization of drought and failure of food systems in the Greater Horn of Africa. Environ Hazards 16:71–91

Schlee G (2003) Redrawing the map of the horn: the politics of difference. Africa: J Int Afr Inst 73:343–368

Schlüter T (2008) Geological Atlas of Africa, 2nd edn. Springer, Berlin, p 307

Solomon N, Birhane E, Gordon C, Haile M, Taheri F, Azadi H, Scheffran J (2018) Environmental impacts and causes of conflict in the Horn of Africa: a review. Earth Sci Rev 177:284–290

Vågen T-G, Winowiecki LA, Tondoh JE, Desta LT, Gumbricht T (2016) Mapping of soil properties and land degradation risk in Africa using MODIS reflectance. Geoderma 263:216–225

Viste E, Sorteberg A (2013) Moisture transport into the Ethiopian highlands. Int J Climatol 33:249–263

Wang Y, Shilenje ZW, Sagero PO, Nyongesa AM, Banda N (2017) Rainfall variability and meteorological drought in the Horn of Africa. Mausam 68:463–474

Webb NP, Marshall NA, Stringer LC, Reed MS, Chappell A, Herrick JE (2017) Land degradation and climate change: building climate resilience in agriculture. Front Ecol Environ 15:450–459

White P (2005) War and food security in Eritrea and Ethiopia, 1998–2000. Disasters 29:S92–S113

Williams AP, Funk C, Michaelsen J, Rauscher SA, Robertson I, Wils TGW, Koprowski M, Eshetu Z, Loader NJ (2012) Recent summer precipitation trends in the Greater Horn of Africa and the emerging role of Indian Ocean sea surface temperature. Clim Dyn 39:2307–2328

Wischmeier WH, Smith DD (1978) Predicting rainfall erosion losses—a guide to conservation planning. U.S. Department of Agriculture, Agriculture Handbook 537, 59pp

Woodruff NP, Siddoway FH (1965) A wind erosion equation. Soil Sci Soc Am J 29:602–608

10 Landscape and Landforms of Northern Somalia

Bruno Petrucci

Abstract

The morphology of northern Somalia is substantially influenced by the tectonic activity caused by the rifting between the African and Arabic plates, known as Gulf of Aden rift. The Precambrian African shield with Late Mesozoic and Paleogene sedimentary cover was broken along a main line, commonly by several faults, subparallel to the rift (and the modern coast) and the inner southern section was uplifted and slightly tilted southward. The uplift caused the formation of a long East–West escarpment (about 750 km long, from the Ethiopian border to as the Bosaso valley) that divides Somaliland in a coastal plain and a plateau. This complex tectonic framework provided the region with unique landscapes and landforms that vary from the flatlands of the plateau, to dramatic escarpments dropping from more than 2000 m asl to 500 m of the coastal plain gently descending seaward. The strong erosion, partly due to the still active tectonic, plays an important role in the shaping of the modern landscapes and resulted in the incision of narrow gorges, with steep sides, in mountainous areas or sharp sandy banks in the wadis that cross the plateau and the coastal plain.

Keywords

Uplifting • Escarpment • Wadi • Gorge • Gypsum crust

10.1 Introduction

Northern Somalia includes the Republic of Somaliland and the western part of the Puntland Region of the Federal Republic of Somalia. The state of Somaliland is an autonomous republic, self-declared in the early 1990s, following a political agreement amongst the main clans of the area. Somaliland corresponds to the territories included in the British protectorate at the end of WWII.

Most of the research work on the geology and geomorphology of northern Somalia was carried out before the outburst of the civil war, especially by British and Italian scholars (see Hadden 2007, for a comprehensive bibliography). This chapter describes the main geomorphological features of the northern part of Somalia, which includes the Republic of Somaliland and the western region of the Puntland Region of the Federal Republic of Somalia and represents an update of existing regional geomorphological knowledge based on field work carried out in the region during the last two decades. The pictures and the description of the landscapes are the summary of 22 years of field work, during which I was never subjected to any hostile act and I could appreciate the friendliness of generous people.

10.2 Study Area

The study area includes the northern part of Somalia, commonly also known as Somaliland, and the adjacent part of Puntland region, where the town of Bosaso is located. The whole area faces the Gulf of Aden. Somaliland has constituted an autonomous state (Republic of Somaliland) since the early 1990s, following the collapse of the Siad Barre regimen, and today is still one of the most democratic states of Africa, with regular national elections every 5 years. Unfortunately, despite its democratic spirit and the peaceful behaviour of the population (the population has been always rejecting the influence of any terrorist group) the country is not recognized by the international community. Puntland, after a period of independence, decided to join the Federal Republic of Somalia, but the border between Somaliland and Puntland is still disputed.

B. Petrucci (✉)
Free-Lance Geologist, UN Agencies and Italian NGOs Consultant, Vimercate, Monza, Italy
e-mail: bruno_petrucci@yahoo.com

P. Billi (ed.), *Landscapes and Landforms of the Horn of Africa*, World Geomorphological Landscapes,
https://doi.org/10.1007/978-3-031-05487-7_10

The main structural element of Somaliland is an escarpment, which is the country's main physiographic element separating the relatively flat highlands from the coastal plain (Fig. 10.1). The latter, whose elevation usually does not exceed 500 m asl, is mostly covered by alluvial and aeolian deposits, whereas the plateau is a gently southward sloping plain descending from the top of the escarpment (2500 m asl near Ceerigabo) to the Ethiopian border in the south. Whilst the altitude of the escarpment top varies within a narrow range of elevations, the plateau drops gently towards the Indian Ocean and the Ethiopian border, descending from elevations close to 1600 to almost 650 m asl at the triple border point amongst Somaliland, Ethiopia and Puntland. The great tectonic depression of the Nogal valley that hosts an important and deep sedimentary basin, occupies the central part of the plateau. In the Puntland region, south of Bosaso, the S–N trending Bosaso valley and another important tectonic feature, the Darror valley, oriented WNW–ESE and descending to the Indian Ocean, divide the escarpment edge and the Karkar plateau from the Ahl Miskat and Ahl Bari Mountains that extend up to the Indian Ocean and constitute the true Horn of Africa that from Cape Guardafui faces both seas (Fig. 10.2).

The climate is typically arid or semi-arid, with rainfall concentrated in two seasons, known as Gu, from March to May, in which the main rainy events occur, and Dayr, from September to November. However, minor rains may also occur from June to August. More intense rainstorms are localized at the escarpment top due to both elevation and the barrier effect of the mountains that force the humid air coming from the sea to rise. Here, the annual rainfall can be as much as 500–600 mm, whereas in the coastal plain that is a very arid area, the annual rainfall drops to less than 100 mm and it is not uncommon to record 2–3 consecutive years with no rain at all. In the same area summer temperature reaches peaks of more than 50 °C, whilst on the plateau it seldom exceeds 35 °C. Winter is characterized by strong winds from NE, whilst in summer the main air flow is from SE. There are no permanent rivers in the region, but some of them may exhibit a large discharge in response to intense rainfalls.

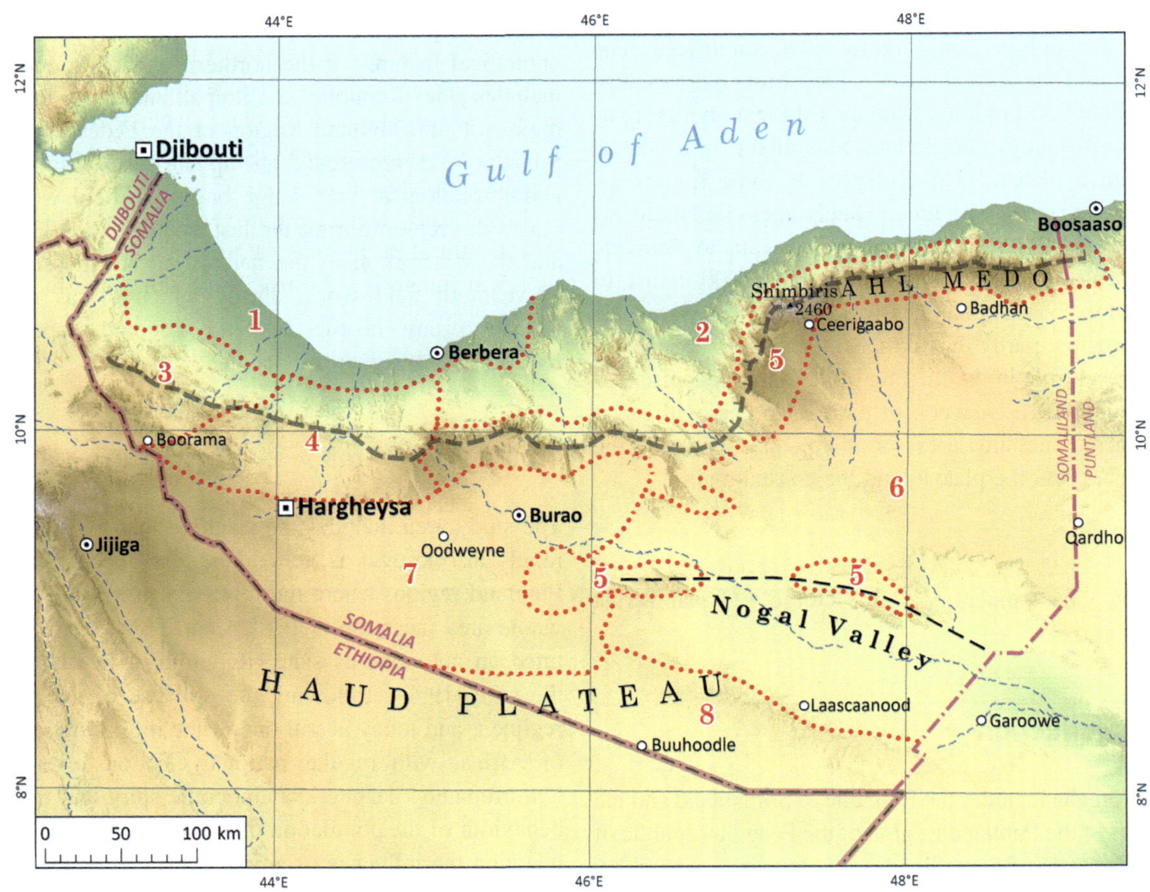

Fig. 10.1 Geomorphological provinces of Somaliland: (1) coastal plain; (2) hilly coastal belt; (3) limestones basement; (4) basement; (5) Auradu/basement; (6) Taleex Karkar; (7) Yesomma; (8) Karkar

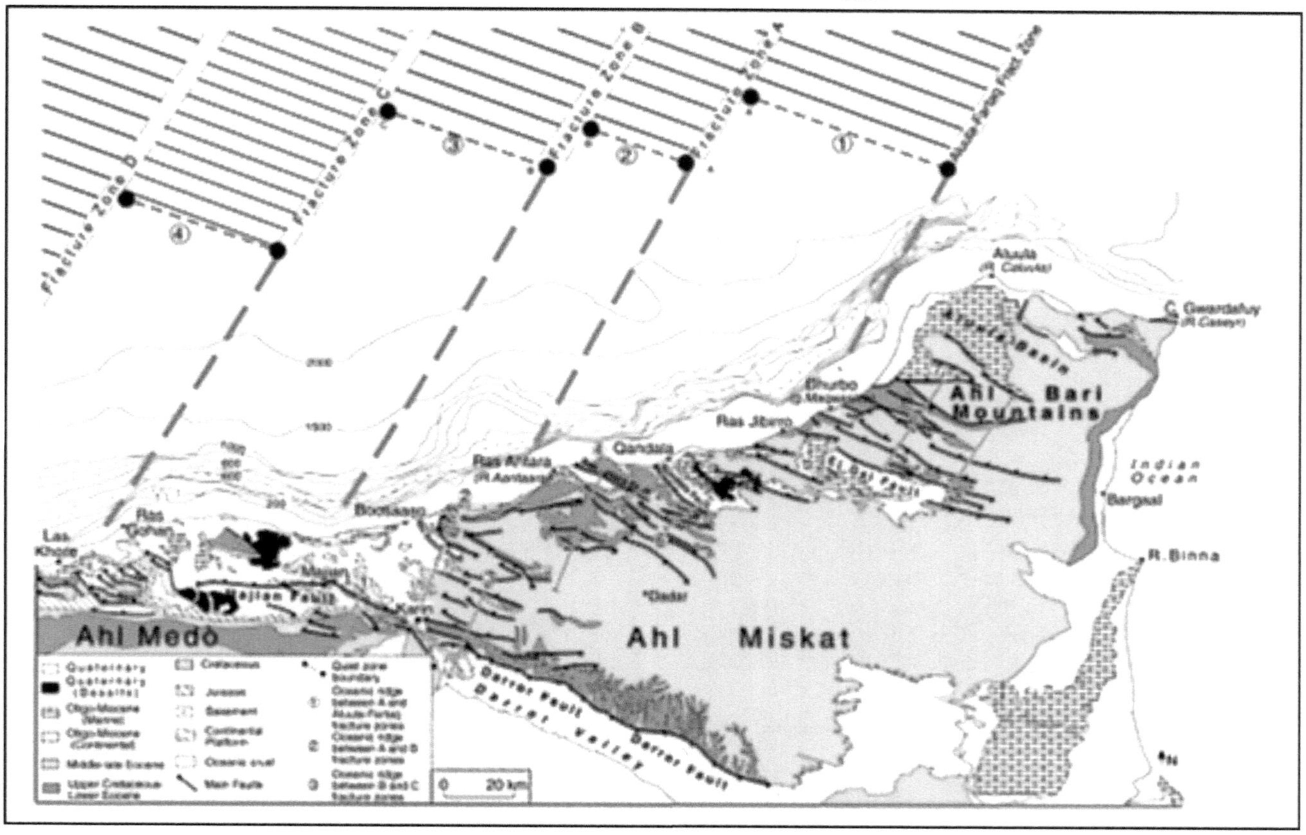

Fig. 10.2 Geological and tectonic scheme of NE Somalia (modified from Fantozzi et al. 2002)

10.3 Landscape and Landforms of the Awdal Region

The Awdal Region takes up the western section of Somaliland, adjacent to the Republic of Djibouti border (Fig. 10.3).

The Precambrian basement crops out at the base of the North Somali escarpment and is covered by different geological units that in the Awdal region are mostly calcareous (Figs. 10.1 and 10.3). The Jurassic Limestones unit forms sub-vertical slopes facing seaward and a gently descending plain towards SSW. The limestones form ridges parallel to the coastline both in the uplifted and in the downthrown blocks, normally separated by minor normal faults, parallel to the escarpment main fault and responsible for seaward lowering of crustal blocks. In this way, the basement—limestone succession is repeated several times (Fig. 10.3). Between the Jurassic limestones and the basement rocks a unit called the Adigrat Sandstones is commonly interposed. It is a basal conglomerate, few dozen of metres thick, which formed at the beginning of the Jurassic marine transgression.

The Jurassic unit mostly consists of stratified limestones with marly levels, which are more abundant at the bottom, at the transition from the Adigrat Sandstones formation. It is considered the best aquifer of the country for the high density of fractures and karstic structures in the upper calcareous portion and for the primary porosity of the Adigrat unit. The best exposed sections are in the Baki district, where a thickness of as much as 200 m of limestones is estimated, and significant springs and the only perennial river of the country (Durdur) occur. The Durdur River descends seaward forming a steep valley, on the bottom of which fruit trees and vegetables are cultivated. In the Xalimale area the escarpment is spectacular and offers a wide view of the lower plain and hill ranges (Fig. 10.4).

In the inner part of the region, such as at Boorama, limestone outcrops never exceed 100 m of thickness (Fig. 10.5). In western basin of the town, ground water exploration led to suppose an underground connection with the Jurassic series beyond the Ethiopian border, whose thicknesses is suspected to be more than 200 m.

Southward, beyond the northernmost ridge built of Jurassic limestones (Gerisa-Gargara), the coastal plain gently descends seaward, with gradients less than 1%. The plain consists of alluvial deposits laid down by the wadis during floods. This sediment is characterized by a downstream decreasing grain size (from gravel to silt), with the formation of small dunes where the sandy fraction prevails.

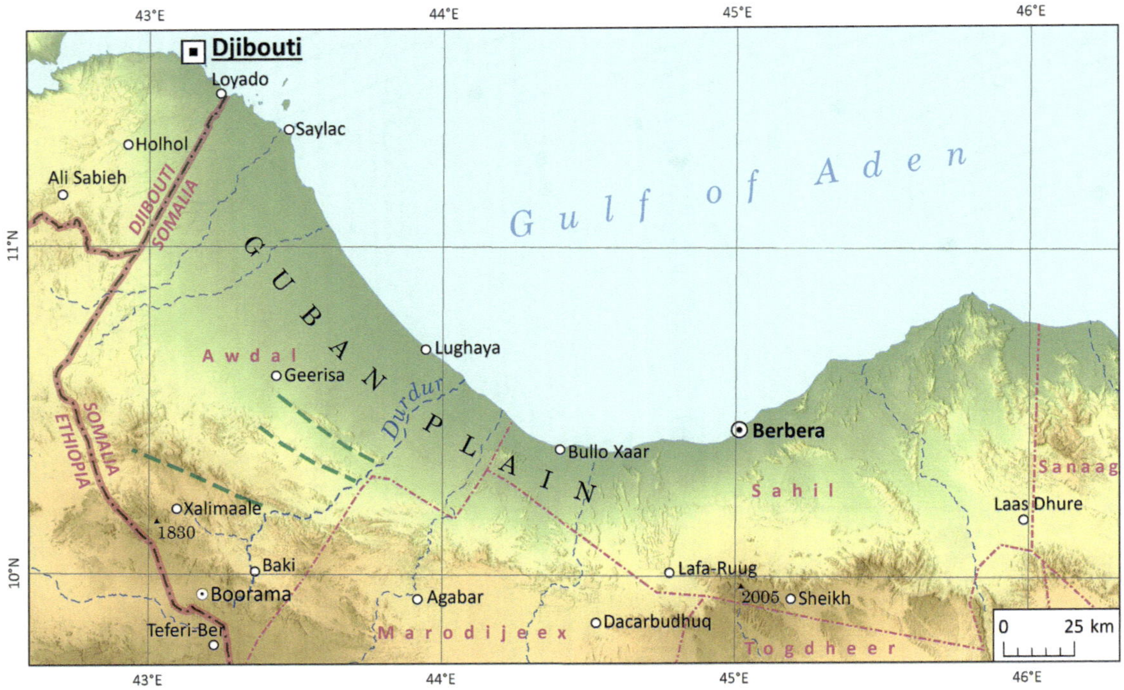

Fig. 10.3 Map of Awdal and Sahil regions. *Note* elongated limestone ridges between Boorama in the south and Geerisa in the north

Fig. 10.4 The great escarpment topped by limestones at Xalimale; notice the steep slopes of the limestone crests (above the blue lines) and gentler slopes within the crystalline basement

Fig. 10.5 Jurassic limestones (above blue line) resting on crystalline basement at Borama (Awdal capital city)

Fig. 10.6 Ruins of the old Turkish mosque made of white coral blocks at Zeila

Near the coastline wide reddish clayey tidal plains, well developed at Zeila (Saylac) and Lugaya (Lughaye), are found. In the area between Loyado (close to the Djibouti border) and Zeila there are small commercial activities that exploit salt lenses embedded within the clay layers. The town of Zeila was built during the Turkish dominion on an ancient coral reef, recently emerged, from which white coral blocks were taken to build the settlement (Fig. 10.6). The town was destroyed in 1990 by the concurrence of a stream flood and the high tide. The coastal plain is crossed by hazardous wadis. They remain dry, with white sandy beds for most of the time but they can unexpectedly resume high flows in response to intense rainfalls on the plateau,

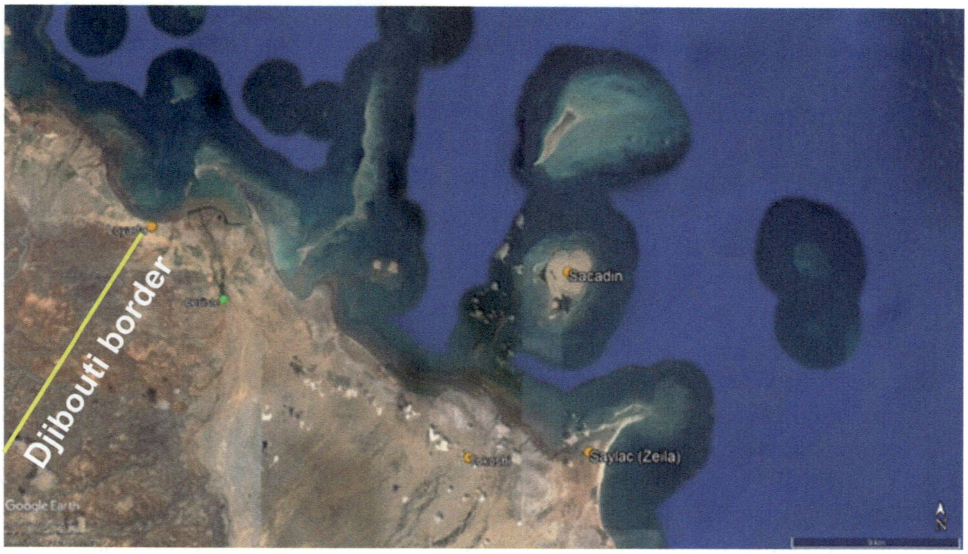

Fig. 10.7 Satellite image of Zeila-Loyado plain and of the Sahadin Islands

sometimes with one or two days of delay. The sea is not easily accessible because of the presence of lagoons, quick sands and, locally, mangrove strips.

Along the Djibouti border and widely beyond it, a volcanic belt less than 20 km wide and mostly made by Pliocene lava and pyroclastic products is present. Between Zeila and Djibouti, several islands of volcanic-coralline origin are present as well (Fig. 10.7). The largest one amongst them is Sacadin (Sahadin) and it is fully surrounded by mangroves and coralline reef.

10.4 Landscape and Landforms of the Marodijeex Region

Moving eastward into the Marodijeex (once Woqoy-Galbeed) region, where the state capital Hargeisa is located (Figs. 10.1 and 10.8), the plateau edge and part of the coastal strip are covered by a dark Pliocene lava flow (Late Miocene—Pleistocene—Abbate et al. 1994) (Fig. 10.8), probably coming from a linear effusive source. A volcanic edifice of similar age

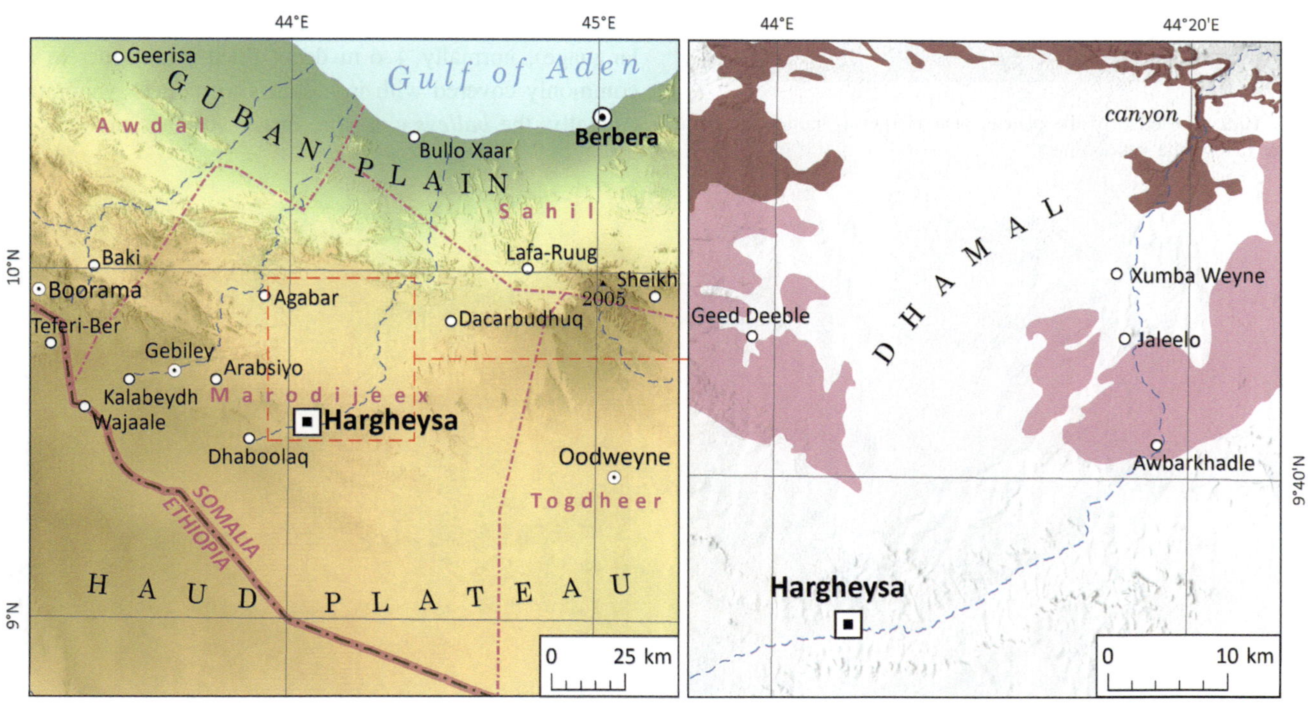

Fig. 10.8 Marodijeex (Woqooyi Galbeed) region and detail of Hargeisa surroundings; the areas coloured in purple mark the basement outcrops, dark areas are covered by basalt flows

and composition, mainly basaltic, Mount Elmis, formed on the coast about 90 km to the north of the Somaliland capital of Hargeisa (Fig. 10.9).

Near Hargeisa, scattered remains of the Auradu limestones form isolated hills. Two of them, located in the eastern suburbs of the town, are called the Hargeisa Breasts (*Naasablod* in the local language) and are considered the brand of the town (Fig. 10.10). The peculiar morphology of these hills results from the occurrence of a hard limestone top, which protects the underlying Yesomma sandstones from erosion. The Late Cretaceous Yesomma sandstones, which constitute most of the surrounding plain south of the town, are of fluvial origin and are the Somali equivalent of the Adigrat Sandstones extensively outcropping in Ethiopia.

In the Marodijeex region, the Yesomma sandstones are exposed over most of the plateau. This unit of continental environment consists of alluvial deposits, mostly sandy and silty that filled many depressions of tectonic nature and contributes, along with the deposition of a thin Quaternary alluvial blanket, to the contemporary flat morphology of the plateau. The unit is part of a widespread formation known in most of the Sahelian area as the Nubian Sandstones and in its thickest sections can reach more than 500 m. In the Hargeisa surroundings the escarpment is split into two main steps, with a deep tectonic depression in between, filled by Quaternary deposits that host some of the best aquifers of the country (Geed Deeble, Dhamal, Jaleelo).

Fig. 10.9 The edge of the plateau near Hargeisa—notice the basalt cover (above the yellow line)

Whilst the areas near the escarpment are characterized by strong erosion, which gives rise to spectacular structures, the Haud plateau (Fig. 10.1) and the coastal plain do not offer many landforms of particular interest. For the most part, the plateau is characterized by an arid, sandy-silty plain that appears slightly convex to the south of the capital or flat and dipping southward, as in the areas of Odoweyne or Burco (Togdheer Region), or northward, as in the coastal plain, everywhere covered by alluvial-aeolian deposits. The landscape is monotonous, only interrupted by isolated, flat-topped hill chains, a few dozen metres higher than the surrounding plain. Vegetation is scarce and mainly consisting of low bushes and scattered acacias. The plateau is gently inclined towards the Ethiopian border and it is crossed by drainage systems, along which many sandy dams (*balleys* in the local language) have been built to supply water for any use (civil, rural or livestock). In an area of many thousands of square km, in fact, deep wells and perennial aquifers are lacking. Here the population tries to extend the water supply beyond the rainy seasons harvesting the running water in wide structures like the *balleys* (Fig. 10.11) or smaller square trenches (*berkads* in the local language), normally 4–6 m deep, lined with concrete and commonly covered with iron sheets to reduce evaporation. Actually, the *balleys* can store larger volumes of water but

Fig. 10.10 Hargeisa twin hills (Naasablod) protruding from the plain: the hills are topped by a layer of Auradu limestones

Fig. 10.11 A dried-up balley at the peak of the rainy season at Botor (Gabiley)

seldom has the water supply lasted beyond the month of January, whereas the larger *berkads* are a reliable resource throughout a normal dry season, up to February, but cannot face prolonged droughts.

The region gets the name from the wadi that originates a few km east of the town. The river runs for about 60 km in the prevailing WSW–ENE direction as far as the settlement of Awbarakhadle, then it turns sharply northward as far as the village of Xomboweyne. In this area the wadi gets the name of Tog Wahen and the streambed becomes very wide, as much as 1 km. The alluvial deposits are abundantly recharged by the river flows and the subsurface water level is found at the depth of less than 1 m. The river subsurface flow is commonly exploited by means of trenches that capture the water and convey it to farms located downstream, mostly cultivating fruit trees. Shortly downstream of Xomboweyne the plain ends abruptly as the river enters the basalt flows that cover the crystalline basement. This section of the valley is deeply cut into the basalts, forming black sub-vertical walls on the flanks and acquiring a canyon-like morphology (Fig. 10.12).

In the rainy season, small ephemeral lakes form a few km downstream of the Xomboweyne village (Fig. 10.13). The surroundings of these small lakes are dangerous for the presence of quick sand.

Downstream of the village, the river descends from elevations of about 1000 m down to 500 m asl in the coastal plain. In this reach, the river has eroded deep and narrow gorges incised into the basalts and the basement rocks (Fig. 10.14). Just upstream of its fan delta, Tog Wahen River forms spectacular narrow gorges in granite and in the Yesomma sandstones (Figs. 10.14 and 10.15). The water flowing in the upstream reaches infiltrates in the fractured basement and, at the escarpment foot, the reddish granite gorge (Fig. 10.14) hosts only a small water flow in the dry

Fig. 10.13 Ephemeral lake near Xomboweyne (Hargeisa) formed by Tog Wahen. In this area the basalt is characterized by reddish colour

Fig. 10.14 Tog Wahen gorge cut into a reddish granite. Water flows can be as much as 10 m deep. In the upper part of the gorge typical weathering forms of granite rocks, such as spheroidal boulders, widened fissures and tors can be seen

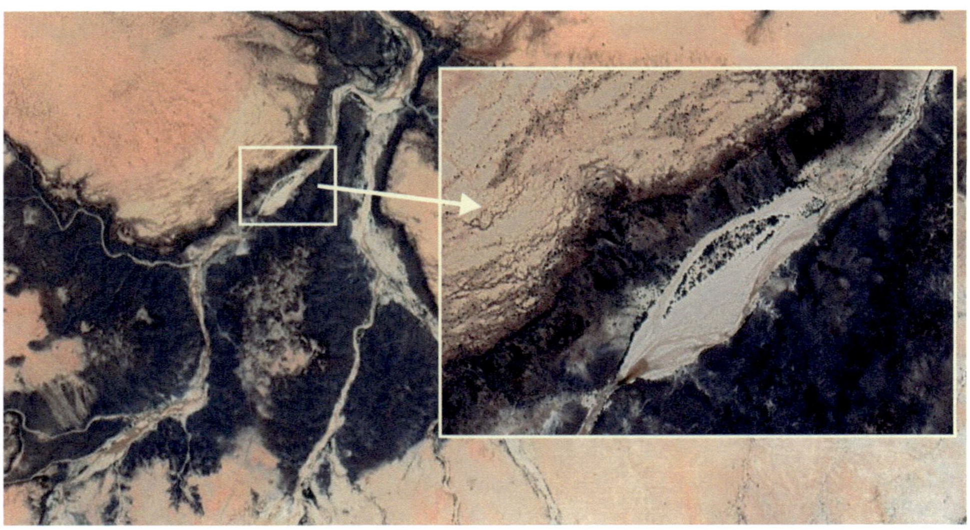

Fig. 10.12 The Tog Wahen river entering the basalt flows and forming a canyon with sub-vertical valley flanks. Tog or Togga means 'a river' in the local language

Fig. 10.15 Pink, criss-cross laminated Yesomma sandstones exposed in one of Tog Wahen gorges

Fig. 10.16 The gorge (G) and fan delta of Tog Wahen

season, with scattered deeper pools. The gorges are not accessible in the rainy season since the risk to be flushed away by sudden flows is very high.

A few hundred metres downstream of the gorge, the Tog Wahen streambed widens and splits into a multiple channel system to form a fan delta, which is about 14 km wide (Fig. 10.16).

10.5 Sahil Region

Sahil region lies at the centre of the country, along the coast of the Gulf of Aden, and the capital is Berbera, once capital of the English protectorate. It extends up to the escarpment and to the northern section of the plateau, where the town of Shiik occurs (Fig. 10.17).

The Sahil geology is the outcome of an intense tectonic activity that formed several blocks with thick sedimentary series along the SW–NE chain (Fig. 10.17) that surrounds the coastal plain of Berbera. The coastal plain consists of Quaternary alluvial deposits with scattered hills built of Miocene-Pliocene deposits, whereas pre-Eocene units are not visible as far as the first hills chain south of the plain. The Jurassic limestones form a series of important blocks in the area south of Berbera, at Bixindhule (Fig. 10.17).

In this location the thickness of the unit is estimated for about 700 m and consists of different sub-units, with the presence of thick marly sections interbedded with medium-fine graded limestones (Figs. 10.18 and 10.19). The Adigrat sub-unit at the base may include thin pillow basalts.

The E–W alignment of limestone blocks is sharply bound, in the south, by a long normal fault that lowers the upper surface of the blocks to the level of the basement. Because of this tectonic structure and the southward dip of limestones, the calcareous sub-units constitute important aquifers that supply several springs (Fig. 10.19) and a permanent stream, running in a narrow valley cut into the basement (Fig. 10.20). The stream disappears in the sandy alluvial deposits of the Togga (wadi) Kalajab at Faradero site, where the well field of Berbera is located.

At Faradero it is possible to see the Precambrian basement and the Boulder Beds unit (Plio-Pleistocene), the latter normally 5000 m above the basement, but juxtaposed here because of a major geological fault that downshifted the eastern section of the valley. From geophysical data collected during oil surveys, the limestones extend widely at depth, between the Paleogene units and the Precambrian basement.

10.6 The Eastern Regions of Somaliland

The eastern section of Somaliland includes three regions: Togdheer, Sool and Sanaag, whose capitals are Burco, Laascaanood and Ceerigaabo (Fig. 10.21). To the east of the last two regions there is the disputed border with Puntland. In this area, on the top of the Cretaceous unit, a younger formation, the Auradu Limestones, marks a new marine transgression in the lower Eocene. The limestones thicken from west to east, from a few dozen metres in the Marodijeex valley to some hundreds of metres in the area east of

Fig. 10.17 Map of eastern part of Sahil coastal region and prominent inland escarpment. (1) Hills built of Cenozoic sedimentary rocks, (2) blocks of Jurassic limestone

Fig. 10.18 Bixindhule limestones interbedded with marly levels (between the green lines)

Burao (Burco). This suggests that the transgression came from the east. The Auradu unit forms spectacular cliffs on the top of the northern escarpment around the town of Ceerigaabo (Sanaag region) (Fig. 10.21), whilst in the central and southern sections of the plateau (Togdheer and Sool regions) it forms long multiple ridges subparallel to the escarpment. The unit consists of massive biogenic limestones, with lenses of sandy, clayey and marly limestones (Fig. 10.22).

Fig. 10.19 Fresh water emerging at the foot of a limestone wall

Fig. 10.20 The stream fed by the Bixindule springs, running between black gabbro rocks and reddish orthoclase dikes, before disappearing in the sands of Kalajab valley

The Gulf of Aden pre-rifting succession continued in the lower to middle Eocene with the deposition of Taleh (Taleex) evaporites that mark a regressive phase, with a vast lagoon environment in which gypsum and anhydrites with limestones and cherty levels formed (Fig. 10.23).

The Taleex unit is dominant in the eastern regions (Togdheer, Sool, Sanaag) (Figs. 10.24 and 10.25) where it covers the older units (Auradu, Yesomma). The first outcrops appear in the headwaters of the Nugaal (Nogal) valley and from there, they extend northward and eastward. From a morphological point of view the Taleex evaporites form a wide plain dotted by rounded hills a few metres high. In the Sanaag region, on the top of these hills stone rings of uncertain origin are found. The local inhabitants declare that their function is unknown. They may be old gravestones or simple nomadic shields against the wind (Fig. 10.24).

Commonly, large gypsum crusts form near the ground surface, where the Taleex formation is present, though they

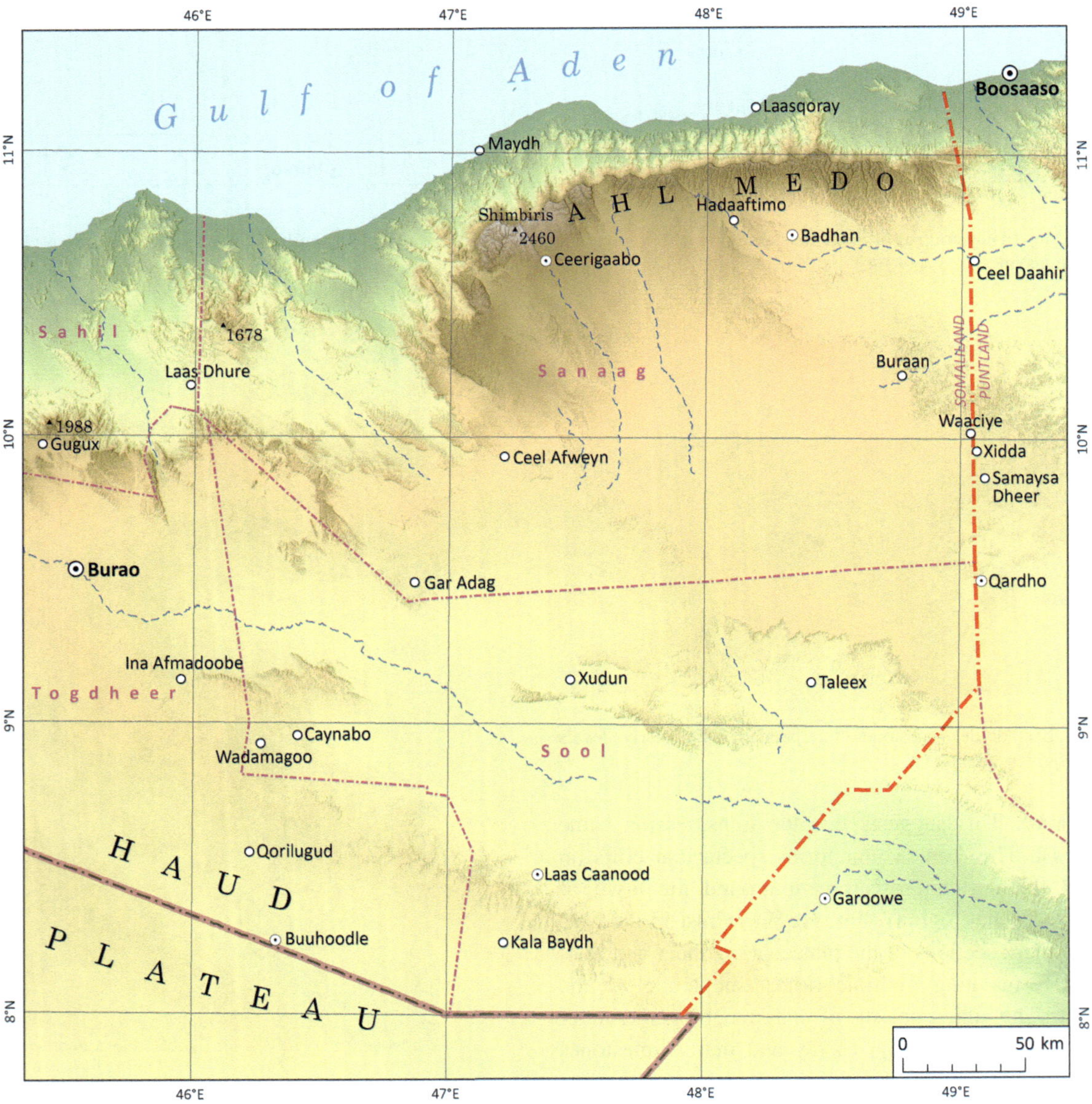

Fig. 10.21 Map of Eastern Somaliland

Fig. 10.22 The cliff of massive Auradu Limestones at Erigavo, facing the coastal plain

Fig. 10.24 Typical gypsum hill in the Sanaag region. Notice the stone ring on the hill top

Fig. 10.23 Spectacular erosive landforms of the sub-horizontal Auradu layers on the escarpment top (aerial oblique view)

are not exposed. In some places, as in the Nugaal valley, the gypsum that forms the crusts comes from far away hills as ions in the water running across the ground surface during floods (Fig. 10.26). The evaporites are responsible for the salinity of shallow and deep waters of these regions and of most of the Karkar region in Puntland. The thickness of the Taleex formation can reach 300 m.

On the northern side of the Nugaal valley a high range rises to high elevation, providing termination of a tectonic trench that extends for most of the valley length and width. The trench bottom occurs a few kilometres below sea level, where the crystalline basement lies. At the northern edge, a big fault has caused uplift of the geological succession that includes deposits spanning a very long interval from more than 500 million (Precambrian) to 4–5 million years ago (Pliocene) and building the massive range, mostly made by Auradu limestones (Fig. 10.27).

10.7 The Bosaso (Boossaso) Plain

At the eastern edge of the Nugaal valley, beyond Laas Canood, there is the disputed border with Puntland, presently part of the Federal Republic of Somalia. From Garowe to the escarpment facing the Gulf of Aden, the landscape consists of a wide monotonous calcareous plain, formed by gently undulated or flat layers of the Karkar Formation, with only a small outcrop of the whitish gypsum of Taleex formation in between; the flat Karkar limestones extend as far as the escarpment at the margin of the Haud Plateau (Fig. 10.1). The road that descends from the plateau to the sea enters a narrow valley excavated by the Dhagan wadi that drains a wide basin. With its headwaters behind the escarpment edge, the course receives a large amount of water because at the high elevation (>1.000 m asl) the rainfall is relatively abundant (300–400 mm yr^{-1}). For this reason, the water flows in the wadi for a few kilometres from the escarpment foot, near the Laag village, where it takes the name of Whadhod (Figs. 10.28 and 10.29). A few hundred metres beyond the Bosaso-Garowe road bridge the water disappears into coarse bed deposits and probably flows underground following the course of the wadi as far as the Kaw gorge that constitutes the river mouth into the sea (Fig. 10.30). The gorge is spectacular if seen from a plane when arriving or leaving the town. This is because during high tide it is occupied by sea water, whereas during low tide

Fig. 10.25 Taleex range at Laas Canood. The town extends in a wide, gently inclined valley between the hills

Fig. 10.26 The central part of the Nugaal Valley: yellow dust and gypsum crusts

Fig. 10.27 The Auradu range at the northern edge of the Nugaal valley

it becomes a muddy swamp, on which flamingos often walk. The site can be reached by a gravel road but, as everywhere in Puntland, an SPU escort and military permission are necessary since a military camp has to be crossed beyond the airport.

The Dhagan wadi, dry for more than 90% of its course, marks the eastern edge of the western escarpment, named as the Ahl Medo range in Puntland, and separates it from the Ahl Miskat range that forms most of the true Horn of Africa. From the Laag village, a wide plain extends seaward. It is formed by older Quaternary alluvial deposits and is punctuated by hills made of Mesozoic geological units. The area is very arid, almost devoid of vegetation, but in the surroundings of Laag, especially westward, there are many oases around springs (Duud Shabel, War War, Hilqad) (Fig. 10.31) and extending along other branches of Wadi Dhagan, where the water running otherwise underground comes to the surface and flows for a few hundred metres (Laag, Gub wadis). At Duud Shabel there are high palms, dwarf palms and several farms where greens are cultivated. A project to bring water from the springs to Boosaso is going on. From this area northward the landscape is rather monotonous, only interrupted by alluvial terraces, locally made of ochre or red coloured sandy-clayey deposits (Fig. 10.32) or outcrops of the Dubhar limestones.

The town of Bosaso rests on a low terrace between two wadis, Balade and Kulule and it is surrounded by a wide plain made of pebbles. A few kilometres to the east, the Ahl Miskat range closes the plain with beautiful cliffs formed by the Auradu Limestones, cut by an important fault and by deep valleys. In one of these valleys, formed by the Byio Kulule stream (Fig. 10.33), there are hot water springs and a thermal station with water pools of different temperatures, built during the early Italian occupation, surrounded by a beautiful oasis.

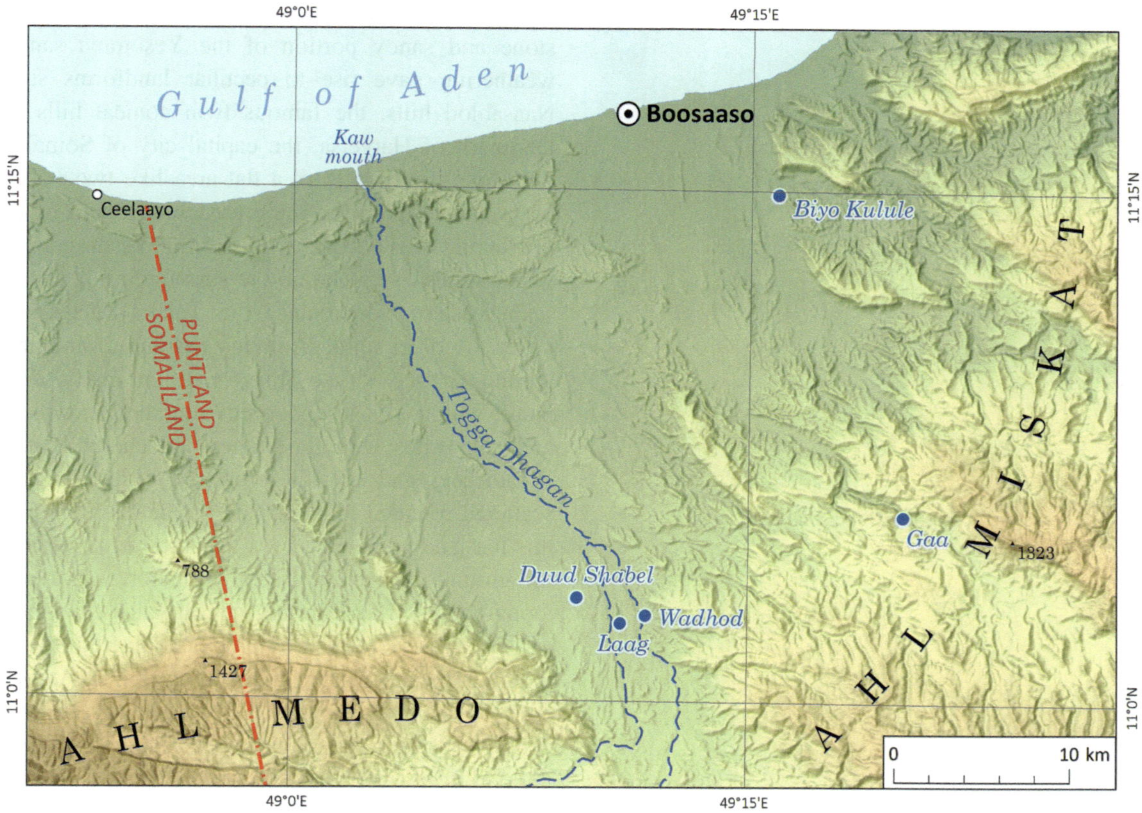

Fig. 10.28 Bosaso plain, the main springs and the course of Wadi Dhagan

Fig. 10.29 Water in Wadi Dhagan under the road bridge near the Laag village

Fig. 10.30 See water entering the Kaw mouth during high tide

10.8 Concluding Remarks

The contemporary geomorphological landscapes of northern Somalia reflect the long sequence of geological and tectonic events that affected this region. Planation of the crystalline basement, marine transgressions and deposition, and uplift associated with the doming of the Horn of Africa created a landscape with marked physiographic differences. The main, large-scale elements of northern Somalia landscape are the plateau, the escarpment and the coastal plain. The contrast between the sedimentary rocks resting on the crystalline basement provides the landscape with spectacular, sharp-crested top of the main escarpment in the Awdal region.

Fig. 10.31 The oasis encircling the Duud Shabel springs

Fig. 10.32 Desert view from Duud Shabel spring: in the background the edge of the plateau escarpment (Ahl Medo) descending from about 2000 to 500 m asl; in the foreground old alluvial terraces

Fig. 10.33 Byio Kulule valley, eastern flank

In the Marodijeex Region, variable resistance of limestone and sandy portion of the Yesomma sandstones to weathering gave rise to peculiar landforms such as the Naasablod hills, the famous twin conical hills that are a landmark of Hargeisa, the capital city of Somaliland. The twin hills protrude from a flat area like two isolated inselbergs and results from the presence of a hard limestone top preventing erosion of the underlying Yesomma sandstones, which otherwise make up the surrounding plain.

In eastern Somaliland, the Taleh (Taleex) evaporites cover the older units (Auradu, Yesomma) and are the prevailing surface rocks. This formation marks a regressive phase within a vast lagoon environment, in which gypsum and anhydrites with limestones and cherty levels formed. Weathering and erosion produced a flat landscape with scattered rounded hills only a few metres high. Commonly, this flat land is covered by a gypsum crust, which forms on the ground even if the Taleh formation is not close to the topographic surface. In the Nugaal valley, for instance, the gypsum crust is formed by precipitation from overland flow draining distant gypsum hills during floods. At the northern edge of the Nugaal valley, a big fault caused uplift of the geological series that includes deposits spanning a very long interval from more than 500 million years (Precambrian) up to 4–5 million years ago (Pliocene) building the massive range, mostly made by Auradu limestones, and forming spectacular cliffs near Ceerigabo.

In the eastern part of the plateau deep gorges, incised by ephemeral streams, occur. The larger of them, such as Wadi Dhagan, have large catchments, whose headwaters are located on the plateau, beyond the escarpment edge. On the plateau the annual rainfall is relatively high (300–400 mm). This geomorphological setting favours the collection of large quantities of water that flows in the wadi downstream of the escarpment foot, for a few kilometres in the coastal plain. This water is a very important resource for the communities living in the dry costal belt, where it can be used for small crops irrigation.

Though this chapter is a summary description of the main geomorphological landscapes of northern Somalia, the peculiar combination of a complex geological and structural evolution of the area and the prevailing dry climate of today resulted in landscapes and landforms, whose generating processes and controlling factors are known only superficially. Unfortunately, the state of permanent civil war and the resulting political instability that characterized the region during the last decades have discouraged geomorphologists from undertaking field investigation campaigns. In northern Somalia, however, many interesting scientific questions are still open and only the main geomorphological characteristics of this region were delineated in this chapter in order to stimulate further research as soon as sufficient security conditions will be restored in the region.

References

Abbate E, Sagri M, Sassi FP (1994) Geological map of Somalia. Somali National University, Mogadishu

Fantozzi PL, Abdirahman HM, Ali Kassim M, Carmignani L (2002) Geological map of Northeastern Somalia. Scale 1:200.000. CNR—University of Siena

Hadden RL (2007) The geology of Somalia: a selected bibliography of Somalian geology, geography and earth science. US Army Corps of Engineers, Topographic Engineering Center, Alexandria, Virginia

Inselberg Landscape of the Bur Area in Southern Somalia

11

Borislav K. Kamenov, Paolo Billi, and Piotr Migoń

Abstract

A great number of spectacular inselbergs (burs in the Somali language) punctuates an otherwise flat landscape of a vast area in southern Somalia known as the Bur Area. These inselbergs are built mainly by granitoids, which represent two types: syn- to late kinematic and late-kinematic plutons. Their mineralogy, petrology and geochronological dating indicate that the youngest granitoids from the Bur Area belong to the final Pan-African magmatic event in the East African Orogen. Long-term surface lowering in the Mesozoic and Cenozoic, probably by a range of land-forming mechanisms, resulted in the formation of a vast plain dotted by inselbergs of various heights and shapes. Post-Miocene uplift of the Bur Area horst, elongated in a NE–SW direction, was moderate and did not trigger considerable incision of the plain. Inselbergs represent a range of types, from massive domes with smooth slopes of variable steepness through castellated forms to conical and crested hills. The contemporary morphology of the inselbergs reflects both structural control (joints, faults, sheeting) and the operation of various surface processes, mainly weathering and mass movements.

Keywords

Inselberg • Peneplanation • Block faulting • Granite weathering • Erosion cycles

B. K. Kamenov (✉)
Faculty of Geology and Geography, Sofia University "St. Kliment Ohridski", Sofia, Bulgaria
e-mail: b.kamenov@gmail.com

P. Billi
International Platfrom for Dryland Research and Education, Tottori University, Tottori, Japan

P. Migoń
Institute of Geography and Regional Development, University of Wrocław, Wrocław, Poland

11.1 Introduction

A semi-arid area in southern Somalia is unusually rich in inselbergs, and it is known as the Bur Region. The physiographic features and petrologic research on the rock varieties of these inselbergs reveal that they are built up by a very hard and erosion-resisting lithology. The Bur Area is therefore an excellent example of residual relief forms, strongly influenced by the nature of the underlying rock.

This paper summarizes the geological and geomorphological background of the Bur Area. Unfortunately, the permanent state of civil war that has been distressing Somalia during the last two decades prevented researchers from continuing field work, thus not allowing one to upgrade the basic geological knowledge or to verify modern theoretical geomorphological models. Though the level of knowledge of the area is mainly based on data collected in the past, the landscape and landforms of the Bur Area are so peculiar that they deserve to be investigated and brought to the knowledge of the scientific community. Very few papers (e.g. Azzaroli and De Angelis 1965; Borsi 1965; Haider 1993; Ali Kassim et al. 2002; Kamenov and Lilov 2012), in fact, focus on this area, the geology and geomorphology of which are therefore poorly known. Given the limitations addressed above, this contribution is mainly descriptive and focused on data available at present. The main aim of this paper is therefore to provide data on the basic characteristics of the Bur inselbergs, to emphasize the causal connection between the lithological compositions of bedrock and the relief and indicate the role of inselbergs in human expansion in the region.

11.2 Physiographic Features of the Region

Somalia differs very much from other equatorial countries. The country is located in the equatorial and subequatorial climatic belts, but one cannot find here the humid climate,

P. Billi (ed.), *Landscapes and Landforms of the Horn of Africa*, World Geomorphological Landscapes,
https://doi.org/10.1007/978-3-031-05487-7_11

large discharge rivers, dense woodlands and high-grass savannas typical for such countries. The climate is semi-arid (see Chap. 2 of this publication or more details) and the landscape is monotonous, strewn with acacia and mimosa trees, thorny bush and blinding sand dunes (along the coast), whereas sunny days predominate. Except for the northernmost part of Somalia, flat land prevails for the most part of the country. Only the northern part of the plateau (Karkaar Mountains) is raised and cut by a fault-generated escarpment having the appearance of a mountainous range. The gradual lowering of the relief from north to south is also a typical landscape feature of the country (Fig. 11.1).

Southwestern Somalia is dominated by the country's only two permanent rivers, the Juba and the Wabe Shabelle (see Chap. 14 of this publication for more details on these rivers). With their sources in the Ethiopian highlands, these rivers flow in a generally southerly direction, cutting wide valleys in the Somali Plateau as it descends towards the sea; the plateau's elevation decreases rapidly in this area. The adjacent coastal zone, which includes the lower reaches of the rivers and extends to the Kenyan border, averages 180 m asl. Favourable rainfall and soil conditions make the entire riverine region a fertile agricultural area and the centre of the country's largest sedentary population.

The region between the Wabe Shabelle and Juba rivers, located in the southern part of the Central Plateau between the towns of Bur Acaba and Dinsoor, stands out from the lowlands with its outstanding morphological features. Isolated, steep-sided residual hills, rocky and usually bare rise conspicuously above the general level of the surrounding flat area, as either solitary hills or clusters of hills. These loaf-shaped small mountains reach elevations of up to 450 m asl, or some 200 m above the southern part of the plateau, although others are less than 100 m high. They are typical inselbergs, or "*burs*" in the Somali language, after which the whole region is named. The Bur Area lies some 180 km to the west of Mogadishu and occupies an area of about 25,000 km^2 (Fig. 11.1). The Bur Area is a flat erosional peneplain about 200 m asl, and occasional rocky domes and castellated hills are the only landmarks in the area.

Fig. 11.1 Main physiographic zones of Somalia and the location of Bur Area. B—Baydhabo, D—Dinsoor. *Source* of image: Landsat/ Copernicus © 2020 Google

The Bur Area is an uplifted horst of Precambrian basement rocks (Stefanini 1918) with a southwest-northeast orientation, about 200 km long and 100 km wide (Fig. 11.2). About 200 burs have been observed on aerial photographs. These are scattered throughout the area, being most numerous in its southwestern portion. Two systems of faults cut across the horst and divide it into minor blocks. One system of normal faults has NW–SE to N–S trend, whereas the second one has a SW–NE to W–E trend.

The most striking inselbergs are mainly built of granitoids and, to a lesser extent, of quartzite. Amongst them, impressive examples are provided by Bur Acaba, Bur Degis, Banoda Boss, Safarnolei, Bur Gulo, Bur Galin, Bur Eibi, Alio Gheleh, Bur Geluai, Bur Dur and others.

11.3 Geological Background

The Somali territory is a transitional zone between the large Eastern African Arch and the deep sedimentary basin of the Indian Ocean. The Eastern African Arch (covering Ethiopia, Kenya, Uganda, Tanzania, Mozambique, Red Sea, Gulf of Aden and their neighbouring countries) is cut by deep penetrating faults of the Great Rift System. The biggest crustal structures were formed in the Precambrian and were influenced by different tectonic movements and igneous processes later on. Somalia is situated just within the eastern shoulder of the large Ethiopian Arch Uplift. Metamorphic and igneous complexes are exposed only in the uplifted blocks, which are, at the same time, the principal structural units:

- Northern Somali Zone, for which the largest gradients of vertical displacements are known.
- Bur Area, which pertains to the wide belt of Precambrian rocks, designated as the Mozambique Belt.
- Mesozoic-Cenozoic epiplatform sedimentary sequences that cover the rest of the country.

Basement exposures occur only in the first two zones, which are referred to the East African Orogen, resulting from the continental collision between East Gondwana and West

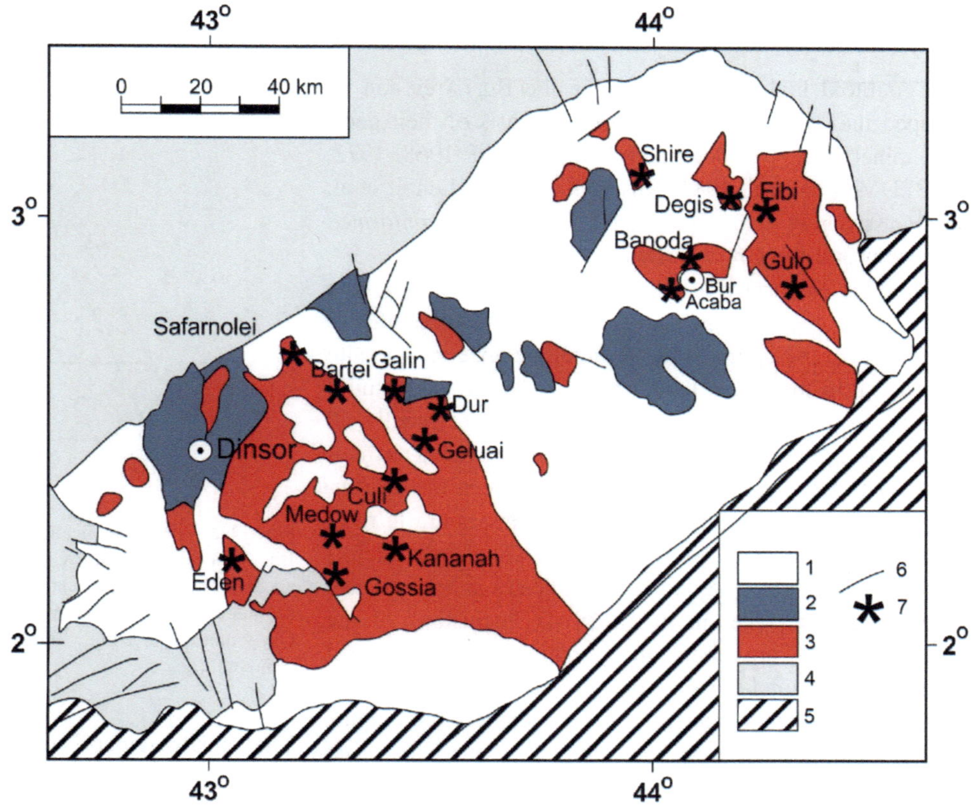

Fig. 11.2 Simplified geological map of Bur Area (after Kamenov and Lilov 2012). 1—Lower Olontaleh series of Precambrian metamorphic rocks, 2—upper Dinsor series of Precambrian metamorphic rocks, 3—granitoids, 4—Jurassic sedimentary rocks, 5—Pleistocene to recent deposits, 6—faults, 7—selected inselbergs (burs)

Gondwana at about 870–550 Ma. The southern part of the East African Orogen is a distinctive geotectonic unit called the Mozambique Belt, overprinted on the Eastern African Arch. The north-eastern branch of the Mozambique Belt in Somalia includes the Northern Somali crystalline basement and the Bur Area. Several major deformational phases were recognized, the last one being termed the Pan-African Orogeny. It is interpreted as the final tectonic-thermal event, culminating at 650–500 Ma ago. The granitoids making up the inselbergs of the Bur Area are products of these Pan-African magmatic events (Kamenov and Lilov 2012).

In the study area, the basement rocks are divided into two series: Lower (Olontaleh) and Upper (Dinsor) Series (Warden and Hörkel 1984) (Fig. 11.2). These series were metamorphosed simultaneously, and both were involved in a process of generation of silicic magma in situ or emplaced at close distance from the melting point. Basement rocks occur also outside the Bur Area, but they are traced by drillings or presumed to be present on the basis of geophysical investigations. It is generally accepted that they can be found at depths deeper than 5000 m. A number of block uplift movements complicate this area, and in places, the gradients of the tectonic displacements reach up to 1:5 (over 10 km). This is the reason why next to the contemporary erosion surface depressions thick sedimentary sequences in the order of over 10 km are found. The vertical movements brought the Precambrian Basement to the surface.

The Lower Olontaleh Series includes schists, gneisses, amphibolites and granulite-like rocks. The total thickness in its eastern part is estimated to be as much as about 3000 m. The metamorphic rocks are strongly migmatized in certain localities. The Upper Dinsor Series consists of several units of silicate rocks (gneisses, amphibolites and quartzites), calc-silicate rocks and marbles. The presence of beds of ferruginous quartzites is a characteristic peculiarity for its top part. The total thickness of the series is estimated as 2500 m.

The Mesozoic-Cenozoic epiplatform sedimentary sequences are mainly Jurassic and, in some places, Cretaceous in age, covering the western, northern and eastern parts of the Bur Area. The rocks are sandstones, argillaceous marls and limestones, dipping gently to the southeast. Along the northern and the western boundary of the Bur uplift zone basaltic dykes was intruded, tentatively dated as Miocene in age. Quaternary and recent deposits occur as marine, colluvial and alluvial sediments.

The igneous rocks in the Bur Area vary from leucocratic granites to gabbro. The prevailing rock types are granitoids and granite-gneisses, cutting the iron-bearing quartzites and assimilating them to a certain degree.

11.4 Granitoid Petrology

The granitoid bodies outcropping in the Bur Area can be grouped into two different types on the basis of their geology, mineralogy, petrology and dating (UNDP 1968, 1972, 1973; D'Amico et al. 1981; Küster et al. 1994; Lenoir et al. 1994; Kamenov and Lilov 2012): the *Thin conformed intrusions* and the *Great discordant plutonic bodies*.

The *Thin conformed intrusions* are emplaced between the schists and gneisses and have variable thickness. Commonly, the granite plutons have a gneissic appearance, with inherited schistosity and banding and are encircled by migmatites (from nebulites to typical anatectites). Usually all gradations between isotropic granites and banded orthogneisses can be observed. The contacts between these intrusions and the surrounding migmatized gneisses are in most cases concordant, diffusive and gradual. The internal deformation is penetrative foliation. It indicates a syn- to late kinematic setting (Lenoir et al. 1994). Inclusions of the basement rocks are abundant. Usually, they are spread within the marginal parts of the plutons. The rather small granitoid inselbergs consist of leucocratic fine-grained monzogranite and syenogranite, rimmed by migmatites and gneisses. It seems that in the southwestern sector of the Bur Area mainly this type of granite is outcropping, supporting Bur Muun, Bur Daxale, Bur Kananah, Bur Gossia, Bur Geel Way and Bur Galin elevations. Quartz, microcline and plagioclase are the principal rock-forming minerals. Green amphibole is the minor constituent. Clinopyroxene and garnet are rare, but they are typical for this type of granitoids. Biotite flakes are found only in the pegmatite dykes. Cataclastic to low-grade metamorphic overprints are typical in some intrusions.

The *Great discordant plutonic bodies* are relatively homogeneous and often characterized by melanosoma inclusions. These granitoids were discordantly emplaced into the high-grade basement rocks, probably at a shallower level than the first group. Lenoir et al. (1994) referred the group to the post-kinematic type. Locally, weak foliation of magmatic origin is manifested by discontinuous layers of biotite or amphibole grains. These sections gradually pass into clear isotropic fabric in granite within short distances, of several metres only. Typical intrusive homogeneous textures are more frequent in the north-eastern sector of the Bur Area. The modal relationships of the principal rock-forming minerals in the sampled burs (Bur Gulo, Bur Eibi, Bur Acaba, Bur Banoda, Bur Degis, Bur Shire) confirm their assignment mainly to the monzogranites, but some samples fall also in the fields of quartz monzonite, granodiorite and quartz monzodiorite (Fig. 11.3). A characteristic feature of these rocks is the scarcity of femic minerals, and some of the samples could be classified as leucogranite (e.g. Bur Acaba).

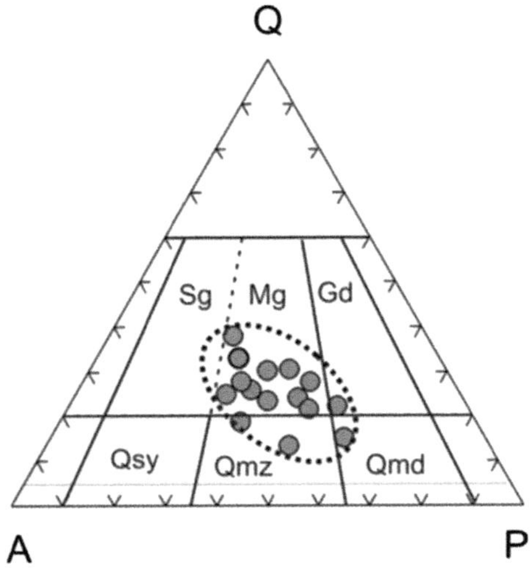

Fig. 11.3 APQ classification diagram of the analysed granitoids in the Bur Area (after Kamenov and Lilov 2012). A—percentage (mode) of alkali feldspar, P—percentage of plagioclase feldspar, Q—percentage of quartz; Sg—syenogranite, Mg—monzogranite, Gd—granodiorite, Qsy—quartz syenite, Qmz—quartz monzonite, Qmd—quartz monzodiorite

The post-magmatic process of "albitization" over the microcline is widespread. It is most intensively developed in the sector close to Alio Geleh. Here, the metasomatic alteration of granites reaches its maximum to the real "albitites" and "alkaline syenites", which include rare-metal economic mineralization of thorium specialization. The weathered granitic crust contains kaolin occurrences around the burs.

Almost all granitoid inselbergs are quite similar in terms of texture and structure, mineral and chemical composition, structural and erosional level of emplacement. Different generations of dykes can be distinguished as well.

The bigger granitoid bodies from the first diffusive group occur exceptionally within the Lower Olontaleh Series, whereas the typical discordant intrusions with sharp contacts can be found in both basement series. The above-described granitoid suites, i.e. "diffusive" and "discordant", were assumed also by Dal Piaz and Sassi (1986) and by Küster et al. (1994).

The earlier foliated suite indicates the presence of $\sim$2.5 Ga crust in the region, with anatexis occurring at 536 $\pm$ 18 Ma (Küster et al. 1994). Younger massive granites contain zircon that indicates post-tectonic emplacement at 474 $\pm$ 9 Ma (Küster et al. 1994; Lenoir et al. 1994). The implemented discriminations (Lenoir et al. 1994) show that the syn- to late kinematic granitoids are referred mainly to Within-Plate Group and less probably to the Syn-Collisional Granites, while the post-kinematic samples fall in the field of Volcanic Arc Granite. Newly published K–Ar data (Kamenov and Lilov 2012) confirmed the age of the younger granitoid magmatism (ranging 440–490 Ma), whereas pegmatites from northern Somalia turned out to be affiliated to the syn- to late-kinematic magmatic event (500–560 Ma, Kamenov and Lilov 2012). Granitoid magmatism and high-grade metamorphism in Southern Somalia are probably both related to post-collisional lithospheric thinning, magmatic underplating and crustal relaxation.

11.5 Landforms and Geomorphological Processes

Morphologically the inselbergs are mainly rounded hills (domes) standing out of the plain (Fig. 11.4). In some areas, larger domes are surrounded by smaller ones and the landscape is like a flat ocean surface with scattered islands, ideally consistent with the original German term coined at the turn of the twentieth century (Fig. 11.5). Sharp pointed peaks are not observed, although crested hills are present. The Bur Area landscape is monotonous as it consists of a flat land covered with sparse bushes not higher than 2–3 m. The average monthly maximum temperature recorded at Bur Acaba is over 30 °C in 10 months a year, and the maximum monthly temperatures peak between 40 and 45 °C in eight months a year (Fig. 11.6). The soil is predominantly reddish-brown, but in some places, dark grey-brownish soils are found. Vegetation is developed only on the quartzite burs, whereas the granitoid burs are bare.

The Bur Area is crossed by a number of drainage lines, but water flow is ephemeral and discharge is present only in response to high-intensity rainstorms. These stream channels are incised into alluvial deposits or basement weathering products (Ferrari et al. 1985). The flat land between the inselbergs consists of past alluvium deposited by several fluvial cycles, modern alluvium and endorheic depressions (Ferrari et al. 1985). The older alluvial deposits are terraced and make a transition to the modern alluvial material and/or to the clayey facies of the present plain (Ferrari et al. 1985). A general scheme of this geomorphological setting, based on the situation of Bur Acaba, is illustrated in Fig. 11.7.

The granites are mainly fine-grained and seldom show a porphyritic structure (Azzaroli and De Angelis 1965). The exposed parts of the burs are considerably jointed. Well-expressed parallel-piped cleavage along three orthogonal joints systems favours rock splitting into large blocks. Layers rich in biotite and poor in quartz and feldspars typically alternate with those including more quartz. Late thin pegmatite veins cut the rocks. In places, next to these veins, metasomatic processes formed aureoles of epimagmatic microclinization.

Foliation is common in all inselbergs, but it does not coincide with the elongated shapes of contemporary exposures. In the Bur Area, fault dislocations are usually oriented along two main systems with the Red Sea one (NNW) prevailing over the Cameroon Line (NE) (Fig. 11.2). The shape of many inselbergs is elongated for it is controlled by these fault line systems. For example, the shape of some burs of Kananah inselberg group is controlled by faults parallel to the Red Sea, while other granite burs such as Bur Finini, Warabe, Banoda, Wagelle and Eibi are strongly influenced by faults almost perpendicular to the former. Another group of granitoid hills shows foliation oblique or, in some places, even transversal to the general elongation of the outcrops. The best examples of this peculiarity can be observed in the group of granitic hills named Bur Gossia.

A typical weathering product of the granite is kaolin. Preliminary investigations and boreholes indicate that the kaolin deposits are located at a distance of not more than 150–200 m from the inselberg/plain junction. Their form and position are genetically related to the fault tectonics in the Bur Area. The kaolin occurrences are usually elongated and oriented along the two fault systems mentioned above. Representative examples of kaolin occurrences parallel to the two fault systems can be seen at Bur Galan, Bur Siabo, Bur Dur and Bur Gosei, respectively. The kaolin

Fig. 11.4 Google Earth transverse view of Bur Acaba, with Bur Eibi in the background

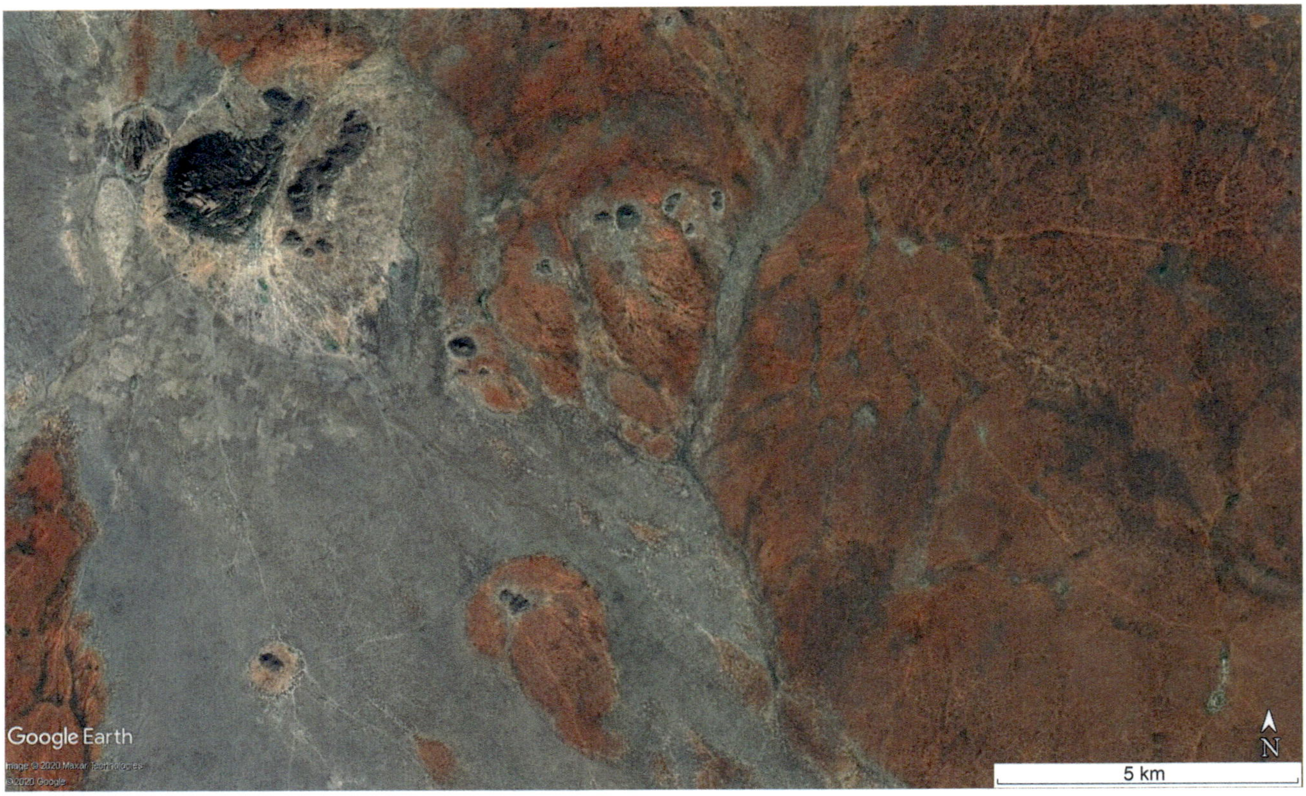

Fig. 11.5 Bur Eibi (upper left) is one of the largest inselbergs in the region, whereas several minor burs occur in the vicinity. *Note* the presence of SW–NE faults dividing Bur Eibi into three smaller elevations (Bur Eibi coordinates: 03° 00′ 01″ N; 44° 18′ 07″ E; for location, see Fig. 11.10). *Source* of image: © Google

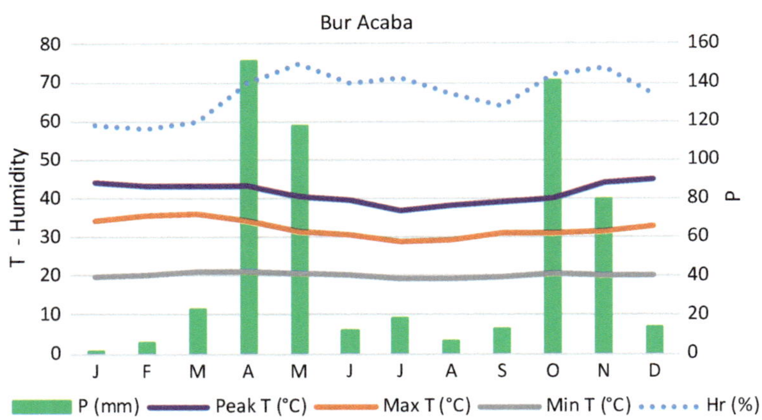

Fig. 11.6 Climograph of Bur Acaba. Observation interval 1961–1990. *Source* of data Deutscher Wetterdienst DWD (2016)

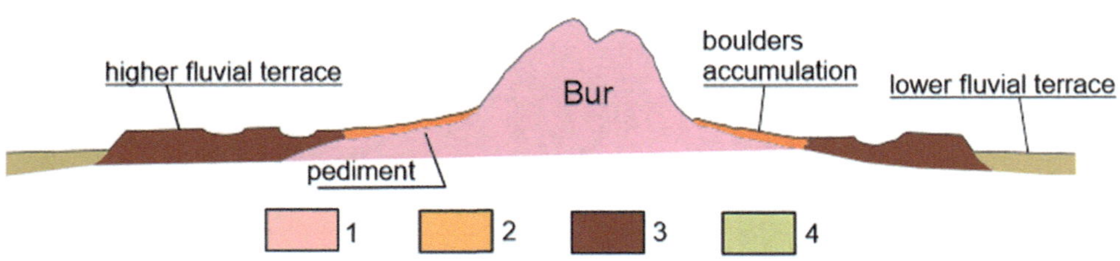

Fig. 11.7 Schematic geomorphological setting and alluvial–colluvial deposits based on the situation of Bur Acaba. 1—granite, 2—accumulation of boulders derived from physical weathering on the inselberg, 3—higher fluvial terrace (older alluvial deposits, consisting of coarse, rounded clasts), 4—lower fluvial terrace (modified from Ferrari et al. 1985)

Fig. 11.8 Temporary water wells at the foot of Bur Banoda (photo by B. Kamenov)

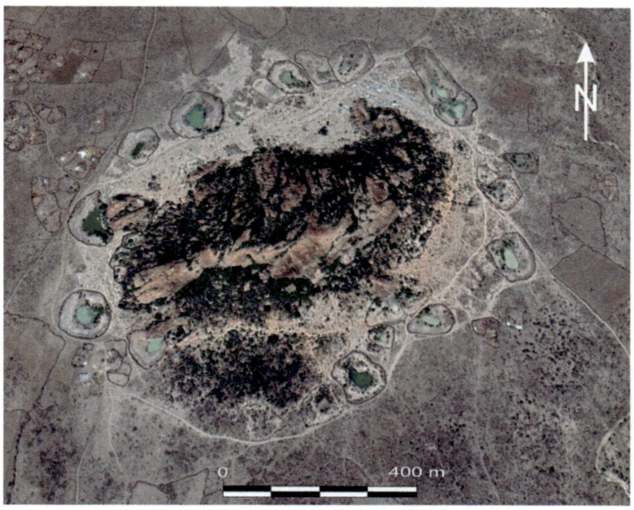

Fig. 11.9 Hand-dug water ponds around Bur Degis. *Source* of image: © Google

occurrences are 50–350 m long, 20–110 m wide and 0.5–10 m thick and lie at a depth to 20 m. Although the hills have a limited surface, along with their footslopes underlain by kaolin materials, they have enough capacity to store a great deal of rainwater. Therefore, inselbergs provide most of water supply to the local communities and commonly the water wells encircle almost completely the slope/plain junction (Figs. 11.8 and 11.9).

11.6 Diversity of Inselberg Morphology

The generally good quality of Google Earth imagery allows one to examine inselberg morphology in more detail. The diversity of form amongst inselbergs in general has long been known, resulting in several attempts to produce a typology of these residual hills (Thomas 1978; Twidale 1981, 1982; Migoń 2006). Referring specifically to granite settings, Twidale (1981) distinguished domes (= bornhardts), block-strewn inselbergs (= nubbins) and castellated hills (= castle koppies), arguing that they are all genetically related in that the latter two are products of disintegration of primary domes. Domed inselbergs may assume different shapes: high, steep-sided cupolas, similar to "sugar loafs" as described from Rio de Janeiro and elsewhere in Brazil (e.g. Fernandes et al. 2010) or low shields (whalebacks, ruwares) with gentle slopes and flattened top surfaces. However, inselbergs may also be conical (Meyer 1967; Kesel 1977). It seems agreed that this diversity of form reflects the variable patterns of jointing and other fractures, bornhardts being formed in particularly massive bedrock compartments, experiencing significant compression.

An overview of inselberg morphology of the Bur Area shows that all morphological types are present, clearly indicating the variety of structural controls and the different pathways of degradation of these landforms. Figure 11.10 shows the location of representative examples of inselbergs, which will be briefly described. Bur Dur is an example of a massive dome, elongated in a N–S direction, c. 1.4 km long and up to 0.6 km wide (Fig. 11.11). Its peak is at 483 m asl, which is more than 200 m higher than the level of the adjacent plain. The dome is cut by one prominent discontinuity in the south that forks into two in the northern part. Vertical cliffs have developed along these structures, whereas the benches below are apparently covered by enough regolith to allow vegetation growth in specific places. The southern perimeter of the dome is punctuated by two large recesses (alcoves), more than 100 m across. Large boulders at their foot suggest the role of rock slope collapses in their origin. Otherwise, talus is scarce around Bur Dur, indicating the limited role of catastrophic slope failures in the evolution of slopes. By contrast, light-tone aprons around the perimeter, partly overgrown, are probably accumulations of products of granular disintegration of granite. Structural control is even more evident at Bur Acaba (Fig. 11.12)—a prominent complex dome in the middle of Bur Acaba town, 1.1 km long and 0.5 km wide. This one is N–S elongated too, rising to nearly 290 m asl, more than 100 m above the plain. Several master fractures trending N345W are exposed within the hill, which therefore consists of a few elongated, upward-convex ridges. Slope inclinations vary between vertical to gently inclined, whereas the summit parts are flattish. Intersections of vertical fractures with sheet joints give rise to minor benches and, locally, overhangs. Talus at the slope/plain junction is scarce, although does occur

(e.g. at the southern end). Shallow channels modelled by sporadic runoff can be recognized, especially on the less steep, eastern side.

Durdur inselberg (Fig. 11.13) has smaller dimensions than the two described above, but provides a good example

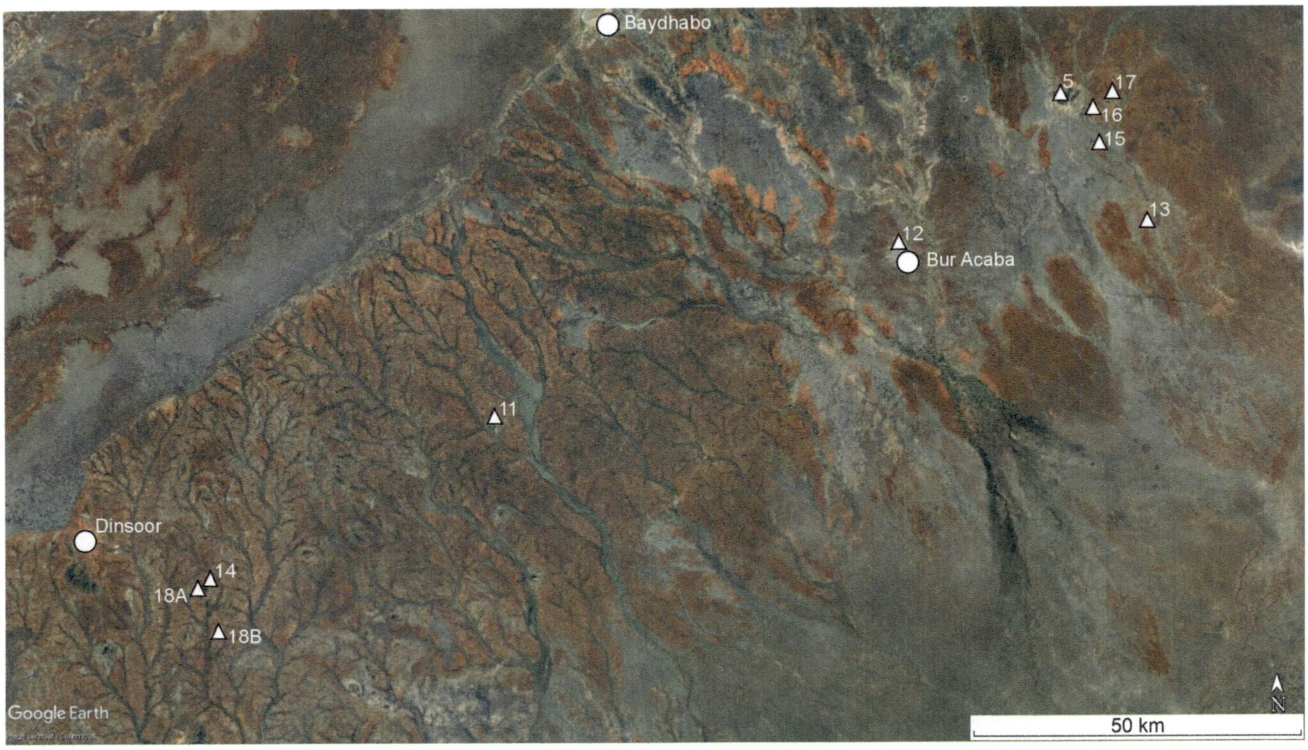

Fig. 11.10 Location of inselbergs described in more detail. Numbers next to triangles refer to figure numbering in this chapter. *Source* of image: © Google

Fig. 11.11 Bur Dur—an example of tripartite massive dome (for location, see Fig. 11.10). *Source* of image: © Google

Fig. 11.12 Bur Acaba—an example of fracture control on dome morphology (for location, see Fig. 11.10). *Source* of image: © Google

Fig. 11.13 Complex morphology of Durdur inselberg, with elements of a dome, castle koppie and nubbin (for location, see Fig. 11.10). *Source* of image: © Google

Fig. 11.14 Castellated inselbergs located 20 km to SE from Dinsoor (for location, see Fig. 11.10). *Source* of image: © Google

of a transitional form between a dome, a castle koppie and a nubbin. It is nearly circular, 300 m long and 250 m wide, reaching the height of 212 m asl (c. 50 m above the plain). Remnants of a dome form the peak of the hill, whereas numerous multidirectional joints separate massive blocky compartments, some more than 20 m long. A cluster of big boulders on the north-eastern side, just below the peak, is probably a talus from rock slope collapse, but otherwise big boulders appear to be in situ or moved only little. No boulders or shield-like bedrock outcrops are present in the vicinity of the inselberg. An unnamed inselberg some 20 km to the SE from the town of Dinsoor (Fig. 11.14) is castellated in the upper part, whereas moderately inclined convex rock surfaces are exposed below. W–E trending vertical joints outline remnant ridges and clusters of cuboid blocks, which give rise to a small rock city on the west-facing slope. The hill is 400 × 300 m and reaches c. 315 m asl, with a relative height in the order of 50 m. Another example of a castellated inselberg is provided on Fig. 11.15. The twin hill of Madhabey (229 m asl) is located c. 8 km to SE from Bur Eibe and consists of a larger southern hill (350 × 250 m) and smaller northern hill (200 × 170 m). NW–SE and NE–SW joint directions control the shape of bedrock outcrops, dividing these into mostly rectangular blocks which occur either in isolation or in clusters. The relative height of the hill is about 40 m. More advanced degradation can be recognized at Heudooliyow hill (283 m asl), 4 km to SE from Bur Eibe (Fig. 11.16). The core of the hill, which still rises some 60 m above the plain, is made of a remnant dome, which is 200 × 120 m. It is surrounded on all sides by a wide belt of displaced talus boulders, some more than 20 m long. Their occurrence strongly suggests frequent rock slope collapses. Reasons for this rather unusual pattern of degradation, standing in contrast with the fairly limited amount of talus elsewhere in the region, remain unclear, although another granite elevation located 3.5 km to the NW shows a very similar garland of talus too (Fig. 11.17).

Inselbergs conical in cross section occur too (Fig. 11.18), as demonstrated by two examples located to the SE from Dinsoor, at a distance of c. 18 and 24 km from the town. They are 900 × 500 m and 850 × 250 m, respectively, with the higher one being more than 100 m high in respect to the plain. In both cases, a narrow rock fin forms the axial part. Degradation patterns appear to vary. On the northern hill (Fig. 11.18a), the north-facing slope is covered by many rectangular blocks—almost certainly products of rock slope collapses, whereas the south-facing ones lack boulders and seem to have mantling materials of smaller size. The southern hill seems to be a hogback ridge, with the form adjusted to the near-vertical position of discontinuities (Fig. 11.18b).

Fig. 11.15 Twin castellated inselberg of Madhabey (for location, see Fig. 11.10). *Source* of image: © Google

Fig. 11.16 Advanced disintegration of an inselberg—Heudooliyow hill (for location, see Fig. 11.10). *Source* of image: © Google

Fig. 11.17 Another example of considerable degradation of an inselberg, as attested by widespread talus blocks (for location, see Fig. 11.10). *Source* of image: © Google

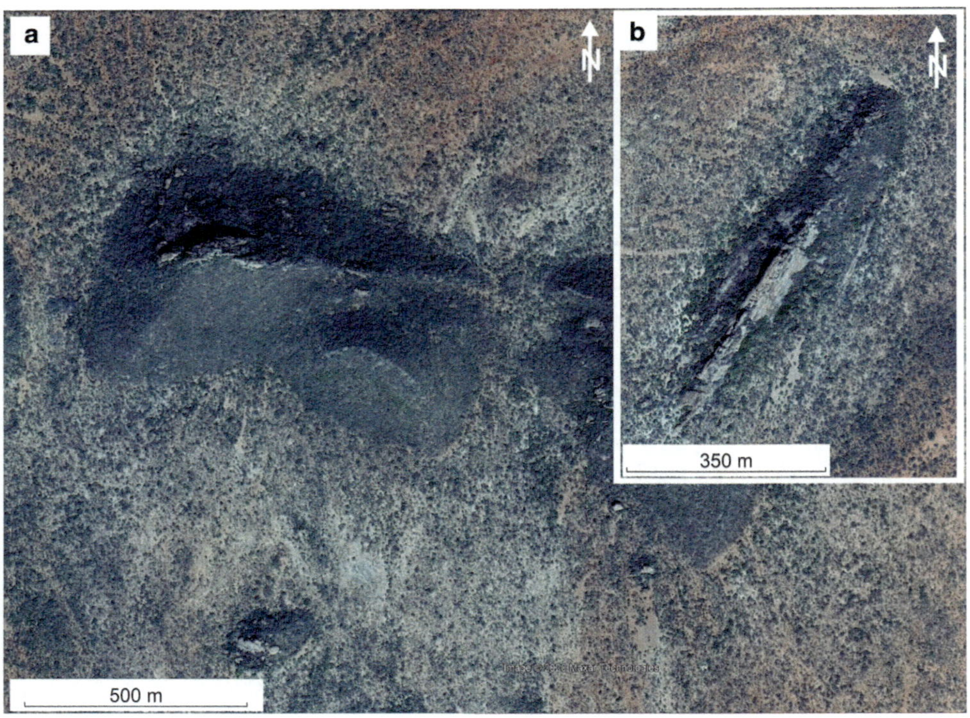

Fig. 11.18 Two examples of elongated inselbergs in the vicinity of Dinsoor, conical in cross section and probably built of quartzite (for location, see Fig. 11.10). *Source* of image: © Google

Fig. 11.19 Rounded top with evidence of exfoliation of Bur Eibi. Notice Bur Acaba in the background and the flat bush land in between (photo B. Kamenov)

Limited ground observations revealed the presence of "onion skin" spheroidal weathering leading to exfoliation (Fig. 11.19), which is probably attributable to diurnal temperature ranges. Large boulders litter the perimeter of inselbergs (Fig. 11.20) and locally extend over the surrounding pediment. The burs also show various large to small-scale weathering forms that typically characterize inselbergs in the crystalline rocks (Twidale and Vidal Romani 2005; Migoń 2006). They include large tafoni (a very big one can be seen in Bur Acaba), alveolar weathering and pitting (examples of different size can be seen on Bur Eibi), exfoliation slabs and balanced rocks (Bur Eibi and Bur Gulo), rounded top and spheroidal blocks, flutings (Bur Acaba and Bur Gulo), plinths (Bur Acaba) and flared slopes.

11.7 Inselbergs and Geoarchaeology

Limited geoarchaeological work undertaken in the region before the onset of civil war and following unrest suggests that inselbergs in the Bur Area played a significant role in the life of prehistoric human populations and may have been catalysts of permanent settlement (Clark 1954; Brandt 1986, 1988; Jones et al. 2018). Archaeological excavations were performed in rockshelters at Bur Eibe (Brandt 1988) and Bur Acaba (Jones et al. 2018), yielding human skeletons, artefacts and animal bones which allowed one to reconstruct hunting patterns during the transition period from hunter–gatherer to more sedentary subsistence and settlement patterns. The record covers the period from more than 20 ka to the Mid-Holocene, when significant environmental changes occurred. Faunal assemblages show that during the Last Glacial Maximum the rockshelters were occupied by mobile groups who hunted big game over very arid grasslands. Climate amelioration allowed a change in strategy since reliable water resources on and around inselbergs continued to attract animals, so that human groups focused mainly on dwarf antelopes. The beginning of a lifestyle probably induced ever-increasing emotional ties with inselbergs, as evidenced by burials and rock art. It is noticeable in this context that relatively large settlements are located at the foot of the largest inselbergs, Bur Acaba and Bur Eibi being particularly good examples (see Fig. 11.12).

Fig. 11.20 Talus of big boulders accumulated at the base of Bur Acaba

Table 11.1 Hypothetical sequence of geologic and geomorphic events that led to the formation of inselbergs in Bur Area

Phase	Events	Time interval
11	Reworking of weathering products from inselbergs by slope and fluvial processes, boulder accumulation, formation of pediments and depositional surfaces around the inselbergs	Quaternary
10	Uplift associated with the East Africa doming and the formation of the Rift Valley Differential weathering and further growth of inselbergs, shaping of their form in detail	Post-Miocene
9	Long period of stability, with the formation of an etchplain as witnessed by laterites and fluvial deposits with rounded pebbles	Cretaceous to Miocene
8	Transgression and deposition of predominantly carbonaceous sediments over a peneplain with isolated inselbergs. Complete submergence probably accomplished by the end of the Jurassic	Jurassic
7	Uplift followed by a new period of peneplanation	Triassic—early Jurassic
6	Subaerial denudation and peneplanation	Palaeozoic
5	Intrusions of post-kinematic granitoids of metaluminous composition	490–440 Ma
4	Large syn- to late kinematic intrusions of granitic masses into the Olontole rocks. Magmatism probably related to both post-collisional lithospheric thinning, magmatic underplating and crustal relaxation	Approx. 550–600 Ma
3	Folding and metamorphosis of the Olontole and Dinsor series towards amphibolite grade rocks. This orogenic event was accompanied by large-scale anatexis and migmatization	Approx. 600 Ma
2	Deposition of arenaceous, argillaceous and calcareous rocks of the Dinsor Series with their interbedded spilite flows	Early Proterozoic
1	Deposition of sediments to form the Olontole series. This deposition may have been multi-cyclic and represented a long time span, possibly including early Proterozoic time	Archean, possibly including early Proterozoic

11.8 Conclusions

The Bur Area is flat land (though very small-scale undulations are present as well) punctuated by a few tens of isolated hills protruding to a height from a few metres to more than 200 m above the surrounding plain. These hills are typical examples of inselbergs, made up by crystalline basement rocks. The majority of these inselbergs occur as individual outcrops and are spaced several kilometres from each other. In places, they occur in groups, including 4–5 hills of different size. Their petrological composition is typically granitic or gneissic, though some variations in the mineralogical composition emerged from the petrographic analysis of a number of samples. The mineralogical and petrographic data and some field observation allowed the reconstruction of the geological history of the inselberg landscape of the Bur Area and its morphological evolution (Table 11.1).

Unfortunately, the state of war that has been devastating Somalia for a long time has prevented scientists (and ourselves as well) from refining their studies and results. The Bur Area is a very interesting and, in some instances, a unique place in the world to investigate the interactions of geological, geomorphological and pedogenic processes. We do hope that peace will be eventually restored in the country and that international scholars will have again the possibility to do field work all over Somalia in collaboration with their Somali colleagues and with the support of what has always been a friendly population.

References

Ali Kassim M, Carmignani L, Conti P, Fantozzi PL (2002) Geology of Mesozoic-tertiary sedimentary basins in Southwestern Somalia. J Afr Geol 34:3–20

Azzaroli P, De Angelis AM (1965) Lineamenti geologici della regione dei Bur in Somalia. Atti Soc Toscana Sci Nat Mem Serie A 72:537–547

Borsi S (1965) Determinazione di età con il metodo Rb/Sr di alcune rocce della Regione dei Bur in Somalia. Atti Soc Toscana Sci Nat Mem Serie A 72:3–36

Brandt SA (1986) The upper Pleistocene and early Holocene prehistory of the horn of Africa. Afr Archaeol Rev 4:41–82

Brandt SA (1988) Early Holocene mortuary practices and hunter-gatherer adaptations in Southern Somalia. World Archaeol 20:40–56

Clark JD (1954) The prehistoric cultures of the horn of Africa. Cambridge University Press, Cambridge

Dal Piaz CV, Sassi FP (1986) The crystalline basement of Somalia: a review. Mem Soc Geol Italiana 31:351–361

D'Amico C, Ibrahim HA, Sassi NP (1981) Outline of the Somali basement. Geol Rundsch 70:882–896

Deutscher Wetterdienst (DWD) (2016) Klimatafel von Iscia Buurhakaba/Somalia. Baseline climate means (1961–1990) from stations all over the world (in German). https://www.dwd.de/DWD/klima/beratung/ak/ak_632470_kt.pdf

Fernandes NF, Tupinambá M, Mello CL, Peixoto MNO (2010) Rio de Janeiro: a metropolis between granite-gneiss massifs. In: Migoń P (ed) Geomorphological landscapes of the world. Springer, Heidelberg—Dordrecht, pp 89–100

Ferrari GA, Balley O, Bruno V, Chaie M (1985) Una sequenza di paleosuoli nelal regione dei Bur (Bay Region) Somalia. Rivista Agricoltura Subtropicale e Trop 79:471–487

Haider A (1993) Preliminary data on the magmatic origin and emplacement context of the Precambrian amphibolites of the Buur region (southern Somalia). In: Abbate E, Sagri M, Sassi FP (eds) Geology and mineral resources of Somalia and surrounding regions 113. Istituto Agronomico per l'Oltremare, Relazioni e Monografie Agrarie Subtropicali e Tropicali, Nuova Serie, Firenze, pp 129–142

Jones MB, Brandt SA, Marshall F (2018) Hunter-gatherer reliance on inselbergs, big game, and dwarf antelope at the Rifle range site, Buur Hakaba, southern Somalia ~20,000–5000 BP. Quat Int 471A:55–65

Kamenov BK, Lilov P (2012) The late post-orogenic granitoid episode from Bur Area, Somalia—petrology and K-Ar data. Geologie 103:57–73

Kesel RH (1977) Some aspects of the geomorphology of inselbergs in central Arizona, USA. Z Geomorph NF 21:119–146

Kűster D, Utke A, Leopold L, Lenoir J-L, Haider A (1994) Pan-African granitoid magmatism in north-eastern and southern Somalia. Berl Geowiss Abh 120:519–536

Lenoir J-L, Liégeois J-P, Kűster D, Utke A, Matheis G, Haider A (1994) Origin and regional significance of late Precambrian and early Paleozoic granitoids in the Pan-African belt of Somalia. Geol Rundsch 83:624–641

Meyer R (1967) Studien über Inselberge und Rumpfflächen in Nordtransvaal. Münchner Geogr Hefte 31

Migon P (2006) Granite landscapes of the world. Oxford University Press, Oxford

Paron P, Vargas R (2007) Landform of selected study areas in Somaliland and Southern Somalia. Integrated landform mapping approach at semi-detailed scale using remote sensing and GIS techniques. FAO-SWALIM. Project Report L-02. Nairobi, Kenya

Stefanini G (1918) Osservazioni geologiche nella Somalia Italiana meridionale. Boll Soc Geol Italiana 32

Thomas MF (1978) The study of inselbergs. Z Geomorph NF 31:1–41

Twidale CR (1981) Granite inselbergs: domed, block-strewn and castellated. Geogr J 147:54–71

Twidale CR (1982) Granite landforms. Elsevier, Amsterdam

Twidale CR, Vidal Romani JR (2005) Landforms and geology of granite terrains. Balkema, Leiden

UNDP 1968 Report on Uranium, Thorium and rare earths at Alio Ghelle, Somalia, Mineral and Groundwater Survey Project

UNDP 1972 Geological map of Somalia D.R. 1:1,000,000 Mineral and Groundwater Survey Project, phase II, 2 sheets (unpublished)

UNDP 1973 Ground water in the Somali democratic republic, Mineral and Groundwater Survey, phase II, Technical Report 3 United Nations, New York

Warden AJ, Hörkel AD (1984) The petrological evolution of the north-eastern branch of the Mozambique Belt (Kenya, Somalia, Ethiopia). Mitt Österr Geol Ges 77:161–184

Fluvial Landscape of the Dabaan Basin, Northern Somalia

12

Paolo Billi

Abstract

The Dabaan is a Middle Eocene to Early Miocene sedimentary basin. It stretches from the plateau margin to the Gulf of Aden and is drained by two main ephemeral streams: the Byoguure and the Kalajab. The climate ranges from semiarid in the headwaters to hyper-arid in the middle and lower reaches. Both rivers have an ephemeral flow regime as they are dry for long periods, and water flows only in response to occasional, very intense rainfalls. Field investigations were carried out to investigate channel morphology and sediment organisation in order to improve our knowledge of dryland rivers hydrology and sediment transport processes. The main morpho-sedimentary units of the streambed were identified and measured. Sediment samples were collected, and detailed observations on bed material sediment organisation, the occurrence of bedforms and sedimentary structures were carried out. The results indicate that, despite the braided channel morphology, the bars do not originate by depositional processes and downstream migration, as in permanent rivers. The bars and the channels, in fact, have no sedimentary structures but horizontal lamination, which implies flows with Froude numbers around one and the formation of a plane bed. During extreme floods, the upper part of the streambed is supposed to move en masse, and the bars are formed by scouring processes during the receding flood flow. Other, much less common bedforms are described. Boulder crescent scours and a new bedform, the cuspate sand waves, are found to be formed in flashy streams and, therefore, are considered as diagnostic of arid environments.

Keywords

Ephemeral streams • Braided channels • Horizontal lamination • Boulder crescent scours • Cuspate sand waves • Dryland

12.1 Introduction

Recently, ephemeral streams have attracted attention of several scientists for their importance as the only surface water resource for irrigation in many drylands (FAO 2010), for their flashy and devastating floods with negative implications such as casualties, property damage and infrastructure destruction (Lin 1999; Billi et al. 2015; Korichi et al. 2016; Pacheco-Guerrero et al. 2017), for their channel dynamics (Demissie et al. 2019), high sediment transport (Billi 2011) and deposition rates (Hooke 2019) and bridge clogging (Demissie et al. 2016) affecting roads and settlements. The sedimentology of modern ephemeral streams is also studied as a valid analogue of ancient dryland depositional systems in the context of oil exploration (Kelly and Olsen 1993; North and Taylor 1996; Billi 2007).

Ephemeral streams have peculiar characteristics that make them very different from perennial rivers of humid and sub-humid areas (Billi et al. 2018). Ephemeral stream channels, in fact, are dry for the most of the time and water flows only in response to sporadic, very intense rainfalls. Such an impulsive nature and the great availability of sediment supply, given sparse or absent vegetation in the headwaters, provide these dryland rivers with high mobility and peculiar geomorphological and sedimentological features that, despite increasing interest from several scientists, are still poorly known.

This chapter reports about the fluvial landscape in the Dabaan basin, an arid zone in northern Somalia, and investigates morphological and sedimentological characteristics of

P. Billi (✉)
International Platform for Dryland Research and Education, Tottori University, Tottori, Japan
e-mail: bli@unife.it

P. Billi (ed.), *Landscapes and Landforms of the Horn of Africa*, World Geomorphological Landscapes,
https://doi.org/10.1007/978-3-031-05487-7_12

two neighbouring ephemeral streams that flow from the margin of the escarpment of the Gulf of Aden trough to the Gulf of Aden coast, near the town of Berbera.

12.2 Study Area

The Dabaan basin is located to the southeast of the town of Berbera, in northern Somalia (Somaliland), and has an approximate area of 600 km² (Fig. 12.1). The basin is underlain by Proterozoic to Cambrian rocks of the basement and a 2400 m thick sedimentary sequence of Middle Eocene to Early Miocene age, formed during the early phases of opening of the Gulf of Aden through (Fig. 12.2) (Abbate et al. 1983; Sagri et al. 1989).

To the south, the Dabaan basin is bound by a major faulting system, associated with rifting of the Gulf of Aden that formed a 1500 m high escarpment. The basin structure mainly consists of a mild asymmetric syncline whose northern beds dip southward with an inclination of 15°–20° that tends to decrease in the central part of the basin. In the southern part of the basin, the beds dip northward with high dip angles and the sedimentary sequence is juxtaposed with

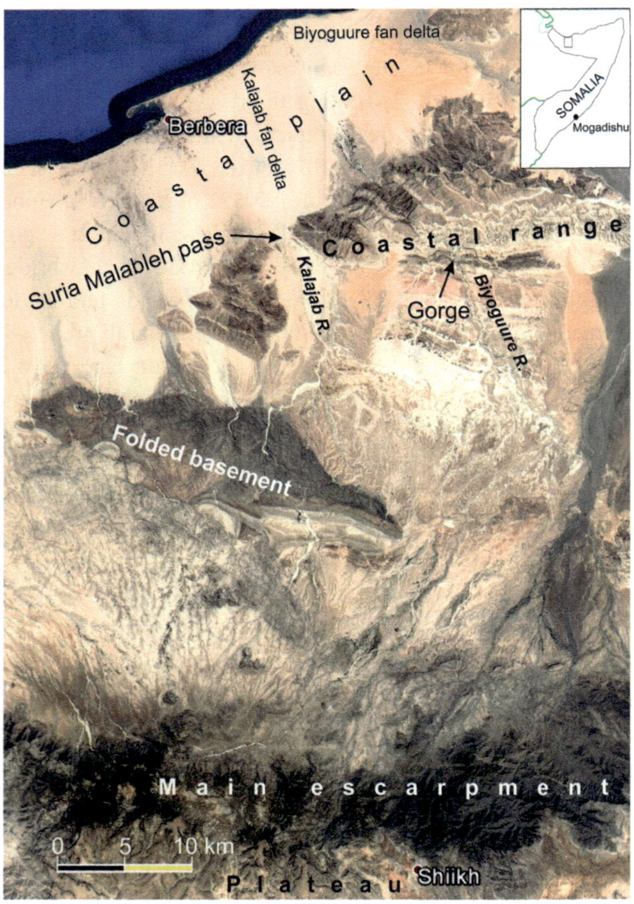

Fig. 12.1 Main physiographic elements of the study area

the crystalline basement along an important fault (Abbate et al. 1983).

The studied rivers, the Biyoguure and Kalajab, originate within the rift escarpment margin near the town of Shiikh (Fig. 12.1) and occasionally reach the sea 10 and 4 km east of the town of Berbera, respectively. The catchment of the Biyoguure is about 696 km², and the main river stem is about 52 km long. The Kalajab River has a catchment of about 356 km² and the length of 33 km (Billi and Tacconi 1985) (Fig. 12.3).

The headwater portions of both catchments are underlain by crystalline rocks of the basement and correspond to the escarpment of the Somali highland plateau. This mountainous part extends from the continental divide to the elevation of 700 m asl and comprises 25% and 10% of the catchment of the Biyoguure and Kalajab, respectively. From the elevation of 700–450 m asl, both rivers flow incised into a gently inclined terrain underlain by the Yesomma Sandstones for about 10 km. Proceeding northward, the two rivers cross the Dabaan Tertiary sequence (Sagri et al. 1989) for a distance of about 20 km, as far as the elevation of 250 m asl. Beyond this reach, both rivers cross the coastal mountain range (the highest peak of which is at 950 m asl) consisting of Mesozoic and Eocene sandstones and evaporates. The Biyoguure River, which probably predates uplift of the coastal range, has incised a deep and narrow gorge (Fig. 12.4), beyond which it enters the coastal plain, where it forms a flat and poorly developed delta and finally debouches into the Gulf of Aden. The Kalajab River crosses the coastal range through a wide saddle (the Suria Malableh pass) (Fig. 12.1), beyond which it forms a large (12 km long and 9 km wide) fan delta between Berbera and the Biyoguure streambed and delta (Fig. 12.5). A large, right branch of Kalajab fan delta joins the Biyoguure about 4.5 km downstream of the coastal range, whereas smaller distributive channels on the left side of the fan delta proceed towards Berbera and its vicinity (Fig. 12.5), which are occasionally flooded.

In both catchments, rock-weathering processes are very active. Slopes are covered by fine rock fragments that are easily washed away by sporadic but intense rainfalls. The flat alluvial paleo-surfaces of the Tertiary deposits in the central part of the catchments are covered by rock fragments, similar to the desert pavement. The larger blocks are not moved because no runoff is generated and are weathered in place; the finer particles are blown away by the wind.

12.2.1 Climate

Unfortunately, very few climatic data are available for the Dabaan basin. Old data, collected by Hunt (1960) for the 1944–1950 interval, were combined with more recent data

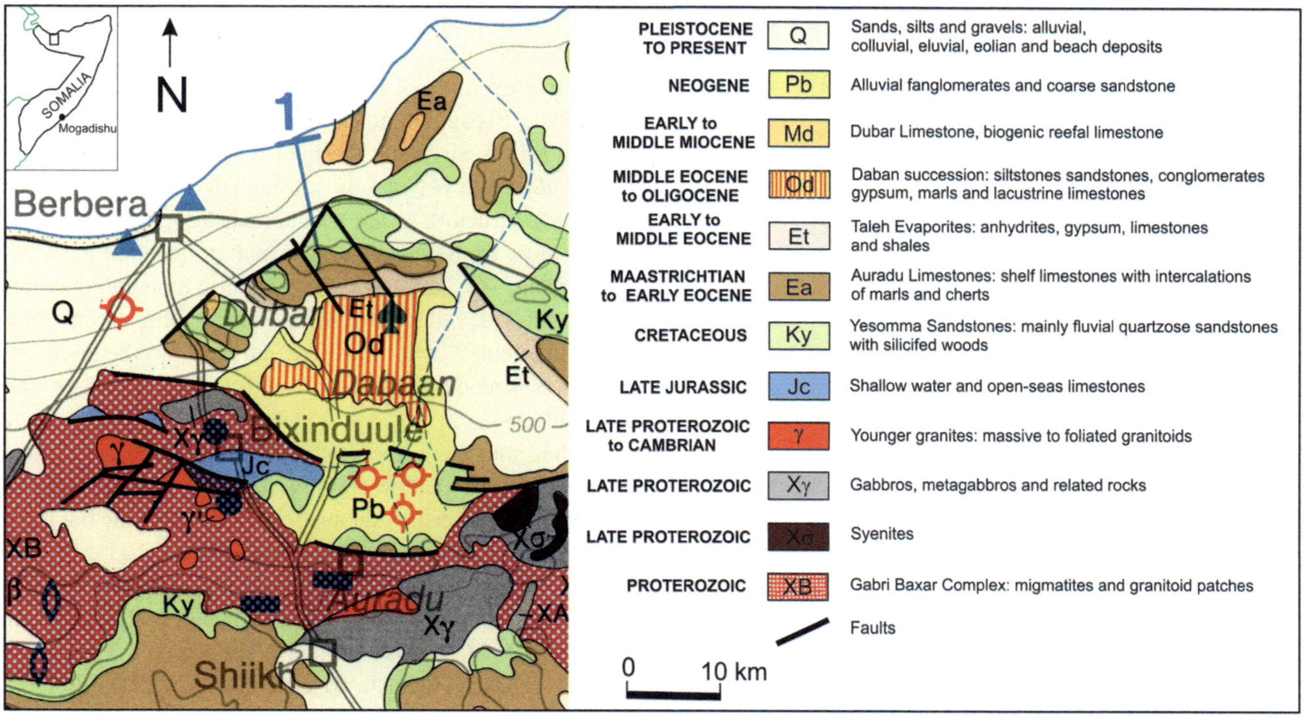

Fig. 12.2 Basic geology of the study are (modified from Abbate et al. 1983 and Sagri et al. 1989). Symbols: red circle = oil well; blue triangle = monazite, zircon, rutile placers; blue circle = Pb, Zn, Ba, F veins; blue rectangle = stratiform Mn mineralization (gondite); blue tears = Nb, Ta, Be quartz-pegmatites; blue spade = lignites and coals

spanning the last decade (https://en.climate-data.org/) for the stations of Berbera, Bixinduule and Shiikh (Fig. 12.3).

As expected, temperatures are higher in Berbera and lower in Shiikh due to their different location: Berbera is on the coastline, whereas Shiikh is on the plateau, at an elevation of 1441 m asl (Fig. 12.6c). Bixinduule is in between these stations, at an elevation of 750 m asl, and shows intermediate values. In Berbera, the average maximum monthly temperature is recorded in July, with 42.1 °C, whereas in Bixinduule and Shiikh the highest temperatures are in August and September, with 36.3 and 29.0 °C, respectively. The average monthly temperature range is wider in Shiikh, with 12.4 °C, and narrower in Berbera (7.9 °C). With such high temperatures, Berbera can be ranked among the hottest inhabited places in the world.

Like temperature, annual and monthly precipitation amounts are largely influenced by the elevation of the meteo-station. In Shiikh, annual rainfall is 461.5 mm, in Bixinduule the respective value is 257.9 mm, and in Berbera, annual rainfall is only 49.7 mm. The monthly distribution indicates that in Boxinduule two rainy seasons occur (Fig. 12.7b). The main one is in April and May, and minor rains occur in October. In Shiikh, the monthly rainfall distribution follows a similar bimodal pattern, but appreciable rains occur also in June and September (Fig. 12.7c). In Berbera, rainfall is so sparse that no evident monthly pattern can be pointed out (Fig. 12.7a). In Fig. 12.7, the highest monthly rainfalls ever recorded are also reported. These data indicate that in Bixinduule maximum precipitation is recorded in April and October (Fig. 12.7b), whereas in Shiikh high rains are recorded in April, May and June and a secondary peak is observed in October (Fig. 12.7c).

It is worth noticing considerable difference between mean and maximum monthly precipitation (Fig. 12.7). The largest one is recorded in Shiikh, with 125.9 mm in June, but the highest ratio is observed in Bixinduule where, in November, maximum rainfall is five times the mean rainfall (Fig. 12.7b and c). Unfortunately, no long time series of rainfall data are available for these meteo-stations, but information obtained from local people, who report occasional, very intense rainstorms, and the large difference between low mean and high maximum monthly rainfalls, with the latter corresponding to 40–70% of the mean annual precipitation, suggest a wide interannual and monthly variability of precipitation. Such high rainfall accounts for the ephemeral nature of the two rivers, in which floods occur only occasionally and seldom the water flow is able to reach the sea. Nevertheless, high floods are described by local people as very impressive, very fast and powerful. Given scarce precipitation, the streambed of the Kalajab is completely dry for most of the time. The Biyoguure would be the same except that, due to the uplifted impervious strata of the coastal

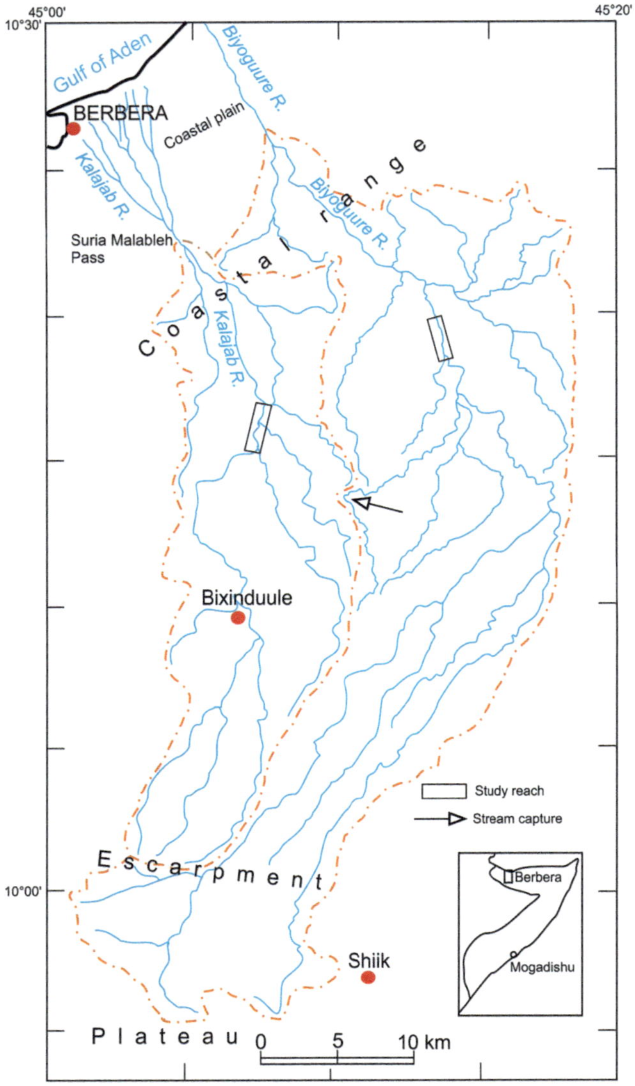

Fig. 12.3 Simplified drainage networks of the studied rivers

range, the subsurface water emerges in the streambed, resulting in a permanent flow of a few litres per second in the gorge and for a few kilometres upstream (Figs. 12.4 and 12.8).

For the high temperatures and low precipitation, the climate of the Dabaan basin can be classified as a hot desert climate (*BWh* according to Koeppen classification), and the environment is typical of a rocky desert devoid of vegetation (Fig. 12.9), except some shrubs in the dry river beds with very sparse vegetation (Fig. 12.10).

12.2.2 The Rivers

Both the studied rivers have narrow catchments, with a similar shape (Fig. 12.3). In the headwaters, parallel drainage network occurs, with a N-NE flow direction as far as 10° 15′ N. Beyond this latitude, the rivers turn to N-NW and the drainage network has a more angular shape. An instructive example of river capture, caused by a left tributary of the Biyoguure at the expense of a right tributary of the Kalajab, is visible in the field (Fig. 12.3).

The drainage network is substantially influenced by the geological structure as are all the main orographic elements of this area. In the headwaters, which correspond to the rift escarpment, the landscape is typically mountainous with deeply incised valleys and steep slopes. The streams have steep gradients; their beds are carved into bedrock and devoid of streambed sediment, with the exception of very big boulders.

In their middle reaches both the rivers flow on the bottom of narrow and shallow (a few tens of metres) canyons incised into the Tertiary depositional sequence, the flat top of which is a preserved alluvial surface (Fig. 12.11). In these middle reaches, the streambed gradient ranges between 0.01 and 0.02, the rivers have a braided channel morphology and there is no real, well-developed alluvial plain. The lateral, alluvial deposits, in fact, do not show marked sedimentological differences with the main channel deposits. The former are topographically higher than the latter and seem to have been deposited during very high floods rather than being a typical fine-grained overbank deposit. In ephemeral streams, large floods are commonly able to mobilise the entire streambed, which is then dissected during the receding flood flows (Billi 2016).

While crossing the coastal range, the Kalajab does not show marked changes in its channel morphology. By contrast, the Biyoguure has cut a deep and narrow gorge, in which the river changes the pattern from braiding into an entrenched meandering channel. Downstream of the coastal range, the Biyoguure resumes a straight braided channel morphology as far as the fan delta, where the main channel divides into a main distributary channel and a few smaller ones (Figs. 12.4 and 12.5).

12.3 Channel Morphology, Bedforms and Sedimentary Structures

For both rivers, field data were collected along four cross sections representative of typical morphology and sedimentology of the streambed, within a middle reach about 6.5 and 5.5 km long for the Biyoguure and the Kalajab, respectively (indicated by rectangles in Fig. 12.3). The broadest cross sections measured 709 m in the Biyoguure and 542 m in the Kalajab. The streambed gradient was measured across a stretch of about 500 m centred on each cross section. The data collection and observations carried out along each cross section included recording of channel geometry (width and depth), bed material grain-size

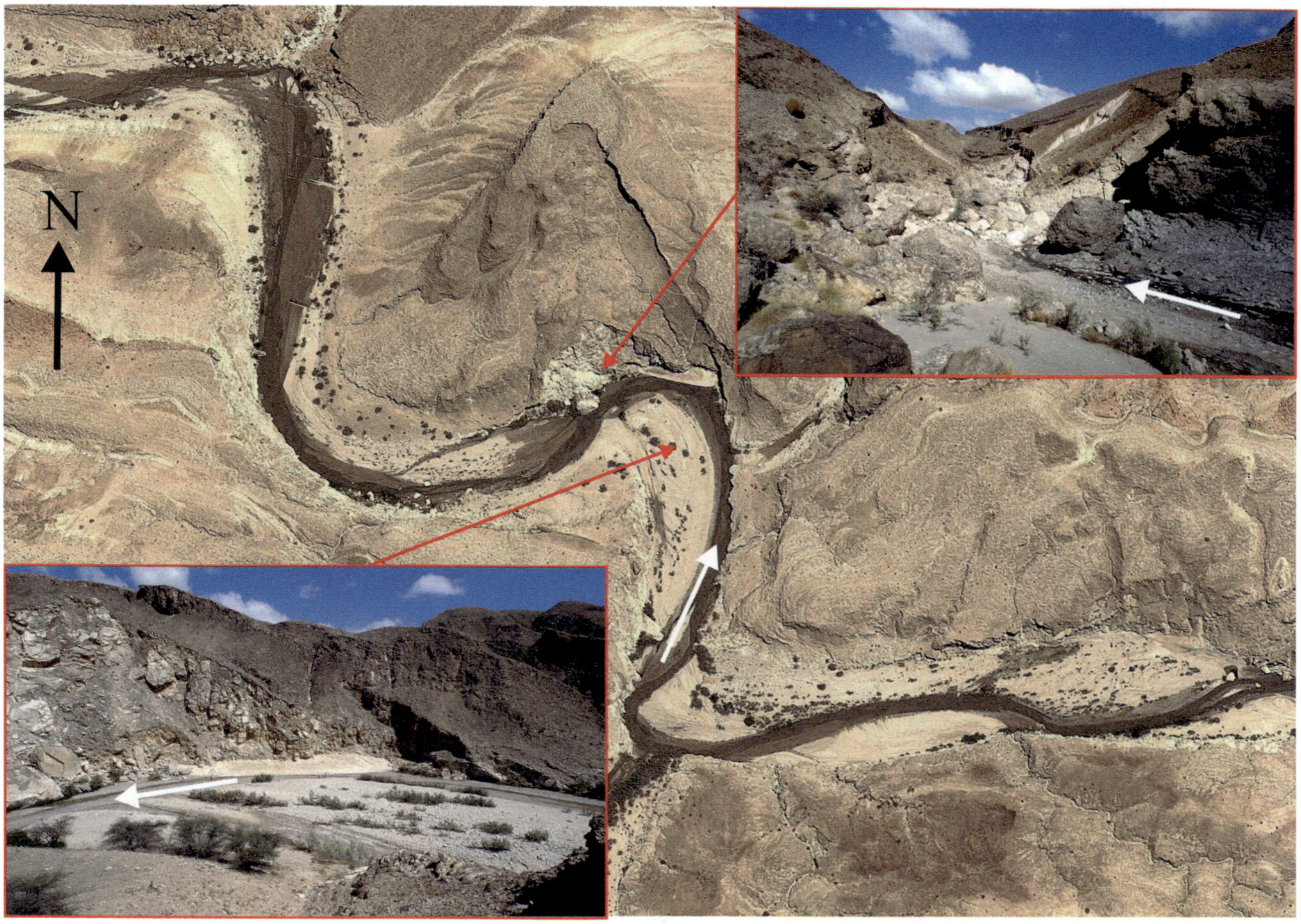

Fig. 12.4 Narrow and deep gorge of the Byoguure River cutting through the coastal range. The white arrows indicate flow direction

distribution sampled using the transect line method (Leopold 1970), the average size of the largest particles, the occurrence of bedforms and their geometry and sedimentary structures of specific morpho-sedimentary units.

12.3.1 Channel Morphology

The channel morphology of both rivers is typically braided (Fig. 12.12). The streambed, in fact, consists of several braided channels separated by median longitudinal bars. The channels are very large and shallow: the average width/depth ratio is 85 (range 12–253) in the Biyoguure and 100 (range 23–400) in the Kalajab; i.e., they show typical values of dryland ephemeral streams (Graf 1988). The braided channels are very shallow (mean depth 0.62 m; range 0.30–1.75 m) and have an average streambed gradient of 0.015.

At the cross-sectional scale, the longitudinal profile is characterised by the alternation of steeper and almost flat segments that could be associated with riffle and pool sequences. The general structure of the streambed consists of flat surfaces placed at different elevation and commonly steeper segments connect a higher surface with a lower one. This is not in agreement with the typical riffle geomorphology of braided rivers in more humid climates as these steeper segments seem to have been formed during the receding flood flows, when most of discharge occurs in the channels and the higher elevated surfaces are washed out by water flowing into the deeper and lower elevated channels. A similar setting was found also by Billi (2016) in boulder bed ephemeral streams.

Similar to braided rivers in humid environments, lateral and median longitudinal bars have typical rhomboid shape, and their maximum width is of the same order of magnitude as the channel width. What makes these bars very different from those of perennial rivers is their internal structure. The Biyoguure and Kalajab bars, in fact, are devoid of any sedimentary structures that would indicate lateral or downstream accretion. In other words, unlike perennial braided rivers, the study reaches do not show any evidence of bar downcurrent migration taking place through bar head erosion and bar tail sedimentation (Bluck 1979, 1982). Moreover, no substantial longitudinal variation of grain size was observed within the bars to point out bar components with

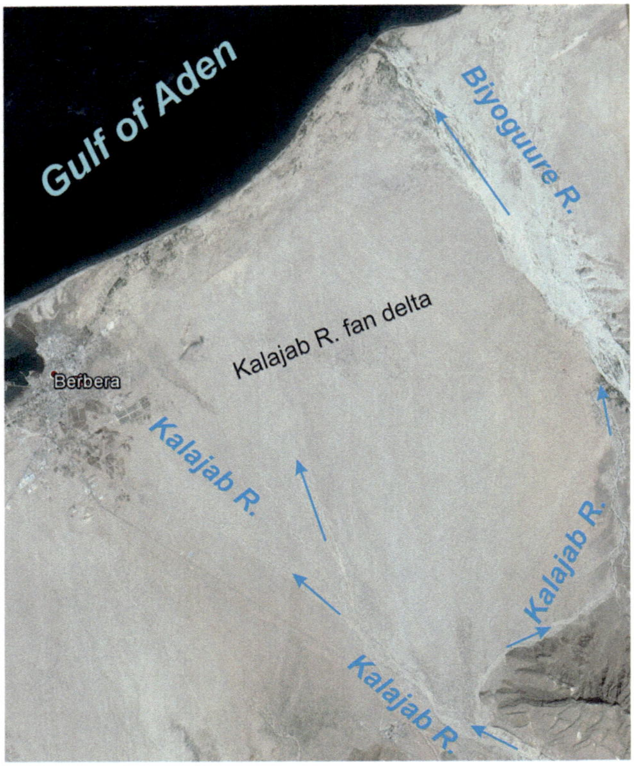

Fig. 12.5 Kalajab fan delta and the Byoguure delta. Notice the larger size of the Kalajab fan delta despite the catchment area of this river is about the size of the Byoguure. Occasionally, the rightmost distributary channel of the Kalajab fan delta joins the Byoguure

different gran size and morphology. The most common sedimentary structure of the bars is horizontal planar lamination of fine, sandy sediment, locally resting on a massive coarse gravel platform as observed by Bluck (1974) on sandur deposits in Iceland. The lack of both accretionary fronts and typical sedimentary structures of braided streams indicates that the bars in the study reaches are not formed by localised deposition processes but, rather, they are the product of channel incision during the receding flood flows or subsequent smaller floods. This hypothesis is supported by the presence of sedimentary structures and their arrangement that are the same for bars and channels (see also the next sections) and by the spatial continuity of facies architecture between channel and bar deposits. This implies that during large floods the whole streambed is entrained, and when discharge decreases, flow is mainly confined in shallow channels that dissect the streambed and transport finer sediment that is deposited in the pool-like flat stretches. Billi (2016) found a similar uniformity between bar and channel deposits in very coarse boulder bed ephemeral streams and postulated that bedload is transported as massive bedload sheets. Also, in the Biyoguure and the Kalajab, horizontal planar lamination and massive sand and/or fine gravel layers are the most common sedimentary structures.

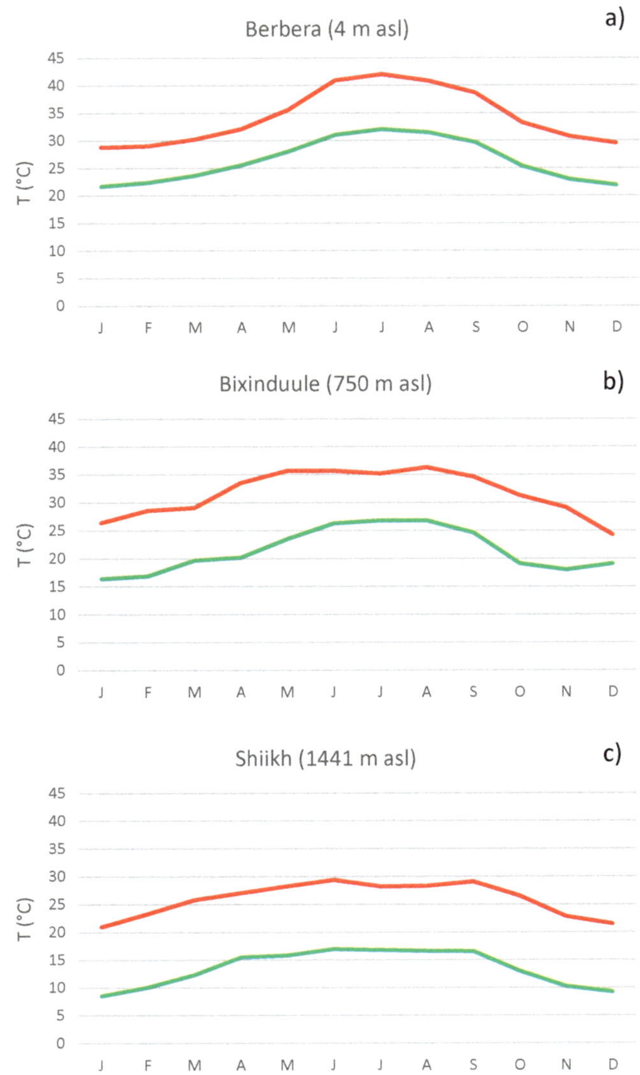

Fig. 12.6 Mean monthly temperature measured at three meteo-stations located at different elevations within the basin: Berbera (**a**) on the coast; Bixinfuule (**b**) in the middle reaches; Shiik (**c**) in the headwaters

12.3.2 Sediment

The bed material grain-size distribution of the Biyoguure River is very similar to that of the Kalajab. Composite samples from both rivers were aggregated to represent the grain size of main morpho-sedimentary units, i.e. channels, riffles and bars rivers (Fig. 12.13). Bars and channels have very similar grain-size characteristics: D_{50} is the same (−0.84 phi; 1.79 mm), sorting is almost the same (2.16 and 2.72 mm for channels and bars, respectively), and sand prevails over gravel. Riffles, instead, are substantially coarser ($D_{50} = -5.58$ phi or 47.8 mm), a little less sorted (standard deviation = 3.17), and 78% of bed material is gravel.

The three main streambed units appear more dissimilar if the size of the largest particles (D_{max} or D_{90}) is considered.

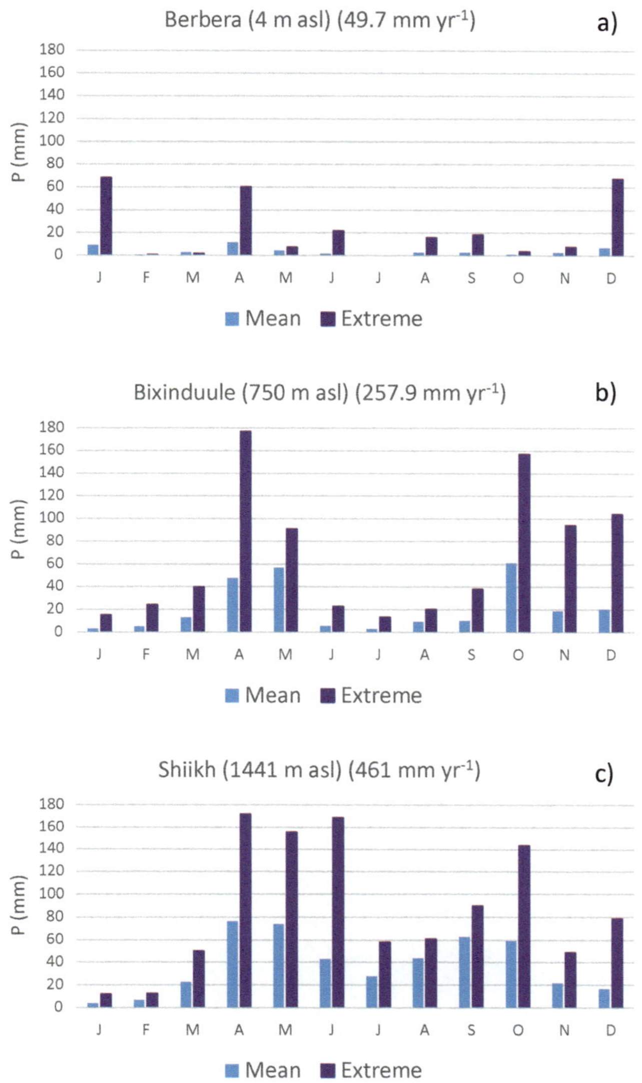

Fig. 12.7 Mean and maximum monthly precipitation measured at three meteo-stations located at different elevations within the basin: Berbera (**a**) on the coast; Bixinfuule (**b**) in the middle reaches; Shiik (**c**) in the headwaters

On bars, mean D_{max} is 500 mm; in channels, it is 225 mm, and on riffles, it is 425 mm. D_{90} follows a slightly different pattern, with the largest D_{90} on riffles (−6.89 phi or 119 mm), whereas D_{90} of bars (−6.49 phi or 89.7 mm) is substantially coarser than of the channel streambed (−5.29 phi or 39.1 mm). The occurrence of larger particles on the bar top suggests that the bar surface experienced higher stream power than the channels and that bars and channels were likely formed in different moments within a flood (or between floods), i.e. when depositional processes occurred with different flow discharges.

Using the equations proposed by Costa (1983) to calculate the flow velocity (v) capable to entrain large boulders,

$$v = 4.63D^{0.455} \tag{12.1}$$

$$v = 5.2D^{0.487} \tag{12.2}$$

and introducing the values of D_{max} measured on top of three main streambeds, the average values of the threshold flow velocity of 3.54, 2.43 and 3.28 ms^{-1} were obtained for the largest particles on bars, channels and riffles, respectively. Moreover, for the average channel depth and gradient a shear stress (τ) of about 92 Nm^{-2} was calculated. This result (in kg m^{-2}) was introduced in the equations of Baker and Ritter (1975), Costa (1983) and Williams (1983), which were reversed to calculate the largest particle entrainment by a given shear stress:

$$D_{max} = (\tau/88.5)^{0.67} \tag{12.3}$$

$$D_{max} = (\tau/24.4)^{0.824} \tag{12.4}$$

$$D_{max} = \tau/17 \tag{12.5}$$

The predicted D_{max} ranges between 220 and 550 mm, i.e. values comparable to those of the largest particle size measured in the field.

The bimodal particle size distribution of riffles may be explained by deposition of coarse particle, transported during higher flows, which are lately infiltrated by finer sediment during the receding flood flow phase.

12.3.3 Bedforms

Commonly, ephemeral streams do not show a wide variety of bedforms. This is probably due to short duration of floods and their high Froude numbers (Billi 2011). The bedforms observed on the streambed of the studied rivers are the following:

Plane bed Plane bed is by far the most common bedform in the Biyoguure and Kalajab rivers. Classical flume experiments (e.g. Simons and Richardson 1961) have shown that the plane bed is formed when the Froude number (Fr = $v/(gh)^{0.5}$ in which v is flow velocity, h is flow depth and g is gravity) is around 1. Taking the flow velocity calculated in the previous sections and the average flow depth of 0.625 m, Froude numbers in the 0.98–1.43 range are obtained. Imposing alternative, though reasonable, flow depths of 1.0 and 0.5 m, the Froude number is still around 1. The hydraulic calculations confirm that flow conditions during floods are suitable for the development of a plane bed. The short duration of floods must be considered as well. In fact, floods of dryland ephemeral streams are typically very intense, but also very short. Flash floods commonly last only

Fig. 12.8 Very small, though semi-permanent base flow of the Byoguure River a few kilometres upstream of the gorge in the coastal range

Fig. 12.9 Arid climate conditions are well reflected by hillslopes devoid of vegetation and covered by large quantities of rock fragments and loose material produced by weathering

a few hours, which is a too short time for the formation of more complex bedforms such as dunes and ripples.

Current ripples These bedforms are not common and are found mainly in restricted areas in the streambed, such as the deepest channels, where shallow flow may have lasted for a longer time after the flood wave has passed by (Fig. 12.14). Two-dimensional ripples have wavelength ranging from 100 to 200 mm, and their height is between 5 and 12 mm. Three-dimensional ripples are smaller, with wavelength between 90 and 140 mm and height between 8 and 15 mm. The size of these bedforms allows one to classify them as ripples as the diagram (Fig. 12.15) demonstrates. Dunes of any kind were not observed.

Wind ripples These bedforms are rather uncommon, probably because their preservation potential is limited. Their geometry is rather constant; wavelength and height range are between 50 and 100 mm and between 5 and 10 mm, respectively.

Fig. 12.10 In the study area, the only vegetation consists of sparse shrubs and very few shrubby trees in the dry riverbeds

Boulder crescent scours These are mainly erosive rather than depositional bedforms. They are formed by the horseshoe vortex flow that develops upstream and on the sides of a standing obstacle like a large boulder and is strong enough to erode the sediment around the standing particle (Picard and Heigh 1973) (Fig. 12.14). Downcurrent of the obstacle stone, a narrow wake of sediment may accumulate. These bedforms are formed during later stages of a flood, when large particles stop moving and very shallow, a few centimetres deep, separation flows are still capable to develop local eddies. Pebble crescent scours are typical of dryland ephemeral streams characterised by low flows and flash floods. In the studied rivers, they were found also in older deposits (Fig. 12.16), confirming that they can be used in paleo-environment reconstructions.

Pebble clusters This coarse-grained bedform is rather uncommon in the studied rivers. Pebble clusters consist of aggregations of smaller particle upstream of a larger stone and the downcurrent formation of a wake made of fine gravel and sand wake (Brayshaw 1984). They are found mainly on steep-gradient surfaces and where bed material is coarse, such as on riffles. Occasionally, they may form also on low-gradient channel bottom.

Transverse ribs Transverse ribs are lines of pebbles/boulders across a small channel. They are normally formed under shallow flow conditions during the receding flood limb and are not typical of dryland rivers as they are more common in perennial rivers (Koster 1978). In the study rivers, transverse ribs have an average spacing of 1.35 m, rest on streambeds with gradients ranging from 0.052 to 0.026, with D_{50} = 70 mm and D_{max} = 350 mm. Applying the relationship between flow depth (h) and transverse rib spacing (L_s) developed from flume data of McDonald and Day (1978),

$$h = (L_s - 0.02)/4.47 \qquad (12.6)$$

to the studied rivers data, a flow depth of about 0.30 m is obtained, which is comparable to the mean diameter of the largest stones forming the transverse ribs.

Mud drape This is a very fine-grained deposit, no more than a few centimetres thick, that is very common and covers large portions of the streambeds. It is deposited during the very last flood phases or by very shallow flood flows. On mud drapes, dissection cracks and flakes are rather common (Fig. 12.17).

Salt crust and raindrop prints Salt crust are very common on the streambeds of the two rivers for two main reasons: (1) high temperatures and evaporation rates; (2) high concentration of salt in the subsurface water which rests on the uplifted Lower to Middle Eocene Taleh evaporates. Raindrop prints are well preserved in the salt crust because often the sporadic rainfalls evaporate a few tens of metres above the ground, which is impacted by only a few drops (Fig. 12.18).

Cuspate sand waves In the streambeds, several examples of very thin (10 mm) cuspate bedforms were observed (Fig. 12.19). Their shape recalls that of linguoid bars (Collinson 1986), but the latter are much thicker and larger and were found mainly in perennial rivers of humid and sub-humid environments. According to Collinson (1970), in fact, "linguoid bars are large enough to have smaller forms superimposed on their stoss sides (though they do not always have them) and … are synonymous with the mid-channel bars". In the Tanana river, linguoid bars are longer than 30 m, but they may also reach a maximum length of as much as 200 m (Ashley 1990), have a sharp slip face 0.5–2.0 m high and are arranged in a generally out-of-phase relationship with one another (Collinson 1970). The cuspate sand waves of the Dabaan rivers do not have smaller superimposed bedforms and are found in groups of a few bedforms

Fig. 12.11 In their middle reaches both the rivers occupy narrow and shallow (a few tens of metres) canyons incised into the Tertiary depositional sequence: **a** Byoguure; **b** Kalajab

(Fig. 12.20) or are isolated. Their length varies between 2 to 10 m, the length/maximum width ratio is around 4, and the length/thickness ratio is around 1000, whereas it is about 100 in the Tanana river. The cuspate sand waves reported in this study have a similar shape and are arranged in a way reminding that of lunate or linguoid ripples, but the latter are much smaller, made of finer sediment and have distinct lee and stoss sides.

The cuspate sand waves of this study consist of a narrow water and sediment supply channel, 0.10–0.30 m wide, from which divergent streaming lineations radially depart to form the main bedform body (Figs. 12.19 and 12.20), reflecting rapid washing away flow typical of dryland ephemeral streams.

Cuspate sand waves were reported in an early report by Billi and Tacconi (1985), but no other paper has investigated

Fig. 12.12 Example of braided channel morphology: the Byoguure River middle reach

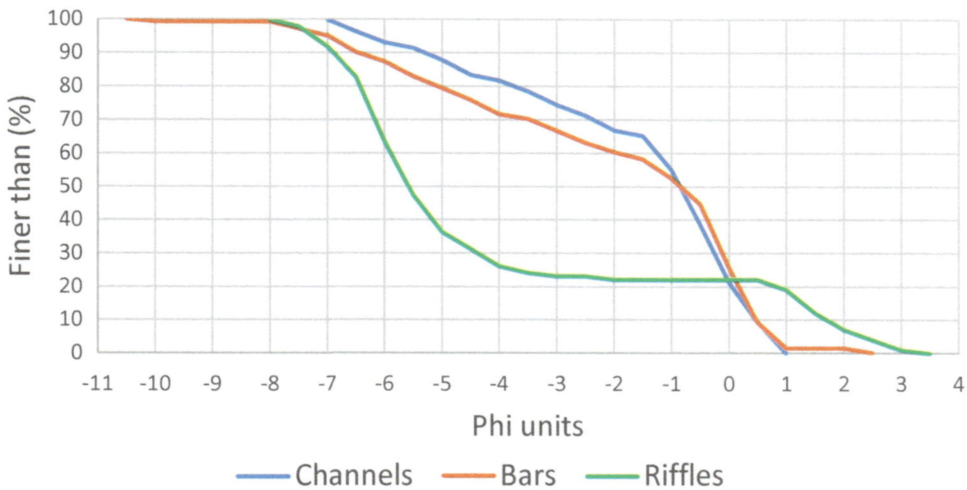

Fig. 12.13 Composite grain-size distributions representative of both rivers for the three main morpho-sedimentary units: channels, bars and riffles

this bedform, which seems to be typical of dryland ephemeral streams. Since it can be considered as a bedform diagnostic of ephemeral streams, further studies are necessary to define the hydraulic conditions of their formation and their relation with horizontal lamination.

12.3.4 Sedimentary Structures

Stratification and sedimentary structures were observed in natural sections and in shallow observation pits. The most common sedimentary structure is horizontal planar lamination, which is present in 90% of field inspection sites. The typical thickness of the laminae ranges between 5 and 15 mm (extreme values are 1–40 mm), and commonly, the thicker laminae are composed of coarser particles. The base of the laminae is often marked by dark heavy metal granules. Such widespread occurrence of horizontal plane lamination can be accounted for by three main factors: (1) short duration of floods, so that more complex bedforms do not have time to develop; (2) flows with Froude numbers commonly around 1, which corresponds to the plane bed; (3) rapid waning of flood flow for infiltration, which causes a rapid decrease in shear stress and high sedimentation rates, thus favouring the formation of the plane bed configuration.

Other authors (Karcz 1972; Picard and High 1973; Graf 1988) did not find such a predominance of horizontal planar lamination over the other bedforms, but this difference can be explained by a more pronounced flashy character of floods in the studied rivers.

Fig. 12.14 Boulder crescent scour with current ripples all around. The white arrow indicates the flow direction

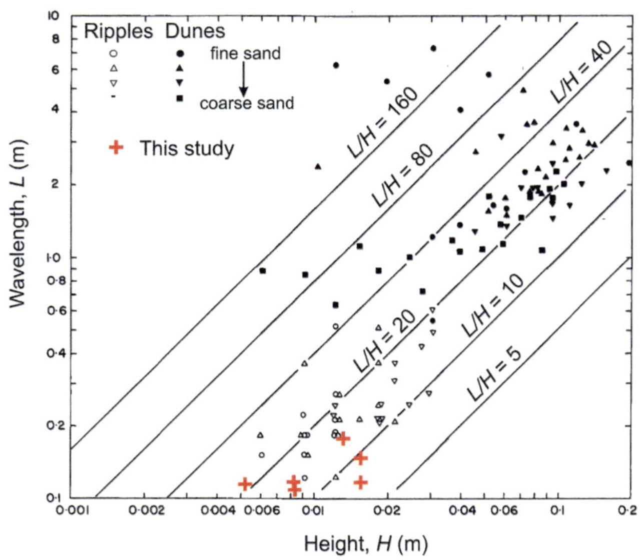

Fig. 12.15 Diagram of bedform hight versus bedform wavelength. Bedforms of the studied rivers plot in the field of fine sand ripples (modified from Allen 1982)

Other sedimentary structures observed are, in a decreasing order of frequency: horizontal massive beds, inclined planar lamination, and trough cross-lamination. Massive sandy or fine to coarse gravel layers are typically found in the channel deposits and their average thickness is 140 mm (extreme values 10–700 mm).

The inclined planar lamination was observed in a few accretionary fronts downstream of small transverse bars and consist of sand and fine gravel. The average inclination of the laminae is 7° (range 4–11°), and their typical thickness ranges from 5 to 20 mm (extreme values 1–70 mm).

The trough cross-lamination is very rare in the studied rivers and was observed only in three cases, downstream of large lateral bars. The scarcity of these sedimentary structures reflects the marked flashy character of floods and accounts for the occasional occurrence of three-dimensional current ripples, the migration of which generates the through cross-bedding.

Fig. 12.16 Example of a boulder crescent scour in the Kalajab River older alluvium

Fig. 12.17 Clay flakes in the Byoguure streambed

Fig. 12.18 Raindrop impact on a salt crust in the streambed of the Byoguure

12.4 Concluding Remarks

The Dabaan basin is subjected to a semiarid to hyper-arid climate. It is drained by two ephemeral rivers, the Byoguure and the Kalajab, which are dry for most of the time. Water flow is sporadic and shallow, but very occasional intense rainfalls may propel powerful floods. Like in other hot desert environments, the Dabaan basin is characterised by the lack of vegetation and active weathering processes producing large quantities of loose material on slopes. The combination of sporadic floods with the high sediment supply provides the rivers with very high width to depth ratios (85–100 on average with maximum values between 250 and 400) and a braided channel morphology. Though braided channels and bars are commonly observed, the bars are not formed by depositional processes as described by various authors (e.g. Bluck 1979) for braided rivers in humid and sub-humid areas. The sediments of both the channels and the bars have the same structure and internal organisation. The bars do not show any evidence of downstream migration or accretionary faces. The streambed does not show lateral differences in terms of the internal organisation of the sediment, which seems to be transported en masse. During extreme floods, the entire streambed is moved in a thickness of a few tens of centimetres and the bars are probably formed by local scouring processes and the concurrent development of braided channels during the receding flood flows. Flow velocities (2.5–3.5 ms^{-1}) calculated by Costa's (1983) Eqs. (12.1) and (12.2) and flow depths inferred from field observations return Froude number values around 1, which are compatible with the ubiquitous occurrence of planar horizontal lamination, very commonly observed in the bed material. Flow depths of 30 cm were obtained from Eq. (12.6), based on the wavelength of transverse ribs, and are consistent with field observations.

Plane bed by far prevails over other classic bedforms reported in the literature. Current ripples are occasionally present on the streambeds, but not dunes of any kind. Boulders and pebble crescent scours are instead very

Fig. 12.19 Example of a cuspate sand wave. Notice the radial flow away from the bedform body. Flow is towards the reader

common and can be considered as bedforms indicative of ephemeral flows in arid environments.

A new bedform, also indicative of flash floods, is the cuspate sand waves. They consist of a narrow water and sediment supply channel, 0.10–0.30 m wide, from which streaming divergent lineations radially depart to form the leaf-shaped bedform body, reflecting a rapidly vanishing flow, as typically observed in dryland ephemeral streams. Apparently, the cuspate sand waves are similar to the linguoid bars of Collinson (1986), but the former are much smaller (maximum length is between 2 and 10 m, whereas the linguoid bars can be longer than 30 m), thinner (a few centimetres compared to 0.5–2.0 m of linguoid bars), have no superimposed bedforms nor any downstream accretionary faces. Further field investigations on the cuspate sand waves are, however, necessary to better understand the hydraulic conditions for their origin and occurrence in other environments.

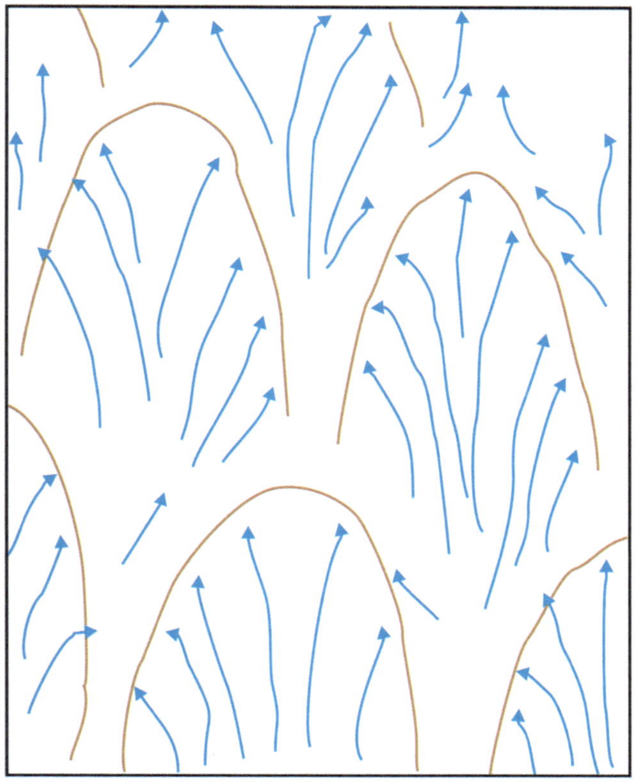

Fig. 12.20 Sketch of the arrangement of cuspate sand waves. The blue arrows indicate flow lines on the bedform body

References

Abbate E, Sagri M, Sassi FO (1994) Geological map of Somalia. Somali National University, S.E.L.C.A, Firenze

Abbate E, Bruni P, Fazzuoli M, Sagri M (1983) Le facies di transizione e continentali nel bacino Terziario del Daban, Somalia settentrionale. Dati preliminari. Quaderni Di Geologia Della Somalia 7:7–37

Allen JRL (1982) Sedimentary structures. Their character and physical basis, volume I. In: Developments in sedimentology 30A, Elsevier, Amsterdam

Ashley GM (1990) Classification of large-scale subaqueous bedforms: a new look at an old problem. J Sediment Petrol 60(1):160–172

Baker VR, Ritter DF (1975) Competence of rivers to transport coarse bedload material. Geol Soc Am Bull 86:975–978

Billi P (2007) Morphology and sediment dynamics of ephemeral streams terminal reaches in the Kobo basin (northern Welo, Ethiopia). Geomorphology 85:98–113

Billi P (2011) Flash flood sediment transport in a steep sand-bed ephemeral stream. Int J Sedim Res 26(2):193–209

Billi P (2016) Channel processes and sedimentology of a boulder-bed ephemeral stream in the Western Afar margin. Z Geomorphol 60 (1):35–52

Billi P, Tacconi P (1985) Morfologia, sediemntologia e dinamica fluviale di due corsi d'acqua effimeri della Somalia settentrionale. Quaderni Di Geologia Della Somalia 8:43–83

Billi P, Alemu Y, Ciampalini R (2015) Increased frequency of flash floods in Dire Dawa, Ethiopia: change in rainfall intensity or human impact? Nat Hazards 76(2):1373–1394

Billi P, Demissie B, Nyssen J, Moges G, Fazzini M (2018) Meander hydromorphology of ephemeral streams: similarities and differences with perennial rivers. Geomorphology 319:35–46
Bluck BJ (1974) Structure and directional properties of some valley sandur deposits in southern Iceland. Sedimentology 21:533–554
Bluck BJ (1979) Structure of coarse-grained braided stream alluvium. Trans R Soc Edinb 70:182–221
Bluck BJ (1982) Texture of gravel bars in braided streams. In: Hey RD, Bathurst JC, Thorne CR (eds) Gravel-bed river. Wiley, Chichester, pp 339–355
Brayshaw AC (1984) Characteristics and origin of cluster bedforms in coarse-grained alluvial channels. In: Koster EH, Steel RJ (eds) Sedimentology of gravels and conglomerates. Memoir 10, Canadian Society of Petroleum Geologists, Calgary, pp 77–65
Collinson JD (1970) Bedforms of the Tana river, Norway. Geogr Ann Ser A Phys Geogr 52(1):31–56
Collinson JD (1986) Alluvial sediments. In: Reading HG (ed) Sedimentary environments and facies, 2nd edn. Balckwell, Oxford, pp 20–62
Costa JE (1983) Paleohydraulics reconstruction of flash-flood peaks from boulder deposits in the Colorado front range. Geol Soc Am Bull 94:986–1004
Demissie B, Billi P, Frankl A, Haile M, Nyssen J (2016) Excess river sedimentation at bridges in the Raya graben (northern Ethiopia). Land Degrad Dev 28:946–958
Demissie B, Nyssen J, Billi P, Haile M, Frankl A (2019) Land changes in relation to stream dynamics in a marginal graben along the northern Ethiopian Rift Valley. Phys Geogr 40(1):71–90
FAO (2010) Guidelines on spate irrigation. FAO Irrigation and Drainage Paper 65, FAO, Rome
Graf WL (1988) Fluvial processes in dryland rivers. Springer, Berlin
Hooke JM (2019) Extreme sediment fluxes in a dryland flash flood. Sci Rep 9:1686
Hunt JA (1960) Report on teh Geology of the Berbera - Sheik area. Report No. 4, Geological Survey, Somaliland Protectorate, Eyre and Spottiswoode, Margate
Karcz I (1972) Sedimentary structures formed by flash floods in southern Israel. Sed Geol 7(161):182
Kelly SB, Olsen H (1993) Terminal fans—a review with reference to Devonian examples. Sed Geol 85:339–374
Korichi K, Hazzab A, Atallah M (2016) Flash floods risk analysis in ephemeral streams: a case study on Wadi Mekerra (northwestern Algeria). Arab J Geosci 9(11):589
Koster EH (1978) Transverse ribs: their characteristics, origin and paleohydraulic significance. In: Miall AD (ed) Fluvial sedimentology. Canadian Society of Petroleum Geologists, Memoir 5, Calgary, pp 161–186
Leopold LB (1970) An improved method for size distribution of stream bed gravel. Water Resour Res 6(5):1357–1366
Lin X (1999) Flash floods in arid and semiarid zones. UNESCO-IHP, Paris
McDonald BC, Day TJ (1978) An experimental flume study on the formation of transverse ribs. Canada Geological Survey Paper 78-1A:441–451
North CP, Taylor KS (1996) Ephemeral-fluvial deposits; integrated outcrop and simulation studies reveal complexity. AAPG Bull 80 (6):811–830
Pacheco-Guerrero A, Goodrich DC, González-Trinidad J, Júnez-Ferreira HE, Bautista-Capetillo CF (2017) Flooding in ephemeral streams: incorporating transmission losses. J Maps 13(2):350–357
Picard MD, High LR Jr (1973) Sedimentary structures of ephemeral streams. In: Developments in sedimentology, vol 17. Elsevier, Amsterdam
Sagri M, Abbate E, Bruni P (1989) Deposits of ephemeral and perennial lakes in the tertiary Daban basin (Northern Somalia). Palaeogeogr Palaeoclimatol Palaeoecol 70:225–233
Simons DB, Richardson EV (1961) Forms of bed roughness in alluvial channels. Proc ASCE J Hydraul Div 87, HY 3, paper 2816:87–105
Williams GP (1983) Paleohydrological methods and some examples from Swedish fluvial environment, I—cobble and boulder deposits. Geogr Ann 65:227–243

River Hydromorphology and Flood Hazard in Somalia

13

Paolo Billi and Mesenbet Yibeltal Sebhat

Abstract

The Juba and Wabe Shabelle are the largest permanent rivers of Somalia and of the whole Horn of Africa. Though these rivers have neighboring catchments of almost the same size, their hydrology and channel dynamics are rather different. Such differences are investigated in the modern rivers, and a comparison with old (Quaternary?) avulsion channels is made. Basic geomorphic parameters of the old channels are measured from satellite images and bankfull discharge is calculated by simple equations using meander wavelength as the main entry parameter. Geological information and satellite images analysis revealed the occurrence of old, presently inland, deltas. Presently, the Shabelle R. is not entering the ocean north of Mogadisho as it would be expected for the regional gradient, but proceeds southwestward parallel to the coast for a few hundreds of kilometers before to reach the Juba R. A new hypothesis to explain this apparent anomaly is presented. In the last two decades, both the study rivers experienced an increased frequency of high, devastating floods causing several fatalities and affecting thousands of people. The causes of such increase in flood hazard are manifold and include both a climate worsening and human impact.

Keywords

Hydromorphology • Avulsion • Meandering • Aggradation • Bank breach

P. Billi (✉)
International Platform for Dryland Research and Education, Tottori University, Tottori, Japan
e-mail: bli@unife.it

M. Y. Sebhat
Bahir Dar Institute of Technology, Bahir Dar University, Bahir Dar, Ethiopia

13.1 Introduction

In Somalia, there are only two big permanent rivers: the Juba and the Wabe Shabelle (in the local language, "*Wabe*" means river and "*Shabelle*" leopard). There is also another river, the Byoguure (in the local language this name means "running water") in the Daban basin of northern Somalia, but this river has a small catchment and a very limited base flow of a few liters per second (see Chap. 12 of this publication for more details).

Throughout the whole Horn of Africa, the Juba and Shabelle rivers have distinctive geomorphological and hydrological characteristics, and though their catchments are neighboring and with similar size, orientation, source area (the Bale mountains in Ethiopia) and both rivers cross similar terrains, they show different channel morphology dynamics, runoff volumes and flow patterns. Repeated, impressive channel avulsions and evidence of very long paleo/old channels are very common in the Shabelle and provide unique examples to investigate this process.

The Juba and Shabelle rivers play a very important role in Southern Somalia as they are the main source of water for irrigation in a country where the other two-thirds of the territory is dryland unsuitable for agriculture. Though floods supply the surrounding plain with water and nutrients and provide the local farmers with good opportunities for flood recession cultivations, the progressive abandonment of almost all the flow regulation structures and irrigation schemes after the onset of the civil war in 1991 resulted in an increased frequency of devastating floods that affected a large number of people and caused several casualties (Basnyat and Gadain 2009). In the last decade, the frequency of overbank floods significantly increased in response to higher than usual rainfalls in the lower Juba and Shabelle rivers flood plain. However, this already hazardous situation has been exacerbated by uncontrolled river bank breaches by local farmers in an attempt to replace the function of the former irrigation schemes, presently no longer operating.

P. Billi (ed.), *Landscapes and Landforms of the Horn of Africa*, World Geomorphological Landscapes,
https://doi.org/10.1007/978-3-031-05487-7_13

The observed increase in precipitation, however, has not decreased the frequency of droughts (SWALIM 2016) that became with floods a major hazard in Somalia (see also Chap. 1 of this book for details).

In this chapter, the hydromorphology characteristics of these two rivers and their recent geomorphological evolution are examined, and the reasons for the increased frequency of devastating floods in their lower reaches are investigated.

13.2 Geographic Setting

The Juba and Shabelle rivers have their source in the Bale and the Chercher and Ahmar mountains in Ethiopia, respectively, (Billi 2015) (Fig. 13.1) at elevations higher than 4000 m asl (Fig. 13.2). In Ethiopia, the Juba is named as Genale river. Both rivers flow from the Ethiopian Plateau-Rift Valley margin and maintain a south-east direction as far as the Ethiopia-Somalia border. Beyond the border, the rivers flow southward but, while the Juba reaches the sea near Kisimayo, the Shabelle turns to south-west around the town of Jowhaar and flows parallel to the coastal dunes running out in the swamps of the Balli plain (i.e., wet plain in the local language) (Vicinanza 1910; Basnyat 2007)) downstream of Afgoye (Fig. 13.1). Beyond these wetlands, the river resumes a poorly defined channel and a very low and intermittent flow. The Shabelle joins the Juba upstream of Jamame only during occasional, extreme floods.

The Juba results from three main rivers (Genale, Dawa and Weyb) joining at Dolo, just before to cross the Somali border, and about 80 km upstream of Luuq, where the most upstream flow gage in Somalia is located. This river has a catchment of about 221,000 km^2 (the catchment portion in Somalia is about 50,000 km^2) and a total length of 2078 km of which 1204 are in Somalia. The catchment area of the Juba almost doubles and becomes about 450,000 km^2 if also the Laag Dheera river, which occasionally joins the Juba a few kilometers upstream the river outlet into the Indian Ocean, near Kisimayo, is included.

The Shabelle has no large tributaries in the Ethiopian headwaters, with the exception of the Fafan river. This river is geomorphologically a tributary of the Shabelle, joining it near Ferfer, at the Somali border (about 30 km upstream of Beled Weyne, where the most upstream flow gage of the Shabelle river in Somalia is located), but actually its flow is very erratic and reaches the Shabelle only very occasionally during extreme floods (Fig. 13.3). The Shabelle has a catchment of about 296,000 km^2 at the confluence with the Juba (about 245,000 km^2 upstream of the Balli swamps) and a length of 2041 km at the confluence with the Juba. In Somalia, its catchment area is about 103,000 km^2 for a length of 803 km Basnyat and Gadain (2009) . The Juba-Shabelle river system is the largest of the Horn of Africa with a total catchment area of about 750,000 km^2.

13.2.1 Climate

The climate of both catchments is very similar and highly variable since it is much more humid and cool in the highlands and rather hot and drier in the lowland (Fig. 13.4). In the Bale and the Ahmar mountains in Ethiopia, annual precipitation is over 1500 and 1000 mm, respectively, whereas in the middle and lower river reaches, it can be as low as 350 (Kebri Dehar) and 250 mm (Beled Weyne), respectively, (Muchiri 2007; Fazzini et al. 2015). The monthly rainfall pattern shows a main wet season (*Gu* in the local language) from April to June and a minor rainy season (*Deyr* in the local language) from October to November (Muchiri 2007: Chap. 2, this volume) (Fig. 13.5). In the catchment portions within Somalia, the highest annual rainfall is recorded at Baidoa (566 mm) and the lowest at Dolo (250 mm). The *Gu* rains account for 43–56% of annual precipitation. In eight out of the 12 meteo-stations considered the month with the highest rainfall is April, with a mean monthly rain of about 93 mm. The onset of the *Gu* rain is rather abrupt since the mean monthly rain of March is about 16 mm, that is, about half of the average monthly precipitation (33.9 mm) which, on its turn, is slightly more than one third of mean April rain (92.8 mm).

Annual precipitation is highly variable in the study area (Table 13.1). On the base of very few representative meteo-stations data, annual rainfall variation coefficient is lower in the headwater of both catchment (e.g., 0.20 in Goba, 0.27 in Negelle and 0.39 in Jijiga) and tends to increase in the downstream portion of the catchments (e.g., 0.49 in Kebri Dehar, 0.54 in Beled Weyne) to decrease again along to the coast (e.g., 0.32 in Afgoye, 0.42 in Kisimayo) (Basnyat 2007; Fazzini et al. 2015).

Rainfall intensity is a very important parameter to predict floods and soil erosion. Unfortunately, hourly data are very scarce, whereas some daily intensity data are available for two meteo-station in the headwaters of both study rivers. In the Juba headwaters, the maximum daily rainfall intensities ever recorded are 112 and 137 mm day^{-1} (mean 43 and 66 mm day^{-1}, respectively) measured by the rain gages of Robe Bale and Negelle, respectively, (Figs. 13.1 and 13.6). In the Shebelle headwaters, the highest daily rainfalls ever recorded are 118 and 128 mm day^{-1} (mean 54 and 59 mm day^{-1}, respectively) at the meteo-stations of Haramaya and Kebri Dehar (about 80 km West of Jijiga), respectively, (Figs. 13.1 and 13.6). It is worth noticing that

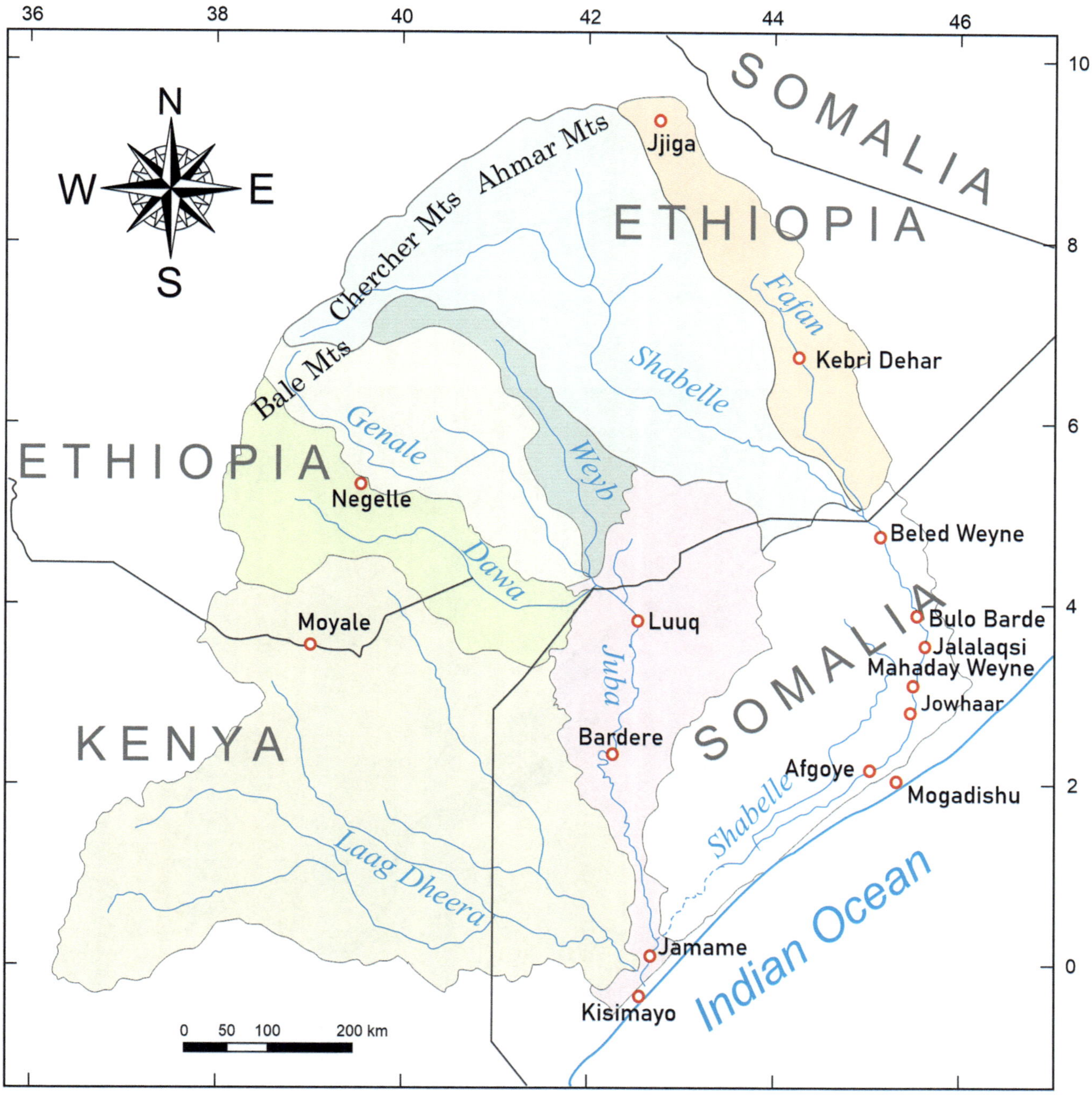

Fig. 13.1 Main catchments and sub-catchments of the Juba and Shabelle rivers. Modified from Balint et al. (2010)

commonly such daily rain amounts are reached in a shorter time (typically two–three hours) rather than in 24 h. Figure 13.6 shows that in these meteo-stations, daily rainfall with 5–10 years return time may range from 50–60 to more than 90–100 mm day^{-1}, i.e., values high enough to produce flooding (Diakakis 2012).

In the 18 meteo-stations considered within Somalia (see Chap. 1 of this volume for more details), mean monthly temperature is rather high in every month ranging from 14.1 to 36.6 °C. The town with the highest mean annual temperature is Luuq (30.7 °C), and the cooler one is Ceerigabo with 17.2 °C. The hottest months typically coincide with the *Gu* rains (the main rainy season from April to June), whereas the relatively cooler ones are from November to February. Mean monthly temperature does not change much through the year. The maximum range, in fact, is recorded at Berbera (11.7 °C) and the minimum at Kismayo (2.9 °C). More detailed information about the climate of Somalia in general and the study area in particular can be found in Muchiri (2007) and in Chap. 1 of this volume.

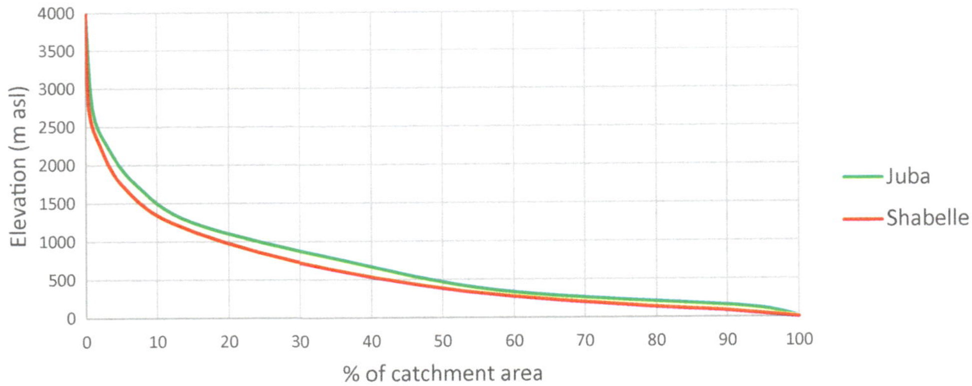

Fig. 13.2 Hypsometric curves for the Juba and Shabelle rivers. Data from Basnyat and Gadain (2009)

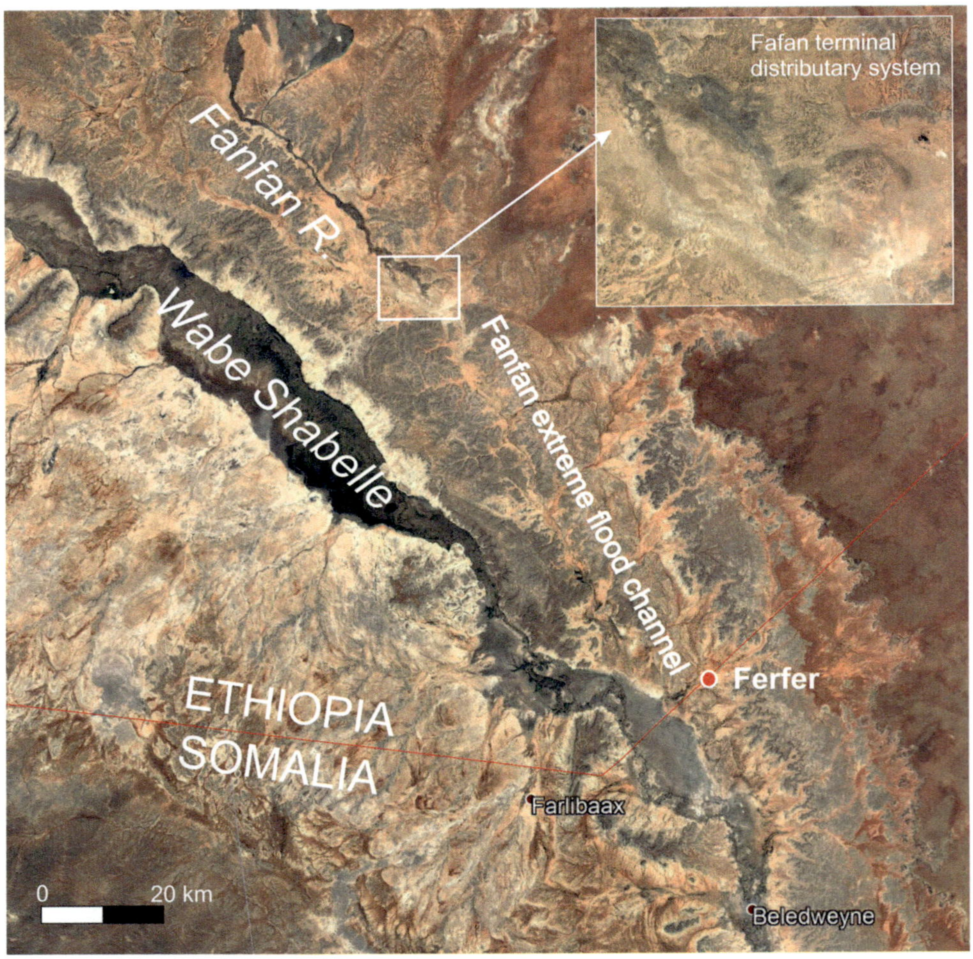

Fig. 13.3 Confluence area between the Shabelle and the Fanfan rivers

13.2.2 River Hydrology

Both the Juba and Shabelle rivers have flow data dating back to the mid-1950s, measured at a few flow gages in Ethiopia and Somalia (the main flaw gages are reported in Fig. 1), but some sparse measurements were undertaken by Italian authorities since 1925 (Italian Government 1925). Though these data sets are affected by several gaps, the situation was much better than in 1990s when almost all the recording stations were not operating because of the civil war. In the early 2000s, the FAO-SWALIM project restored the most important flow gages, flow measurements were resumed, and flow data are available since recent (Muthusi and Gadain 2009). Of course, some differences in the rating curves between pre- and post-war were expected (for many reasons, including streambed deposition and scouring, channel narrowing/widening, etc.). Some calibration to make the two data set comparable was performed by SWALIM (Muthusi

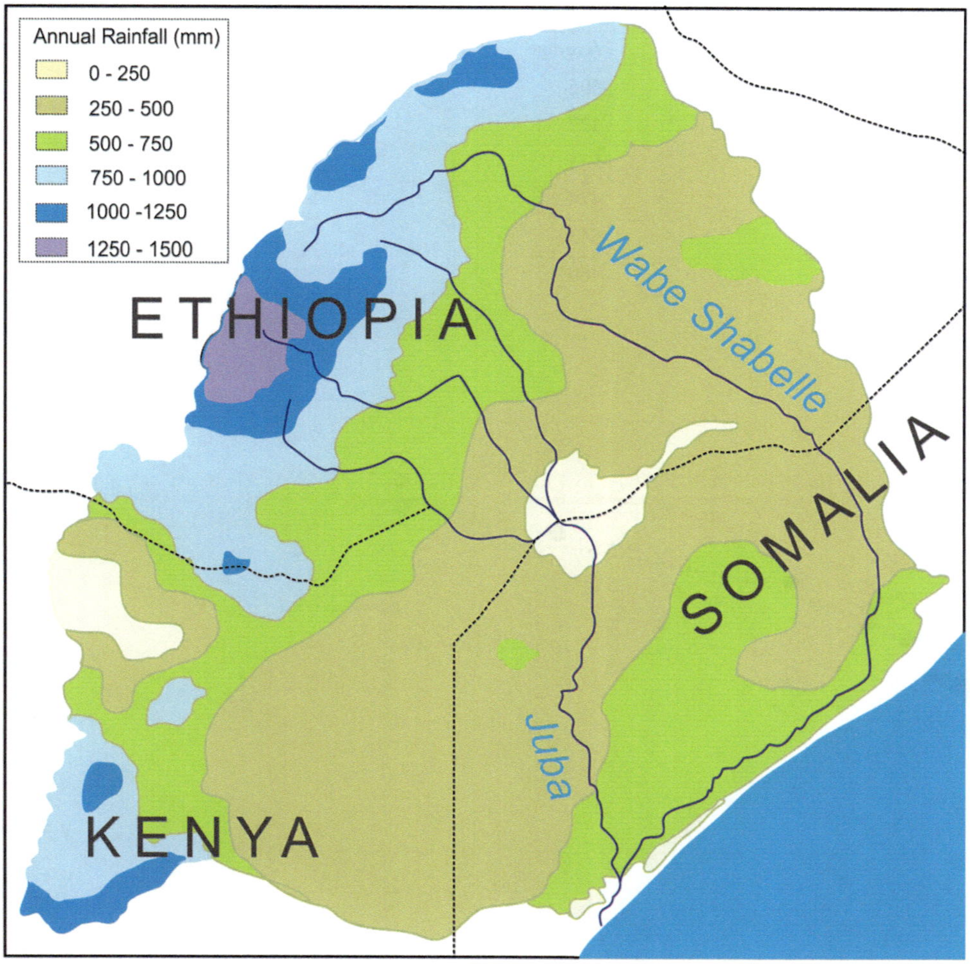

Fig. 13.4 Precipitation distribution across the Juba and Shabelle catchments. Modified from UNEP (2010)

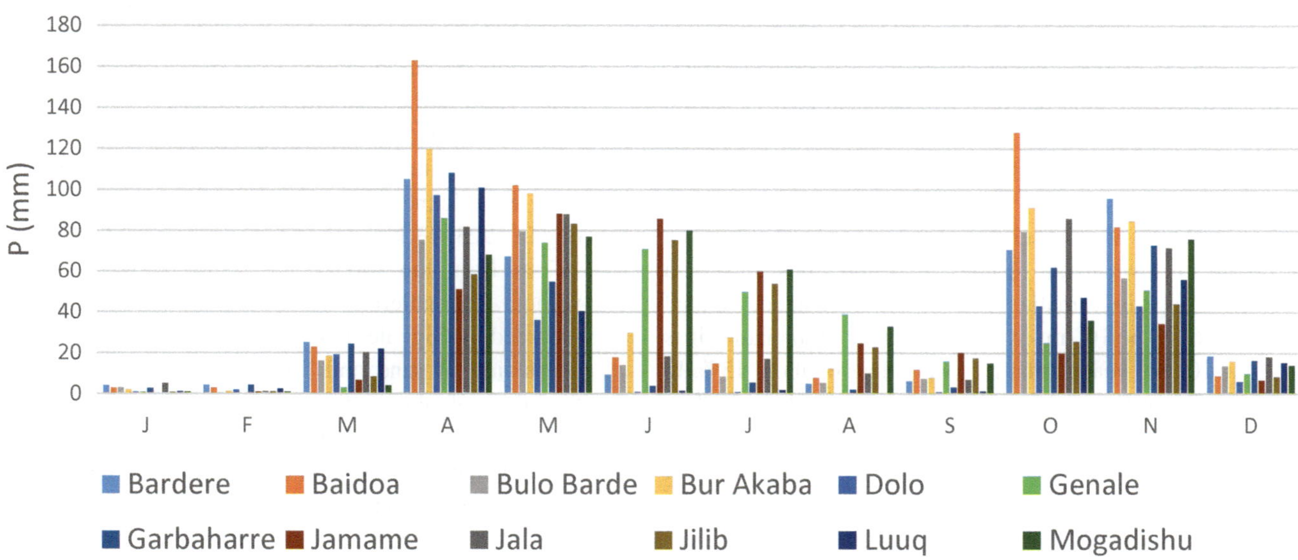

Fig. 13.5 Monthly rainfall variation for a few selected meteo-stations in the study area

Table 13.1 Average annual rainfall (mm) over the study area

Catchment	Average	Maximum	Minimum
Juba (whole basin)	595	1275	239
Juba (within Somalia)	427	704	279
Laag Dheera (whole basin)	534	1355	279
Laag Dheera (within Somalia)	478	571	332
Shabelle (whole basin)	543	1129	266
Shabelle (within Somalia)	460	651	279

Data from Basnyat (2007)

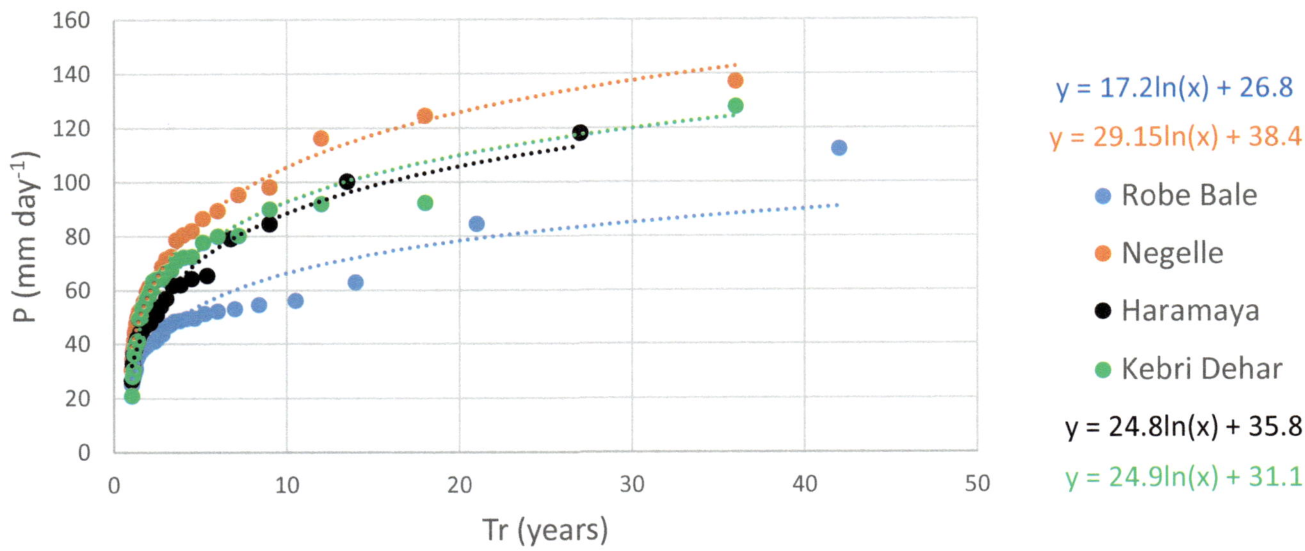

Fig. 13.6 Daily rainfall frequency curves of a few representative meteo-stations in the Juba and Shabelle river headwaters

and Gadain 2009), and new rating curves are still under construction. An important result of the work of Muthusi and Gadain (2009) is that, though the rating curves have changed, the discharge ratio between two adjacent flow gages did not substantially change.

Though the Shabelle has a larger catchment than the Juba (without the Laag Dheera), its average annual runoff volume is less. In both rivers, discharge decreases remarkably from upstream to downstream reaches. At Luuq (most upstream Juba flow gage in Somalia—Fig. 13.1), for instance, annual runoff is 6128 × 10^6 m^3 $year^{-1}$, whereas at Jamame, it is 5145 × 10^6 m^3 $year^{-1}$, that is, a water loss of about 15%. In the Shabelle, at Beled Weyne (most upstream Shabelle flow gage in Somalia—Fig. 13.1), annual runoff is 2580 × 10^6 m^3 $year^{-1}$, but it becomes as low as 126 × 10^6 m^3 $year^{-1}$ at Afgoye, which corresponds to a reduction of about 95%. The reasons for such a different hydrological behavior of this river are manifold, the most important of which include:

(a) the lack of tributaries of the Shabelle in its lower course. The Fafan river is the most downstream tributary joining the Shabelle at the Ethiopia-Somalia border. But, this river has a very low discharge and only occasionally; during very large flood discharges, its water reaches the Shabelle (Fig. 13.3);
(b) water diversion for irrigation purposes;
(c) streambed infiltration and ground water recharge;
(d) the great thickness of the alluvial deposits. A borehole dug in 1923, 13 km east of Jowhaar, reached a depth of 107 m from the surface, i.e., 3 m below sea level, without encountering the bedrock. The stratigraphy recorded by Gigliale (in Crema 1923) consists of a basal alternation of sand and fine and coarser gravel for a thickness of about 40 m, overlain by alternations of sand, sometimes with fine gravel and dark red clays. Groundwater was found at a depth of 68 m in the basal gravels, but discharge was rather modest: 2.5 m^3 per hour;
(e) bedrock characteristics. The middle to lower reaches of the Shabelle are underlain by limestone and gypsum formations within which karst cavities may be present at depth (Fig. 13.7).

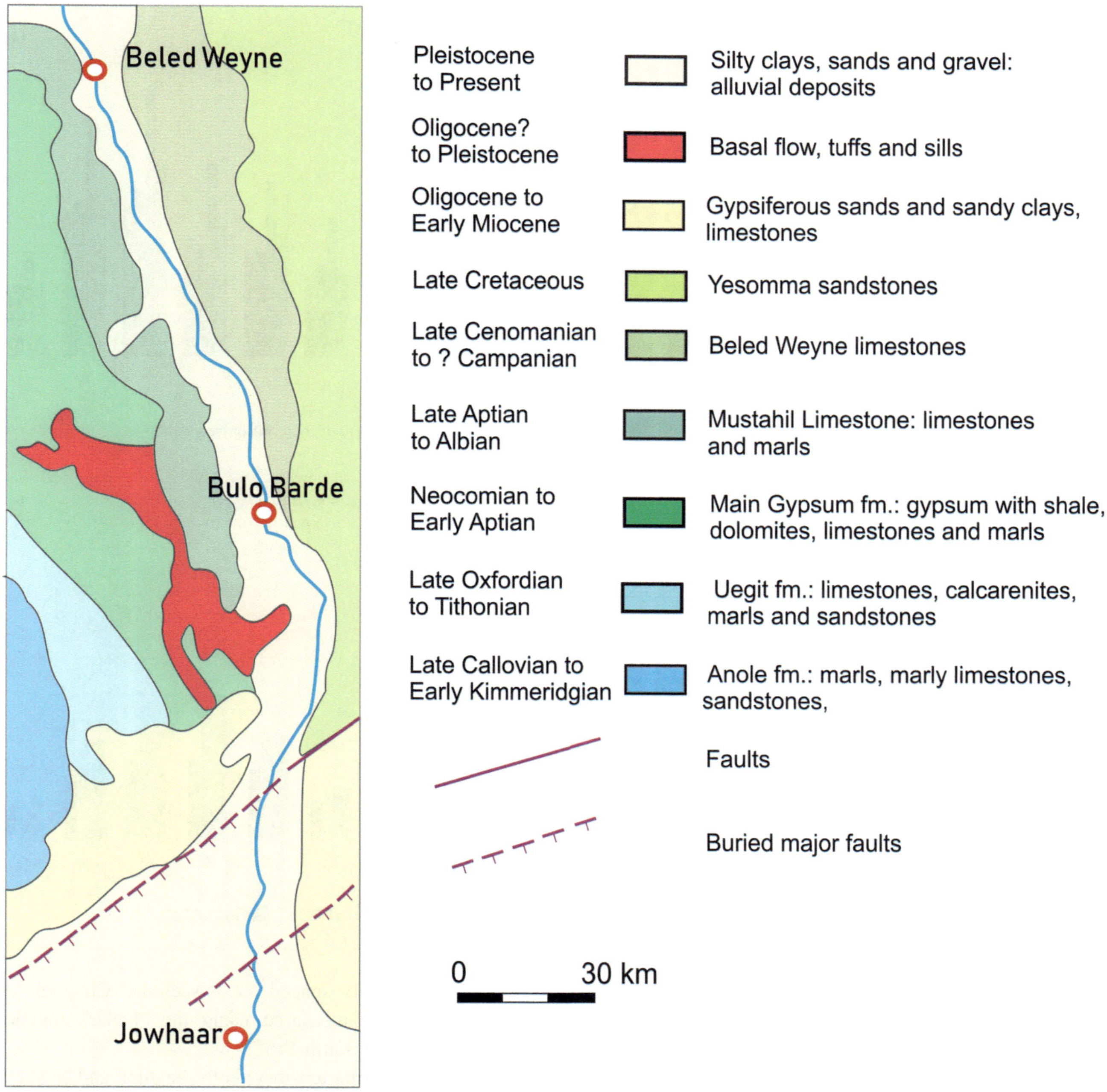

Fig. 13.7 Geological map of the area around the Shabelle river from Beled Wayne to Jowhaar. Modified from Abbate et al. (1994)

The monthly flow pattern follows that of monthly precipitation (Figs. 13.8 and 13.9). The largest mean runoff volumes and discharges are recorded in October–November and May at Bardere for the Juba and at Beled Weyne for the Shabelle (Figs. 13.8 and 13.9). The flow peaks in May follow the *Gu* rains, whereas those of October and November coincide with the *deyr* rains. The latter are propelled by the monsoon type rains that have their maximum from July to September in the Ethiopian headwaters (Fazzini et al. 2015).

Only the flow time series recorded at Luuq on the Juba and at Beled Weyne on the Shabelle are long enough to try some considerations on flow variability. Not only the annual runoff of the Juba is much larger (about 58%) than in the Shabelle, but this latter river is subjected also to a wider runoff variability with a variation coefficient of 0.48, whereas it is only 0.28 in the Juba (Fig. 13.10). The highest flow variability is recorded in March (1.73) in the Juba as runoff volume is substantially influenced by sporadic early beginning of the *Gu* rains in this month. The lowest flow discharges are recorded from January to March in both rivers with values ranging from 1 to 6 m^3s^{-1} in the Juba and from 0.76 to 2.00 m^3s^{-1} in the Shabelle. In this latter river,

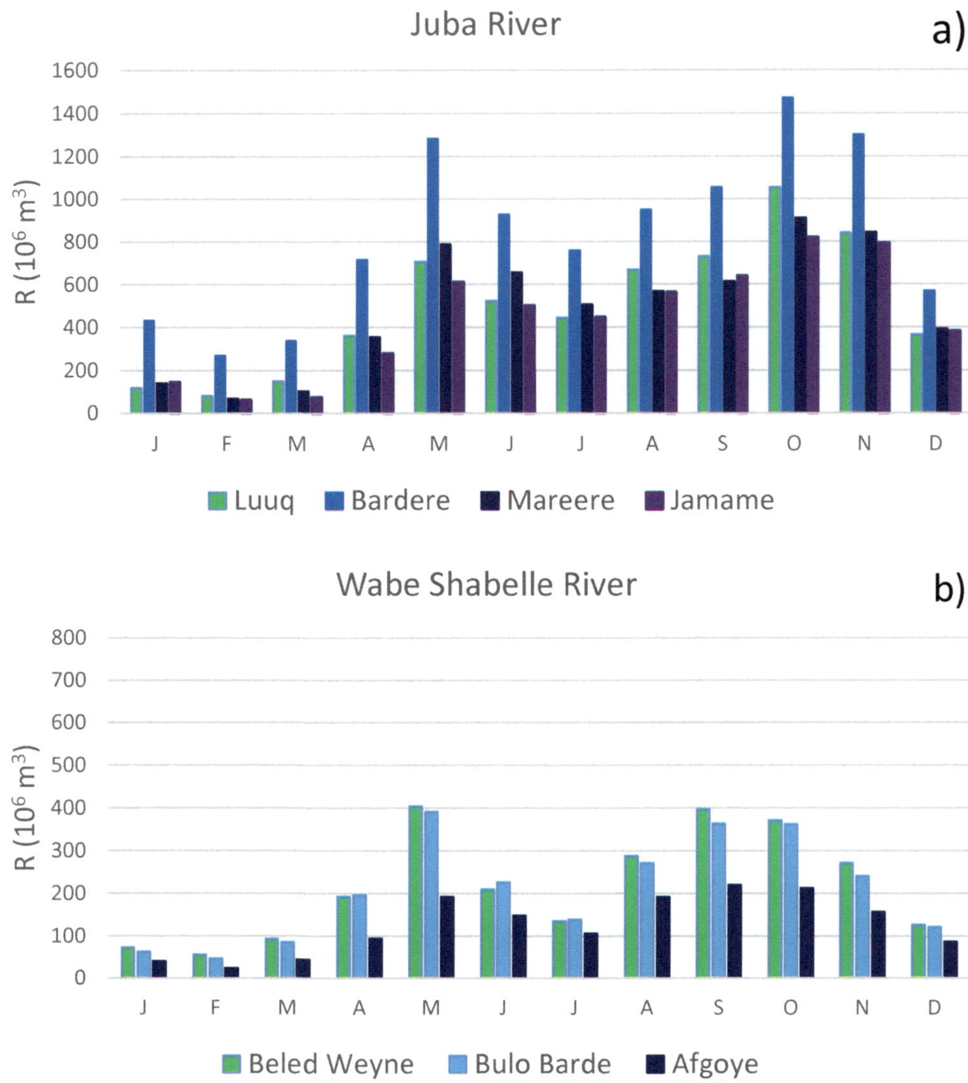

Fig. 13.8 Runoff monthly distribution of Juba **a** and Wabe Shabelle **b** rivers. Notice, the runoff volume of the former is about three times larger than the latter

however, during particularly long periods of low or no rain, the streambed may become partially or even completely dry (Fig. 13.11), especially in the middle-lower reaches, from Mahaday Weyne to Balcad, as it happened twice in one year in March 2016 and in December 2016-March 2017 (World Bank-FAO 2018).

13.3 River Morphology

All the geomorphic parameters used in this study were obtained from measurement on Google Earth Pro® satellite images. In order to increase the accuracy of meander wavelength (*Lw*) data, measurements were repeated two times on both side of the river reach. Elevation data for gradient calculation were obtained from averaging 4–5 values measure in the vicinity of the reach extremities. Average channel width was calculated by measuring the width at not less than 20, equally spaced, cross-sections. Channel and valley length were measured using the "Ruler" function available in Google Earth Pro®.

Within the Somalia territory, both the Juba and Shabelle river courses can be divided in two main portions with different geomorphological characteristics. In the upstream part, both rivers are entrenched and flow within a very narrow alluvial plain incised into a generally flat landscape 10–60 m higher. In the Juba, entrenched reaches (Fig. 13.12) alternate with bedrock reaches in which there is no flood plain and the river is bound by the valley slopes (Fig. 13.13). In the Shabelle, the flood plain of the entrenched part is wider (Fig. 13.14) and the river maintains such a morphology as far as Jalalaqsi. In these upstream reaches, both rivers show a moderate sinuosity. In the Juba, channel sinuosity is mainly controlled by bedrock structure as a fault system is present (Abbate et al. 1994), especially in the bedrock reaches between Luuq and Bardere. The Shabelle,

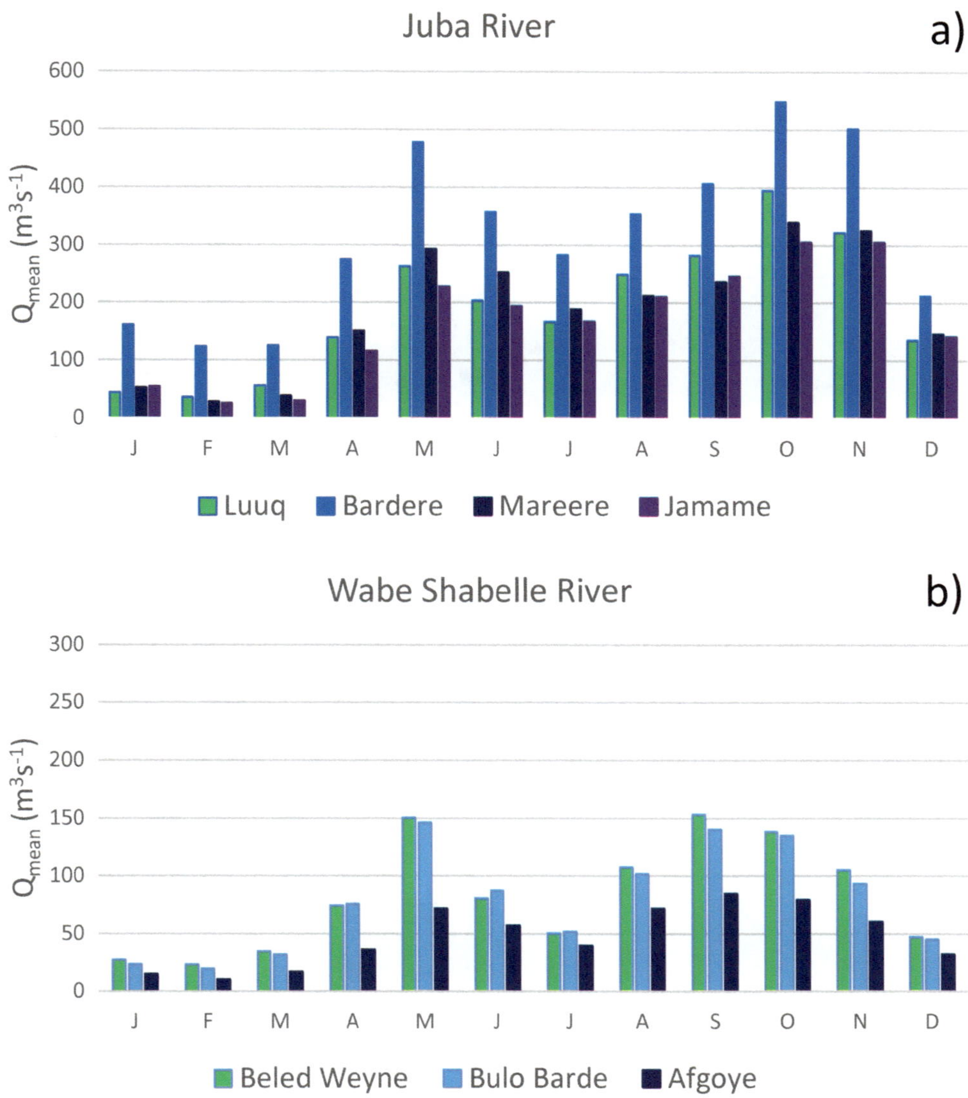

Fig. 13.9 Mean monthly discharge of Juba **a** and Wabe Shabelle **b** rivers. Notice, the discharge of the former is about three times larger than the latter

instead, flows across a wider alluvial plain, and the river is not laterally constrained. Though its sinuosity is low (1.60), examples of channel dynamics such as meander neck cut offs, abandoned channels, ox-bow lakes, point bar scrolls and lateral meander migrations can be observed (Fig. 13.15).

In the downstream reaches, both rivers enter a wide flood plain (from a few kilometers upstream of Dujuma in the Juba and from about 20 km downstream of Jalalaqsi in the Shabelle), without any lateral confinement (Fig. 13.16). In these downstream reaches, channel sinuosity increases to 1.80 in the Shabelle and 1.90 in the Juba and both rivers show all the typical morphological elements of meandering rivers (Figs. 13.17 and 13.18).

The Juba river shows an increase in channel dynamics in the downstream reaches, whereas the Shabelle is more dynamic than Juba between Jalalaqsi and Jowhaar. In its most downstream reaches, this river seems to be rather stable with very few cases of lateral migration or meander cutoff (Table 13.1).

A characteristic that the two study rivers have in common is a tendency to anastomosing (Fig. 13.19) and channel avulsion. In the Shabelle, this latter process is more pronounced and extensive (Fig. 13.20). Some basic geomorphic parameters of the modern Shabelle river and of the old avulsion channels between Mahaday Weyne and Afgoye (as indicated in Fig. 13.20) were measured and reported in Table 13.2. These data show clearly that the old channels were larger than that of the modern Shabelle. However, such a difference is even more striking if we consider bankfull discharge.

The discharge of a river may be highly variable, from drought to devastating floods, depending on local climate and catchment physical characteristics. Among the very many flow conditions experienced by a river, geomorphologists and hydrologists (e.g., Wolman and Leopold 1957; Leopold et al. 1964; Ackers and Charlton 1970) have identified bankfull flow as the most appropriate discharge to

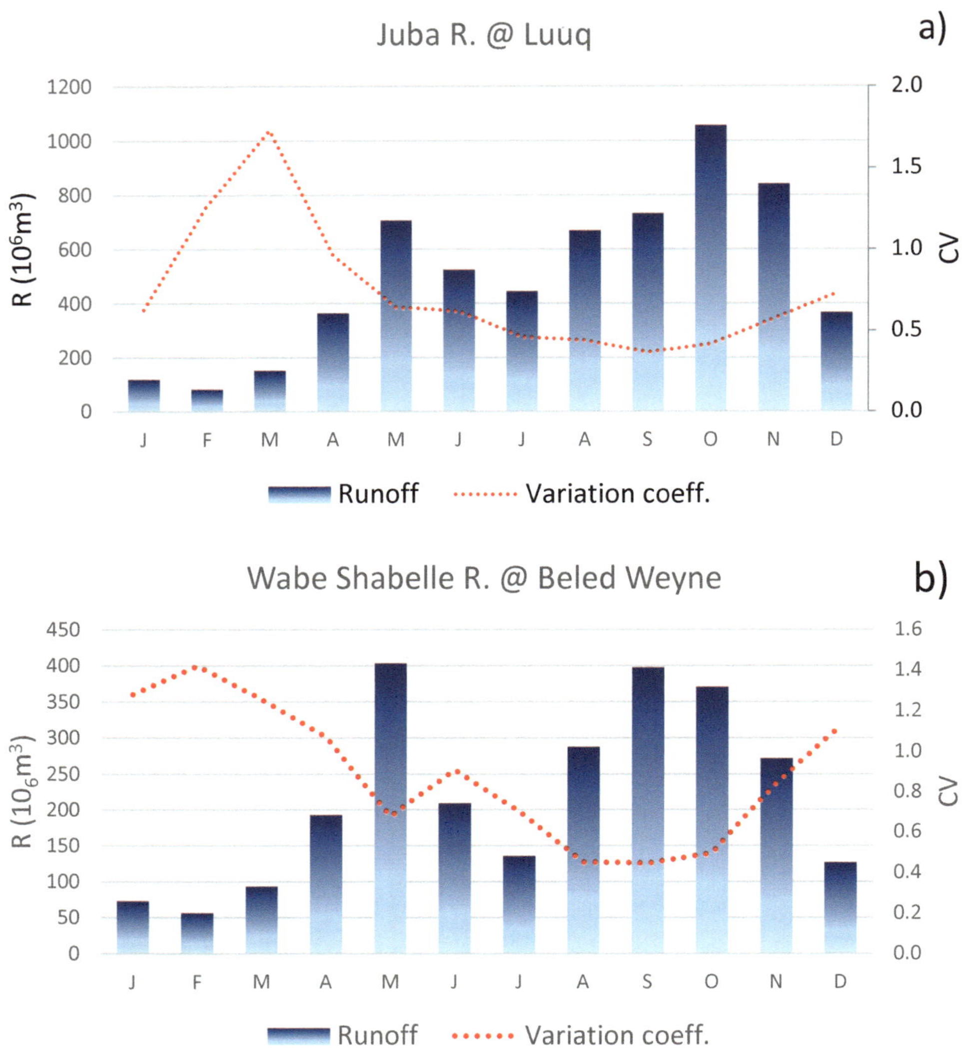

Fig. 13.10 Mean monthly distribution of runoff and of variation coefficient observed at Luuq on the Juba (**a**) and at Beled Weyne on the Shabelle (**b**)

Fig. 13.11 Very low discharge (26/02/2018) and dry bed in the Shabelle river between Jowhaar and Balcad

Fig. 13.12 Entrenched reach of the Juba river at Luuq

compare different river systems in terms of channel pattern and morphology, flow hydrology and sediment yield. In fact, bankfull discharge is associated with the "dominant discharge", i.e., the flow which, in the long term, performs most of the work in terms of sediment transport and yields the largest part of the sediment load (Wolman and Miller 1960). For these reasons, bankfull flow has a relevant geomorphologic significance, since some channel parameters, such as channel hydraulic geometry or meander wavelength, are adjusted to it. In other words, the morphology of a river channel is largely determined by bankfull flow. Hence, in paleohydrologic analysis, this is the best, unambiguous parameter to connote a river flow since it can be inferred from the paleochannel geometry and sediment characteristics and can be used for comparison with modern fluvial systems. Though there is some uncertainty about which discharge best approximates bankfull flow, especially for rivers under different climate and flow regime conditions, many authors (e.g., Leopold et al. 1964; Andrews 1980; Torizzo and Pitlick 2004) restrict bankfull flow to a discharge with a return time ranging from 1.58 to 2.33 years.

For the Shabelle at Beled Weyne, the time series of discharge is not very long (only 47 years), but it can be considered sufficient to construct a reliable flood frequency diagram (Fig. 13.21). The discharges with a return time

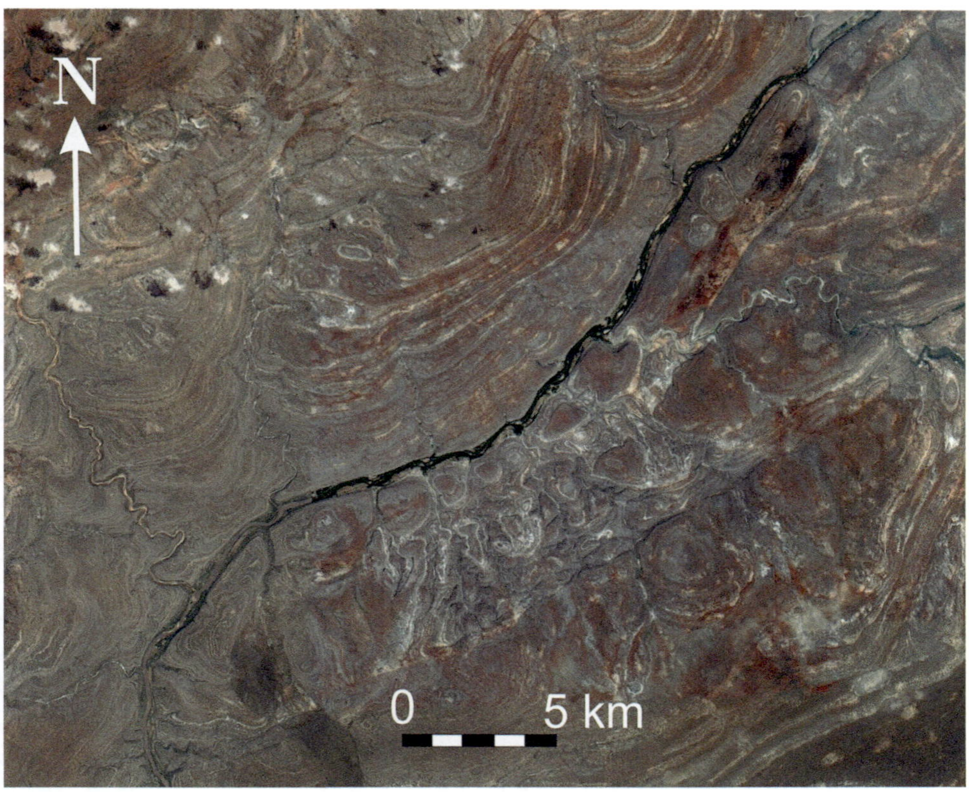

Fig. 13.13 Bedrock reach of the Juba river about 100 km downstream of Luuq

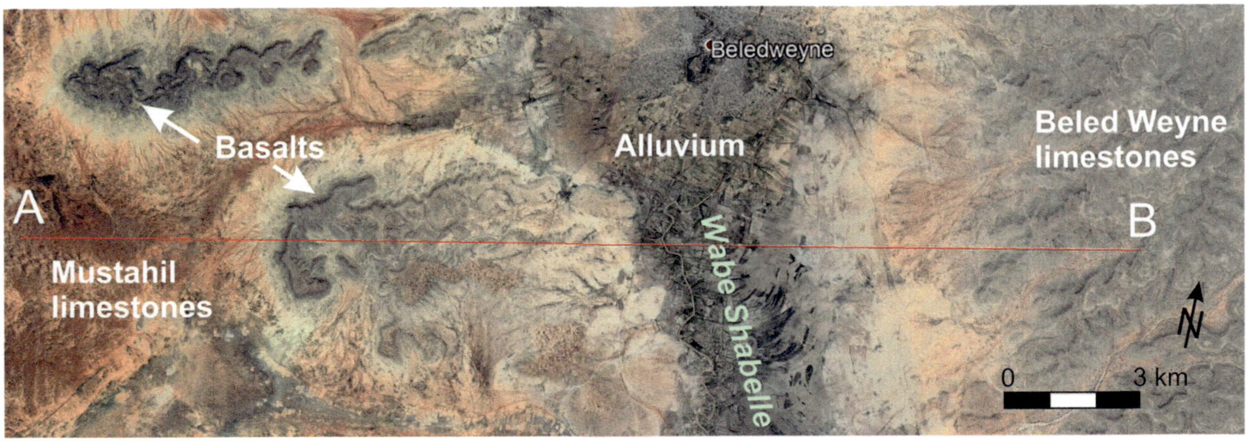

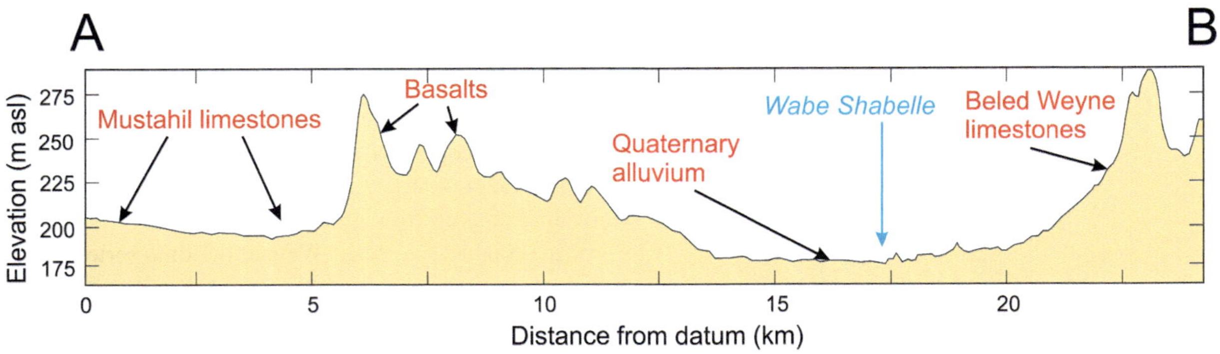

Fig. 13.14 Cross-valley profile of the Shabelle river near Beled Weyne. The alluvial plain is narrow and confined

Fig. 13.15 Channel dynamics evidence in the Shabelle river about 12 km downstream of Beled Weyne: mlm = meander lateral migration; *s* = point bar scroll; ac = abandoned channel; obl = ox-bow lake

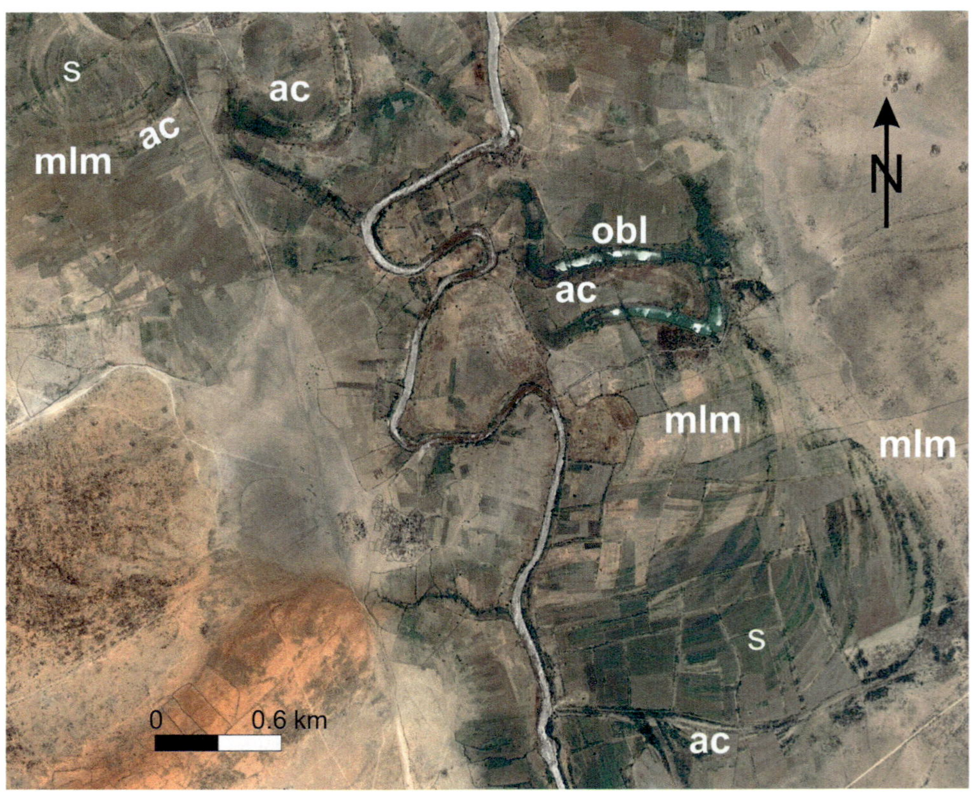

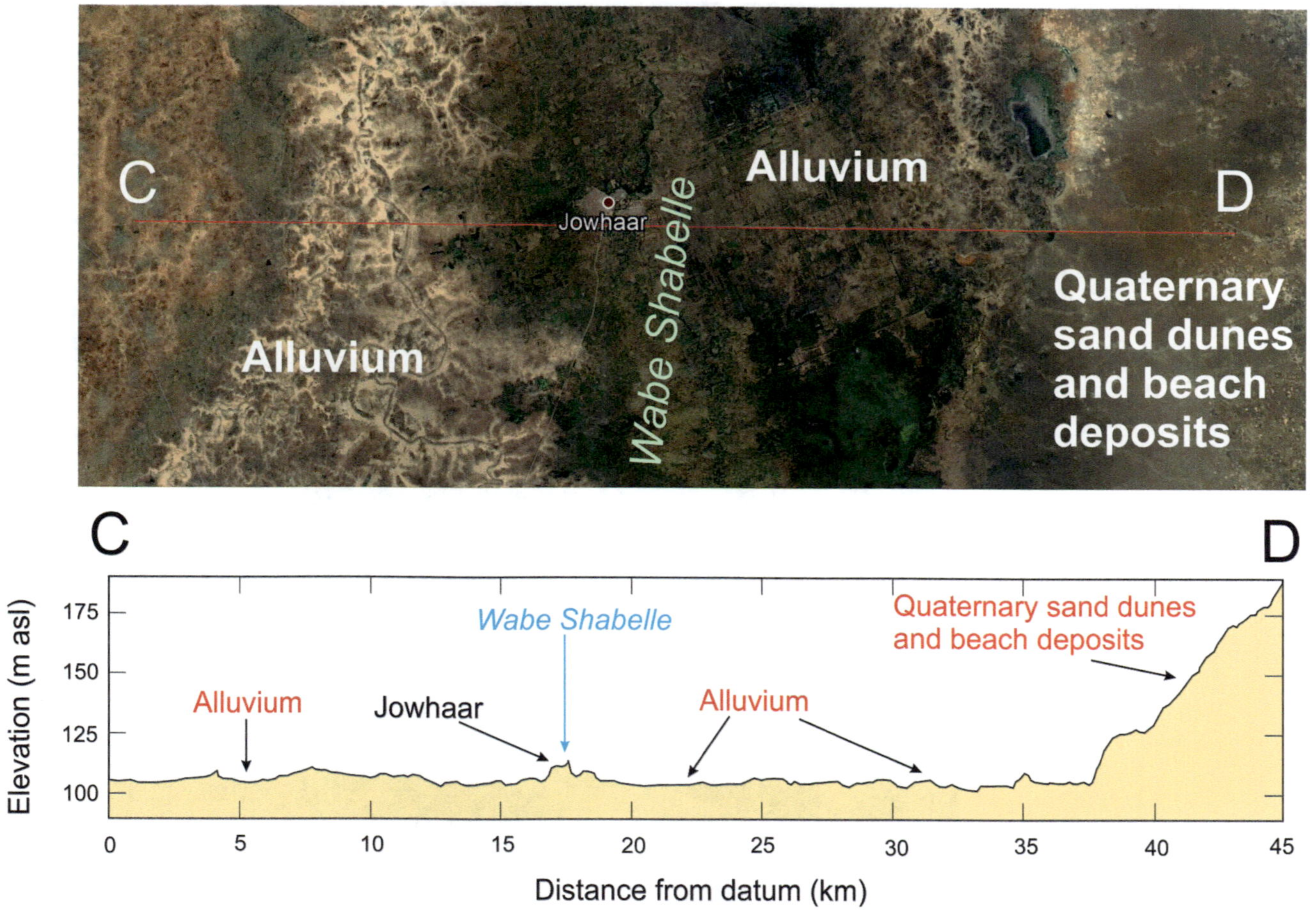

Fig. 13.16 Cross-valley profile of the Shabelle river at Jowhaar. Notice the wide alluvial plain (more than 35 km), the lack of lateral constraints and the perched position of the river, which is more elevated than the surrounding flood plain

interval of 1.58 and 2.33 years calculated with the semi-log diagram of Fig. 13.21 and with the Gumbel extreme values method range from 209 to 269 m^3s^{-1}. For the downstream reach, the only data available are those measured at Mahaday Weyne (Basnyat 2007). Here, the time series is much shorter (only 26 years), but though its limitation is recognized, for completeness of approach, the same calculations were adopted and bankfull discharge resulted to range from 143 to 150 m^3s^{-1}. Notwithstanding the limited data, the lower values obtained at Mahday Weyne are not surprising since both mean monthly discharge and maximum discharge at this measuring station are about 45% smaller than at Beled Weyne.

In order to infer bankfull discharge of the old channels, the empirical relationship of Dury (1976) (in Knighton 1998), which is based on meander wavelength, was used:

$$Q_b = (L_w/32.86)^{1.81} \quad (13.1)$$

in which L_w is the average meander wavelength (Table 13.2). Equation (13.1) was first calibrated on the data of the modern Shabelle. For the reach downstream of Mahaday Weyne, Eq. (13.1) returns a bankfull discharge of 184 m^3s^{-1}. This value is slightly higher than the calculated ones, but the small number of data and the very small standard deviation may lead to consider the values of 143–150 m^3s^{-1} probably a little lower than the actual ones. To further analyze this assumption, the equation of Williams (1978),

$$Q_b = 4.0A_b^{1.21}J^{0.28} \quad (13.2)$$

in which A_b is the cross-sectional area at bankfull flow and J is gradient which was also used. To calculate the cross-section area, information was derived from the cross-sections and the bankfull depth data reported by Basnyat (2007). Equation (13.2) returned a bankfull discharge of about 168 m^3s^{-1}. This value falls in between the bankfull discharge obtained by the flood frequency distribution and the empirical Eq. (13.1) of Dury (1976). The difference between the result of Eqs. (13.1) and (13.2) is small, and since bankfull discharge of the modern river at Mahaday Weyne was calculated from a short time series with values of maximum annual discharge very similar (Fig. 13.21), it can be supposed that such a short time series is not able to

Fig. 13.17 Meander neck cutoffs on the Juba river near Jamame

capture the natural variability of high flow and likely the actual bankfull discharge is a little larger than the calculated values of 143–150 m^3s^{-1}. Moreover, given that the result of Eq. (13.2) is very close to that of Eq. (13.1) and considering that for the old channels, the parameter that can be measured with the best accuracy is meander wavelength, bankfull discharge of these avulsion channels was calculated by means of Eq. (13.1), and the results are reported in Table 13.3. The old channel OC R1.3 is straight and only a couple of meanders are present. Its sinuosity, in fact, is 1.26, which is typical of straight channels, so, for this river, another equation derived by Dury (1976) on the base of theoretical considerations and field data, which is more suitable for non-meandering rivers, was used:

$$Q_b = (0.0396/J)^{1.35} \tag{13.3}$$

in which J is gradient.

The results of Table 13.3, which were obtained from meander wavelength (and gradient as in the case of channel OC R1.3), confirm the difference already observed considering the geomorphic parameter but also point out that old channel OC R1 was formed by a river much larger than the modern one. With the exception of the straight channel OC R1.3, which is an avulsion of channel OC R1 (Fig. 13.20) and does not show evidence of channel lateral dynamics, thus likely indicating a short occupancy of this channel; in all the other old channels, examples of the typical dynamics of a meandering river (cutoffs, point bar lateral migrations, ox-bow lakes, etc.) are present (Fig. 13.22) and suggest that in the old channels bankfull discharges much larger than present were maintained for a time long enough to develop these characteristic meandering river morphologic units.

Unlike the Shabelle, in the Juba, old avulsion channels are not so common and restricted to the lowermost reach

Fig. 13.18 Crevasse splay on the Shabelle river (Google Earth coordinates: 3°09″53″ N 45°32″ 30″ E)

between 20 km upstream of Jilib to Jamame. In this reach, the modern river has an average bankfull width of 80 m, a sinuosity of 1.97, mean meander wavelength of 1102 m, mean radius of curvature of 263 m, and the gradient is about 0.000012. All these geomorphic data indicate that the Juba is a larger river with a larger runoff (Figs. 13.9 and 13.10), larger mean meander wavelength and curvature radius than the Shabelle in its middle to lower reach between Mahady Weyne and Jowhaar (Table 13.2). In the Juba, avulsion channels are much less in number, length and continuity. For these reasons, it was not possible to measure meander wavelength and curvature radius for a sufficient number of bends. The average channel width of avulsion channels (about 60 m) was measured on different short reaches that, however, cannot be assumed as part of the same avulsion channel given the fragmentation of their pattern. This is also confirmed by the standard deviation which is 8.9 m in the modern river and 31 m in the old channels which, in any case, are smaller than the modern Juba.

The assessment of bankfull discharge with Eq. (13.1) returned a value of 577 m^3s^{-1}, whereas bankfull discharge calculated by measured data is in the range of 424–449 m^3s^{-1}. Apparently the difference with the result of Eq. (13.1) seems large, but at the closest gaging station in Jamame, only 16 years of reliable data measured before 1990 are available, and the variation coefficient of the yearly maxima is only 14% in spite of the high annual variability of rainfall (SWALIM 2016). In the work of Basnyat and Gadain (2009), a longer list of yearly maxima is reported, but the value of 477 m^3s^{-1} recorded in 12 years out of 26, with also this same discharge recorded in five years in a row, opens serious question of uncertainty about the bankfull discharge obtained from these field data. These authors report that for both the Juba and the Shabelle rivers, flood peaks are flattened because of bank breaching for irrigation purposes and indicate discharges of 500 and 160 m^3s^{-1} as the bankfull discharges of the Juba at Jamame and of the Shabelle at Mahaday Weyne, respectively, i.e., values not too far from those obtained by Eq. (13.1) (577 and 187 m^3s^{-1} for the Juba and Shabelle, respectively).

13.4 Flooding

In the last two decades, an increased frequency of high floods has been recorded in Somalia (Fig. 13.23) the large majority of which occurred in the middle to lower Juba and Shabelle rivers reaches (SWALIM 2016). In this region, according to SWALIM (2016), the number of people affected by the five- and ten-year flood is about 382,000 and

Fig. 13.19 Anastomosing reach of the Juba river (Google Earth coordinates: 0°12′14″ N 42°46′ 04″ E)

808,000, respectively. Most devastating floods were recorded at the end of the 1990s and in 2006–2007, but recurrent inundations are a serious threat with fatalities and property damage. SWALIM (2016) associates the high floods of 1991/1992, 1994/1995, 1997/1998, 2002/2003, 2006/2007 and 2009/2010 to el Niño episodes, that were particularly marked also in 2014/2015 and 2015/2016.

The maps of the rainfall anomalies over Somalia (FSNAU-FAO 2018) for the 2000–2017 interval (Fig. 13.24) clearly show an increase in the frequency of higher than average rains during the minor rainy season (*Deyr*, i.e., October and November), paralleled by a similar pattern of the wetter *Gu* rainy season (April-June), though with lower positive anomalies and the increased frequency recorded predominantly in April. In most of the cases, the highest rainfalls are concentrated in the southern portion of Somalia, that is, in the middle to lower reaches of the Juba and Shabelle rivers. Very long time series of *Deyr* and *Gu* rains are not available for the study rivers headwaters. Those with the longest record and as least as possible data gaps are Gode and Jijiga for the Shabelle and Moyale and Goba for the Juba. Unfortunately, these time series have different lengths and cover different time intervals. Nevertheless, though with caution, they can be used to investigate if in the river headwaters, there is some evidence of increasing trends. Figure 13.25 shows that with the exception of the *Gu* rain at Jijiga which is characterized by an increasing trend, all the other trends are negative or there is no trend at all.

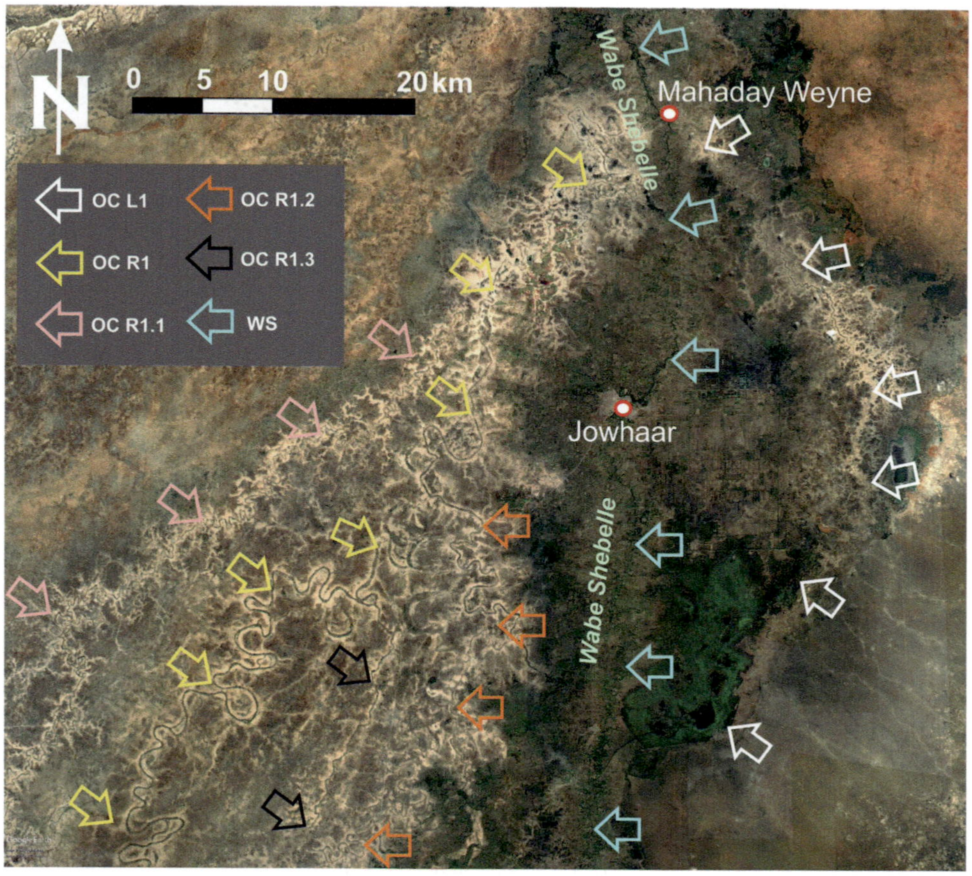

Fig. 13.20 Avulsions and old channels of the Wabe Shabelle downstream of Mahaday Weyne. OC = old channel; *R* = right; *L* = left; 1 = main avulsion; 1.2 = secondary avulsion of main avulsion; 1.3 = secondary avulsion downstream of 1.2; ws = Wabe Shabelle

Table 13.2 Main geomorphic parameters of the Shabelle and the avulsion channels between Mahaday Weyne and Jowhaar (Fig. 17.20)

	S	J (m/m)	W (m)	R_c (m)	R_c max (m)	R_c min (m)	R_c/W	L_w (m)	L_w max (m)	L_w min (m)	L_w/W	L_w/R_c
W S	1.93	0.000246	25.1	84.7	184	41	3.37	587	1018	326	23.4	6.93
OC R1	2.64	0.00013	122.8	434.2	905	198	3.54	2108	3956	1041	17.2	4.85
OC R1.1	2.36	0.00020	67.8	133.3	234	81	1.97	656	1192	235	9.7	4.92
OC R1.2	3.19	9.68E-05	78.6	274.6	507	123	3.49	1118	1987	643	14.2	4.07
OC R1.3[a]	1.26	0.00033	89.4		358	164						
OC L1	1.74	0.00017	44.4	117.8	207	47	2.65	548	958	181	12.4	4.66

[a] This channel is straight (S = 1.26), and no meander parameter was measured. WS—modern Wabe Shabelle river, OC R1, OC R1.1, OC R1.2, OC R1.3—avulsion old channels on the right of the present river (see Fig. 17.21); OC L1—avulsion old channels on the left of the present river (see Fig. 17.21), S—channel sinuosity, J—channel gradient; W—width; R_c—meander curvature radius, L_w—meander wavelength

Long-term daily discharge data are available only for the flow gages of Luuq and Beled Weyne on the Juba and Shabelle, respectively. The time series of the Shabelle is rather continuous, whereas that of the Juba is interrupted by two wide gaps (Fig. 13.26). In spite of a general decreasing trend of precipitation in the headwaters, the Shabelle daily discharges show an increasing trend, especially after 2002, whereas in the Juba, there is no evidence of significant change throughout the last six decades.

In their middle to lower reaches, the river pattern and the flood plain of the Shabelle and, to a much lesser extent, of the Juba are characterized by several avulsion episodes and channels (Figs. 13.19 and 13.20). According to Aslan et al. (2005), avulsion is favored when: (a) The river is rapidly aggrading, thus resulting in a channel perched over the flood plain (Fig. 13.16) and creating a cross-valley gradient. The aggradation process is typically driven by high sediment transport and deposition rates which, in rivers with cohesive

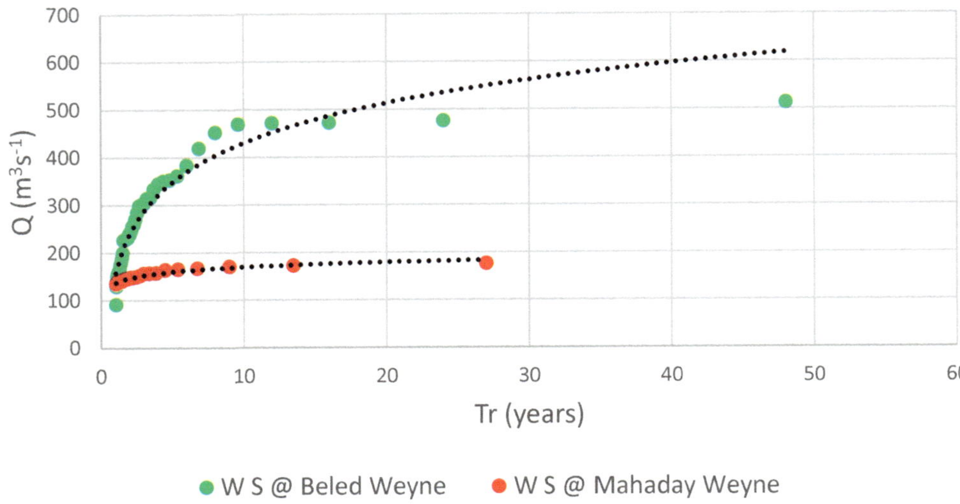

Fig. 13.21 Flood frequency curves for the Shabelle river (WS) at Beled Weyne and Mahaday Weyne

Table 13.3 Bankfull discharge calculated by different methods for the modern Shabelle between Mahaday Weyne and Jowhaar and the avulsion, old channel in the same area

	Actual	Dury (1976)	Williams (1978)
	Q_b (m^3s^{-1})	Q_b (m^3s^{-1})	Q_b (m^3s^{-1})
W S	143–150	184	167.8
O C R1		1867	
OC R1.1		226	
OC R1.2		592	
OC R1.3[a]		654	
OC L1		163	

[a] This channel is straight (S = 1.26), so no meander wavelength was measured; discharge was measured with Eq. (13.3) (see text). *WS*—modern Wabe Shabelle river; OC R1, OC R1.1, OC R1.2—avulsion old channels on the right of the present river (see Fig. 17.21), OC L1—avulsion old channels on the left of the present river (see Fig. 17.21). OC r1.3 was not included because it is a straight channel and no meander parameter was measured

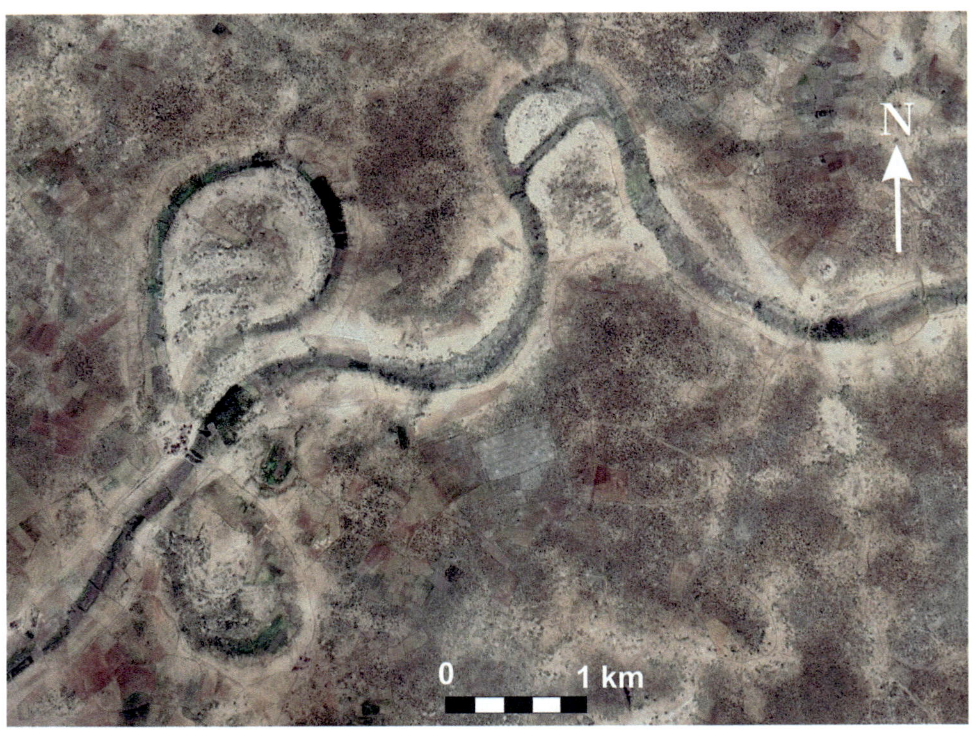

Fig. 13.22 Neck and chute meander cutoffs in the old avulsion channel OC R1 (Google Earth coordinates: 20°21′32″ N 45°06′40″ E)

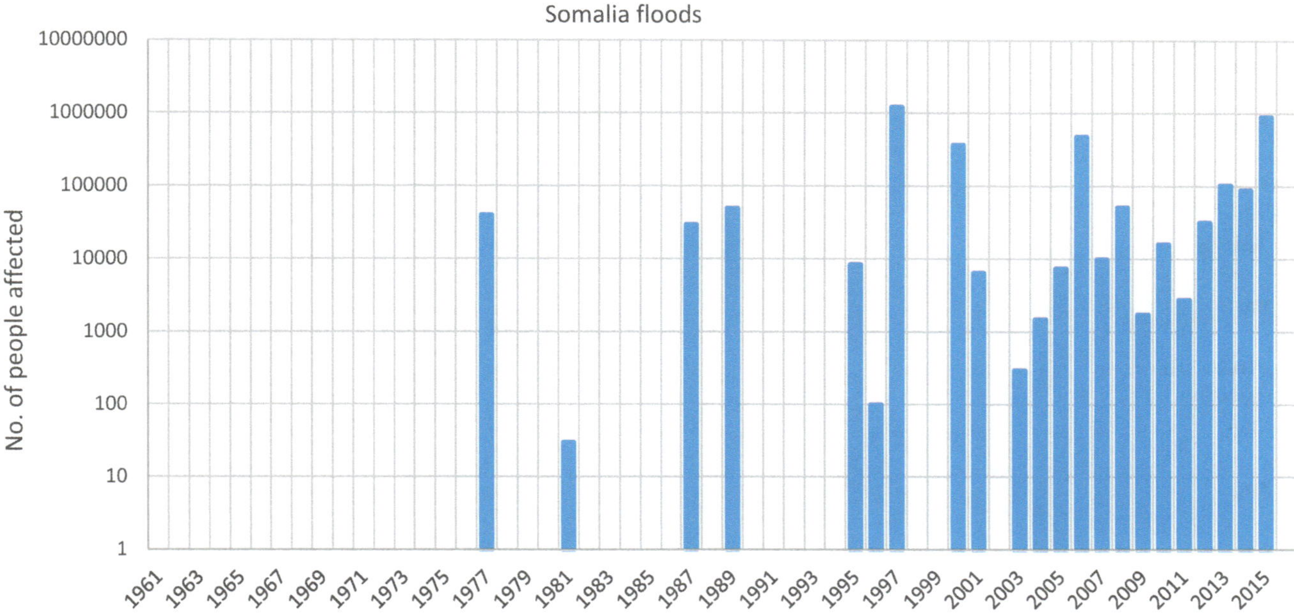

Fig. 13.23 Number of people affected by the increased frequency of floods in Somalia

banks, may result in a marked reduction of channel conveyance capacity; (b) substantial discharge variations; (c) subsidence; (d) in-channel human impact. Both the Juba and Shabelle transport high concentrations of suspended sediment; however, the only suspended sediment concentration data available were obtained by Omuto et al. (2009) by a two-year (2007–2008) field measuring campaign. Though their data are temporally limited, they cover a wide range of rates (from 2 to almost 3000 kg s^{-1}) and these authors were able to construct a suspended sediment rating curve, which is valid for both rivers:

$$Q_{ss} = 0.087Q^{1.9645} \quad (13.4)$$

in which Q_{ss} is the suspended sediment transport rate in kg s^{-1} and Q is discharge in m^3 s^{-1}.

The rating curve of Omuto et al. (2009) shows that an appreciable amount of suspended sediment is transported even with very low discharges, as low as 4 m^3 s^{-1}. Such a result was anticipated by Crema (1923) in one of his very first reports about the Southern Shabelle, described as transporting fine sediment in suspension all year round.

A remarkable reduction in flow discharge is observed in the Shabelle, whose annual runoff decreases from 2580 × 10^6 m^3 $year^{-1}$ at Beled Weyne to 1507 × 10^6 m^3 $year^{-1}$ at Afgoye, corresponding to a water loss of about 42% throughout a river length of about 422 km. Water loss affects also the Juba, but at a much lesser extent. In fact, runoff is 6097 × 10^6 m^3 $year^{-1}$ at Luuq; it increases to 8787 × 10^6 m^3 $year^{-1}$ at Bardere to decrease again at Jamame, where it is 5145 × 10^6 m^3 $year^{-1}$ and records a water loss of 16% in the 1204 km long reach between Luuq and Jamame.

For Beled Weyne, a rather long series of flow data is available to construct an average flow duration curve, whereas for the same river at Afgoye, unfortunately, the data are not enough. The flow duration data for this flow gage were therefore obtained from the Beled Weyne data by simply reducing them by 42%. Two suspended sediment transport duration curves were then obtained by applying Eq. (13.4) to both station flow duration data (Fig. 13.27). The result is that at Beled Weyne the annual suspended sediment yield is 28 × 10^6 t $year^{-1}$ (equivalent to 147 t km^{-2} $year^{-1}$) and at Afgoye is 9.7 × 10^6 t $year^{-1}$ (equivalent to 50 t km^{-2} $year^{-1}$).

For lowland, low-gradient rivers bedload is commonly in the range of 1–10% of suspended load with the most common range between 2 and 5% (e.g., Schumm 1977; Babinski 2005; Cantalice et al. 2013; Joshi and Xu 2017). Using these latter, precautionary values, the difference between the bedload yield in Beled Weyne and Afgoye is 651,651 and 23,266 m^3 $year^{-1}$ considering a bedload proportion of 5 and 2%, respectively. Assuming that these volumes are trapped as bedload sediment (Joshi and Xu 2017), an average aggradation rate of about 0.05 and 0.002 m $year^{-1}$, results, respectively, for the reach between Beled Weyne and Afgoye. Repeating the same procedure for the Juba between Bardere and Jamame, similar values of aggradation rates of 0.09 and 0.003 m $year^{-1-}$ are obtained and a sediment yield of 543 and 158 t km^{-2} $year^{-1}$ at Bardere and Jamame, respectively. At Bardere, the bedload yield calculated in this study as 5% of suspended sediment is about 13 × 10^6 t $year^{-1}$. This value is not too far from that of 20 × 10^6 t $year^{-1}$ reported by Pozzi (1982), which corresponds to a

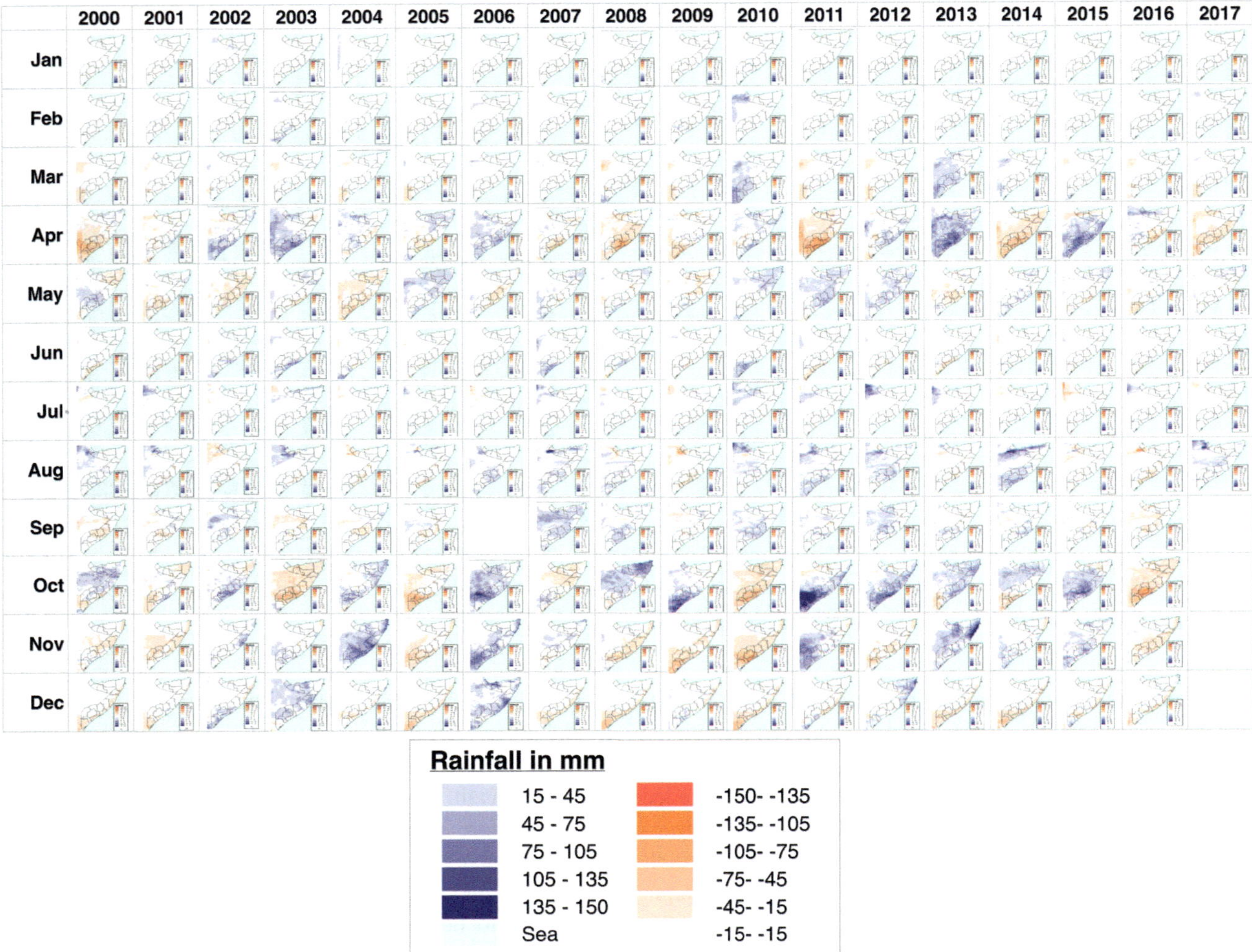

Fig. 13.24 Monthly rainfall distribution across the last two decades in Somalia (FSNAU-FAO 2018)

proportion of 8%. This author affirms that his results were presented in an unpublished technical note (by Technital) and were obtained on the base of FAO data, but no citation is reported nor any information is given about the method used to calculate the bedload yield. Pozzi reports also that at Soblale (150 km SW of Afgoye), the Shabelle has no sediment transport, its water is permanently clear, that there is no evidence of in-channel sedimentation and the channel width measured in 1960 was exactly the same measured about 70 years before by captain Vittorio Bottego (1895).

13.4.1 Human Impact on Flooding

According to Mbara et al (2007), before the onset of the civil war in 1991, irrigation schemes were rather efficient in Somalia. They relied on the Juba and Wabe Shabelle river water and largely contributed to the development of agriculture. Irrigation water was diverted from the rivers by weirs (Fig. 13.28) and conveyed to primary supply canals (see Mbara et al. 2007 and Basnyat and Gadain 2009 for a detailed description of all the irrigation schemes of southern Somalia). Irrigation schemes are more numerous in the Shabelle with 135,000 ha of potentially served land, against the 25,000 of the Juba (Basnyat and Gadain 2009). The long-lasting state of civil war resulted in the abandonment of many of these irrigation schemes that forced the farmers to breach the river banks, thus weakening them and causing uncontrolled flooding and water loss (Fig. 13.29). Due to lack of maintenance, deposition occurred at the gates and in the canals reducing substantially their water conveyance efficiency.

From satellite images (Google Earth 2018), 88 points with bank breach/overtop were identified in the Wabe Shabelle between Beled Weyne and Afgoye, though the most of them is concentrated in the reach between Bulo Barde and Mahaday Weyne. Out of these 88 bank breach/overtop points, 15 are clearly unofficial man made, whereas the other

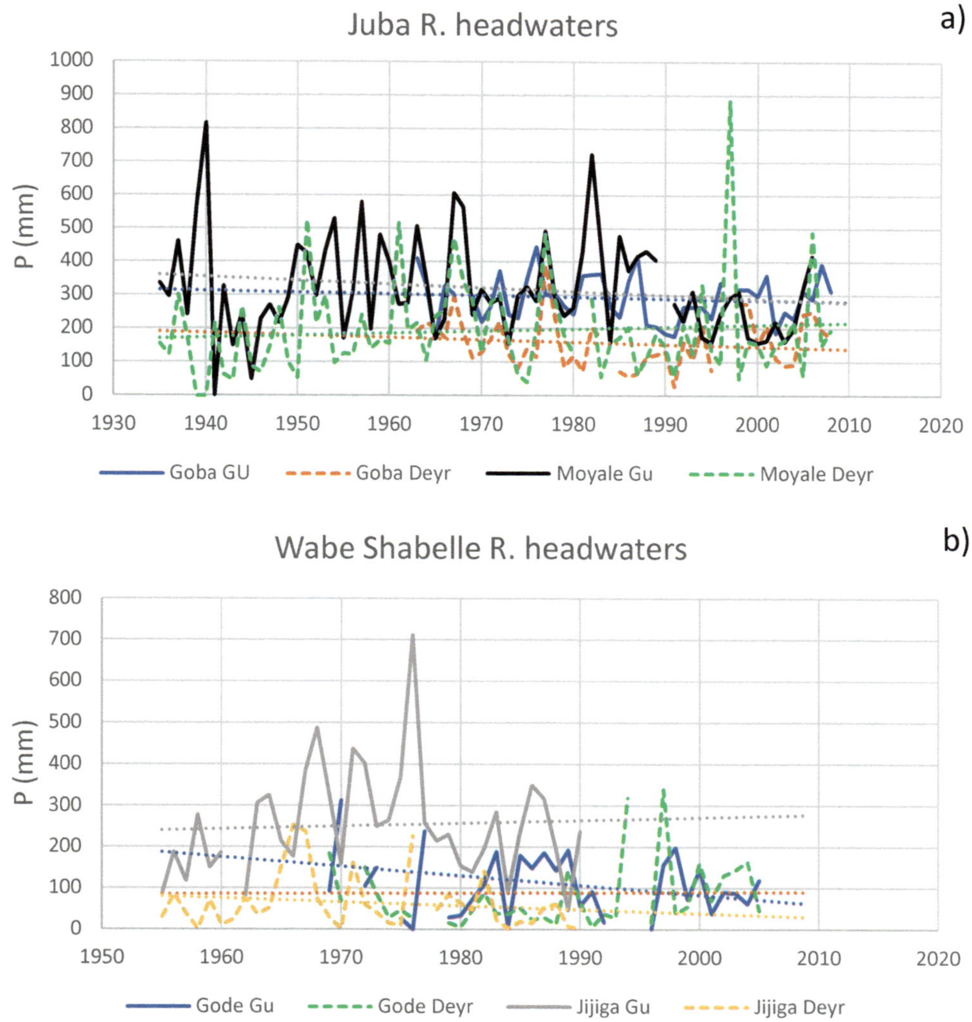

Fig. 13.25 Precipitation time series in the Juba **a** and Shabelle **b** headwaters

73 consist of natural bank breaches/overtop with which evidence of recent or repeated sedimentation is associated. Some of the bank breaches are used by villagers simply to access the river, but can work also as conveying waterway to irrigate the cropland near the village. This situation is particularly common on the Shabelle, whereas on the Juba breaching by farmers is rare.

Basnyat and Gadain (2009) already warned that "unregulated practices of river bank breaching have increased the vulnerability of the riverine communities to progressively smaller peak flows". In some of its lower reaches, the Shabelle is perched (Fig. 13.16) thus exacerbating the risk and the consequences of flooding. The vertical accretion of a lowland, low-gradient flood plain river is commonly associated with high rates of fine sediment transport. The finer component (silt and clay and, subordinately, sand) is deposited close to the bank by overbank flows, whereas the coarser fraction (typically sand and small proportions of fine gravel and grains) is deposited on the streambed because the transport capacity decreases as the river flow spills over the banks, thus resulting in the streambed aggradation (Brierley and Fryirs 2005). The greater the height of the streambed and the natural levees above the flood plain and the higher is the probability of crevasse splays and flood channel formation. Sediment has often been neglected in flood risk analyses, but changes in sediment dynamics and supply may result in channel instability and may have important negative repercussions affecting people and properties.

Both the Juba and the Shabelle transport a lot of sediment. Omuto et al. (2009) have measured concentrations ranging from 1 to 40 gl^{-1} in both rivers. In both the study rivers, the remarkable downstream water loss, the large volumes of water diversion for irrigation (about 12,000 $m^3\ ha^{-1}\ year^{-1}$, which mostly affects the downstream flow decrease of the Shabelle given its lower runoff) (Yibeltal 2015) and the additional water loss by unregulated withdrawal (especially in the Shabelle) are all conditions leading to a marked increase of the already observed high potential for streambed aggradation as showed in the previous section. Hailemariam et al. (2016) studied the land

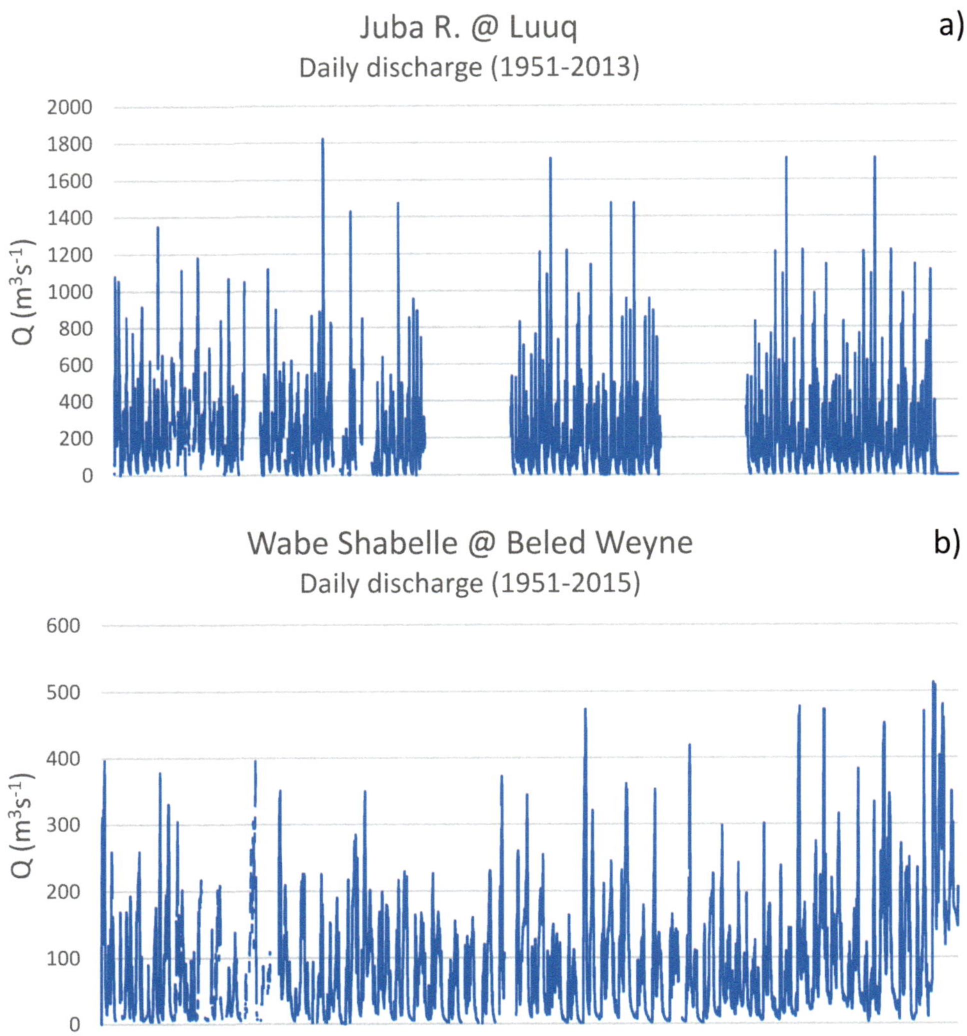

Fig. 13.26 Daily discharge time series for the Juba at Luuq **a** and the Shabelle at Beled Weyne **b** in the last six decades

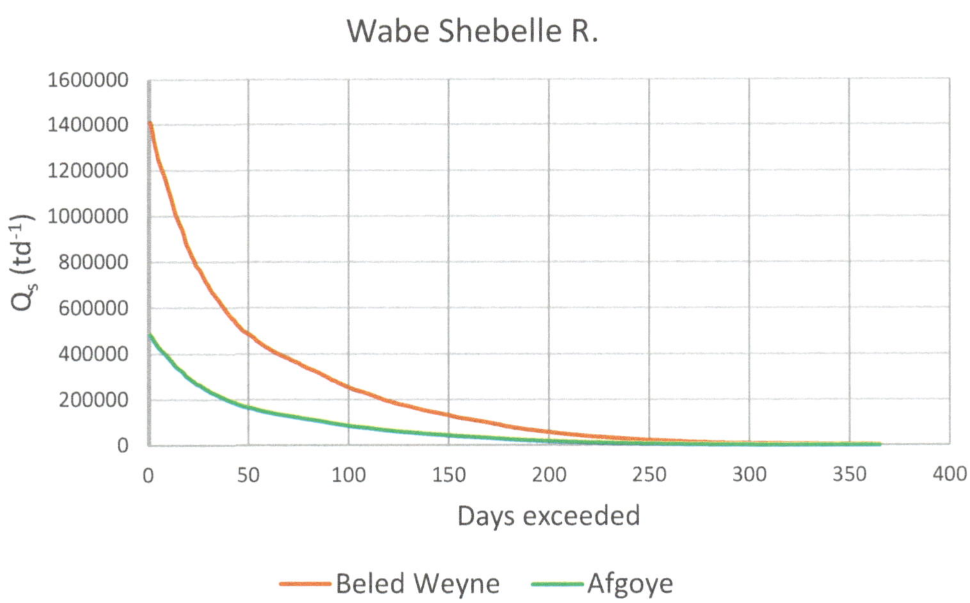

Fig. 13.27 Suspended load duration curve of the Shabelle at Beled Weyne and Afgoye

Fig. 13.28 Weir and irrigation scheme on the Shabelle (Google Earth coordinates: 2°53′14″ N 45° 32′37″ E)

Fig. 13.29 Manmade bank breaching on the Shabelle

Table 13.4 Land use/cover distribution (%) and change in the Bale mountains (Hailemariam et al. 2016)

Land use/cover	1985	1995	2005	2015	2015–1985
Afroalpine	0.9	0.9	0.9	0.8	−0.1
Grassland	21.5	21.0	20.7	19.2	−2.3
Woodland	43.7	43.5	41.6	40.3	−3.4
Shrubland	18.2	17.6	16.0	15.8	−2.4
Farmland	15.4	16.8	20.5	23.2	7.8
Urbanized	0.02	0.1	0.1	0.7	0.68
Total	100	100	100	100	

use/land cover change in the Bale mountains, which are the source of both the study rivers, through the 1985–2015 interval. Their results are summarized in Table 13.4 and clearly indicate that farmland is the only land use constantly increasing during the study interval. Woodland, grassland and shrubland altogether decreased with the same proportion of farmland increase. This is an additional human activity factor leading to an increase in sediment supply, and presumably, it is contributing to the streambed aggradation of the study rivers and to the reduction of their flow conveying capacity.

13.5 Discussion

13.5.1 Channel Changes

A primeval fluvial system of the Shabelle flowing along the same regional slope as today was probably active since long time ago as witnessed by some meandering flood basalt flows (Oligocene? to Pliocene—Abbate et al. 1994) that filled the valley bottom and that today offer typical examples of relief inversion (Sembroni and Molin 2018) (Fig. 13.30).

High sediment transport rates and/or a marked change in discharge and the creation of a transverse valley gradient are among the main causes of river avulsion (Aslan et al. 2005). All these factors are present in the modern Shabelle river. The many avulsion channels that are present in the flood plain downstream of Jalalaqsi are very evident in satellite images. Unfortunately, there is no information about their age, but the very good preservation of their geomorphological characteristics and of the channel dynamics processes traces and the fact that old avulsion channels are present also in the Balli plain as far south as Baraawe suggests a much.

Younger age (probably Holocene) than the meandering flood basalts of Fig. 13.30. The data of Table 13.2 and the bankfull discharges obtained from Eq. (13.1), which are based on the average meander wavelength (i.e., a parameter that can be measured with greater accuracy for both old and modern channels), demonstrate that the old avulsion channels were larger and were formed by a dominant discharge larger than that of the modern Shabelle. Moreover, the variety of well-preserved evidences of channel processes suggests that higher than today forming discharge conditions were maintained for a relatively long time.

Pollen analysis data (Jolly et al. 1998), the existence of lakes now disappeared in East Sahara, and lakes with higher water levels than experienced at present (Hoelzmann et al. 1998) in Eastern Sahel and in Eastern Ethiopia (Lake Abhe) (Mayewski et al. 2004) indicate that between approximately 9.000 and 6.000 years ago, rainfall was much higher than present in Eastern Sahara (100–200 mm $year^{-1}$—Haynes and Haas 1980; Kropelin 1987) and in Eastern sub-Saharan Africa, probably in response to a strong enhancement of the African monsoon (Street-Perrott et al. 1990). Two–three thousand years of higher precipitation and discharges (as much as ten times larger) are enough to provide the old avulsion channels with a larger size than the modern Shabelle and to propel their channel dynamics.

The occurrence of a transverse valley gradient is another important factor favoring channel avulsion (Aslan et al. 2005). By means of the "Elevation Profile" function of Google Earth Pro®, a west–east cross-section was traced a few km north of Mahaday Weyne (Fig. 13.31). Notwithstanding some inaccuracy of the profile due to excess exaggeration of the elevation scale, a general convex upward shape of the surface can be easily identified. The modern Shabelle is located in the most elevated point, and very old avulsion channels are located at a slightly lower elevation, but also the latter are on top of a secondary convex upward surface. This morphology and the lateral spreading of the old avulsion channels recall a delta or a fan delta. The stratigraphic data of the exploration well drilled near Jowhaar, presented by Crema (1923), are not very detailed since only the grain size and color of the different layers are reported. The boring is 107 m deep, and the stratigraphy shows three upward coarsening depositional cycles starting with reddish clays and followed by sand and fine gravel. In the uppermost cycle, the dark red clays include intercalations of gypsum, sand and fine gravel, and occasionally, the clay is mixed up with pebbles. The author affirms that the bore was entirely within alluvial deposits and no bedrock was encountered.

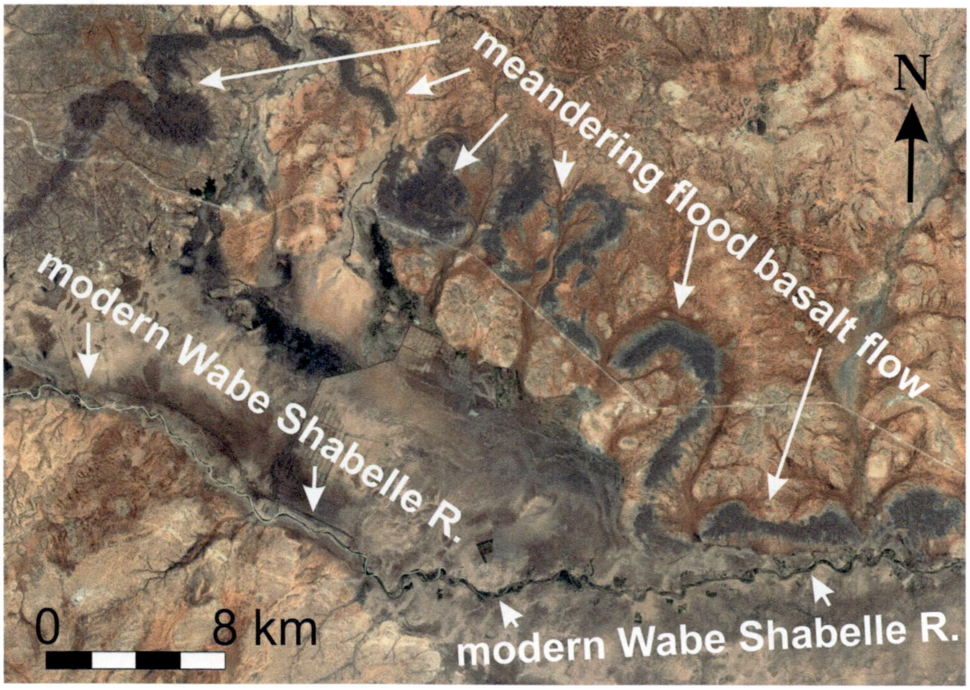

Fig. 13.30 Ancient channels of the Wabe Shabelle filled with basalts. Erosion resulted in a relief inversion with the meandering basalts higher than the surrounding terrains

This information, though poorly detailed, is very important because lets us to suppose the occurrence, downstream of Mahady Weyne, of Quaternary(?) shallowing upward cycles, commonly associated with a coarse grained delta or a fan delta (sensu Nichols 2009). The bore upper part is about 4 m below the present ground surface, whereas the lowermost level (reddish-greenish clays) reached by the boring is 4 m below the current sea level.

Old deltas are still evident in the Juba around Dujuma (Fig. 13.32) and in the Laag Dheera river about 85 km NW of its confluence with the Juba. Abbate et al. (1994) assign a Neogene to Pleistocene age to these deltaic deposits, whereas in the Shabelle, more modern (Pleistocene to present) sediments probably cover the older deltaic deposits. Figure 13.32, and the studies of Mège et al. (2015) and of Sembroni and Molin (2018) indicate that a fluvial system almost coinciding with the modern Shabelle existed (Figs. 13.30 and 13.33) when the old delta of the Juba (Fig. 13.32) was formed.

The occurrence of a narrow graben almost parallel to the modern coast and the continental slope (Abbate et al. 1994) explains why the older delta of the Shabelle is not visible and may also account for the deviation to the SW of the river. Dainelli (1943) assumes that the Shabelle was prevented from entering the Indian Ocean north of Mogadishu due to the presence of the coastal dunes, which forced the river to the SW. Today, the coastal dunes elevation is around 200 m asl, nevertheless, during wetter intervals of the past the combination of hydrometeoric erosion processes and higher flood discharges should have been able to breach the coastal dunes and to outflow into the Indian Ocean forming a delta. Bathymetric maps of the Somalia coast do not show any evidence of a submerged delta. Though south of Merca the 70 km long coastal dunes belt becomes very narrow (less than 4 km) and it is about 70 m higher than the inland flood plain, the Shabelle was never able to cut through it to reach the ocean. This suggests that the recent structural arrangement of the Somalia coastal belt with the formation of the longitudinal depression slightly inclined to the SW may have forced the Shabelle to follow the local slope preventing it to proceed straight into the Indian Ocean (Fig. 13.33). That is confirmed also by the ephemeral streams originating in the Bur region which are perpendicular to the depression axis in their upstream reaches and turn slightly to the south as they enter the depression.

13.5.2 Increased Risk of Flooding

Southern Somalia has seen an increase of flood frequency throughout the last two decades. As reported in the previous sections, it seems there is a decreasing rather than an increasing trend for precipitation in the Juba and Shabelle headwaters. In spite of that, the Shabelle shows an increasing trend in annual runoff for both the rainy seasons, whereas in the Juba, the main rainy season (*Gu*) shows an increasing trend and the small rainy season (*Deyr*) shows a moderate decreasing trend (Fig. 13.34). In the Shabelle annual, seasonal and monthly maximum runoff data indicate a marked increase in the last three decades, with the *Gu* season

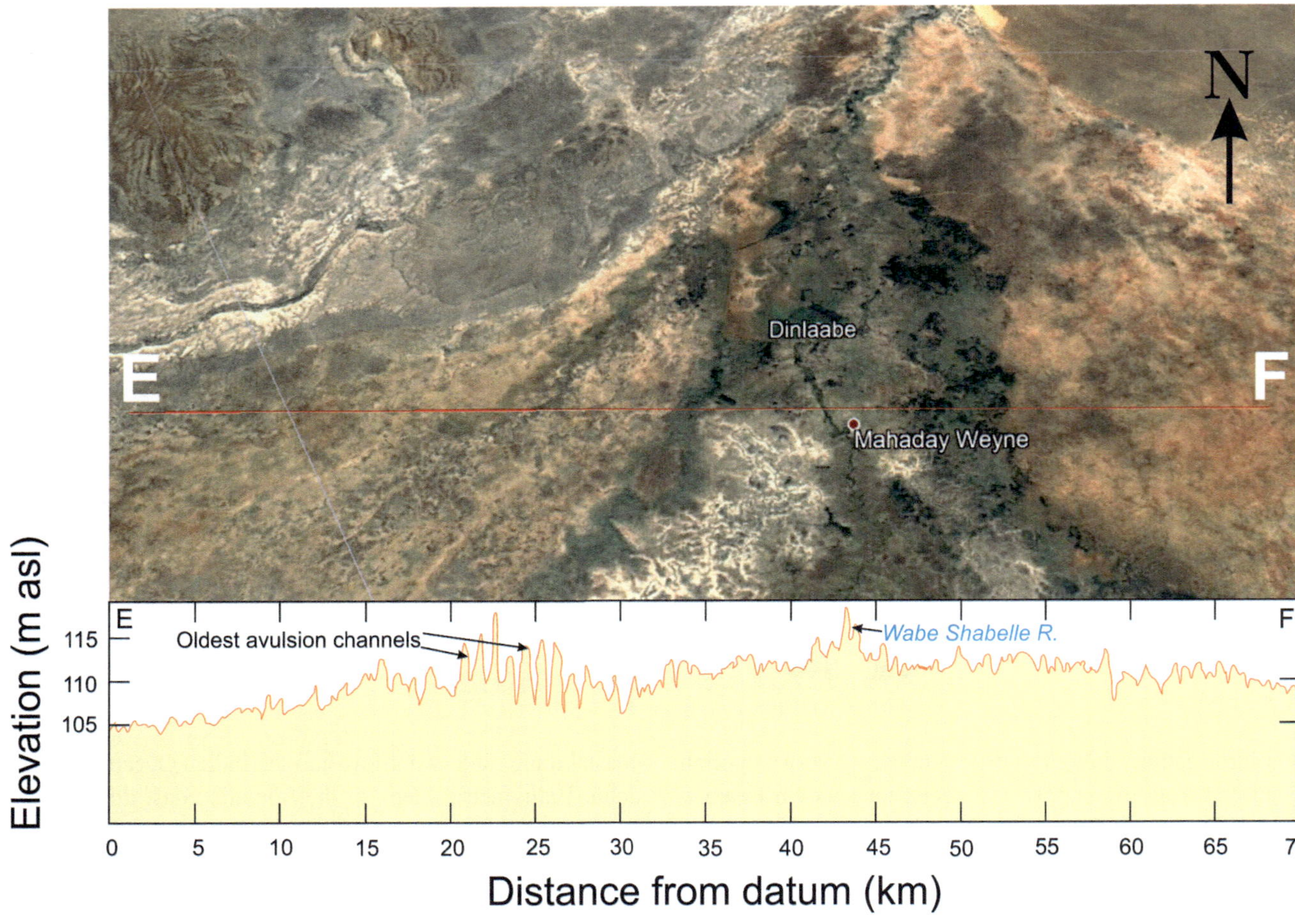

Fig. 13.31 Terrain profile traced across the Wabe Shabelle flood plain a few km north of Mahaday Weyne. Notice, the convex upward shape of the ground surface and the modern Shabelle on its highest point

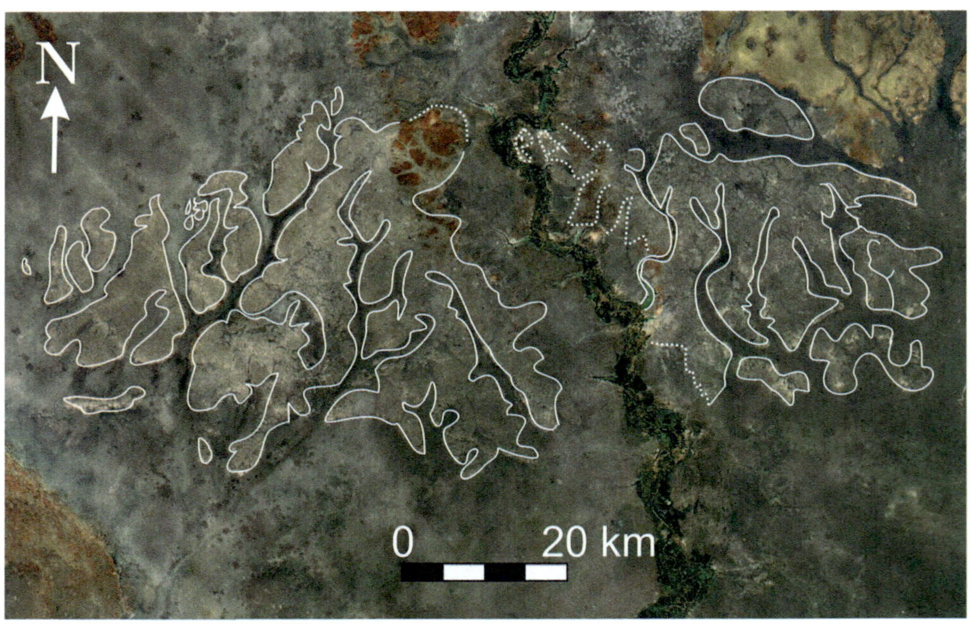

Fig. 13.32 Neogene to Pleistocene (Abbate et al. 1994) delta of the Juba river

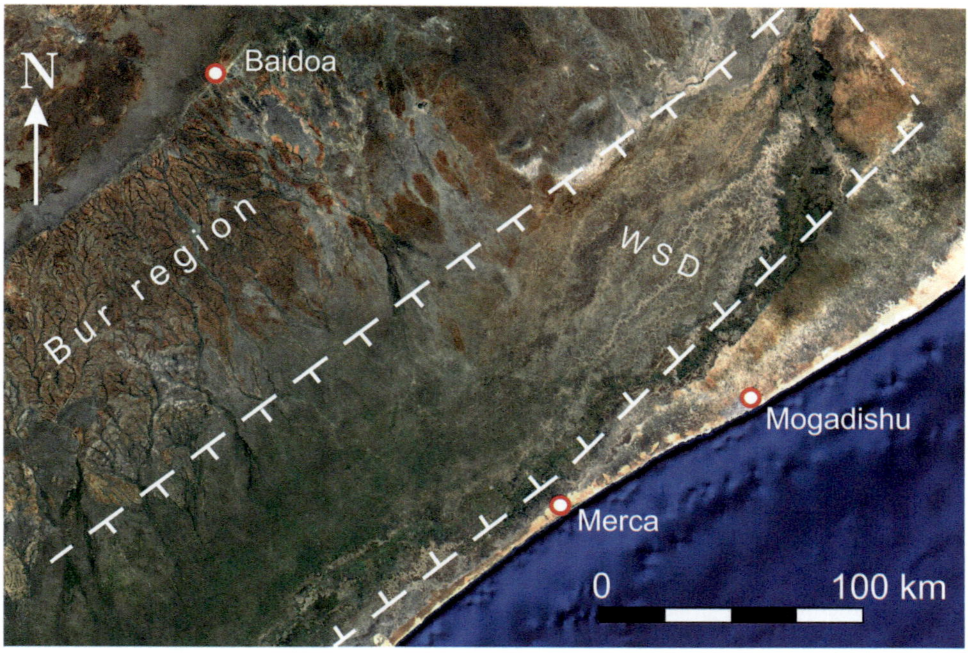

Fig. 13.33 Simplified structural setting of the lower Wabe Shabelle. WSD = Wabe Shabelle Delta

characterized by an increase of 47%. In the Juba, the increase of runoff in the *Gu* season is paired by the decrease in the *Deyr*. These data show a contrasting situation between the two rivers and clearly explain the increased frequency of flooding of the Shabelle, but not that of the Juba. Also, the annual maximum discharge shows a marked increase for the Shabelle at Beled Weyne and a moderate increase for the Juba at Luuq (Fig. 13.35).

Figure 13.24 depicts how torrential rains became more common throughout the last two decades in the lower reaches of both rivers. Under such circumstances, flooding may occur also without a substantial contribution from the monsoon rains on the Ethiopian highlands. Moreover, the perched position of both rivers on the flood plain extended the flooding duration before the inundation waters take their way back into the rivers.

The change in land use/land cover, characterized by a substantial increase in farmland, in the headwaters of both rivers likely resulted in an increase of the sediment supply which, given the downstream decrease in discharge by water loss, resulted in higher sedimentation rates and streambed aggradation. Omuto et al. (2009) report a mean suspended sediment concentration of 11.8 and 0.93 gl^{-1} measured for the Juba at Bardere and Buale, respectively, during a 2007–2008 field campaign. For a location no longer identifiable (Kaitoi, according to the Italian spelling) in between these two localities, Pozzi (1982) reported a maximum suspended sediment concentration of 11.3 gl^{-1} on the base of some FAO data (source not cited in the reference list). These data, though limited in number and in time coverage, suggest that the suspended sediment concentration of the Juba has increased, at least in the last four decades. The same author reports also a few bankfull width data measured in 1960 at Beled Weyne (65 m), Mahaday Weyne (38 m) and Afgooye (47 m) on the Shabelle. Unfortunately, no information is provided about the exact positions of the cross-sections. Nevertheless, repeating some bankfull width measurement from Google Earth on a long reach of a few kilometers upstream and downstream these locations, the average bankfull width of today is substantially smaller: 43 m at Beled Weyne, 28 m at Mahaday Weyne and 22 m at Afgooye. The bankfull widths measured on today's river have a variation coefficient ranging between 12 and 23%, and none of the values is larger than those of 1960. These results confirm that the Shabelle decreased in width in its lower reaches, i.e., those more often affected by devastating floods. This conclusion seems to be confirmed also by the comparison of a postcard pictures of the Shabelle at Jowhaar and Afgoye taken in the early 1920s with pictures of today (Fig. 13.36).

The demand for more water for irrigation, that can be fulfilled only by the rehabilitation of abandoned irrigation schemes and the construction of new ones (Yibeltal 2015), may reduce drastically the practice of uncontrolled river bank breaching by farmers. However, only after an accurate study on the effects of water diversion on sedimentation and streambed aggradation and an accurate plan for water management, the worsening of the situation in terms of the flood hazard can be kept under control and possibly reduced.

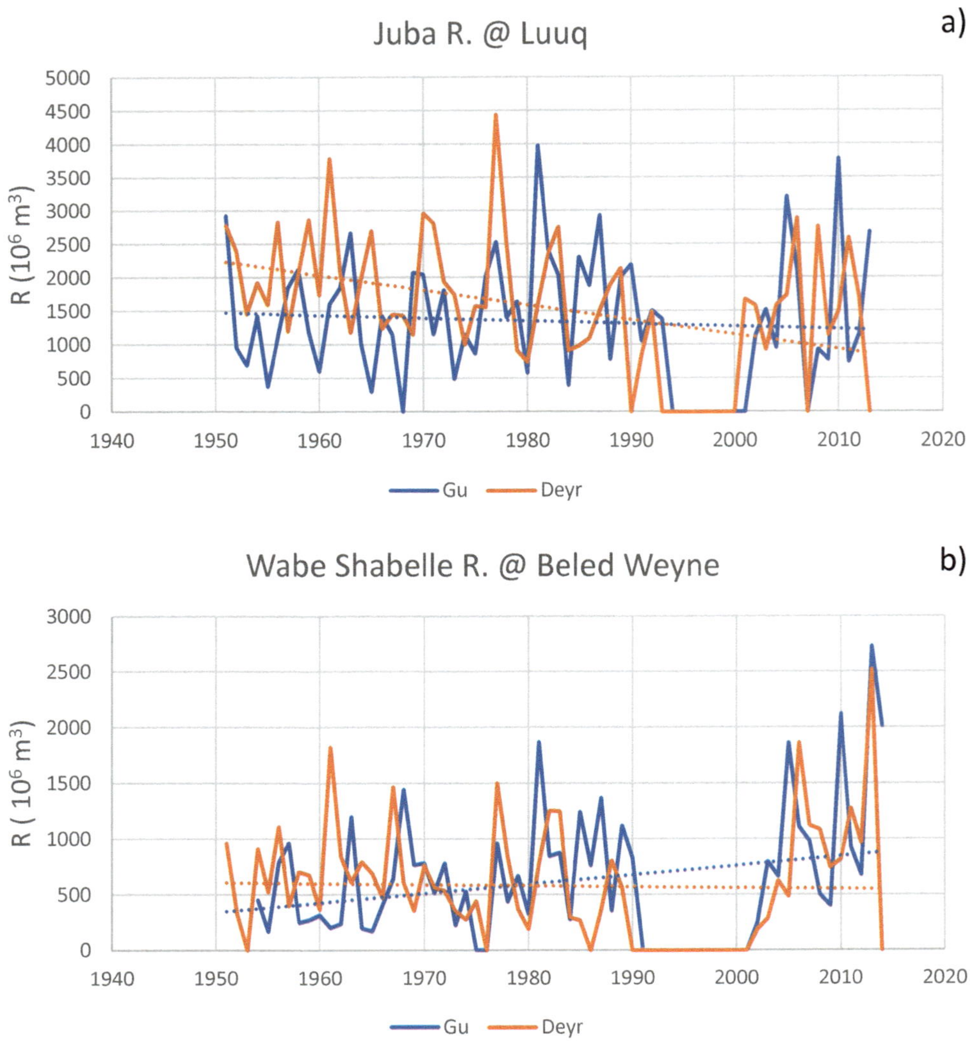

Fig. 13.34 Runoff time series for the Juba @ Luuq (**a**) and the Shabelle @ Beled Weyne (**b**)

13.6 Conclusions

The Juba and Wabe Shabelle rivers are the only permanent rivers of Somalia and the largest of the Horn of Africa. Both rivers are characterized by active channel dynamics that is witnessed also by old, avulsion channels still clearly visible in the lower alluvial plains. Shabelle geomorphic and hydrologic parameters, such as channel width, meander wavelength, sinuosity and bankfull discharge demonstrate that the old avulsion channels were larger and were formed by a dominant discharge larger than today. In the old channels of both Shabelle and Jubabankfull, discharge was maintained for a time long enough to develop the typical geomorphological units of high sinuosity meandering rivers. Between approximately 9.000 and 6.000 years ago, rainfall was more abundant (100–200 mm year^{-1}) than today in Eastern Sahara and in Eastern sub-Saharan Africa. Two–three thousand years of higher precipitation may have resulted in larger discharges (as much as ten times larger in the Wabe Shabelle) capable to form much larger rivers than present and to propel their channel dynamics. The old (Quaternary) rivers formed also large deltas or fan deltas that are still evident from satellite images though in the Shabelle, modern sediment is partly hiding the older deltaic deposits. The recent structural arrangement of Southern Somalia coastal belt with the formation of the longitudinal depression slightly inclined to the SW may have forced the Shabelle to follow the local slope and prevented it from proceeding straight into the Indian Ocean as supposed by a few authors (e.g., Mohr 1962; Sembroni and Molin 2018). This hypothesis is also supported by the lack of any evidence of submerged deltas in bathymetric maps of the Somalia coast.

Both the Juba and Wabe Shabelle are a very important resource for agriculture, but many of the irrigation infrastructures are degraded after several years of civil war. In the last decade, such difficult situation has been exacerbated by a marked increase in the frequency of devastating floods whose driving factors are to be found in climate worsening and several human impacts including an increase of farming

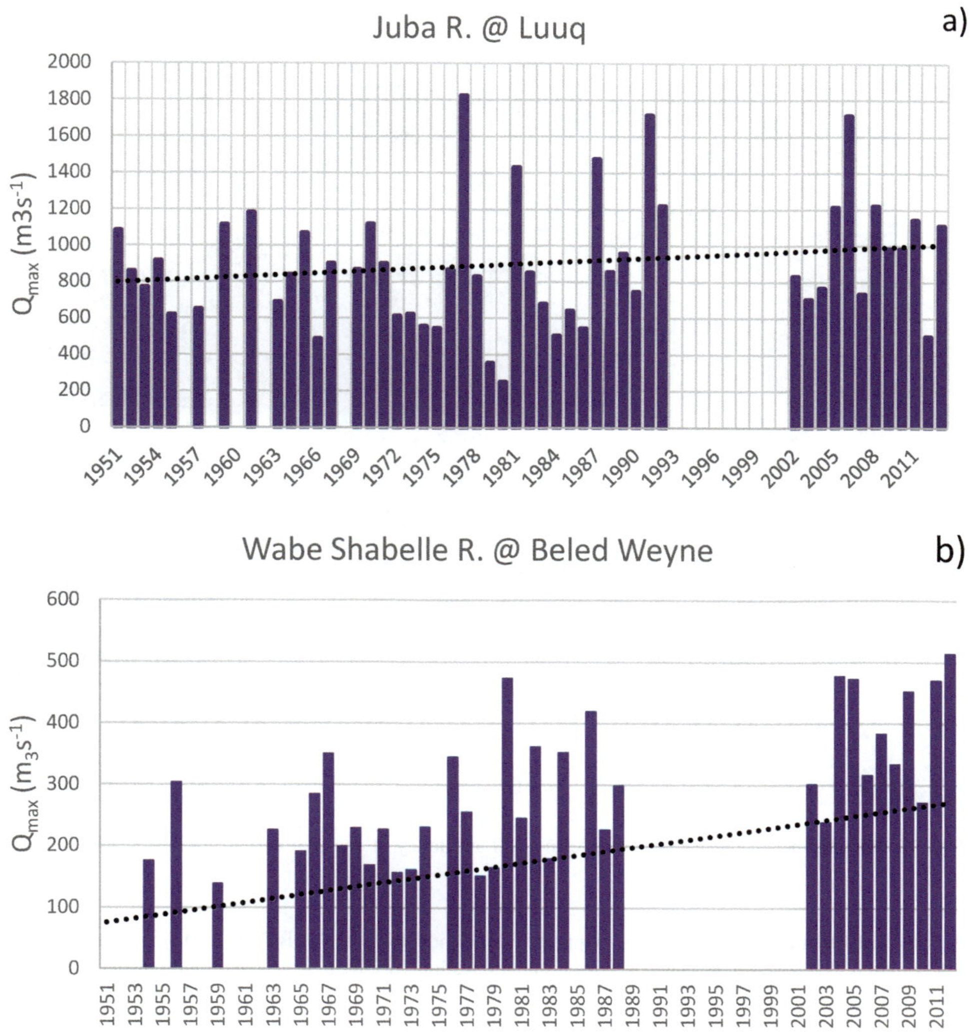

Fig. 13.35 Annual maximum discharge time series for the Juba @ Luuq (**a**) and the Shabelle @ Beled Weyne (**b**)

in the rivers headwaters, manmade bank breaches for irrigation that, associated with high rates of sediment transport and deposition, have negatively affected the conveyance capacity of the rivers. In the Shabelle between Beled Weyne and Afgoye, 88 bank breach/overtop points were found, of which 15 are clearly unofficial manmade, whereas the other 73 consist of natural bank breaches/overtop.

The change of the land use/cover in the headwaters of both rivers, characterized by a substantial increase in farmland, resulted in an increase of sediment supply, higher sedimentation rates and streambed aggradation. An average aggradation rate of about 0.05 and 0.002 m year^{-1} for the Shabelle at Beled Weyne and Afgoye, respectively, was calculated. Moreover, from the comparison of some channel width measurements of the Shabelle made by Pozzi (1982) in 1960 with modern river data, a channel narrowing of about 40% is evident.

All these factors, combined with an increased frequency of heavy rains during the small rainy season (especially in the lower Juba and Wabe Shabelle) resulted in several, devastating floods causing fatalities and affecting several thousands of people. In the Shabelle, in fact, annual, seasonal and monthly maximum runoff data indicate a marked increase in the last three decades with the *Gu* season characterized by an increase of 47%. In the Juba, the increase of runoff in the *Gu* season is paired by the decrease in the *Deyr*. These data show a contrasting situation between the two rivers and clearly explain the increased frequency of flooding of the Shabelle, but not that of the Juba. Yet, the annual maximum discharge time series shows a marked increase for the Shabelle at Beled Weyne and a moderate increase for the Juba at Luuq. Under such circumstances of channel narrowing, aggradation and bank breaches, flooding may occur also without a substantial contribution from the monsoon rains on the Ethiopian highlands.

Sediment has often been neglected in flood risk analyses, but changes in sediment dynamics and supply may result in channel instability (mainly aggradation, degradation and bank breaches) that, combined with substantial change in the rainfall regime, may have negative effects on people and properties.

Fig. 13.36 Channel of the Wabe Shabelle in 1920 and today at Jowhaar and Afgoye. Both the old reaches look wider than today

Acknowledgements This study was supported by funds from the International Platform for Dryland Research and Education (IPDRE), Tottori University.

References

Abbate E, Sagri M, Sassi FP (eds) (1994) Geological map of Somalia. UNDP/United Nations Department of Technical Cooperation of Development (UNDTCD)

Ackers P, Charlton FG (1970) Dimensional analysis of alluvial channels with special reference to meander length. J Hydraul Res 8:287–316

Andrews ED (1980) Effective and bankfull discharges of streams in the Yampa River Basin, Colorado and Wyoming. J Hydrol 46:311–330

Aslan A, Autin WJ, Blum Md (2005) Causes of river avulsion: insights from the Late Holocene avulsion history of the Mississippi River, Y.S.A. J Sediment Res 75:650–664

Babinski Z (2005) The relationship between suspended and bed load transport in river channels. In: Walling DE, Horowitz AJ (eds) Sediment budgets 1, vol 291, pp 182–188. IAHS Publication

Balint Z, Njeru L, Pappas C, Paron P (2010) Atlas of the Juba and Shabelle Rivers in Somalia. FAO, Rome

Basnyat DB (2007) Water resources of Somalia. Technical report no W-11, FAO-SWALIM, Nairobi

Basnyat DB, Gadain HM (2009) Hydraulic behaviour of the Juba and Shabelle Rivers: basic analysis for irrigation and flood management purposes. Technical report no W-13, FAO-SWALIM, Nairobi

Billi P (2015) Geomorphological landscapes of Ethiopia. In: Billi P (ed) Landscapes and landforms of Ethiopia, world geomorphological landscapes. Springer, Dordrecht, pp 3–32

Bottego V (1895) Viaggi di scoperta nel cuore dell'Africa: il Giuba esplorato. Società Geografica Italiana, Loescher & co, Torino

Brierley GJ, Fryirs KA (2005) Geomorphology and river management applications of the river styles framework. Blackwell Publishing, Oxford

Cantalice JRB, Filho MC, Stosic BD, Piscoya VC, Guerra SMS, Singh VP (2013) Relationship between bedload and suspended sediment in the sand-bed Exu River, in the semi-arid region of Brazil. Hydrol Sci J 58(8):1789–1802

CRED (2019) International Disaster Database. https://www.emdat.be/emdat_db/

Crema C (1923) Osservazioni sulla geologia del medio Scebeli (Somalia Italian) in base a materiali raccolti da S.A.R. il Principe Luigi di Savoia, Duca degli Abruzzi (1919–1920). Geologia. Rendiconti Della Reale Accademia Dei Lincei 32:180–184

Dainelli G (1943) Geologia dell'Africa orientale, Reale Accademia d'Italia, Roma

Diakakis M (2012) Rainfall thresholds for flood triggering. The case of Marathonas in Greece. Nat Hazards 60:789–800

Dury GH (1976) Discharge prediction, present and former, from channel dimensions. J Hydrol 30:219–245

Fazzini M, Bisci C, Billi P (2015) The climate of Ethiopia. In: Billi P (ed) Landscapes and landforms of Ethiopia, world geomorphological landscapes. Springer, Dordrecht, pp 65–87

FSNAU-FAO (2018) Somaliarainfall performance, 2001–2019: deviation from long term mean (LTM). http://www.fsnau.org/

Hoelzmann P, Jolly D, Harrison SP, Laarif F, Bonnefille R, Pachur HJ (1998) Mid-Holocene land-surface conditions in northern Africa and Arabian peninsula: a data set for AGCM sensitivity experiments. Glob Biogeochem Cycles 12:35–51
Italian Government (1925) Monografie delle Regioni della Somalia. Regio Governo della Somalia Italiana, De Agostini, Torino
Hailemariam AN, Soromessa T, Teketay D (2016) Land use and land cover change in the Bale Mountain eco-region of Ethiopia during 1985 to 2015. Land 5(4):41
Haynes CV, Haas H (1980) Radiocarbon evidence for Holocene recharge of groundwater, Western Desert, Egypt. Radiocarbon 22:705–717
Jolly D, Prentice IC, Bonnefille R, Ballouche A, Bengo M, Brenac P, Buchet G, Burney D, Cazet JP, Cheddadi R, Edorh T, Elenga H, Elmoutaki S, Guiot, Laarif F, Lamb H, Lezine AM, Maley J, Mbenza M, Peyron O, Reille M, Reynaud-Farrera I, Riollet G, Ritchie JC, Roche E, Scott L, Ssemmanda I, Straka H, Umer M, Van Campo E (1998) Biome reconstruction from pollen and plant macrofossil data for Africa and the Arabian peninsula at 0 and 6000 years. J Biogeograp 25:1007–1027
Joshi S, Xu YJ (2017) Bedload and suspended load transport in the 140-km reach downstream of the Mississippi River avulsion to the Atchafalaya River. Water 9:716
Knighton D (1998) Fluvial forms and processes: a new perspective. Arnold, London
Kropelin S (1987) Palaeoecological evidence from early to mid-Holocene playas in the Gilf Kebir (SW Egypt). Palaeoecol Afr 18:189–208
Leopold LB, Wolman MG, Miller JP (1964) Fluvial processes in geomorphology. Freeman, San Francisco
Mayewski PA. Rohling EE, Stager JC, Karlen W, Maasch KA, Meeker LD, Meyerson EA, Gasse F, van Kreveld S, Holmgren K, Lee-Thorp J, Rosqvist G, Rack F, Staubwasser M, Schneider RR. Steig EJ (2004) Holocene climate variability. Quat Res 62:243–255
Mbara CJ, Gadain HM, Muthusi FM (2007) Status of medium to large irrigation schemes in Southern Somalia, technical report no W-05, FAO-SWALIM, Nairobi
Mège D, Purcell P, Pochat S, Guidat T (2015) The landscape and landform of the Ogaden, southeast Ethiopia. In: Billi P (ed) Landscapes and landforms of Ethiopia. Springer, Dordrecht, pp 323–348
Mohr PA (1962) The Geology of Ethiopia. Addis Ababa Univ. Press, Addis Ababa
Muchiri PW (2007) Climate of Somalia. Technical report no W-01, FAO-SWALIM, Nairobi
Muthusi FM, Gadain HM (2009) Somalia flood forecasting system. Technical Report No W-16, FAO-SWALIM, Nairobi, Kenya
Nichols G (2009) Sedimentology and stratigraphy. Wiley, Chichester
Omuto CT, Vargas RR, Paron P (2009) Soil erosion and sedimentation modelling and monitoring framework of the areas between rivers Juba and Shabelle in southern Somalia. FA0-SWALIM technical report no. L-16, Nairobi, Kenya
Pozzi R (1982) Lineamenti della idrogeologia della Somalia. Quaderni Di Geologia Della Somalia 6:281–321
Schumm SA (1977) The fluvial system. Wiley, New York
Sembroni A, Molin P (2018) Long-term drainage system evolution in the Wabe Shebele River basin (SE Ethiopia—SW Somalia). Geomorphology 320:45–63
Street-Perrott F, Mitchell J, Marchand D, Brunner J (1990) Milankovitch and albedo forcing of the tropical monsoons: A comparison of geological evidence and numerical simulations for 9000 yBP. Trans Roy Soc Edinb Earth Sci 81(4):407–427
SWALIM (2016) Disaster management plan for the Juba and Shabelle basins in Somalia. FAO-SWALIM project report No W-27, Nairobi
Torizzo M, Pitlick J (2004) Magnitude-frequency of bed load transport in mountain streams in Colorado. J Hydrol 290:137–151
UNEP (2010) Africa water Atlas. Division of early warning and assessment (DEWA). United Nations Environment Programme (UNEP), Nairobi
Vicinanza G (1910) La Somalia Italiana. De Rosa & Polidori, Napoli
Williams GP (1978) Bankfull discharge of rivers. Water Resour Res 14:1141–1158
Wolman MG, Miller JP (1960) Magnitude and frequency of forces in geomorphic processes. J Geol 68:114–142
Wolman MG, Leopold LB (1957) River flood plains: some observations of their formation. US Geological Survey, Professional Paper 282-C, Washington
World Bank-FAO (2018) Rebuilding resilient and sustainable agriculture in Somalia, vol 1. International Bank for Reconstruction and Development/The World Bank and the Food and Agriculture Organization of the United Nations
Yibeltal M (2015) Assessments of water demands for the Juba and Shabelle Rivers in Somalia. J Agric Environ Int Dev 109(2):165–177

Recent Landscape Change in Somalia Monitored Through the Use of Repeat Photography

14

Robert Neil Munro and F. Jane Madgwick

Abstract

Using archival vertical and oblique aerial photography (1945–68) and thematic maps, we present observations on landscape change in Somalia, on the Juba and Shabelle River Inter-Riverine, along the coastal sand belt, and among the vegetation arcs and landscapes of the north. Our work does not cover all of Somalia but presents case studies, based on our work in agricultural development, food security and conservation projects that, in the Inter-Riverine area, the authors worked on in the 1980s. We also examine changes in Somaliland and Puntland, repeating Trimetrogon photography of vegetation arcs and desert scenes and photography from a USAID project. The online grey literature archives of WOSSAC at Cranfield University and FAO-SWALIM have provided legacy materials, and Google Earth has been central in making repeats, where one cannot go on the ground today. Google Earth can be adopted as a useful and free tool to make meaningful repeat shots of older oblique aerial photography as a monitoring tool in change detection. Conclusions are drawn on each study, and proposals are made for future investigations to follow up on the assessments and make a more systematic use of the archival aerial photography.

Keywords

Repeat photography • Trimetrogon aerial photography • Juba Shabelle Inter-Riverine • Forest and land use change • Sand dune stabilisation

14.1 Introduction

This chapter has adopted repeat photography methods (Webb et al. 2010) to examine change in land cover, land use and the physical landscape in parts of Somalia (taken here to include the countries now designated as Somalia, Somaliland and Puntland). Repeat photography/repeat imagery methods aim to return to the same position and assess change on the ground from that location. We have made this assessment at numerous locations in Somalia by utilising archive materials. A poster, presented at the 2011 AIG Conference in Addis Ababa, highlighted one of the main archives, a collection of Somalia aerial photography (AP) mosaics made by Hunting Surveys Ltd (HSL) and its possible uses (Paron and Munro 2011). In this chapter, we provide examples of uses.

Munro (RNM) first used repeat photography of landscapes in Jordan (for 1953–1990 period). In Darfur, Sudan, land cover changes have been compared for the 1994–2020 period (work in preparation). Later he used georeferenced older AP and mosaics to assess sand movement in Sudan (Munro et al. 2018; Munro and El Tom 2009) and Arabia (HTS 1999; Munro and Wilkinson 2007; Munro et al. 2012, 2013). In Tigray, Ethiopia, he has assessed changes to land cover over 45 years (Nyssen et al. 2007, 2008, Nyssen et al. 2009a, b, 2010a, b, 2014; Munro et al. 2008 and 2019; De Mûelenaere et al. 2014).

The present work makes a preliminary review of selected images from development studies and air surveys made in Somalia between 1945 and 1985 to determine what was

R. N. Munro (✉)
Institute of Climate and Society, Mekelle University, Tigray, Ethiopia

Ethiopian Geospatial Information Institute, Addis Ababa, Ethiopia
e-mail: neilmunrodirleton@gmail.com

F. J. Madgwick
Wetlands International, Wageningen, Netherlands

P. Billi (ed.), *Landscapes and Landforms of the Horn of Africa*, World Geomorphological Landscapes,
https://doi.org/10.1007/978-3-031-05487-7_14

assessed, what was planned in the studies we worked on and how has land use changed over the intervening period to 2020 as far as this can be followed from the published and grey literature and also an important source to gauge change, the remarkable twenty-first-century armchair geographer's tool, Google Earth ©. We have been able to substitute Google Earth for the aerial camera to repeat oblique and vertical images. The Google Earth repeats are not perfect of course—higher-quality imagery acquired from a research budget and taken at the same repeat times would be necessary—but we regard them as a good introduction.

This chapter does not cover all of Somalia but presents case studies, most drawn from data of development projects and studies the authors worked on in the 1980s, and include reports, maps and our own terrestrial and aerial photography. The examples in this paper are drawn from archived aerial photography (AP) and mosaics and selected from 210 terrestrial and oblique aerial photographs made by the authors (Fig. 14.1).

14.2 Data and Methods

14.2.1 Sources of Archival Aerial Photography and Mosaics

Archival aerial photography (AP) remains the best material to examine past landscapes over a long interval; the high resolution of AP can identify even a tussock grass, and under the stereoscope and using other photogrammetric tools, the structure of vegetation can be quantified. In 1972, on soil surveys in the Awash Valley of Ethiopia RNM used colour AP that showed shrubs <50 cm diameter, and while ERTS (Landsat) MSS false colour imagery acquired from NASA was tremendously exciting to use in the field, it had low resolution (Parry 1974); satellite imagery of any use in soil surveys only became very roughly comparable to AP with the introduction of SPOT imagery in the 1986, but SPOT still had only a 10 m resolution. Today high-resolution imagery is almost taken for granted, but AP remains the only

Fig. 14.1 Location map

available material for detailed examination of landscapes before the 1980s. Fortunately, AP archives exist for Somalia.

We believe, though unverified, that the earliest aerial photography of the Inter-Riverine area—the lands within the Juba and Shabelle basins in Somalia—was flown by the Italian Air Force for Istituto Geografico Militare (IGM) during the 1930s, and is in an, so far undocumented, IGM archive in Florence, not included on the IGM's (now IGMI) online digital repository. The British Royal Air Force (RAF) flew parts of Somalia in the 1940s for wartime use, but most of the East African campaign materials are thought to have been destroyed at the end of WW II. During 1944–45, the United States Army Air Force (USAAF) flew 1:40,000 scale Trimetrogon (TMG) AP over Somalia using photo-reconnaissance B-24 aircraft, fitted with three K-17 cameras (one vertical and two oblique). These flew at a height of 20,000 ft (6096 m) on east–west aligned paths, approximately 40 km apart (Livingstone 1964). Hunt (1952) noted that a nearly complete set of 1945 Trimetrogon photographs was handed over to Director of Agriculture and Veterinary Services (DAVS), Hargeisa, in 1951. It is unlikely to be there, but Trimetrogon photography is available through the National Archives of the USA (information from the National Collection of Aerial Photography (NCAP), Edinburgh, UK, in 2019). From 1946 to 55, the RAF flew blocks of Somaliland that were used in the compilation of the Directorate of Colonial Surveys DCS39 series of topographical sheets at an approximate scale of 1:125,000.

For this study, we have used a few Trimetrogon images from 1945, terrestrial and aerial photography taken between 1958 and 1984, and compared these to recent satellite imagery. Our principal source has been the uncontrolled air photo mosaics constructed by Hunting Surveys Ltd. (HSL) in the 1960s and early 1970s from 1958 Royal Air Force AP. Very considerable efforts and skills went into making the semi-controlled mosaics by HSL staff: the central parts of aerial photos were used, photos were laid down over existing topographical maps, and coordinates of observable points on topographical maps were identified on AP to give as accurate a location as possible. Though they were not constructed as orthophoto maps, the workmanship makes them satisfactorily accurate for modern change detection studies.

The HSL Somalia mosaics used 1958 Royal Air Force AP, and these mosaics were constructed periodically at intervals for prospecting oil companies from 1960 to mid-1970s. The RAF cover provides a quality timeline view of the country at 1958: a period before satellite imagery became available and before imagery could start to match the resolution of AP. In the early 1970s, Landsat (originally called the Earth Resources Technology Satellite—ERTS) imagery became available that could be rectified, but resolution was small: comparing Google imagery and a 1:500,000 scale colour composite ERTS image of the upper Shabelle Valley in Somalia acquired on 26 January 1973 and made by HSL, it is difficult to see the vegetation arcs that cover this region (see Sect. 14.9). Later, as satellite imagery was of higher resolution, and became digitally multispectral, it was cheaper per km^2 to acquire than AP. But importantly, the accuracy of the old mosaics could be checked. As resolution increased, AP on a grand scale was phased out, but remains as a superlative tool to compare with modern digital and multispectral formats. It is also essential for stereo interpretation. These AP mosaics cover almost all of Somalia as shown in Figs. 14.2 and 14.3.

Table 14.1 is a provisional list of Somalia AP and mosaics and provides details, as best as can be produced, of the former HSL AP negative mosaic archive and other collections. The HSL Somalia mosaic archive (>450 large negatives) was saved from destruction by RNM in 1986 and, after many years of private storage, now in the care of NCAP, to whom inquiries for further use should be addressed (http://ncap.org.uk/). This table is incomplete as other materials remain unseen: the IGM archive believed to be extant in Florence and Royal Air Force cover from WW2. The RAF flew much of Somalia from 1947 to 1958, and their photography was used several times by HSL to make mosaics. HSL had copies of the RAF films to make prints. Other AP used in mosaics was flown 1959–60 for the Standard Vacuum Oil Co and Mineraria Somalia. These surveying companies used a wide range of AP cameras including RC 5a, RC 8, RC 9 and Eagle IX. HSL also flew some 1:30,000 colour AP, in January and February 1970 of the Bur Region for Sinclair [HSL-SOM-COL-70:2–10], but prints and films were destroyed around 1985. In 1983–84, new photography was flown by Air Survey and Development (ASD) Gmbh (Frankfurt) at 1:30,000 scale, and 1:100,000 scale mosaics, for the National Tsetse and Trypanosomiasis Control Project (MMP 1983) in the Inter-Riverine area, were delivered to Mogadishu in 1985. However, inquiries at this time have not determined where this 1983–84 AP archive might be. From the 1960s up to 1985, a large AP archive existed at the Survey and Mapping Services of Somalia in Mogadishu. Its current status is not known, but it is likely it did not survive the civil war; the status of similar archive of old AP in Hargeisa is unknown.

14.2.2 Baseline Sources on Natural Resources Assessment

From 1962 to 65, Lockwood Survey Corporation made for FAO the *Agriculture and Water Survey of the Somalia Republic* (FAO/SF:42 SOM; Lockwood/FAO, 1969). The FAO soil studies used the HSL 1:60,000 scale mosaics, and the study produced useful data with a colour soil map at 1:500,000 scale (Fig. 14.4). Slightly later, Hunting

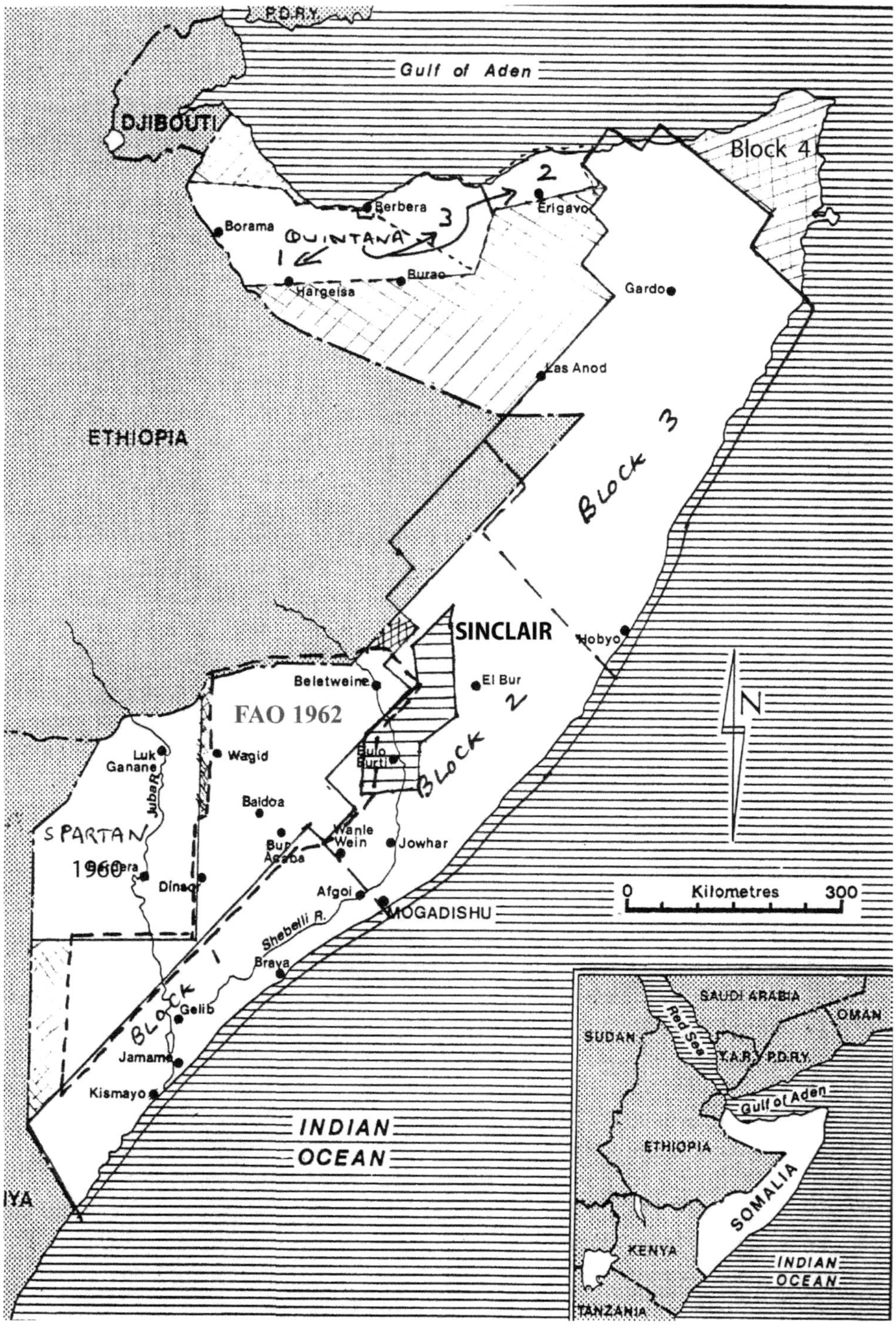

Fig. 14.2 Aerial photo mosaic cover of Somalia (Hunting). Refer to Table 14.1. Details made by Munro at Hunting Survey Records 1986 Original drawings donated to NCAP, Edinburgh 2020

Technical Services (HTS) and Sir Murdoch MacDonald and Partners (MMP) conducted the *Project for the Water Control and Management of the Shabelle River* (HTS-MMP 1969). These were both key baseline efforts that remain very useful. Comparative studies on the Juba River had been made by Selchozpromexport in 1965 and were updated by Technital (1975). On the irrigation and rainfed projects planning studies described in this chapter, we utilised the maps and the soil data from a series of feasibility studies made from 1977 onwards by HTS and MMP in the Inter-Riverine area (HTS 1977a, 1977b, 1978a. 1978b, 1979, 1982, 1984; MMP, 1978a, 1978b, 1980, 1981, 1983, 1985). In the north, Somaliland and Puntland, RNM refers to a 1960s USAID-funded agricultural development project (Wixom 1963, 64; Erdmann 1993; McCarthy et al. 1985); to Trimetrogon oblique AP from 1945 and 1958 Royal Air

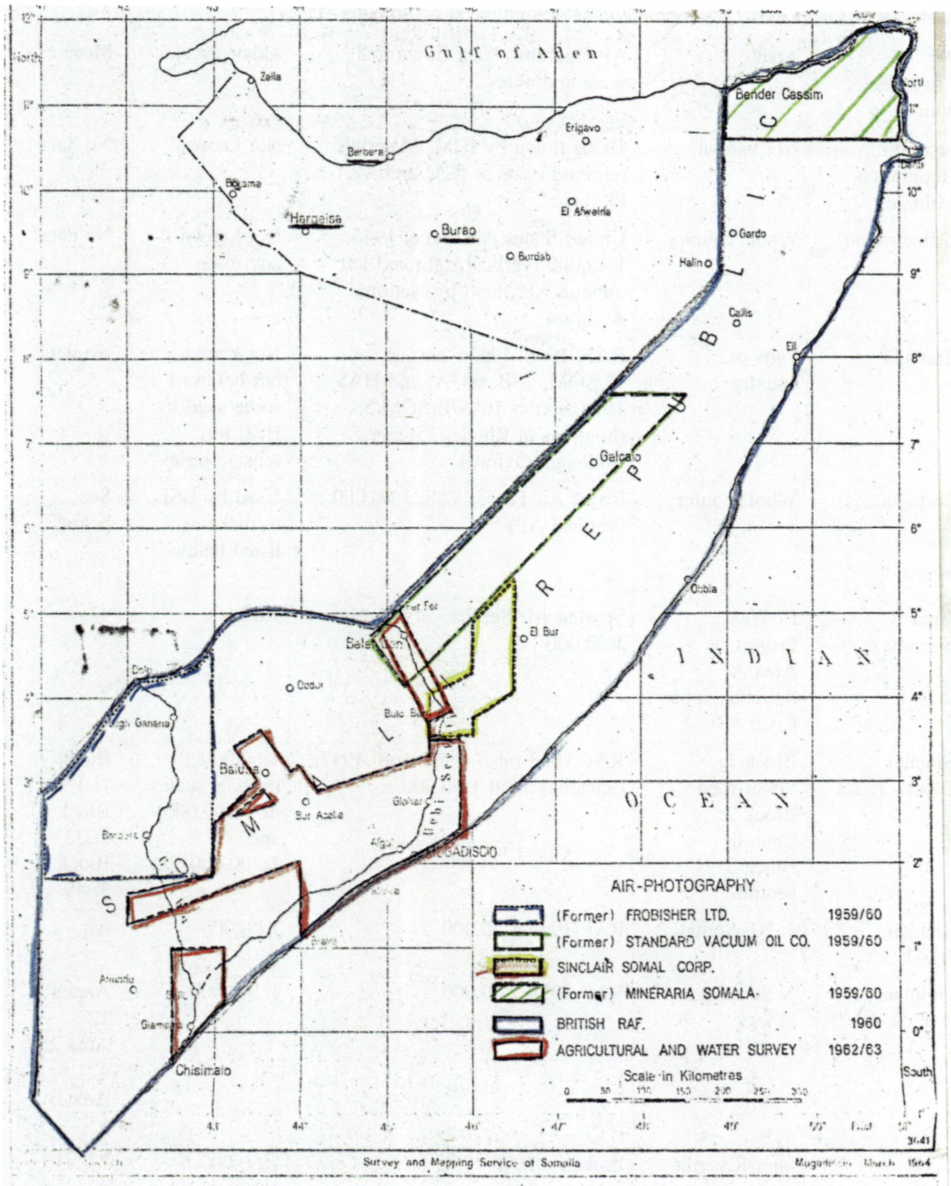

Fig. 14.3 Aerial photo mosaic cover of Somalia. Refer to Table 14.1. A 1964 sketch map from the *Survey and Mapping Services, Mogadishu.* It was salvaged from Survey Records at Hunting Surveys Ltd in 1986. Original drawings donated to NCAP, Edinburgh 2020

Force AP made into mosaics by HSL. For the 1975–76 Somalia Livestock Sector Review, a hydrogeology map at a scale of 1:2 million was made of the entire country (HTS/Gunn Rural Management 1978) and most usefully shows all known springs, wells and boreholes at that time.

Our assessment has been aided by documents held in WOSSAC archive in the UK (www.wossac.com) and the SWALIM archive in Nairobi (www.faoswalim.org). Other useful sources can be found in the Earth Resources bibliographies compiled by Kalb (2000) and Hadden (2007) and the coastal and marine bibliographies of Morcos and Varley (1990) and PERGSA/GEF (2002).

The spelling of place names in Somalia often uses English, Italian, older and modern Somali names and Russian derivations on their maps at 1:100,000 and 1:200,000. Apart from spelling of Mogadishu, we use the convention that names are as placed in the reports at the time of our studies: Hadden (2007) agreed and provided pertinent guidance for the wide range of place name spellings in Somalia.

14.3 Agricultural Development in the Inter-Riverine and Mudug

14.3.1 Cycles of Drought and Flood

Drought and flood conditions have long been regular features of life in southern Somalia. In 1941–42, as Allied forces pushed into Somalia from Kenya and dislodged the

Table 14.1 Older aerial photography and mosaic cover of Somalia

AP Cover/block name	Area	AP Operator. Year flown and scale and notes	Mosaic scale	Mosaics	Hunting mosaic series no.	Notes
Instituto Geografico Militare	Not known	1930s flown by IGM. Materials reported to be at IGM archive, Florence	Not known	No data	Not applicable	Not yet seen
Trimetrogon	Whole country	United States Air Force, 1945, 1:40,000 (vertical right and left oblique AP), held by National Archives	Not known if any made	No data	Not applicable	Available at US National Archives
Somalia	Parts of country	RAF, 1946–1953—sorties 13/SOM; 13B; 683A; and HAS 1958, sorties HAS/B/SOM/58. (holdings at Rhodes Library, Bodleian, Oxford)	Not known but believed some used by HSL for mosaic series	No data	Not applicable	Photography at Bodleian archive
Somalia	Whole country	Royal Air Force, V58. 1:60,000 (vertical AP)	Used by HSL to make as listed below	See below	Used in mosaics listed below	Prints at NCAP
Area A Somalia	El Wak Project. Area A Somalia. Juba River	Spartan Air Services. 1959–60. 1:30,000	1:82,600	47	HSL used Spartan mosaics 1–47	Spartan Air Services, Canada for Frobisher Oil Co Ltd. Later used by HTS. Negatives at NCAP
Sinclair Blocks 1, 2,3	Block 1: South-west Block 2: Central Block 3: N central	RAF 1958 (also some from 1947 onwards) all at 1:60,000	1:60,000 (also, at scales of 1:100,000 and 1:200,000)	Block 1: 150 Block 2: 33 Block 3: 49	Block 1: D-5824 Block 2: D-5868 Block 3: D-5874	HSL for Sinclair
Sinclair Block 4	4: NE Somalia	RAF 1958 1:60,000	1:60,000	n/a	Block 4: D-6294	HSL for Sinclair
Quintana	N Somalia 3 blocks	RAF 1958 1:60,000	1:100,000	Area 1: 12 Area 2: 4 Area 3: 5	1: D-6446 2: D-6445 3: D-7536	HSL for Quintana
FAO	Inter-Riverine area not covered by others	Probably RAF 1958	1:60,000 1:100,000	73 42	60 k: D-5890 100 k: D-5896	HSL for FAO
Sinclair	Bur Region	HSL 1960	1:200,000	2	D-5473	HSL for Sinclair
Sinclair	Bur Region	HSL 1960	1:30,000	19	D-6430	HSL for Sinclair
Sinclair	Northern Somalia	HSL 1970	1:30,000	39	D-5738	HSL for Sinclair
n/a	East of Shabelle	n/a	1:100,000 1:60,000	6 6	100 k: D-6183 60 k: D-6476	HSL, likely for Sinclair
Continental Oil Co	North Somalia (within Block 3)	RAF 1958	1:200,000	20	Not applicable	HSL for Continental 1973
Tsetse Control Project 1980s	Inter-Riverine region	ASD Gmbh, Frankfurt (1983–84) 1:30,000	1:100,000	Not known	Not applicable	ASD Gmbh. No information on location

Sources RN Munro AP archives; NCAP, Edinburgh; Bodleian Library, Oxford; Paolo Paron

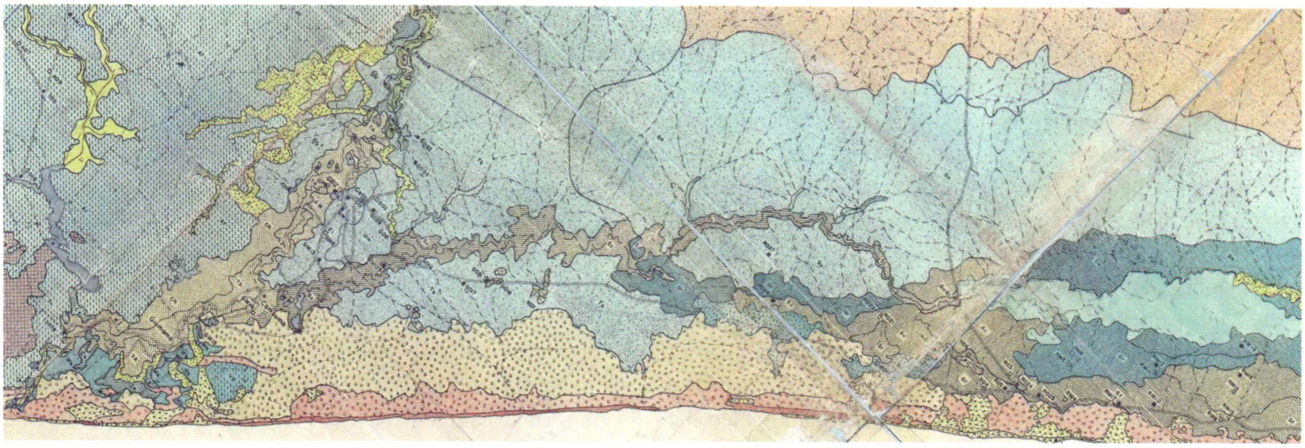

Fig. 14.4 Part of 1:500,000 scale soil map of Juba and Shabelle. Lockwood Landforms and Great Soil Group Map of Somali Republic. FAO Lockwood, 1968. *Archive Source* RNMunro. Also held at WOSSAC

Italians, they crossed first the Dry Juba, an old channel to the west, and then the Wet Juba at Jelib (Fig. 14.5), and in the campaign report commented: "The Juba's floods during the rains are famous. Far upstream, hundreds of miles from the sea, it is capable of drowning the country for miles around" (HMSO 1942).

Flooding, river bank breakages and general fluvial disruption remain common to this day: this is the role floodplains play, and human interference for cultivation or barrages restricts the floodplain. Also, the Dry Juba can still be wet in an unexpected way as we shall show from the 1981 floods study (Sect. 14.4). The economic assessment of Somalia (IBRD 1957) reported on the "extreme paucity of the natural resources, the severity of the environment and the limited extent to which it could be modified"; cropping in the *Gu* (April–May) and *Deyr* (October–November), major and minor rainy seasons, respectively, was always risky from drought. It was felt there were "few possibilities of modifying or of defying the severity of climate" and also noted that the "year ends with the long and punishing dry season—the Gilal". It appears that excessive floods are becoming more common than in 1980: analysis at Belet Weyne by FAO and SWALIM (Relief Web 2020 issued 26 May) showed that for the third year in succession floods

Fig. 14.5 Allied forces crossing the Juba near Gelib, February 1942. From HMSO (1942)

exceeded bank full level surpassing levels long regarded as a 50-year recurrence flood.

The World Bank (WB) agriculture sector study (World Bank 1980) had suggested, somewhat optimistically, that "Somalia is not, in fact, seriously water-short; it has adequate resources to supply its population and to sustain its major economic activities. However, water development in Somalia is expensive and must, therefore, be carefully planned and implemented". What was needed therefore was more coordinated assessments, a strong regulatory framework, land use planning to encompass all users and in general improved management of the land and water resources.

In modern times, a severe series of droughts affected the Horn of Africa and the Sahel region since 1968. Williams (2014a) has noted that an initial drought of 1968 in the Sahel was followed by dry pulses in 1971–73, 1977, 1982–84 and 1987, with another severe drought in 2005. From 1973 to 1975, much of Somalia was under the grip of a severe drought, and during 1974, farming and range conditions had deteriorated as both the *Gu* (April–May) and *Deyr* (October–November) rainfall seasons failed leading to 100% crop failure [see Chap. 2 for more detailed information about the climate of Somalia]. The World Bank (1977) Economic Assessment reported that the drought leading up to 1975 was more widespread and severe since the 1920s and resulted in up to 25% livestock mortality. Sinclair and Fryxell (1985) argued that the Sahelian drought, extending from Senegal through to Somalia, was man-made due to overgrazing and not lack of rain, and all exacerbated by piecemeal development projects. But, while this could have been the case in West Africa, over huge areas of land in Somalia and the Ogaden in adjacent Ethiopia, it was clear that the rains just failed. What water holes there were dried up, livestock populations were decimated and pastoral communities moved *en masse* to locations where the government with donor aid kept them alive and projects such as the planning for irrigation schemes gave hope to displaced peoples.

In contrast, a SWALIM study on preparation of disaster management plans in the Inter-Riverine (SWALIM 2016) recorded that river flooding occurs especially during El Niño conditions, with large floods noted in 1991/1992, 1994/1995, 1997/1998, 2002/2003, 2006/2007 and 2009/2010 periods.

14.3.2 Development Project Assessments in Southern Somalia

14.3.2.1 Physical Setting from the Inter-Riverine to Northern Somalia

East of the Juba and west the Shabelle is the Inter-Riverine of Somalia. From Beled Weyn to Sablaale, an active floodplain is incised into a series of older meander belts, levees and cover floodplains. Below Sablaale, the Shabelle swamps extend to the Juba confluence. To the north, the land rises up gradually as a complex of bedrock pediplains, basement rock inselbergs around Bur, and recent and older alluvial surfaces are dissected by ephemeral streams. Centred on Baidoa, the elevated Bay Region is formed on a thick pile of uplifted Mesozoic calcareous and gypsiferous marine sediments. As the Afro-Arabian dome was uplifted in the late Neogene and Pleistocene, the Shabelle and Juba Rivers cut entrenched courses into these Mesozoic rocks and, as revealed by offshore drilling, deposited large volumes of sediment into the Indian Ocean. Sediments were transported by the Tana River in Kenya to the Indian Ocean also and will have moved north-eastwards by the Somali current. The southward and seaward boundary of the Inter-Riverine zone nowadays is the coastal sand ridge, the so-called Old Red Sandridge (ORS), comprising stabilised and modern active aeolian dunes, that overly a coral and beach rock base over which sit the dunes. From gully sections at Brava examined by RNM in 1985, the ORS is shown to have numerous paleosol layers and a rubefied nature, as well as active dunes derived from rainfed erosion of the sands and onshore dune establishment.

A full chronology of the ancient dunes of the Somali coast remains to be laid out; evidence from coastal sequences in Kenya correlated by Braithwaite (1984) suggests dune sediments from the late Pliocene to Lower Pleistocene, and Somalia is likely to be similar. Age determinations and their meaning in relation to sea levels and Marine Isotope Stages are discussed below in Sect. 14.7 of this chapter. As the ORS is built up, it would have formed a barrier along the coast and the Shabelle would have connected to the Juba. The Shabelle is now confined in this trough, swampy at the distal end and only reaching the Juba, some 1080 river kms downstream from Beled Weyne, when there are large floods.

Bordering the right bank of the lower Shabelle and west of the lower Juba and over the border with Kenya are the so-called Marine Plain lands (FAO-Lockwood 1968), comprising heavy clays with high surface and subsurface sodicity (Solodised Solonetz soils) and also high salinity (Solonchak soils). These were regarded by Lockwood (1968) and HTS (1978a) as ancient lagoonal sediments, but while close to sea level, and taking into account regional uplift, this may well have been their origin as freshwater lagoons with seasonal saline influxes from the ocean perhaps, at higher elevations they represent alluvium derived from volcanic rocks within Ethiopia and Kenya. The Marine Plain lands do not receive run-off from the Juba and had long been considered too expensive to reclaim and only useful for rangeland and wildlife: they remain much thus.

The relatively fertile nature of the alluvial soils of the Juba and Shabelle floodplains was well documented on several studies (FAO-Lockwood 1968; HTS 1969, 1977a, b, 1979; MMP 1978a, 1978b, 1979), but development of lands for irrigation and rainfed farming would require farm management expertise in land reclamation to reduce moderate salinity—and locally some sodicity on levee soils—and improve drainage and tillage issues for heavy clays. Land suitability evaluation in those soil surveys showed that the older the landform, the more limitations were present. For example, in the Janaale-Bulo Marerta area (MMP 1978a, b—Genale–Bulo Marerta Annex I: Soils), the older Shabelle alluvium was given a class 3 suitability with permanent limitations of drainability, profile characteristics and salinity; the semi-recent and recent alluvium was predominantly of suitability class 2 with the same limiting factors with lower threshold values.

Along the course of the Shabelle in Somalia, groundwater that is suitable for irrigation and domestic use is restricted to the present and subsurface connected recent courses of the river and their environs. Beyond these areas, groundwater salinity increases rapidly as effects of direct recharge from annual floods diminish, and there are higher salinity groundwaters derived from bedrock evaporite deposits (MMP 1978a, b—Genale–Bulo Marerta Annex II: Water Resources).

The regional groundwater survey made as part of the Livestock Sector Review and Project Identification (HTS and GRM 1976) with map at 1: 2 million scale summarised the groundwater conditions over all Somalia:

- Hargeisa to Berbera: groundwater is limited to the alluvial wadi aquifers and some fractured metamorphic and karstic rocks;
- Hargeisa to Burao: alluvial and colluvial sediments of Burao basin with deep water of adequate quality in the Yesomma Sandstone;
- Burao to Belet Weyn: extensive area underlain by evaporite deposits. More reliable water supplies occur in karstic rocks;
- Shebelli Valley: groundwater quality along the upper Shebelli is poor but should be adequate for stock watering;
- Inter-Riverine area: groundwater occurs in several aquifers, namely the coastal dunes, Shebelli alluvium, weathered Bur Akaba basement complex and Mesozoic sedimentary rocks;
- Juba Valley: alluvial aquifers border the river and provide adequate watering points along the length of the valley;
- Trans-Juba to Kenya border on the Marine Plains: there is widespread salinity in shallow groundwater in the aquifers in Pliocene to Recent alluvium.

On terms of irrigation possibilities from groundwater, a regional assessment by the United Nations (1989) stated: "the exploitation of the consolidated sub-artesian aquifers for irrigation purposes is considered unlikely. These formations have very poor transmissivity, and this means large drawdowns for small yields. The prospects are better with respect to supplying the people and their livestock, but the boreholes must be installed very efficiently if the desired output is to be obtained with an acceptable drawdown".

The Juba River, that discharges to the Indian Ocean near Kismayu, has good quality water for irrigation but during localised storms the first flush from run-off within its catchment in Somalia is high in silt and total dissolved salts (MMP 1979, 1980). Collier (personal communication 2020) noted that in the 1970s, a visiting specialist to the Inter-Riverine studies was concerned that Juba water was high in salts and unsuitable for irrigation as its waters would salinise the alluvial soils. In fact, the salts come from dissolved gypsum washed off the gypsum outcrops at start of the *Gu* season especially. After the first flush gypsum, soluble salt levels are reduced. The gypsum content of flood waters, such as we noted in 1981 (see Fig. 14.9), will reduce any sodium (alkali) risk hazard in the alluvial soils.

East of the Shabelle and extending to the north-east are the drier Mudug Plains. Rangelands where resilient vegetation arcs are steadily replaced by sub-desert shrubs and grasslands. Further to the north-east in Northern Somalia, the land is dominated by the Haud dissected plateau, on Mesozoic and Cenozoic sediments, that rises to over 2000 m on the northern escarpment. Drainage on the plateau is all ephemeral and to the Indian Ocean. The Haud plateau, partly covered by vegetation arcs (see Sect. 14.8), is terminated at the north-facing escarpment, and there is a sharp descent to the arid coastal lowlands. The eastern part of the escarpment, comprising the Ogo Mountains and Galgodon Highlands, has a mean peak height of 2450 m and is well vegetated as the high grounds catch NE monsoon rains and mist (Sect. 7.2). The western parts of the plateau in Somalia, between Borama and Hargeisa, lie at around 1400 m asl, and the escarpment is less abrupt. Ephemeral streams drain to both the Indian Ocean, mostly through the Nogal Valley some 150 km west of Qardho (Sect. 14.7), and the Gulf of Aden, such as for the Arabsiyo Valley (Sect. 8.2). In the extreme north, raised marine terraces were described by Brook et al. (1996): the highest terrace at 16 m above modern sea level did not provide any datable material, but based on other sites in the region was likely to represent MIS 5e at 132–120 kyr BP. Lower down a terrace at 8 m above sea level was dated at 105 kyr BP (MIS 5c) and a 2 m terrace at 80 kyr BP during MIS 5a. Following the post-glacial transgression, it was considered that sea level was close to its present position by the middle Holocene, at 7 kyr BP.

14.3.2.2 Development Projects

In the 1960s and 70 s, the rangelands, both in the north and also east of the Shabelle, were drought stricken and pastoral nomads lost all their herds. From the late 1960s through to late 1980s, numerous feasibility studies were initiated for irrigation on the main rivers, rehabilitation of rangelands and for rainfed agriculture. Assessments of the livestock situation were made in 1978 by HTS and Gunn Rural Management (GRM), while Murray Watson of Resource Management and Research (RMR) made airborne surveys of livestock and wildlife based on ecological strata (Watson et al. 1979; Watson 1982, Watson and Nimmo 1985). We draw attention to but do not make assessments of these studies, nor for other rangeland projects and studies made in Somalia since the 1960s (e.g. FAO 1980; Hemming 1966, 1971; Le Houérou 1972; Iannelli 1984; Pratt 1972; World Bank 1979; IUCN 2000). We can comment briefly on the National Tsetse and Trypanosomiasis Control Project (NTTCP) that was funded by the United Kingdom's Overseas Development Authority (ODA) during the1980s in the middle and lower Shabelle and aimed to eradicate along the Shabelle the host and its disease. After a planning phase (Bourn 1977), the tsetse control project ran from 1985 to 89 implemented by HTS and Agricair of Zimbabwe. The project has been reported by MMP (1983), Hendy (1985), HTS (1988) and others. The project adopted aerial spraying (Lee 1969), with endosulfan sprayed onto infected areas. Tsetse fly traps were used to monitor results but most were stolen. Many of the project documents are held at the WOSSAC archive, and a more multidisciplinary assessment could reveal much concerning its impact.

In the late 1970s, a large migration of displaced people had moved to the Shabelle and Juba Rivers from traditional grazing areas that had become too dry to sustain any livelihood. Many came from the Ogaden. East of the Shabelle River, in the Central Rangelands area, some 600,000 people were affected, and there were major losses to crops and livestock. To overcome food shortages, 20 relief camps were established, and by 1975, as conditions improved, the government decided to close the camps and resettle people at a number of agricultural and fishery schemes in the south of the country, between the two rivers. These were at Dujuma, Sablaale and Kurten Waarey, and by mid-1975, some 245,000 had been resettled at these schemes, mostly former nomadic livestock owners whose herds had been depleted. It was recognised though that the schemes had specific problems relating to the availability of cultivable land and also irrigation water, and World Bank missions to Somalia in May and July 1975 recommended that future resettlement programmes needed to be based on a sound technical and economic base for rural development: this was the *Inter-Riverine Agricultural Study* (the IRAS), placed within the Somali Government's Settlement Development Agency (HTS 1977a). The study started in mid-1975 and covered the lands between the Shabelle and Juba Rivers, some 200,000 km^2 in all. IRAS provided a comprehensive assessment of previous natural resources and socio-economic data; assessed the water resources of the Juba and Shabelle basins including a review of the hydrogeology of all Somalia; assessed the potential for dryland agriculture irrigation in this region; and determined the suitability of the soils at the three settlements and other areas that might be identified. Extensive soil studies were made along the Juba River for the Fanoole Settlement project (HTS 1978), where a state sugar farm was later established. The Dujuma area survey, made for the Settlement Development Agency, found the soils were on old alluvium partly of non-Juba origin, were derived from adjacent terrains and having variably high salinity and moderate sodicity, and were regarded as not suitable either for investments in irrigation or rainfed cropping (HTS 1977b). This honest appraisal was accepted and became, for Somali specialists anyway, a classic example of the value of making soil surveys prior to implementing development! The Soil Atlas of Africa (Jones et al. 2013) confirms that the soils on the gently rising plains bordering the Juba, that include the Marine Plain soils noted above, are variably saline and sodic.

The IRAS project produced two maps at 1:500,000 scale showing the locations of some 78 Existing and Proposed Development Schemes up to 1977. Parts from sheet 3B (WOSSAC ref 7132) are shown in Figs. 14.6 and 14.7, with selected projects listed in Table 14.2. These maps—perhaps the most useful starting point for new studies to look back at—are held at WOSSAC since 2008 and long before that available from HTS, but remain un-digitised.

Post-IRAS, a series of development projects were made along the Juba and Shabelle plains. These examined lands for soils, irrigation and agriculture for the Fanoole reach on the Juba (HTS 1978), and at Genale-Bulo Marerta (MMP 1978a) and Homboy (HTS 1979) on the Shabelle. Subsequently, the Genale scheme was partly developed, but Homboy was not initiated (SWALIM Report W-05: Mbara et al. 2007). At Jowhar, the problems of drainage and reclamation were intensively studied (MMP 1978b) prior to the development of the Jowhar Off-stream Storage Reservoir. On the uplands between the Juba and Shabelle, the World Bank gave support to the Bay Region Agricultural Development Project (BRADP) assessed by HTS (1982, 1983). Due to lack of older oblique photography for repeating, the status of BRADP is not covered in this paper.

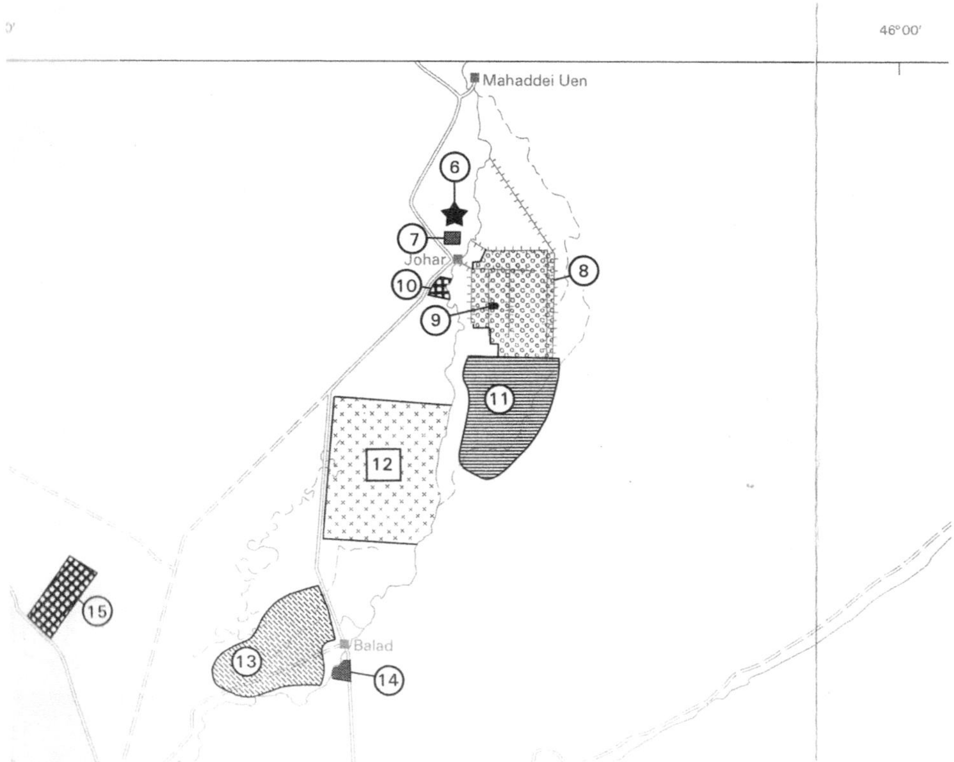

Fig. 14.6 Development studies status as at 1977. Shabelle: Balad to Mahaddei Weyn reach. From Sheet 3B, Inter-Riverine Study, HTS, 1977a, b. Selected projects listed in Table 14.2. *Source* RNM Somalia archives. Also held at WOSSAC Ref 7132

These are the older studies that for the most were available to us in our Inter-Riverine projects made 1980–86 and re-examined in the following sections of this chapter.

14.4 Case Study: Irrigation Designs at Luuq and Jalalaqsi

14.4.1 Background

In 1980, RN Munro was part of the MMP/HTS team making irrigation project plans for the UNHCR at Jalalaqsi on the Shabelle, and Luuq on the Juba, where there was a need to develop permanent agriculture based on irrigation for the large numbers of displaced nomads from the Somalia and Ethiopian Ogaden (MMP 1980). We look at what was proposed and the situation now.

14.4.2 Juba at Luuq

On the Juba River, the reconnaissance survey concluded that the Halba plains on the right bank should be assessed further to a pre-feasibility level of irrigation design. Aerial photography available to the team in 1980 was limited to a 1960 Spartan negatives, made from 1959 to 60 AP. The scanned negative mosaic (Fig. 14.8a), part of a set of the whole Juba donated recently to NCAP, shows the landscape then. We found a previous irrigation system had been installed on the Halba plain, at some date after the 1959 image was taken, but was defunct in 1980. A Google Earth image dated 7 February 2015 (Fig. 14.8a) shows that almost all the forest lands around Luuq and along the Juba have disappeared, likely cut down by the refugees in early 1980s. One can observe that on the alluvial floodplain, forest is now mostly occupied by strips of pump-irrigated lands. These small-scale irrigation farms will be inundated at times of high floods, but pumps can be removed. A larger pump station, such as proposed for Halba, could probably be inundated and lost. Some forests remain upstream, seen at extreme right in the Google Earth © image.

A cross section of the soils at Halba made on this study identified a trough with a deep fill of old alluvial clays (Fig. 14.9). The Dolo series clay soil (Lockwood-FAO 1968) is seen as the dark red linear patches in the Google image (Fig. 14.8b). The Dolo clay surface soil passed down into >5 m of clayey alluvium, and we concluded it was likely a relict Juba channel infill: its highest part is 4 m above present maximum flood level, and the old channel was estimated to extend down to the level of the present Juba River bed. There are other similar linear patches of alluvium on the plateau on both banks, suggesting an anastomosed river. The alluvium-filled troughs were cut into the Main Gypsum Formation (Merla et al. 1979). The adjacent lands

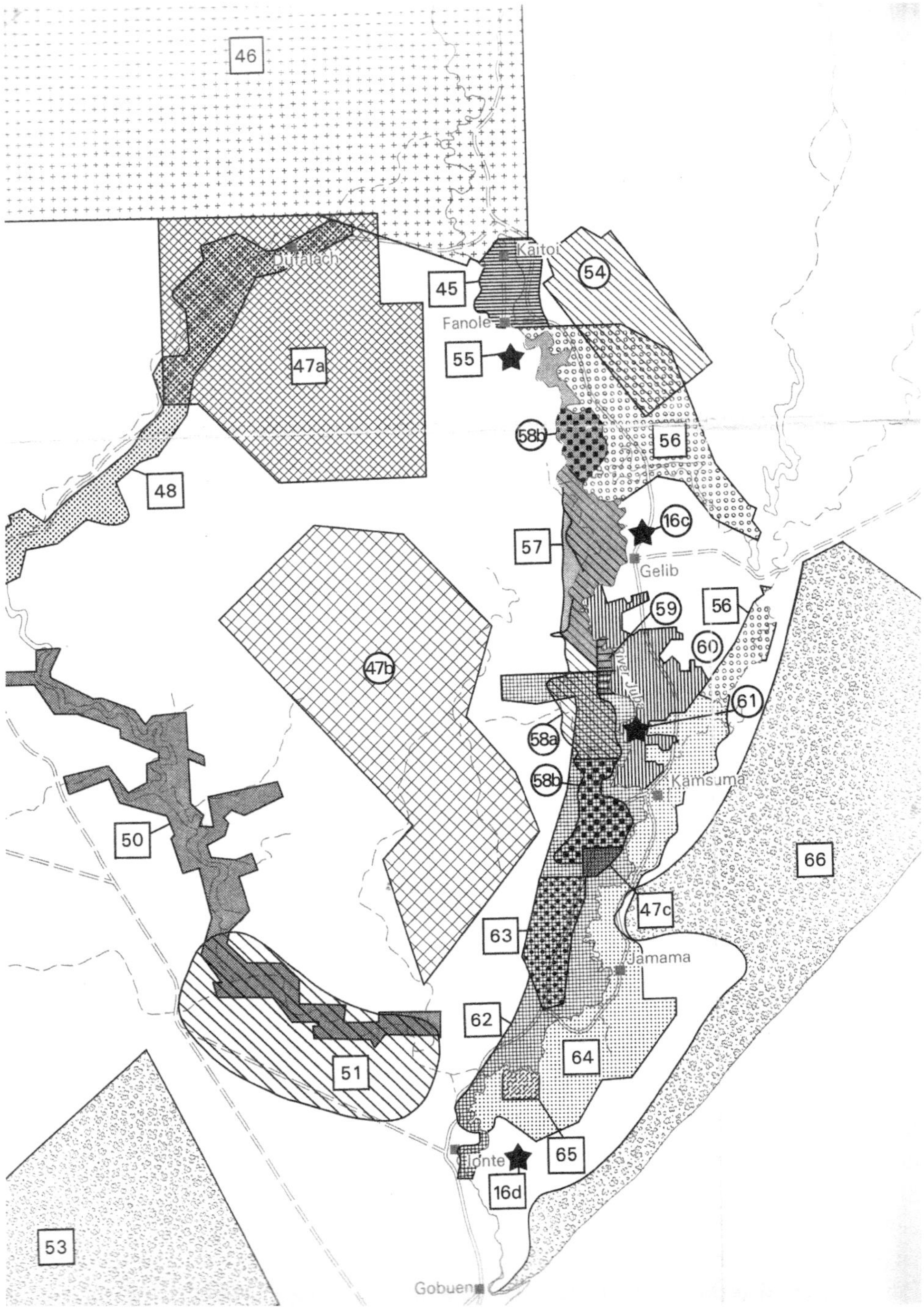

Fig. 14.7 Development studies status as at 1977. Juba: Fanoole to Gobuen. From Sheet 3B, Inter-Riverine Study, HTS, 1977a, b. Selected projects listed in Table 14.2. *Source* RNM Somalia archives. Also held at WOSSAC Ref 7132

beside the Juba are all highly gypsiferous (>70%); the ground in fact is riddled with cavities and gypsum karst sinkholes. The MMP study noted that gypsum was common in all the soils, and care would be needed when designing structures that would not collapse if gypsum was dissolved by irrigation waters.

Later, vegetation and soil studies were made at Luuq by a charity, Inter-Church Response (IRR) (Wieland, 1983; Wieland and Werger 1985) to assess ecological change due to refugee influx. This could be a useful study to re-examine the vegetation and any ecological change in the future. The IRR study referenced the MMP (1980) study somewhat absurdly as "Anonymous", but made no mention of the study in the text. However, they stated that the Dolo clay soils lay on a higher terrace of the Juba, which agreed with our 1980 findings.

In 1986–88, a USAID-funded team made a comprehensive environmental assessment (EA) of the Baardheere Dam Project on the Juba upstream of the proposed Baardheere Dam (Brandt and Gresham 1989; Brandt 2000). In terms of

Table 14.2 Key projects shown on Figs. 14.6 and 14.7

Number on	Name and extent	Reference
6	Johar Experimental Rice Farm 100 ha	Agric Res. Institute. See HTS, 1977a, b
7	Burei, Johar Experimental Sugar Farm. 100 ha	Govt. See HTS 1977a, b
8	Johar Sugar Estate, net potential 8300 ha	Govt
9	Johar Drainage and Reclamation Study	MMP/HTS 1975
10	Johar Crash Programme Farm. 230 ha	See HTS, 1977a, b
11	Johar Off-stream Storage Reservoir. 11,000 ha	HTS/MMP, 1969; MMP, 1973
12	FAO Livestock Fattening Project. 14,500 ha	See HTS, 1977a, b
13	Balad Controlled Irrigation Project. 1000 ha in 1977	HTS/MMP,1969
14	Balad Irrigated Feedlot (M.L.F.R.). 400 ha	See HTS, 1977a, b
15	Warmahan Fodder Farm (M.L.F.R.). 3800 ha	See HTS, 1977a, b
46	Dujuma National Park. 380,000 ha	Technital, 1975
47a and 47b	Trans-Juba Livestock Project. 160,000 ha	See HTS, 1977a, b
51	Descek Uamo National Park. 40,000 ha	Govt
53	Forest Reserve. 140,000 ha	Abel and Kille, 1976
54	Gelib Livestock Holding Ground. 21,500 ha	Govt, and EDF
56	Fanole-Gelib Irrigation District 6. Net area 26,400 ha	See HTS, 1977a, b
58a and 58b	Lower Juba Sugar Project. 23,900 ha	Booker McConnell, 1976—reports now at WOSSAC
59	Kalanji Banana Farm, Pilot 250 ha	National Banana Board. See HTS, 1977a, b
62	Ionte Irrigation District 32,000 ha	Govt. See HTS 1977a, b
63	Mogambo Irrigation Project. 7000	TAMS/FINTEC, 1977; MMP, 1979
64	Jamame Irrigation district. 20,050 ha	Govt. See HTS 1977a, b
65	Torda Irrigation Project. 2400 ha	Govt. See HTS 1977a, b

archaeology, the EA discovered 686 sites within the 420 km^2 of the proposed reservoir; the oldest sites were Middle Stone Age (ca 280–25 ka BP) and included Neolithic rock art cave sites and early Islamic cairns and graves. Geomorphologist Martin Williams found that there were no high terraces in the Juba Valley such as they are commonly found in semi-arid terrains within the Horn of Africa, and there was only clayey alluvium—slack water deposits—in side valleys and low-lying areas (Williams 1995, 2014b, 2016). This accords with our explanation of the clay-rich Dolo soil and alluvial fill found in the Halba area (Fig. 14.9). He found that both paleo- and contemporary floods leak rapidly into the surrounding karstic terrains, and that a dam at Baardheere would never reach full supply level in such limestone karst country. Ali Kasim et al. (2002) also noted the lack of any high terrace on the Juba in this area.

Additionally, Williams (2014b) noted the reservoir would flood all the fertile alluvial lands up to Luuq and destroy the farming livelihoods of thousands of people, as there was almost no soil on the higher grounds—a common result of dams proposed or built on rivers such as the Main Nile in Sudan and upper Egypt that pass through rocky deserts with a narrow and extremely fertile floodplain. Also, from gastropod shell evidence in Holocene alluvium, it was clear that the two main vectors of Schistosomiasis could lead to this parasitic disease becoming endemic in the reservoir and all along Juba Valley.

Another factor that we noted in 1981 on the flood study, and confirmed on the 1:2 million scale Hydrogeology Map of Somalia (HTS 1976), was that directly at the proposed dam site the bedrock comprised a 10-km-wide outcrop of gypsiferous marls and clays of the Warendab (also called Anole) Formation (HTS 1976; Merla et al. 1979; Abbate et al. 1994).

Before the USAID EA study, the feasibility study made by Technital (1975) had stated a future reservoir would be "watertight": it would appear that it would not have been so.

Fig. 14.8 **a** Luuq, 1959. Luuq Area. Sheet D7/C7 of Spartan Negative made in 1960. From former HSL archives donated by RN Munro to NCAP in 2019. Negatives © NCAP. **b** Luuq, February 2015. The Juba flows to the left. Pale tones of gypsiferous bedrock. The Halba area where irrigation was designed is shown to right of meander loop. Google Earth ©

Further, the World Bank's 1981 agriculture sector review (World Bank 1980) stated that "*the Baardheere Dam does not look like an attractive investment in relative terms*". Following the EA, the World Bank agreed to the next phases of the EA in this area and to evaluate the significance of prehistoric/historic sites, conduct excavations and prepare a full Cultural Heritage Management plan. Civil war put a stop to any further work. However, the implications of building a dam on limestone and gypsum karst terrains, the socio-economic factors relating to loss of livelihoods, the loss of known cultural heritage and the likely spread of Schistosomiasis weigh heavily against any future plans for approval of an investment design for a dam at Baardheere. That it all was taken so far towards a final acceptance in the past seems to indicate geopolitical overriding of environmental and cultural heritage concerns.

The Halba scheme that MacDonald's designed to pre-feasibility level in 1980 was not initiated. Google images show canal lines that have not been connected to the Juba for a long time, if ever; the dark red deep clay soils are utilised for rainfed farming only and are partly covered by secondary bush and termitaria. Collectively, the irrigation strips cover far more land than the proposed Halba scheme and are nourished most years by silt-laden floods.

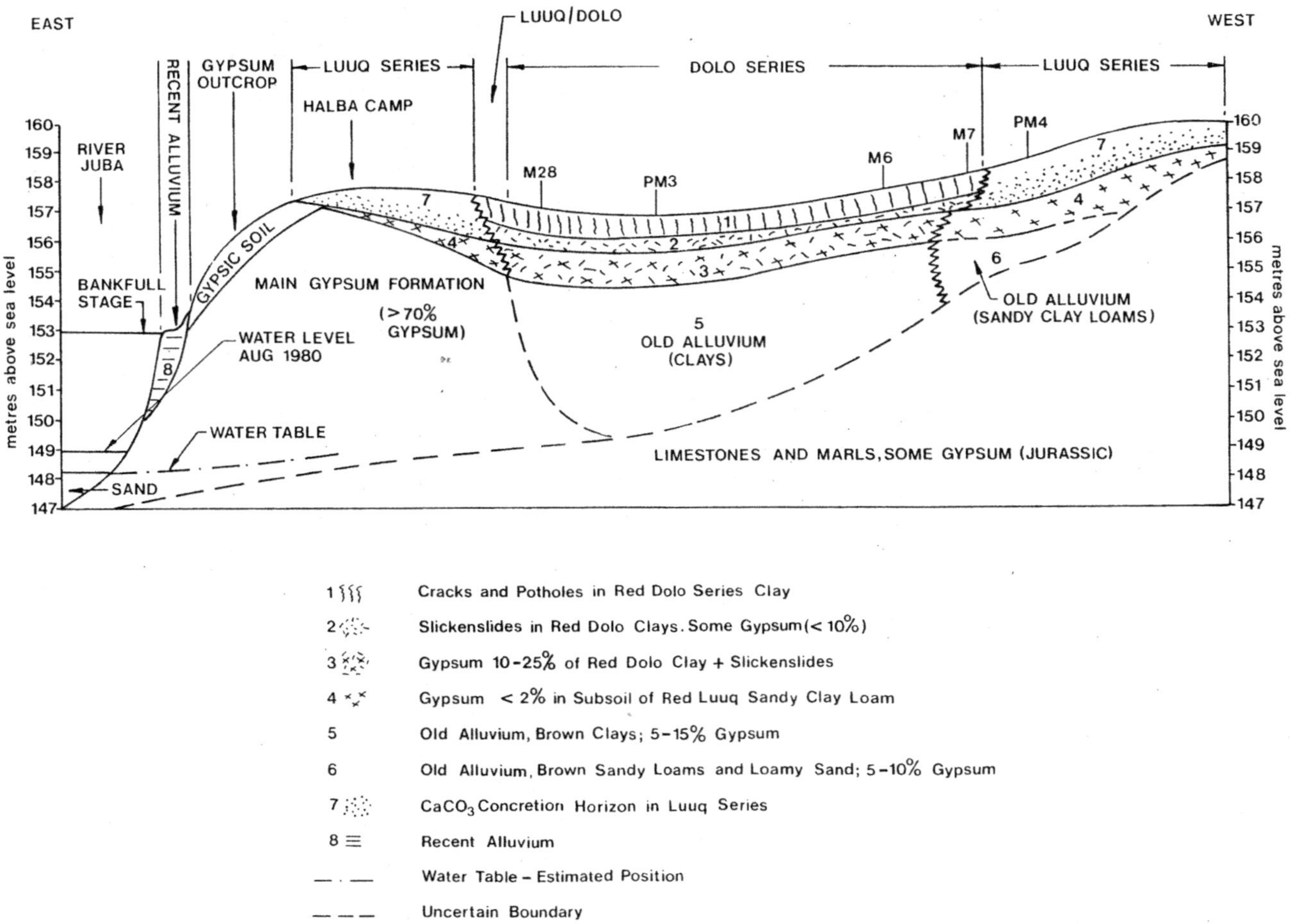

Fig. 14.9 Soils cross section—Halba area. From MMP (1980). *Author* Munro

14.4.3 Shabelle at Jalalaqsi

The Jalalaqsi studies in 1980 were made on the right bank of the Shabelle downstream of Jalalaqsi, on old Shabelle River alluvium. In Fig. 14.10a, an aerial photograph from 1960, Jalalaqsi is at top left, and the pre-feasibility study area lies downstream of the town at far right. Examination of the area now using Google Earth imagery (Fig. 14.10b) shows that various inlet canals extend to east from the Shabelle, and irrigation canals have been laid out to supply fields. These do not look as if they were in use at time of the imagery (24 January 2014). A meander loop downstream of the newer pump station has been cut off since 1960. As refugees moved back to the Ogaden areas, the need for a larger scheme became redundant. There is however considerable small-scale irrigation along the Shabelle alluvial floodplain using portable pumps that can be removed at times of floods. The Jalalaqsi scheme designed to pre-feasibility level in 1981 was never taken to a detailed design level. Google images show several canal lines from the Shabelle but most appears to be rainfed agriculture. Irrigation, as at Lugh on the Juba, is restricted to small-scale pump schemes using alluvium of the floodplain.

The 1980 soil studies, summarised in a cross section (Fig. 14.11), were guided by the Lockwood-FAO (1968) and Inter-Riverine study (HTS 1977a, b). Subsurface gypsum concentrations up to 70% were found in the subsoils, derived it is thought from gypsum-rich groundwaters, and raised a concern when designing engineering structures that could collapse from solution of gypsum: the land suitability classification took this into account.

14.4.4 Conclusions

The schemes designed to pre-feasibility level at Jalalaqsi and Luuq were never taken further by UNHCR: the refugee crisis was downgraded and people returned into the Ethiopian Ogaden. Also, by the mid-1980s there was a period of more favourable rainfall. In 1981, the Agriculture Sector Review (World Bank 1980) concluded that the approach to drought proofing that emphasised resettlement of nomads and

a

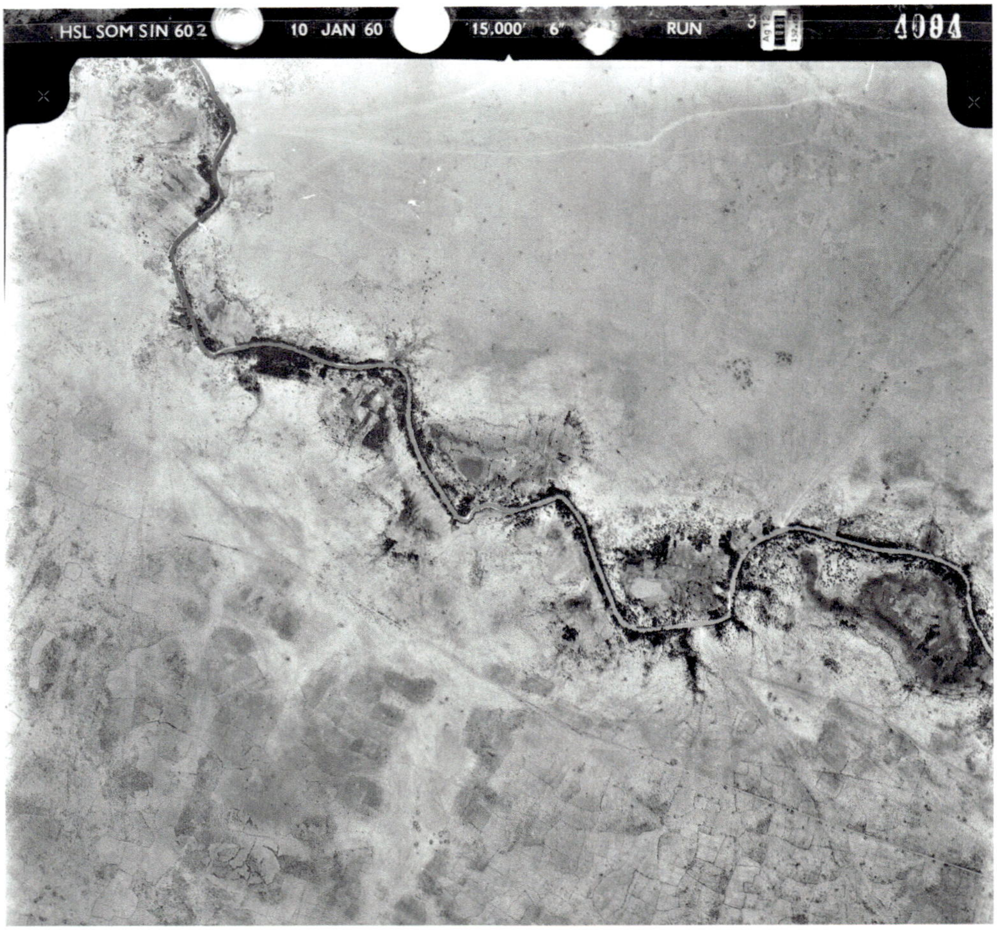

b

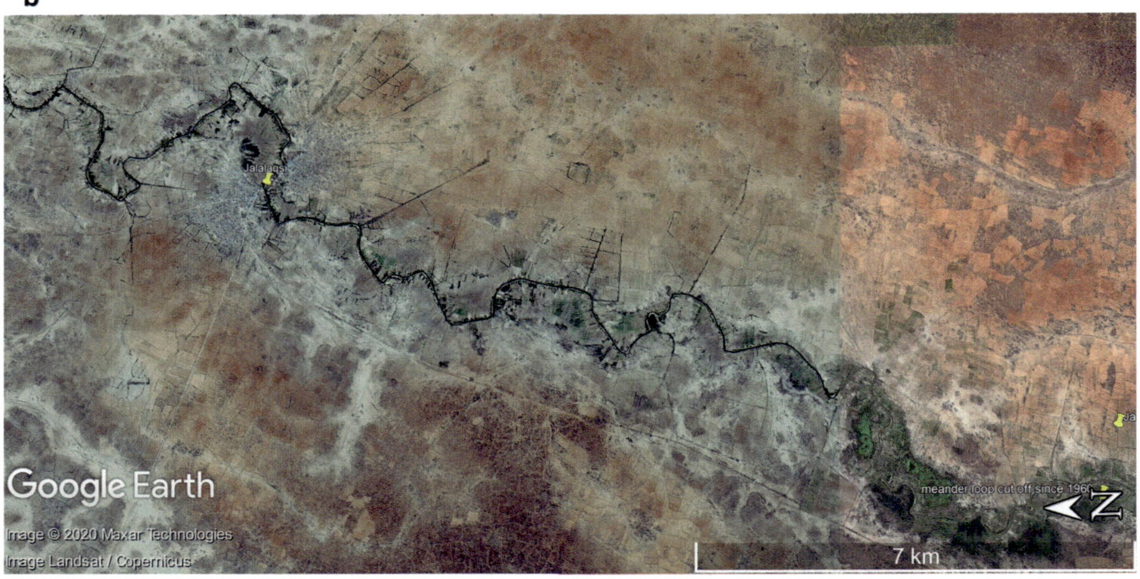

Fig. 14.10 **a** Jalalaqsi is the small settlement with a road bridge at top left. HSL 1:30,000 AP flown 10 January 1960. From RNM photo archives to be donated to NCAP. **b** Jalalaqsi Area today: Some irrigation layouts and pump stations on older alluvial plains. Google Earth 2014 and 2017. Google Earth ©

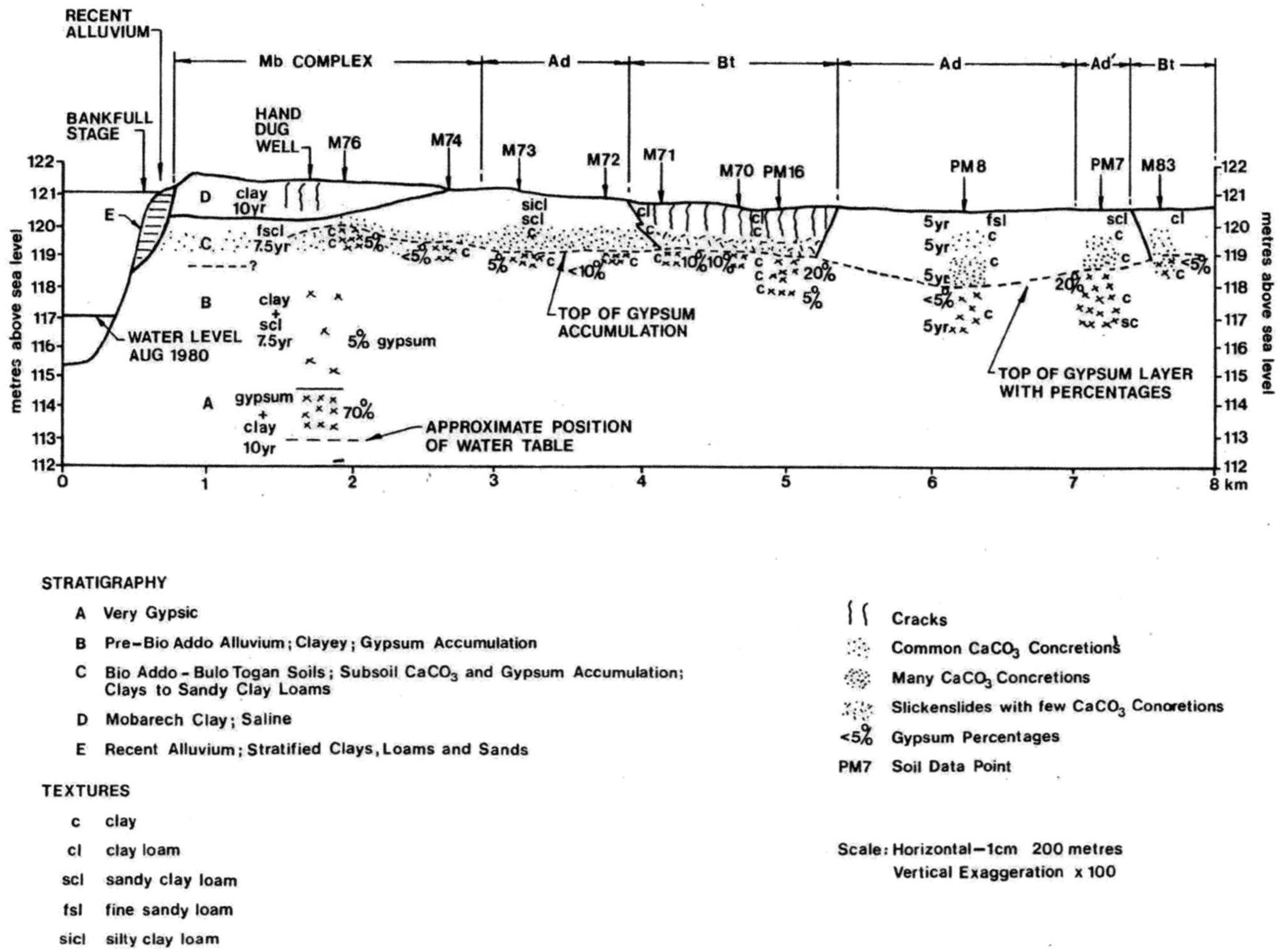

Fig. 14.11 Soils cross section—Jalalaqsi Area. From MMP (1980). *Author* Munro

horizontal expansion of irrigated agriculture was: "judging from the results to date, not the right approach to resettling nomads under Somali conditions, and that the settlement programme, started in 1975, had not succeeded in achieving its objectives of creating self-sustaining communities for former nomads".

However, while the modest schemes that were planned did not materialise on the higher terraces along the Juba and Shabelle, the resulting decades have seen a huge increase in the employment of pumps to irrigate small-scale schemes along both rivers utilising the fertile alluvial lands. These have benefitted many thousands of small holders, and given the success of early warning systems transmitted down to the communities by social media, pumps can be removed easily at times of floods up onto higher ground. Some larger fixed pump stations exist at key places and take water onto the higher elevations: these are more vulnerable if and when a major flood occurs. Rehabilitation of the larger schemes, for example Jowhar and Juba Sugar schemes, awaits a more secure period.

The IRAS project maps at 1:500,000 scale showing the locations of some 78 "Existing and Proposed Development Schemes" are due for an update with a review of the status of the older projects, but access to many areas will remain very difficult at this time.

14.5 Case Study: Flood Damage Study of the Juba and Shabelle Inter-Riverine

14.5.1 Background

In April and May 1981, huge floods swept down the Shabelle and Juba from Ethiopia, impacting heavily on the alluvial lands of the Inter-Riverine area. Locally, high rainfall compounded the disaster.

It is a region that has always suffered from either too much or too little rainfall and either damaging floods or insufficient irrigation water in the two rivers. Successive appraisal reports by UNOCHA now indicate that the *Gu* or

Deyr rains look promising but then later have to report crop failure is possible. SWALIM has archived numerous such similar reports over the past ten years. In 2019, heavy rainfall in the upper catchments of the Juba and Shabelle led to high discharges with floods in Somalia and reported on breakages in the rivers with flooding into back-swamps or old channels, and there is an active early warning system that delivers information on floods and damages to decision-makers and aid agencies. There is little new in these contrasting events in 1981 or 2019 as the hydrology records show.

In 1981, the Somali authorities made their own damage assessment and with FAO requested a detailed engineering (MacDonald's) and agricultural (Hunting's) consultancy to examine all aspects: Mark Brett and Munro, Engineer and Geographer, respectively, mobilised rapidly in mid-June for the FAO Flood Damage Study. It sounds an apocryphal tale, but was true, that in early 1981 an FAO drought assessment mission was mobilised some months earlier than the Flood Damage Study. The former, we were told, took a while to mobilise, and their arrival coincided with the rains and floods, so they had no option but to depart immediately. To us this summed up the rather unpredictable Somalia situation of drought or flood.

The team had use of a 1970 map made from aerial reconnaissance by MMP of the June 1970 Shabelle River floods that covered a huge area of the Shabelle—lower Juba area (MMP 1970) with cross-country floods utilising both old Shabelle channels and back-swamps. In 1981, floods and heavy rainfall had also covered the region. During May and June 1981, FAO hydrologist Brian Gemmell had made a number of flights, drawn preliminary flood maps, and commenced to repair the largely destroyed network of flood level indicators and gauges.

14.5.2 Remote Sensing Methods

FAO Rome had worked at high level with NASA to acquire recent 14 Landsat scenes of the area. Despite the close cooperation, the images arrived after fieldwork ended and while of great interest were of marginal use due to the low resolution of the MSS imagery, the extensive cloud cover and acquisition dates in May and June well after the flood peaks. FAO recognised this as an important test case to acquire imagery rapidly and get it to the people in the field. NASA could not deliver the imagery while the mission was in the field but FAO considered that it was a very useful exercise, and a lesson learnt for future policy; the images also helped us refine our maps back in the office. It is all very different nowadays where imagery is readily available within a short time from various satellite platforms.

On the mission, firstly we examined the flooded lands from the air in a twin-engine Cessna over three days (28–29–30 June) up and down both rivers. Then we made the same journey by car to examine the situation on the ground, where we could visit as flooding was still present over huge areas, and cost up the damages. On our flights, we used Gemmell's maps and marked up revisions of flood limits, as best one could at about 300 km/h. The base maps were the excellent Russian 1:200,000 topographical maps. These amalgamated lines were later redrawn onto fair copies of the maps, with boundaries augmented using the many oblique aerial photos taken by Munro. Four of the final maps for the Shabelle and Juba are shown. Each map is then followed by samples of the oblique shots. Normally, such flights are recommended to be made after a ground reconnaissance, when one knows the terrain intimately and can pinpoint features on maps more easily. Fortunately, RNM had made quite extensive ground journeys in 1980 that covered the Shabelle and Juba Rivers and the Baidoa area. The 1981 flights and motor tours more or less repeated the aerial excursions and showed clearly many interesting points that often would have been difficult to understand on the ground, if indeed access had been possible. The government had made its own study of costs of the floods, and these were compared with the MMP work. Four of the Shabelle and Juba flood maps from 1981 maps are given here and show the extent of the floods over parts of the Juba and Shabelle (Figs. 14.12, 14.13, 14.19 and 14.20).

14.5.3 Assessments on the Shabelle

The extent of the 1981 flooding downstream of the Ethiopian border and in the reaches to Jowhar Sugar Project are shown in Figs. 14.12 and 14.13. Gauge measurements on the Shabelle at the border showed that the flood exceeded 1000 m^3/s on 30 April, more than double of the previous record flood of 400 m^3/s. Inevitably, it covered the entire extent of the floodplain up to the dune ridges on the left bank. The Shabelle had flooded Bulo Barde, and breaches were made to reduce flooding in the town. Water flowed into an old back-swamp leaving an impressive crevasse splay feature (Fig. 14.14a) but had bypassed the flood-relief channel. By 2005, the flood-relief channel had been improved (Fig. 14.14b), and it takes excess flood off the Shabelle at Beled Weyne and then returning it to the river further downstream.

Further downstream, almost all back-swamp areas along the main course of the Shabelle were flooded, as one might expect. Most settlements were on the levees or the higher ground of the adjacent sandy plains that slope upwards away from the river, as this sort of flooding is frequent and acts swiftly.

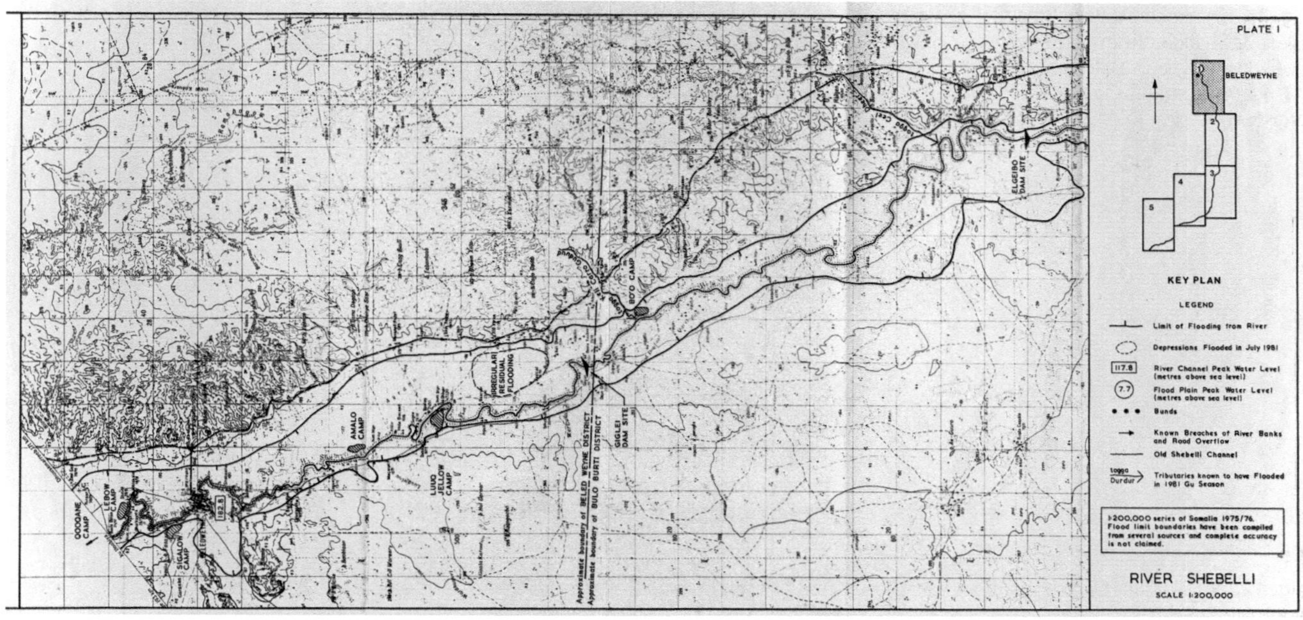

Fig. 14.12 Flood Damage Study Sheet 1: Shabelle from Ethiopian border to El Geibo (MMP 1981). Report held by WOSSAC

On the right bank, the floodplain of the Shabelle opens out just above Jalalaqsi, and there are numerous old meander belts and accompanying back-swamp depressions that extend south-west towards, eventually, the Juba (Fig. 14.15a). By 2017, much of the woodland/forest in this reach had been removed (Fig. 14.15b). Wanle Weyn on the road to Baidoa lies at the limit of the old mega-floodplain. A flood broke through at Dobale and entered one of these low-lying areas at Mahaddayweyne and then went cross-country in a south-west direction, broke through the Baidoa road, flooding the FAO Seed farm, and continued onwards, where, on 28 June, we photographed the margins of this by now quite slow-moving or even static flood—it was difficult to judge, of course, movement from the air (Figs. 14.16 a–b, Fig. 14.17).

Downstream, floods had entered part of the Jowhar sugar plantation, but flood channels worked well and most of the flood was channelled into the storage reservoir from the Sabun Barrage offtake (Fig. 14.18a). The Off-stream Storage Reservoir (MMP 1978b) lies on the Shabelle left bank at Jowhar, on old clay cover floodplains. The Jowhar irrigation project today is only working partially. The reservoir appears to have been functioning in recent past (Mbara et al. 2007), but a Google Earth image taken in 2019 shows limited irrigation activity in the plantation with the reservoir a large swamp (Fig. 14.18b).

Between Jowhar and Dobale, the study recommended that for future flood control the low-lying alluvial lands on the left bank east of Mahaddayweyne and the Sabun Barrage could be utilised as a flood-relief channel between the sugar plantation and the sand ridges. Recent Google Earth imagery shows that this linear depression has been utilised by flood waters in recent years and also that long embankments have been constructed across the valley and to the west of Mahaddayweyne, starting some 27 km downstream of Jalalaqsi and appear to be flood control measures to deflect floods into old channels and protect the main town. On the Shabelle left bank below Sabun, as seen on recent Google Earth imagery, farmlands previously irrigated have been inundated by floods, and swampy low-lying areas are fringed by wide belts of saline efflorescence in the soils.

Downstream of Jowhar and as the river flattens out, the Shabelle is flanked by a series of old channels, some with former meander belts quite clearly defined and others less so. In 1981, there was only limited flooding here all the way to the Shabelle swamps far downstream, and the *Gu* rains allowed rainfed farming to thrive that year: the common pattern of devastating floods next to undamaged lands where agriculture has a normal year.

14.5.4 Assessments on the Juba

On the Juba, the floodplain was filled by flood waters all the way from the Ethiopian border to Fanoole—Maps from the 1981 study are shown in Figs. 14.19 and 14.20. All rainfed or pump-irrigated annual crops on lower-lying lands were flooded and had mostly rotted by the time the flood receded. This was not unusual as the floodplain is narrow, and people were able to remove most pumps to higher ground. On higher grounds of levees, bananas often survived. Below Fanoole, the Juba Valley was constricted by flood protection

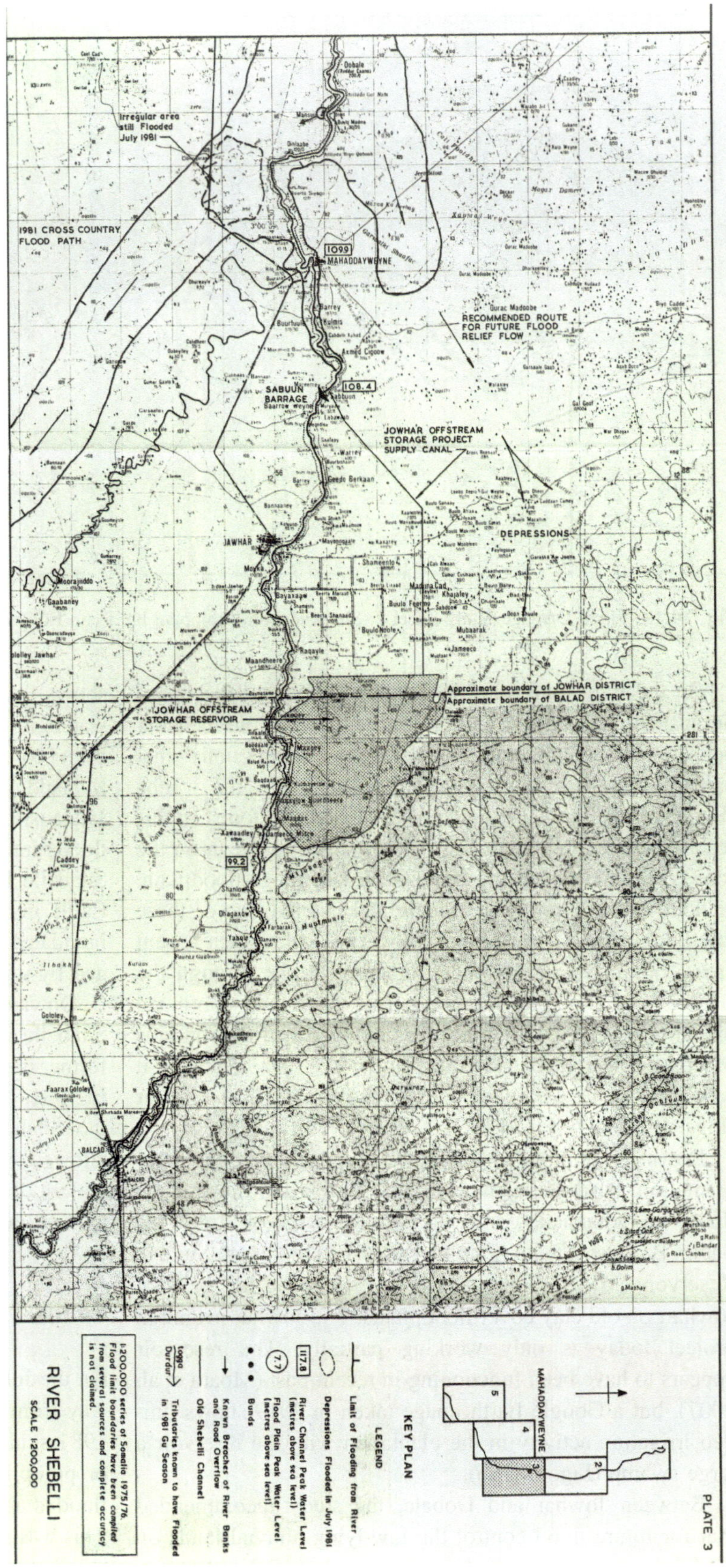

Fig. 14.13 Flood Damage Study Sheet 3: Shabelle from Mahaddayweyne to Balcad (MMP 1981). Report held by WOSSAC

Fig. 14.14 **a** Beled Weyn crevasse splay and flooding in the town, bypassing the flood-relief channel. Downstream is to right. *Photo* RN Munro ©, 28 June 1981. **b** Approximate same view generated from Google Earth ©. The town has expanded; the flood-relief channel had been constructed at the crevasse splay point before 2004 has an improved intake. It takes excess flood to re-enter the Shabelle 2.8 km to the south where an old channel network connects to the Shabelle, but the entire town is at risk from any future mega-flood. Google Earth ©

a

Fig. 14.15 **a** Shabelle, North of Jalalaqsi. View downstream. The 1981 cross-country flood visible at right background. *Photo* RN Munro ©. 28 June 1981. **b** Approximate same view using Google Earth © 4 March 2017. Most of the riverine forest, woodlands and scrublands on terrains at left (on stabilised relict sand dunes) have been degraded and enclosed

bunds around irrigation schemes, such as Fanoole, Juba Sugar and Mogambo, but the flood was able to overflow into the Far Shabelle, an old Juba channel to the west, and travelled for some 50 kms southwards (Fig. 14.21a, b). Then, and much to the project's surprise, the flood entered Juba Sugar on the west unprotected side via Scorpion Lake (Fig. 14.22a, b)—a case of the Scorpion's sting. The repeat images are 35 years apart.

The Somali Government reports suggested that the overflow was due to siltation of the Juba River but our mission discounted this. At Juba sugar, the flood came in, and through the plantation at such a pace though there was not time for stagnant conditions to develop, losses were modest. The old channels, clearly visible on archival aerial photography, satellite imagery, and also depicted on the Russian topographical maps, had been forgotten, but came back to life: this example of fluvial geomorphology in action was most interesting to observe from the air (see Chap. 13), and our study led to a better appreciation in government and the Juba Sugar Project that geomorphology should always be assessed. This right bank flood then continued southwards and came back onto the Juba floodplain near Bur Koy. The subsequent flood control on the Little Juba due to bund constrictions resulted in a concern by the Flood Damage Mission that long-lasting damage would occur to many of the huge mango trees along the Little Juba (Webi Yaro), since flood waters along it had been diverted back into the Juba at Kumbararee, thus leaving the Little Juba dry downstream from there. This would become a disaster for the hundreds of mature and very valuable mango trees along its banks that were already showing signs of drought stress. The Flood Damage Mission recommended that the blockage at Kumbararee be opened soonest, and it was: recent Google Earth imagery shows that the same mature mango trees, up to 50 m high, continue to thrive. A barrage on the Juba now uses the old channel as an emergency flood-relief conduit, and irrigated lands further downstream are protected, though the Juba Sugar factory and its project infrastructure was destroyed in the civil war period.

The Mogambo irrigation project, only partially functioning before (MMP 1979, 1987), was reported by Mbara et al. 2007 (SWALIM W-05) to be inactive. A mosaic of

b

Fig. 14.15 (continued)

Mogambo based on 1963 AP (Fig. 14.23a) can be compared with Google Imagery (Fig. 14.23b, dated January 2016). A close-up shows the main canal completely defunct, and the scheme remains only partly used by small-scale pump operations (Fig. 14.24).

Further south, two sets of repeats show the degradation of a fertile land: firstly, some 14 km SW of Jamaame (Fig. 14.25a, b) and secondly over a part of the Ionte Irrigation District flooded in June 1981 (Fig. 14.26a). The Ionte area lies some 20 km SW of Mogambo scheme, and the main road to Ionte and Kismayu is at the base of the photo. A more recent oblique image generated from Google Earth © dated 7 February 2011 (Fig. 14.26b) illustrates the general degradation of agricultural efficiency in this region: irrigated lands are only close to the Juba, and the riverine forest of 1981 is almost completely removed, while other parts are poorly drained, and terrains not irrigated are reverting to bush.

14.5.5 Conclusion

There remains much scope to rehabilitate the lands on both the Juba and Shabelle Rivers that were flooded in 1981 and of course often before and since. The FAO's 1981 Flood Damage Mission, by combining aerial survey with groundwork and seeking the geomorphological and hydrological explanations and an understanding of why, for example, one area was flooded and another was not, sets a useful standard and precedent for future flood monitoring and damage limitation in Somalia. Security issues in the Inter-Riverine make supporting fieldwork difficult, but it is hoped this will change.

One aspect the Flood Damage Mission noted was the accuracy of estimating the floods. The maps made by Gemmell, then revised by the Mission, and finally verified by ground observations, could be compared with the estimations in each district made by the Somali Flood Committee:

a

Fig. 14.16 a Cross-country flood at main road to north. *Photo* RN Munro ©, 28 June 1981. **b**. NEW Google Earth © 28 February 2013 imagery. The flood occupied the grey Vertisol soils of a former Shabelle back-swamp. Much of the bushland has disappeared and land is now more intensively cropped

- Shabelle. We estimated that out of a gross area of 128,000 ha flooded, some 26,000 ha was agricultural cropland. The official Somali Flood Committee assessed estimated 84,200 ha net.
- The cross-country flood covered 77,000 ha gross and 12,000 ha net. The Somali assessment underestimated at 2700 ha.
- Along the Juba, we estimated that out of a gross area of 144,000 ha gross flooded, some 53,650 ha was agricultural cropland. The official Somali Flood Committee Assessment estimated 66,952 ha net.

It is all much easier today with high-resolution imagery and cloud penetrating radar imagery that can be acquired quite quickly, and operational services/systems, such as the Copernicus Emergency Management Service (EMS) can provide in a few hours much of what the Flood Damage Mission took several weeks and flights by plane and long car journeys to complete. Security difficulties mean that sort of mission is not possible at the present. The EMS is a very useful service that uses imagery from earliest available suitable sensors and resolution be it Sentinel or very high resolution such as WV2 or 3, Pleiades and can provide detailed maps within 2–12 h for first estimate to damage grading assessment maps ready to print: all very more advanced than what was conceivable in 1981. In Somalia, where topographic maps exist at 1:100,000 scale, the infrastructure base could be in place before a flood event, and extent of floods depicted using freely available imagery such as Sentinel or Landsat, and then groundwork would need to follow up to assess crop and infrastructure damage: this is reported from districts. As it is, the assessments in Somalia reported by FAO-SWALIM (funded by UK Aid, USAID, World Bank, Somalia Humanitarian Fund, Swiss Confederation, Italian Cooperation, and EU) use for instance of WV2/WV3 Digital Global imagery that is readily available from archive. It is very useful to assess risk and plan emergency relief through reference and pre-disaster situation maps also available as EMS service:

[https://emergency.copernicus.eu/mapping/ems/emergency-management-service-mapping];

b

Fig. 14.16 (continued)

The assessment of such floods and their damage can be facilitated in part by rapid acquisition of imagery and change detection techniques to show, delineate and measure damaged areas. Hydrologic gauges and current metres on the rivers and routine establishment of rating curves are essential to maintain and are more robust and automatic compared to 1980s and no doubt will be in place for key stations already. In 1981, the gauges established by FAO (Gemmell 1982) following the 1981 floods were soon damaged or destroyed by floods in a following season. Civil war caused more damage, and a recent SWALIM study provided the status of many hydrologic stations and irrigation structures. Repair of structures and scheme rehabilitation in the Inter-Riverine will come with more peaceful times. The swamp maps made by Morgan and Thorn (MMP 1983) were based on the 1983 photography and could be re-examined to assess change. It would be useful also if more extensive use was made of the FAO model *Aquacrop* that uses real-time crop and climate data, together with soil, management and known groundwater conditions to predict the biomass and ultimately the yield of a calibrated crop. It should be in use as a tool.

In the future, it is likely that flood events will continue to cause losses to crops and infrastructure. With global change, these may well become much more frequent and more severe in destructive strength than hitherto. While dams on the

Fig. 14.17 Shabelle cross-country flood moving to the SW with its limit in distance. Difficult to repeat but about 15–20 km SW of the road crossing shown in Fig. 14.17a, b Illustrates the slightly uneven nature of the ground surface. *Photo* RN Munro ©, 28 June 1981

rivers could regulate floods, the investments needed are unlikely to be forthcoming, and the environmental and physical impacts are substantial. Better perhaps that a rethink is made of planning in the entire Inter-Riverine basin in view of these scenarios. One could, for example, abandon thoughts of rehabilitation of large schemes on the lower Juba and Shabelle and allow forests to return, biodiversity to recover; improve rainfed farming; and consider relocation to safer sites of frequently flooded major settlements on the Shabelle and Juba.

14.6 Case Study: Land Evaluation at Farjano

14.6.1 Background and Purpose of the Project

Improvements to the traditional rainfed agricultural sector in southern Somalia were initiated by Andrew MacPherson of Bingle Pty., who established rainfed schemes at Sablaale and Kurtun Waarey adjacent to the Shabelle River in the late 1970s (Citaco 1976; Bingle and Ltd. 1981–84; MacPherson 1983; World Bank 1984; Agrisystems Overseas Ltd 1984). Funded by the World Bank, the projects thrived. An advocate of rainfed farming over irrigation (as long as there is sufficient rainfall of course) MacPherson promoted rainfed farming for decades: on the clay belt rain lands of central Sudan, he considered that improvements in semi-mechanised rainfed farming—using precision levelling, opportunistic planting, improved cultivars, low tillage and good management (with farmers deciding on the spot rather than waiting for decisions from absentee landlords)—could be as productive as irrigation for numerous crops (Newtech-HTSPE 2008), and this has proved so.

On the sand ridge soils along the southern Somali coast, rainfed cropping gradually replaced woodland, but the soils were of low fertility. In 1984, Agrisystems, following assessments by GTZ (1982 1984) on the existing rainfed farms in the middle and lower Shabelle, made an appraisal report of the Farjano area for UNHCR. This proposed a rainfed scheme at Farjano, adjacent to Bingle's

a

Fig. 14.18 a View upstream into Jowhar Sugar Project in background, with excess flood waters entering the Jowhar Off-stream Storage Reservoir. *Photo* RN Munro ©, 28 June 1981. **b** Jowhar Off-stream Storage Reservoir, 19 August 2019. Acting as a large swamp. The plantation above it has patchy cropping. Google Earth ©

semi-mechanised rainfed farming project Sablaale and next door to the Sablaale irrigated farm project that was initiated in 1976 by the Settlement Development Agency. The project commissioned by UNHCR was an intervention to assist the huge refugee problem in the mid-1980s and aimed to determine if the Bingle model could be applied there. The Farjano area was bordered on its north side by the Shabelle, by this point a rather feeble channel and on the south side by the rising ground of the aeolian sand ridge belt. The Farjano soils were on older Shabelle alluvium and generally free of salts or sodic conditions, and it was considered that if used carefully and utilising residual soil moisture, the lands could be developed using the moderate rainfall, though in drought years the soil moisture deficit would be critical for crops. Fieldwork was made by Munro during January and February 1985.

Figure 14.27 is an HSL print-lay-down mosaic constructed using 1958 RAF aerial photography, and it shows there was little development in this part of the lower Shabelle at that time. The Shabelle swamps lie on the left and right of this reach.

At the extreme right of the area, Abel and Kille (1976) proposed a Shabelle Swamp Wildlife Reserve. These swamps remain largely undeveloped. A study by Mostyn Morgan and David Thorn of MacDonald's (MMP 1983) for the UK Aid Tsetse control project (LRDC 1983; Hendy 1985) considered that the swampy terrains were very likely to be responsible for the more elevated rainfall in the lower Shabelle and thus should not be drained.

In the 1985 Farjano study, the author used the 1:30,000 AP from February 1983 (Table 14.1): this appears to have been lost and thus was not available for this review. The Farjano rainfed farming study examined the suitability of soils for rainfed agriculture, settlements, forestry and small-scale irrigation. The 1985 soil studies considered carefully the issue of residual moisture, as advocated by Whiteman (1975, 1984) elsewhere, and concluded it was feasible under careful management and using precise land levelling to reduce run-off. Further, it concluded:

- Settlement planning needs to consider that sandy soils on levees are most suitable for agriculture and the heavy

b

Fig. 14.18 (continued)

cracking nature of the Vertisol clays makes them the least suitable;

- For forestry, better coordination on research was suggested when choosing appropriate species;
- Rainfed farming will always be a risky business due the unpredictable nature of the rainfall.

14.6.2 Assessments and Present Situation

The situation now as seen on Google Earth indicates that the Farjano scheme remains under bush and the Sablaale rainfed farm has been converted to irrigated lands or returned to bushland with settlements within a wooded landscape. Figure 14.28 shows the rainfed farms in these changed conditions: the original irrigated farm at Sablaale remains active, and a large off-stream storage reservoir has been constructed in the Shabelle swamps to the north (dark green area at top of image). The main track from Sablaale to the coast cuts through the remains of the Sablaale semi-mechanised rainfed farm now reverting to bushland. At left centre is a former channel of the Shabelle with Chromic Vertisol soils and ground cover (in 1985) of *Sporobolus-Dactylococtinium-Achinochloa-Enteropogon* grassland and an invasive growth of *Acacia spp*. The lands on either side of the channel comprised mixed bushland of 4–5 m high *Securinega, Acacia, Lawsonia, Boscia, Dobera* and *Phyllanthus spp* on cover plains, also of Chromic Vertisols. This bushland appears to have been cleared completely yet the rainfed farm that we assessed and designed was not initiated. It is apparent that unreliability of the rains combined with onset of a civil war led to closing of the funded rainfed schemes. In an assessment of the rainfed schemes in Somalia, Rees et al. (1991) concluded that while reasonable Gu season maize production can be achieved in most years, fallow soil

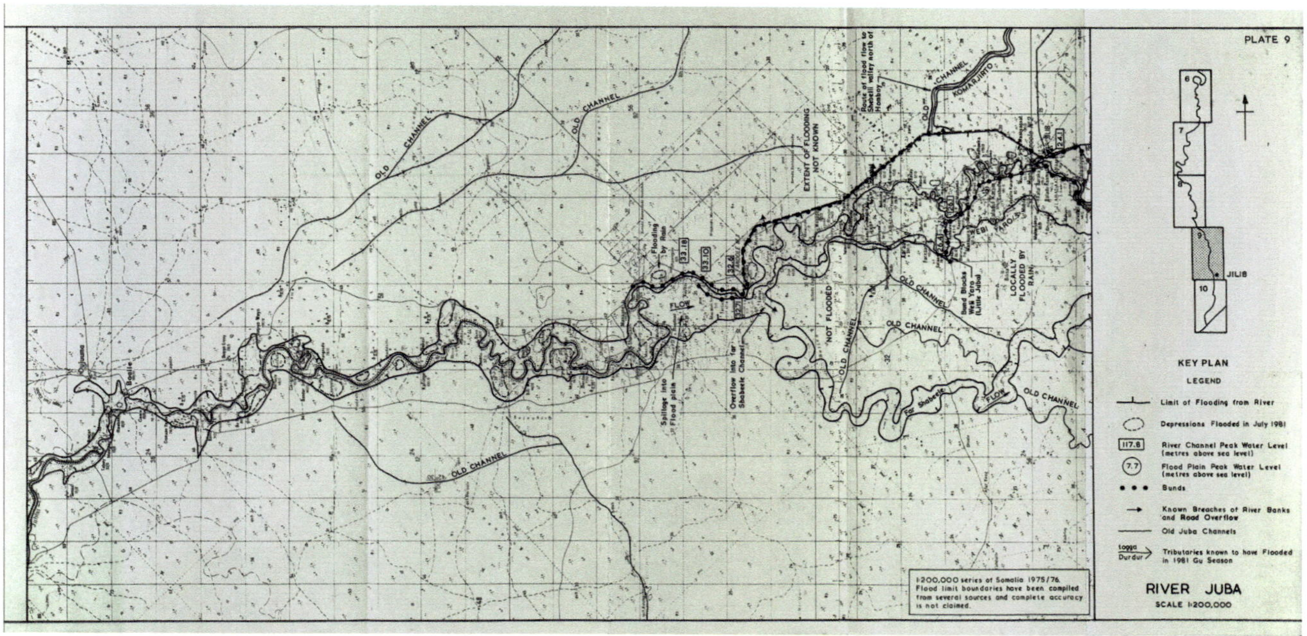

Fig. 14.19 Juba. Flood Damage Study Sheet 9. Juba from Dujuma to Jelib (MMP 1981). Report held by WOSSAC

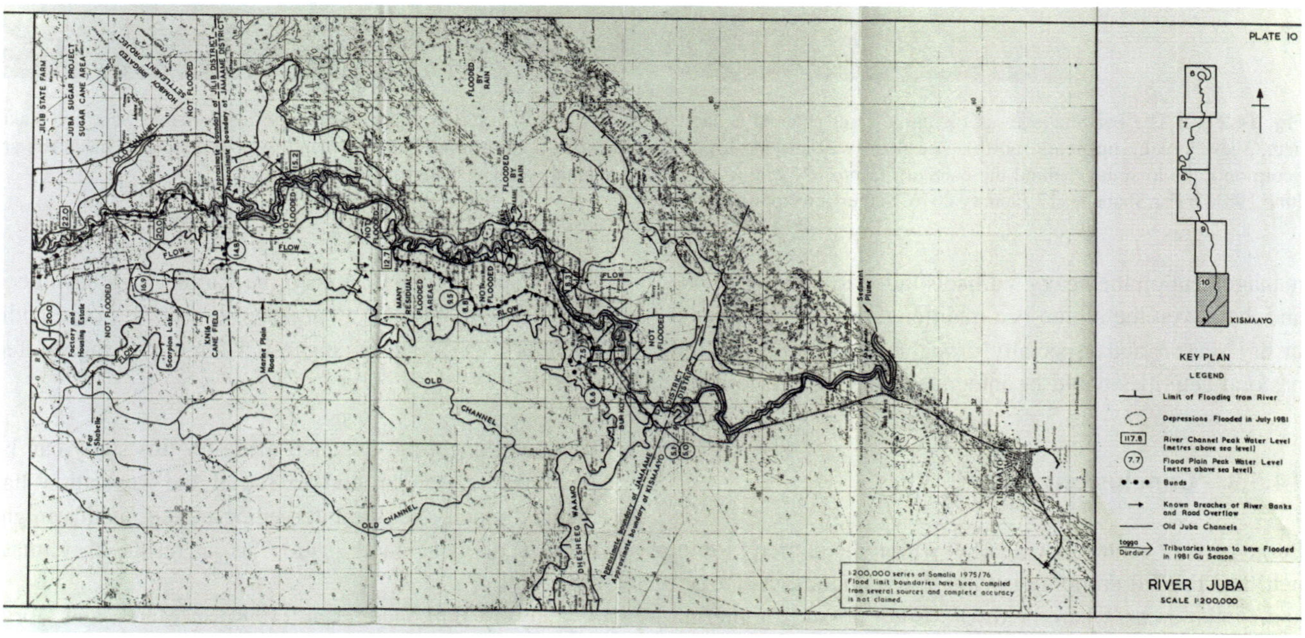

Fig. 14.20 Juba. Flood Damage Study Sheet 10. Juba from Juba Sugar to Indian Ocean (MMP 1981). Report held by WOSSAC

a

Fig. 14.21 a The Far Shabelle, an old Juba River channel is shown here. View looking upstream (north). The flood was moving towards Scorpion Lake to south (behind the camera). *Photo* RN Munro ©, 29 June 1981. **b** Far Shabelle, 17 January 2016. Same view approximately as Fig. 14.21a. Flow is from bottom to top left into Scorpion Lake. View to south. Marine Plain bushland is more degraded; meanders are perennial wetland swamps. Google Earth ©

management on the heavy Vertisols failed to increase yields, and that given the region is a traditional refuge for livestock in dry season and especially drought periods (Fig. 14.29), pastoral activities could be integrated with rainfed farming.

14.6.3 Conclusions

Large parts of central Somalia fall within the rainfed farming belt, and rainfed cropping is vital to livelihoods. At Farjano and similar areas, where fieldwork is not possible at present, it must be hoped that conditions will change for the better so that modern techniques for improving rainfed farming can be introduced, such as improved seeds, opportunistic planting, weed control, integrated pest management including desert locust control and utilisation of residual soil moisture.

Recent surveys that examined the Farjano area have been made by FAO-SWALIM (Vargas et al. 2007). Some additional future work can be suggested for improving rainfed agriculture:

- The land cover/land use map made for this study can be examined on the ground to see if the vegetation has changed much over the intervening period, as this might show ecological modification as a result of climate change.
- A study by Barkhadle et al. (1994) in Marka and Qoryooley Districts upstream of Farjano found that increased population pressure, overgrazing and fuelwood collection had led to a general degradation of the vegetation, and this has upset an equilibrium in traditional land use where the vegetation was determining land use: in 1994, this

b

Fig. 14.21 (continued)

a

Fig. 14.22 **a** Oblique photo of the Far Shabelle channel flood in Juba Sugar Project that was partly flooded. *Photo* RN Munro ©, 29 June 1981. **b** Scorpion Lake. More elevated oblique view than Fig. 14.28. Image from Google Earth © 17 January 2016 shows the degraded state of this area, hardly recognisable from the 1981 photo. The entire Far Shabelle channel and Scorpion Lake are swamps. Sugar fields are returning to bush, and considerable rehabilitation effort would be required

was reversed, and overgrazing is determining type of vegetation. The present situation is unknown.

- Use of the FAO model *Aquacrop* could help substantially to predict crop failure or success when rains are patchy.
- The Farjano soil sites, and those of other baseline surveys, can be georeferenced from the maps and relocated with reasonable accuracy: a soil sampling programme could examine, for example, if there are significant variations to soil organic carbon (SOC) in the rainfed agricultural lands.
- The Middle Shabelle Flood Control Study (MacDonald 1996) should be updated.
- This study has not examined the status of the rangelands east of the Shabelle, but regarding the tsetse control project of the late 1980s, Matthiessen and Douthwaite (1985) noted that the NTTCP was an exception in that it made extensive land use planning and environmental assessments first, but the overall results were not yet clear —the paper was published as the endosulfan spraying programme was starting. A global ban on the manufacture and use of endosulfan took effect during 2012, and the current status of the environment along the Shabelle due to the use of the endosulfan is not known, and an update on this and the tsetse situation is wanting.

14.7 Case Study: Assessment of Forest Ecology Along Middle Juba

14.7.1 Survey of the Forests in 1986

Under the patronage of Professor Lewis and Sir James Lighthill of University College London, the Somalia Research Project (SRP) made studies on the Juba during 1986. Led by Ecologist Jane Madgwick, the expedition spent several months studying the ecological status of the remaining areas of riverine forests along the Juba, between Bu'aale and Maname. The study of the Juba forests between Saacow and Fanoole had been proposed as a National Park and had the overall aim to assess the potential for establishing a reserve in terms of both human use and wildlife

b

Fig. 14.22 (continued)

conservation and to preserve the riverine forests (Madgwick 1984). The Barako and Shoonto area had been declared forest reserves in 1984 following assessments that showed a loss of over 50% between 1960 and 1984 (GTZ 1984; Madgwick et al. 1986). The results of the expedition were reported by Madgwick et al. (1988) and Madgwick (1989, 1990, 2000). The SRP expedition also studied the zoology of the forests (Varty 1988, 1990; Varty and Hill 1988). Other studies were made by Douthwaite (1987) who called them gallery forest: they are probably better termed riverine forest (see Wickens 1976, for a definition of gallery forest).

In terms of flora, the SRP found some 50 species of trees and shrubs in the levee forest, some up to 30 m high. There was a dense understory, and natural regeneration was common. Away from the river edge was a more open woodland, with large crowned *Acacia*. The transition between forest, woodland and bush was relatively sharp. The forest fauna included some 20 species of birds, specialists of riverine forest, who would be likely lost to Somalia if the last forest blocks would be cleared. Many trees found there rely on bats for pollination. The SRP also observed some 32 species of large and small mammals, some of the biggest concentrations in Somalia. Some large mammals visited during the long dry season moving in from the surrounding bushlands, but others, such as the red forest duiker, blue monkey and greater bushbaby, were restricted to these forests. Forest loss was quantified using aerial photographs by Deshmukh (1987) and Madgwick et al. (1988) for three different dates. Between 1960 and 1987, forest was reduced in area from 93.5 to 9.0 km^2. By 1986, at time of the SRP, the residue was fragmented and sizeable areas of riverine forest only existed in the Middle Juba between Fanoole and Bu'aale, but here over 8.5 km^2 of forest was cleared between 1983/4 and 1987. The SRP recommendations are shown in Fig. 14.30. In the two remaining blocks of forest, the main issue was to prevent cultivation encroaching further into the forests, with provision of forest guards and notice boards. However, the team of Madgwick et al. (1988) concluded that the two largest remaining forest blocks (Barako Madow and Shoonto) of the middle Juba remained relatively intact at that time, whilst all other stretches had become fragmented and degraded. These two areas were gazetted by the National Range Agency as forest reserves and had survived since, being valued and protected by local communities. Madgwick et al. (1988) recorded their importance for gathering wild honey, the source of medicines and fibre as well as affording

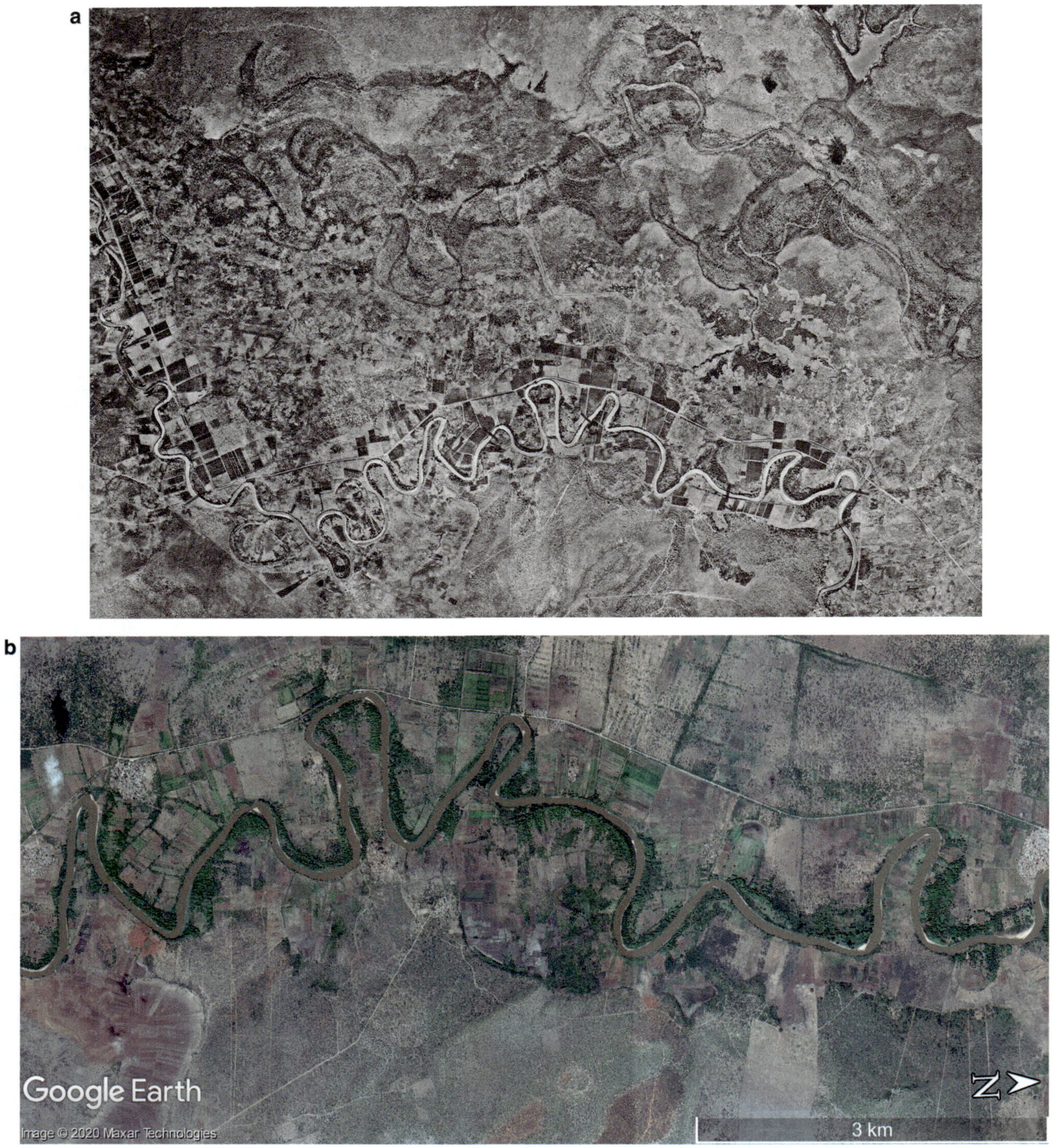

Fig. 14.23 **a** Mogambo Irrigation Project (MIP) area. Mosaic of 1963 AP—prior to development. *Note* old channels of Juba up to 11 km west of current channel. *Source* RN Munro AP archives. Print donated to NCAP, UK). **b** Mogambo. 2016 scene from Google Earth ©. The extent of irrigated lands is similar, though the MIP scheme is essentially not operating, main canal abandoned and all reverting to bush

Fig. 14.24 Enlargement of the MIP scheme in January 2016 showing inactive state. North at top. Google Earth ©

villages some protection from floods. The status of the Somali forests was reviewed by Madgwick and colleagues in an IUCN appraisal of coastal forests (Burgess and Clarke 2000). It was stated that little riverine forest remains in Somalia as most has been cleared for farmland and considered that while in the past forest would have existed along both rivers, on the Shabelle only a small and degraded fragment remained at the Balcad Reserve near Mogadishu. The legal status of the forests is that they remain unprotected.

14.7.2 Present Status

The status of the forests and need for their conservation was further noted by IUCN (1991). A more recent assessment of the forests in Somalia was prepared for the 2015 Global Forest Resources Assessment by Somali foresters (FAO 2014), but this report was limited in providing a sound baseline since fieldwork was not possible and remote sensing was not utilised: most of the report is based on estimates made from extrapolating old data. Unfortunately, they did not study the SRP works noted above that could have provided a baseline for comparative assessment. They reported that no national forest inventory has been made since 1980; all government documents were destroyed in the civil war; overall Somali forest and woodland resources represent a small percentage of the land area; recurrent drought is a serious issue; deforestation is prevalent but even so can barely meet local demands for fuel and wood products.

In 1958, aerial photography showed there was much more forest cover in this area. Figure 14.31a is from the 1958 RAF photography that was constructed into 1:60.000 scale negative mosaic by HSL in 1960 for the FAO Land and Water Surveys. The contrast between the bushlands of Marine Plain (dark) and Juba alluvium (light) defines the floodplain with clarity. Dry season AP shows high reflectance of grass and fields on floodplain soils, likely due to overexposure in mosaic making. What is not so clear are the forests shown in the 1986 surveys and the amount of forest cover: one needs to make stereo interpretation of the individual APs held at NCAP to assess canopy and forest structure. Figure 14.31b, from Madgwick (1988), shows the extent of the residual riverine forests in the SRP study area in 1986. Figure 14.31c is a Google Earth image from 2017 of the same reach of the Juba.

Comparing these by zooming in with Google Earth, one can observe that substantial remnants of the riverine forest and woodland mapped in 1986 still appear to remain

a

Fig. 14.25 **a** Oblique photo in the lower Juba area. Clear that some parts received no overbank floods. Large depressions between Juba and coastal dunes in far distance remained inundated for many months backing up against the dunes. *Photo* RN Munro ©, 29 June 1981. **b** Lower Juba 14 km SW of Jamaame, the 1981 flooded area remains a back-swamp basin lying between Juba and an old channel. The forest cover is considerably reduced along the Juba, but not disappeared. The tall lines of trees as windbreaks have all been cut down. Irrigation is patchy, and some canals appear to be used as tracks. Altogether a rather degraded picture is apparent. Google Earth © imaged on 17 January 2016

relatively intact in this reach, especially in and around Shoonto Forest Reserve, but parts have been converted to mango plantations and irrigated farmland. The dense, tall evergreen levee forest on the big meander of Barako Madow forest reserve has been partially cleared. There has been encroachment of farmland along the levees close to the Juba, but significant dense forest remains on the lower-lying cover plains that surround the back-swamps and old meander loops —the *desheega* basins that remain flooded throughout the year and are not suitable for agriculture. This is all very encouraging, though of course it is difficult to examine the biodiversity in any detail without a field study. The status of the fauna is completely unknown.

Further south towards Gilib, downstream of the CRP area, and onwards to the Indian Ocean, similar small patches of dense riverine forest remain, often adjacent to groves of huge mango trees along levees.

14.7.3 Conclusions

While parts of the original riverine forest remain, the status of faunal and floral biodiversity in the forest of the Juba is elsewhere seriously degraded by deforestation and encroachment of farmland into arable alluvial lands. The Juba Valley is a river system with an active meander belt that periodically cuts new channels. There are also numerous old channels that are still extant and will remain so for hundreds of years. As seen from studying recent imagery though, it is clear these old channels and associated back-swamp basin areas are perennial wetlands and their fringes support riverine forest. A new study to assess all this on the ground is overdue but difficult to achieve at the present time when there is a poor security situation.

In the past, IUCN (2006) reported that biodiversity conservation had a low priority in Somalia. This is

b

Fig. 14.25 (continued)

changing. In 2014, Somalia contributed an important assessment to the Convention on Biological Diversity (Ministry Fisheries and Marine Resources 2014). This covered Somalia, Somaliland and Puntland, and stated that there had been no effective biodiversity resource management and protection for areas since 1991, when the civil war started. In preparing for the 2020 biodiversity target, the report stated that "the most serious concern is the lack of effective legislation and management of protected areas, and absence of a functioning conservation infrastructure. There are fourteen protected areas, representing 0.8% of total land area, with only one measuring more than 100,000 hectares, namely Lag Badana National Park. Eleven wildlife areas had been declared since the 1970s, but only two were thought to be functional. In practice, there has been no formal protection offered to any of these sites since the early 1990s". The concerns over Somalia's biodiversity and environmental status continue to be raised: SWALIM has been making numerous surveys; UNEP made an environmental desk study that led from the impact of the 2004 tsunami to a countrywide survey with a useful bibliography (UNEP 2005) and produced a useful map showing the traditional ecological classification (Barkhadle 1993); biodiversity was further discussed by Ulla and Gadain (2016); the Ministry of Planning, Investment and Economic Development prepared a drought impact and needs assessment (MoPIED, 2018), and this complements the *Somalia National Action Programme for the United Nations Convention to Combat Desertification* (UNDP 2016) and the most recent assessment by the World Bank and FAO (2018) "Rebuilding Resilient and Sustainable Agriculture—Somalia Economic Memorandum". There is thus no shortage of willing and also plans and programmes for Somalia to rebuild its landscape.

Fig. 14.26 **a** Part of Ionte Irrigation District flooded in 1981. This lies 20 km SW of Mogambo scheme. Main road to Ionte and Kismayu at base. *Photo* RN Munro ©, 29 June 1981. **b** Same area shown on Google Earth © dated 7 February 2011. Dry season image. Irrigated lands that lie near the Juba have mostly replaced riverine forest. Other parts are poorly drained, a perennial wetland swamp; other terrains are not irrigated and appear to be reverting to bush

14.8 Case Study: Observations on the Coastal Dunes and Northern Escarpment

14.8.1 The Old Red Sandridge

The Old Red Sandridge, also termed the "Big Somali Dune" (Pignatti et al. 1993) and the "Merka red dune" (Carbone and Accordi 2000; Carbone et al. 1981), is up to 100 m high in places, up to 15 km width (Clark 1972), and is partly a relict feature stabilised and rubefied earlier in the Quaternary during moister phases but with active dune formation from reworked red sands and with dunes moving inland from sands blown onshore at beaches. Sands move north-eastwards along the Somali coast and are derived from offshore sediments of the Juba/Shabelle and Tana Rivers.

Angelucci et al. (1983, 1994, 1995), who examined the area between Mogadishu and Merka, considered the red sands were derived from basement complex metamorphic rocks in the Bur area of Somalia and that the red sands were mostly fluvial deposits. Their stratigraphy showed, from base: basal brecciated limestone bedrock, red sand with caliche (2 m), aeolian red sand with cross-lamination (10 m), red quart-rich sands with bands of white sand (25 m) and active dunes (4 m). The red sands and beach rock calcarenites were regarded as partly heteropic, derived from different rock types. These authors reported that the sands could be used for

b

Fig. 14.26 (continued)

a glass industry, and in the laterised red sands, extraction of Co, Ni, Sc, R, V, Cr, Ti, zircon, ilmenite, kyanite and rutile was possible, though the resource was not quantified nor was any environmental impact assessment provided on how such an extraction would affect dune stabilisation, biodiversity, pastoral activities, water pollution and rainfall.

Brook et al. (1996) and Brook (2002) have used a date from coral reef to indicate the age of the easternmost dunes in the Mogadishu area where they overlie coral reefs. This was in an 11-m-thick sequence of the Ras Aw Maki calcarenite at a quarry near Mogadishu Airport that comprised bioclastic calarenite, reefal bodies and coralgal facies with beach sand and cross-laminated sands above. A coral-rich unit 4 m above sea level was dated by ^{230}Th/^{234}U at 82 ± 6 ka: this was during Marine Isotope Stage MIS 5a when sea level was approximately 20 m below present, and thus tectonic movement has occurred since then. But dune sediments have been found in boreholes well below present sea level and suggest aeolian activity during at least MIS 6 when sea level was at −140 m (Brook 2002; Carbone et al. 1984; Angelucci et al. 1995). This was likely related to movement on two major faults (and others offshore) that lie subparallel to the coast and mark the location of the Somali Coastal Basin (Ali Kasim et al. (2002). The Brava sands may be partly contemporaneous with the beach calcarenite aeolian sands but most accumulated later than 80 ka and were brought onshore from exposed offshore sands at times of low sea level. Brook (2002) showed that post-depositional changes included weathering of feldspars and hornblende leading to formation of kaolinite (with up to 10% clay content) and iron oxides coatings on quartz grains, down to a depth of at least 20 m; later on, erosion of the fixed dunes

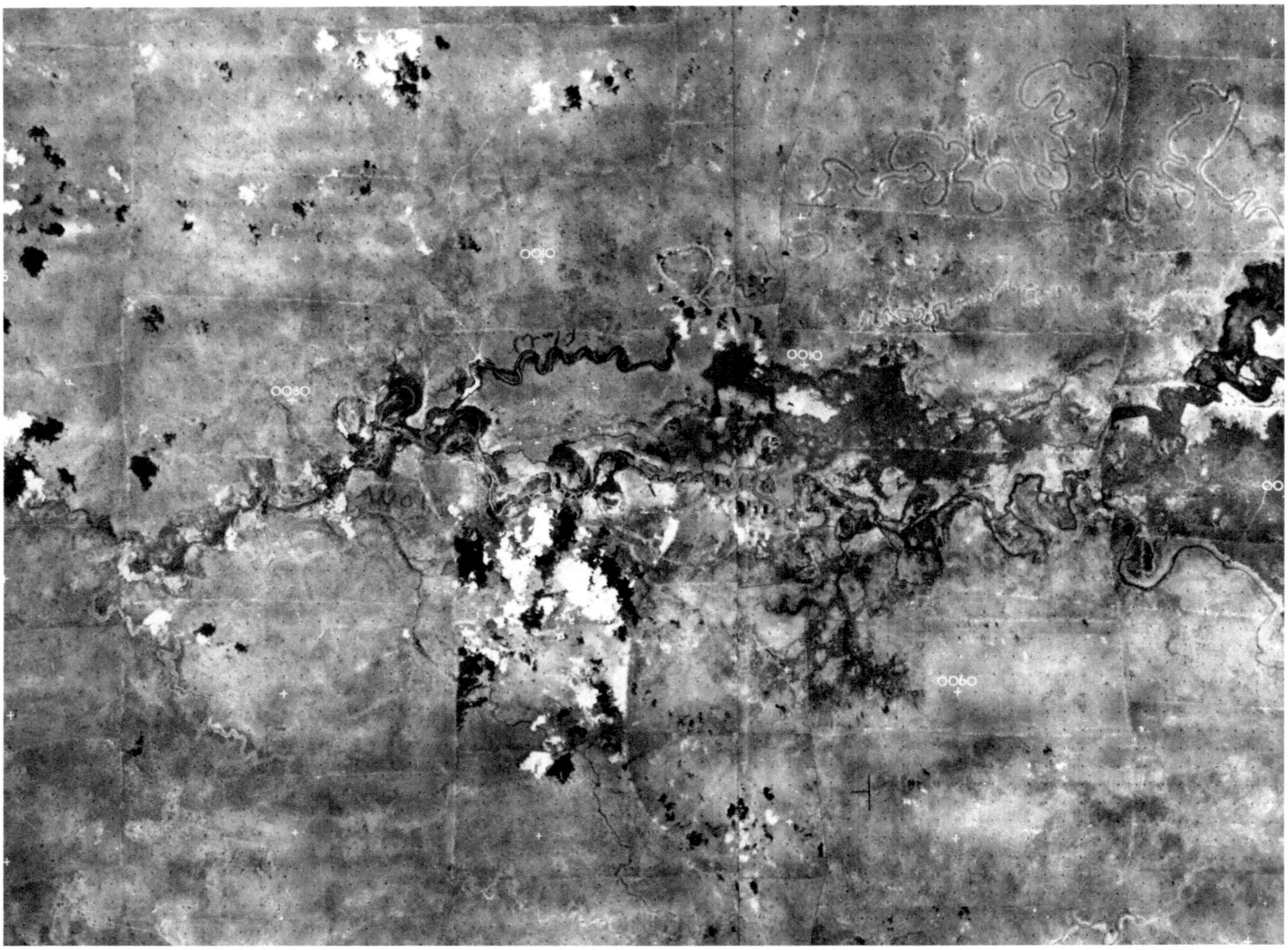

Fig. 14.27 Avai-Farjano-Sablaale Area/ HSL Mosaic at 1:200,000 made from 1958 RAF AP. Ref Block 1, Sheet 9. *Source* RNM AP archives. Original negative now with NCAP, Edinburgh

produced deep gullies on upper slopes, yardangs formed on the lower slopes from residual parts of rills, and gullies cut into the stabilised sands.

Our study of the imagery and aerial photography shows that parabolic dunes spread inland and built up the ridge, probably repeatedly. Given the extent of the sand ridge into Kenya, the mouth of the Tana River and longshore drift towards the NE, we think provenance could well be from a larger catchment area than just the Bur Region. Sands appear to have originated from the Tana, Juba and Shabelle Rivers as they debauched sediments along the coast. In addition, if the red sands were fluvially deposited, it is not clear how this could build up into a sand ridge formation in places 100 m above sea level. The ridge is underlain by beach rock and coral formations well exposed at Brava and Merka that probably constitute the seawards edge of an ancient lagoon. In fact, Artini (1915), 1926) concluded that the mineralogical composition of the dune sands, from Kenya to Mogadishu, corresponded to Juba River sands. Mussi (1975) developed this further to suggest that the dunes built up during periods of low sea levels during glacial epochs, similar to what RNM found on the Yemen Tihama (Munro and Wilkinson 2007).

Stabilisation of dune sands will take place whenever there is rain, but permanent fixing and development of a vegetation cover require a wet season that permits vegetation to cover the landscape and a soil to form. Airborne dust will assist in this process of soil formation. While annual crops (eg millet) can grow on dunes, the stabilisation is lost once soil moisture is utilised and the stover is removed.

At Brava, the chemical changes to the sands as described above will have necessarily been later during repeated moist intervals in the Pleistocene and early Holocene. Williams (2014a, b) discussed the ^{14}C and U-series ages findings of Brook et al. (1997) to show that wet phases in the Horn of Africa occurred at 260–250, 176–160, 116–113, 87–75, 13, 10, 7.5 and 1.5 ka. Soil formation and rubefaction of soils would have transformed active dunes into stable fixed dunes

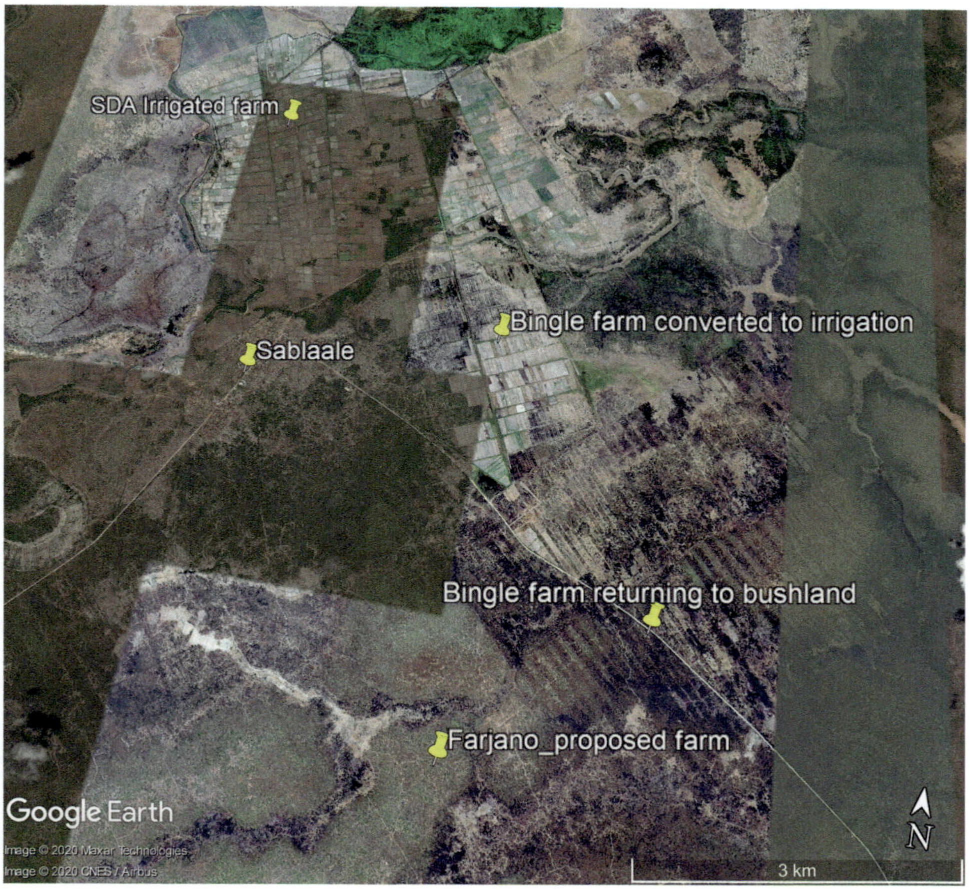

Fig. 14.28 Enlargement of the Farjano area. Google Earth ©. Imagery dates 29 January 2012 and 7 December 2014. The Farjano rainfed farm was never initiated, and 1985 trace lines have disappeared as land reverts to bush. The Bingle Pty Sablaale rainfed farm is partly converted to irrigation and rest reverting to bush

during such wet phases, much like the stabilisation of vast areas of *Qoz* in Darfur and Kordofan, Sudan (Parry and Wickens 1981).

The dunes of Somalia were examined by Lockwood-FAO (1968) who described the soils formed on the stabilised sands. Mainguet (1984) made a useful study of the sands for UNSSO and noted that wind directions and sand movement relate to the seasons: in the Gu (December to March), winds are from east to north-east; in the Der (May to September), from south-west to south. IUCN (1997) examined land degradation and coastal dunes in El Dheer area some 300 km NE of Mogadishu: barchan dunes form on the El Dheer coastal plain (*deh*) as source-bordering dunes from the stream beds (*doh)* and move in zigzag fashion to WSW in Jilal season, and then to NNE in monsoon season.

Examination of Google Earth and older photography (Table 14.1) shows that the landwards edge of the dune belt has characteristic AP features consistent with parabolic dune fronts moving onshore and inland from east to west indicating strong Gu season winds.

Along the Somalia coastline though, sands stream north-eastwards subparallel to the coasts, with dunes emerging from the sea and then entering it again, as at Xaafuun far to the north (see below). Longshore drift brings sands along the Somali coast towards the tip of the Horn, and at intervals—where the geography is appropriate—onshore winds bring these sands onto the beaches where they are then blown inland. It is likely the dunes and drifting sands blew much further inland in the past as faint ghostly traces of parabolic forms are seen in places, such as SW of Galkayo. These may indicate wind directions that no longer operate. There are modern active and ancient stabilised sand dunes that occur along a greater part of the coastline of Somalia from Kismayu to Ras Xaafuun and its lagoon, almost to the tip of the Horn of Africa itself.

The coastal dunes are stabilised in southern Somalia from Kismayu northwards and for several hundred km beyond Mogadishu and have a surface loamy to sandy surface soil layer and buried paleosols. The dune ridges are up to 110 m high and have a complex history that remains to be deciphered. Modern blowout areas are common, and the sands have pale tones and appear to be moving into the farmlands of the Shabelle. Closer inspection shows that they are relatively stable, the fringes have been protected and protective vegetation has recovered, and the active parts have retained a similar form for over 50 years. In addition, much work has been made to stabilise the dune margins where they impact on irrigable lands and infrastructure.

Fig. 14.29 Just a Dry Season Condition or Example of Drought? Munro in the dry Shabelle, Farjano area, 1985. *Photo* RN Munro ©, 1985

The climax vegetation of *Acacia tortilis* and *Acacia busia* (Pignatii and Warfa 1983; Pignatti et al. 1993) is subject to clearance nearer the settlements. Both local remobilisation of sands and onshore drift of sands appear to be taking place. The lack of stabilisation of mobile sands by vegetation has resulted in numerous blowouts that extend inland over the surface of the sand ridge.

Mainguet's (1980) study had recommended that fragile areas should be closed to animals and afforestation made on the sand ridges where degradation has resulted in reactivation of dunes that threaten the roads and irrigated lands inland. This was a common good practice and timely messages in extension and conservation: they were heeded. At Brava, gully erosion into the old sand ridge was halted by planting a graminaceous cover with trees and *Opuntia sp*. on its upper margins to halt erosive run-off from the crest of the ridge. Within the gullies, check dams were planned to permit accumulation of sediment, but it is not known if these were constructed.

Figure 14.32 a shows a large part of the coastline and interior Shabelle plains of southern Somalia. The mosaic was made from 1958 RAF photography by Hunting Surveys. Brava lies at the promontory at the right-hand side of the image. The ancient dunes, whose inland limit is marked by the irregular difference from dark to light tones, are stabilised by iron-stained soils and $CaCO_3$ concretions throughout. The ancient dunes, deeply gullied in places as at Brava, generally show dark tones on the photography from the shrubby vegetation cover. Figure 14.32 b is a Google Earth image (2011–2017) of the same area that indicates only small changes to the large coastal blowouts over 50 years.

At the same time, large gullies have cut down into the ridge. The date of incision of these gullies is not known, but, at Brava, they are older than the 1958 RAF photography. Elsewhere along the coastal sand ridge, there are numerous examples of old gullies now completely re-vegetated, for example 40–48 km SW of Brava, and some 10 km SW of Merca. Closer to Merca and extending for some 40 km to the NE, there are many rills and gullies cut into the paleosol surface of an old sand massif. Some 5 km SW of Merca, from a well-vegetated coastal zone 1 km wide, the old sand ridge rises sharply up to 50–100 m asl and is characterised by many rills and gullies. The convex top of the massif

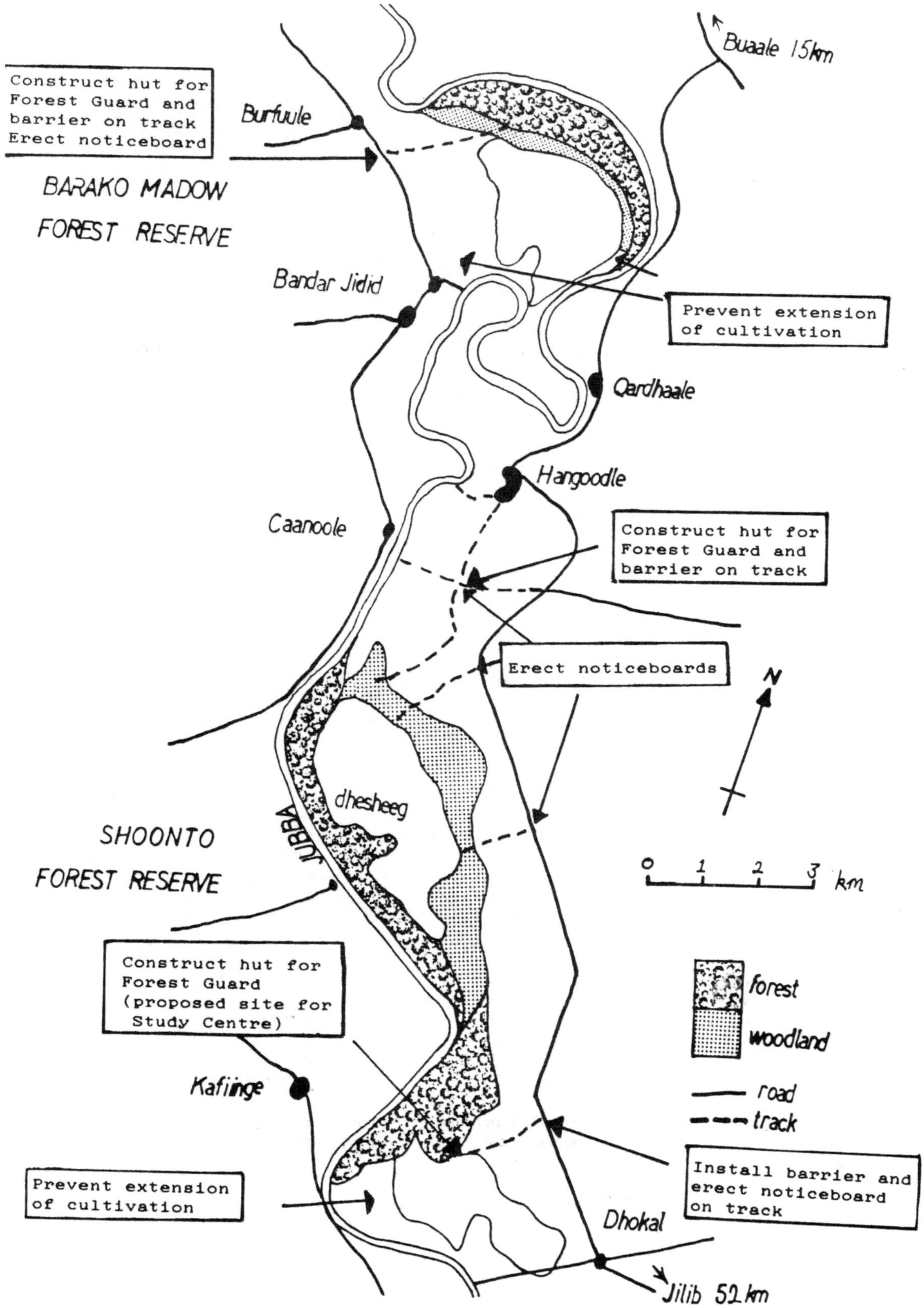

Fig. 14.30 Recommendations of the Somali Research project (Madgwick et al. 1987)

a

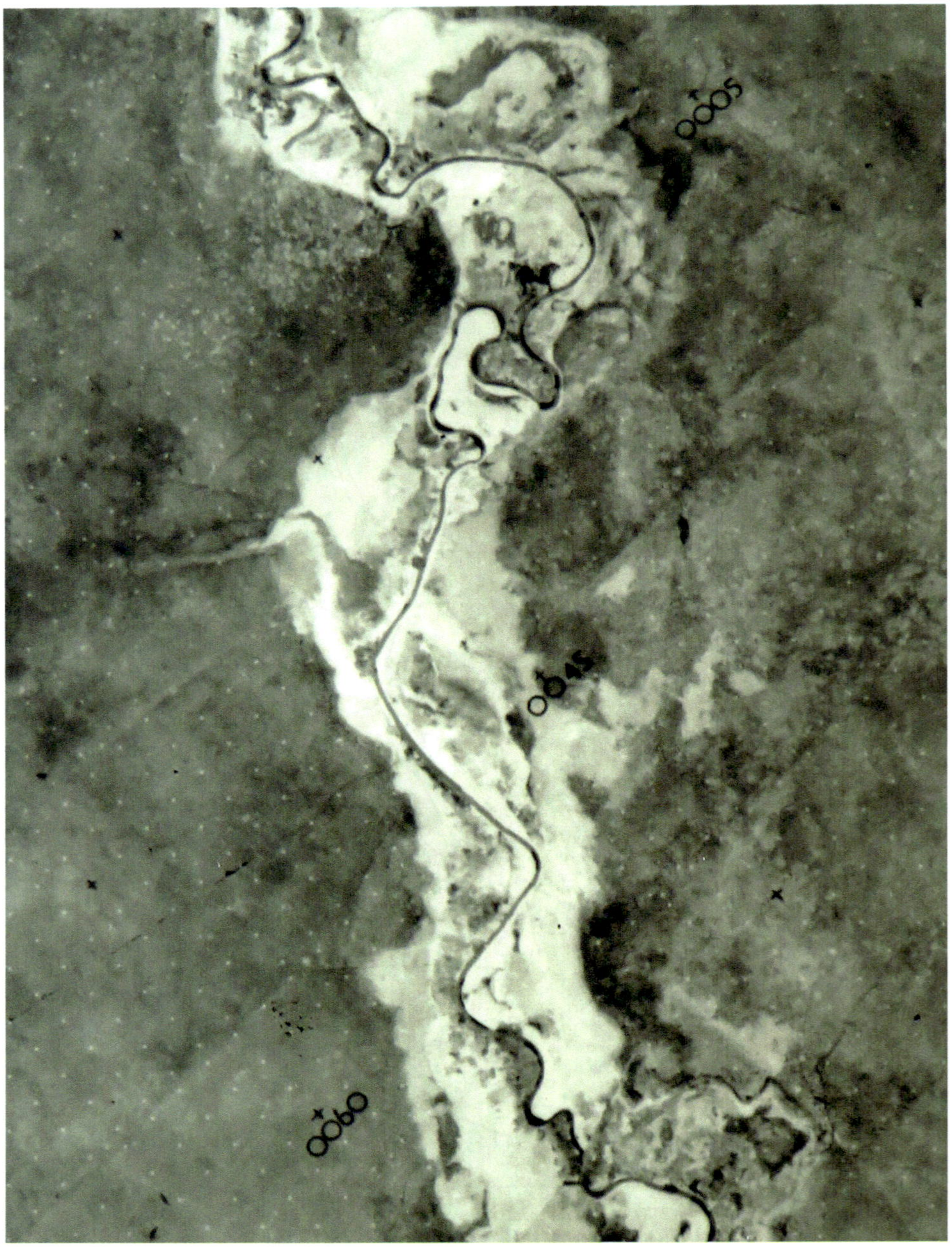

Fig. 14.31 a Juba forests between Buale and south of Maname. 1958 RAF photography. Copied from 1:60.000 scale negative mosaic, made by HSL in 1960, converted to positive using Adobe Photoshop Elements. The photograph of the original print lay down was overexposed making differentiation of alluvium and forest difficult. Sheet 37 of 1:60,000 mosaic series D-5890. North is at top. *Source* RN Munro AP archives. During 2020, the original negative will be donated to NCAP, UK. **b** Forests on the Juba (from Madgwick et al. 1988). Shows the two main forest blocks studied on the SRP. **c** Juba forests between Buale and Maname. Same areas as Photo 30. North is at top. Shows that some forest remains and small changes to river morphology. Google Earth ©. Imagery date: 31 March 2017

extends a further 2 km inland and is gradually buried by active dunes that are transverse ridges formed by SW winds with linear ridges and lobes formed by east and south-east winds. The upper part of this relict dune massif is bare without vegetation cover and run-off from rainfall from rills and gullies that are steadily wasting the entire area.

A close-up view of part of this area is given in Fig. 14.33 a, taken in 1968. The image shows parts of the coastal dune ridge 18 km south-west of Brava. The dunes on this December (1968) AP indicate two wind directions from west and north-east. The winds from the west realign sands that stream to the east, modifying the underlying more massive transverse dunes that have south-west facing slip faces. The Google Earth imagery (Fig. 14.33b) taken in December 2014 shows barchan dunes moving onshore and westwards to fuel the transverse dunes. The barchans pile into the transverse dunes that form most of these active massifs. There appears little change in the 1958 mosaic and 1968 vertical photography. By 2014 though, there is evidence of minor adjustment of the loose sands: the road alignment has been shifted some 400 m to the north and west as sands engulfed the road. The edge of the dunes though is now protected by vegetation. The dunes appear to show a complex pattern of different wind directions with regional winds from west and onshore from east at same time: this can be explained by fact that the dunes reach up to 100 m height in this area, while the Shabelle cover plains inland are at 45 m asl and the onshore winds do not reach far inland. Interior winds in fact

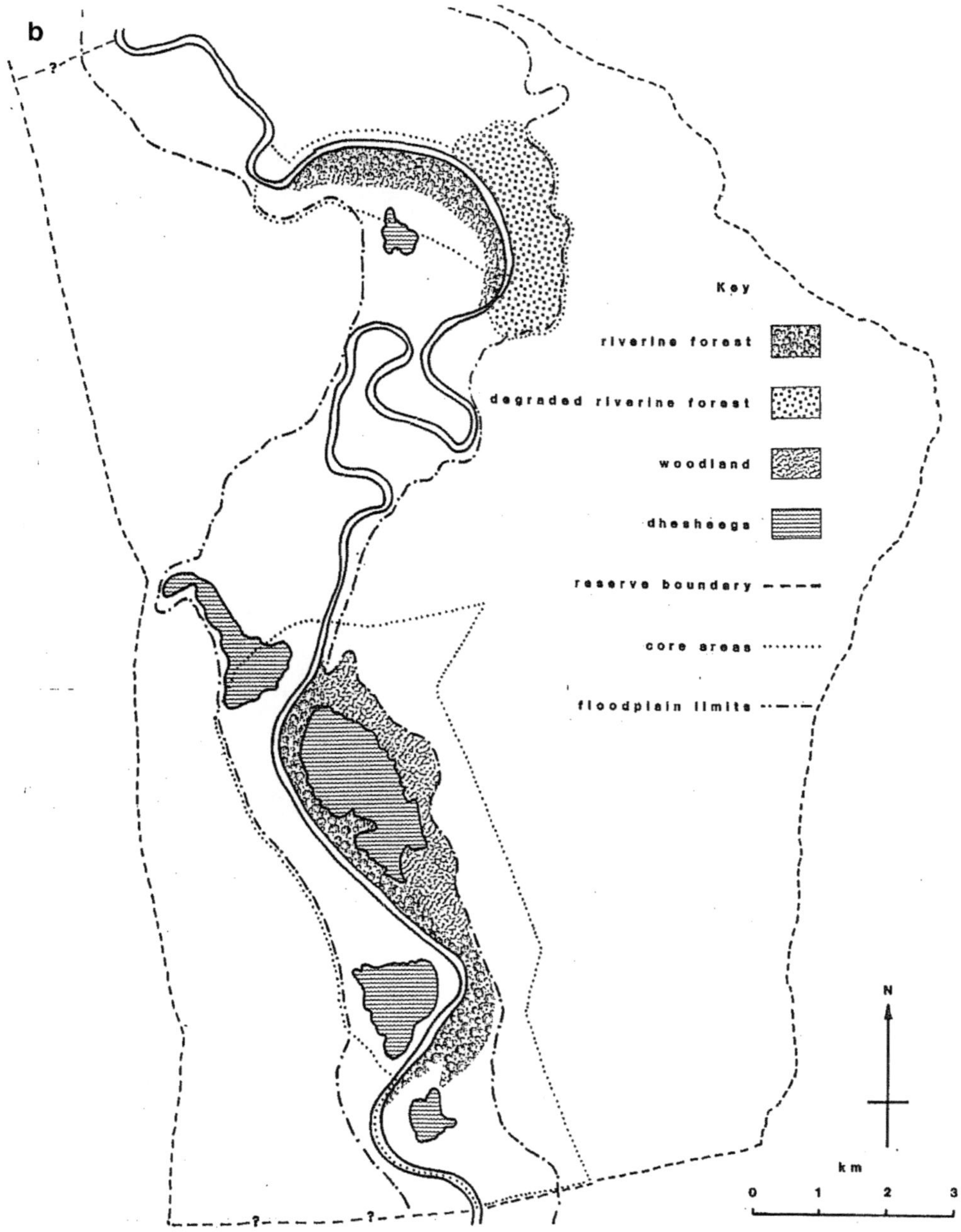

Fig. 14.31 (continued)

c

Fig. 14.31 (continued)

may assist in blocking eastward movement of sands and maintain the dunes where they are.

The situation varies locally: the skilfully made and attractive Italian Military mapping from 1911 provide land cover information and spot heights. The history of Italian mapping in colonial Somalia was documented in excellent monographs by IGM (1939) and Traversi (1964). The Muriole 1:50,000 scale sheet of the El Bacari area, some 31 km ENE of Ione/45 km up the coast from Kismayu, delineates the active dunes (*duna mobile*) and stabilised dunes (*duna stabile*). This identification of active and stable dunes represented an advanced recognition of geomorphic history for the time. Comparing with Google Earth, one can see that *duna mobile* overran *duna stabile* inland and the dune front advanced inland about 1 km (Fig. 14.34a, b). On the 1911 sheet, spot heights made by the topographers show a distinct ridge with elevations reaching 80 m above sea level. Google Earth altimetry here indicates elevations up to 65 m asl: it appears the crest of this dune has been deflated and parts moved inland at some stage between 1911 and 1958: the 1958 photography shows minor changes (not illustrated here due to poor reproduction from the negative), but since then the dunes have moved further inland to the state indicated in the 2011 image.

In a study of the Sar Uanle coast, 20 km south of Kismayu, Vannini et al. (1977) updated the pioneering work by Fantoli (1965) and showed that the winter monsoon winds blow directly off the ocean from the east, but wind speeds in winter were rarely high enough to move sands off the beach.

Yet, the evidence from Google Earth and aerial photography is that strong winds do push sands off the beach. It seems clear that periodically exceptional events, cyclones would seem the likely cause, have occurred to advance the dunes inland. This point was first noted by Mussi (1975).

In Fig. 14.35 a, b, repeat images from 1968 and 2012 show the large gullies (Fig. 14.36) to the immediate north of Brava cut deeply into the ancient (Pleistocene) dune formation massif, but the 2012 Google Earth image indicates that gully headward erosion has been stabilised in terms of further expansion. To the west of Brava and west of the main road (in yellow on Fig. 14.35b), gully stabilisation was initiated on the top of the dunes by the National Range Agency in afforestation schemes in the mid-1980s, notably in Brava area, when RNM assessed the changes. These

a

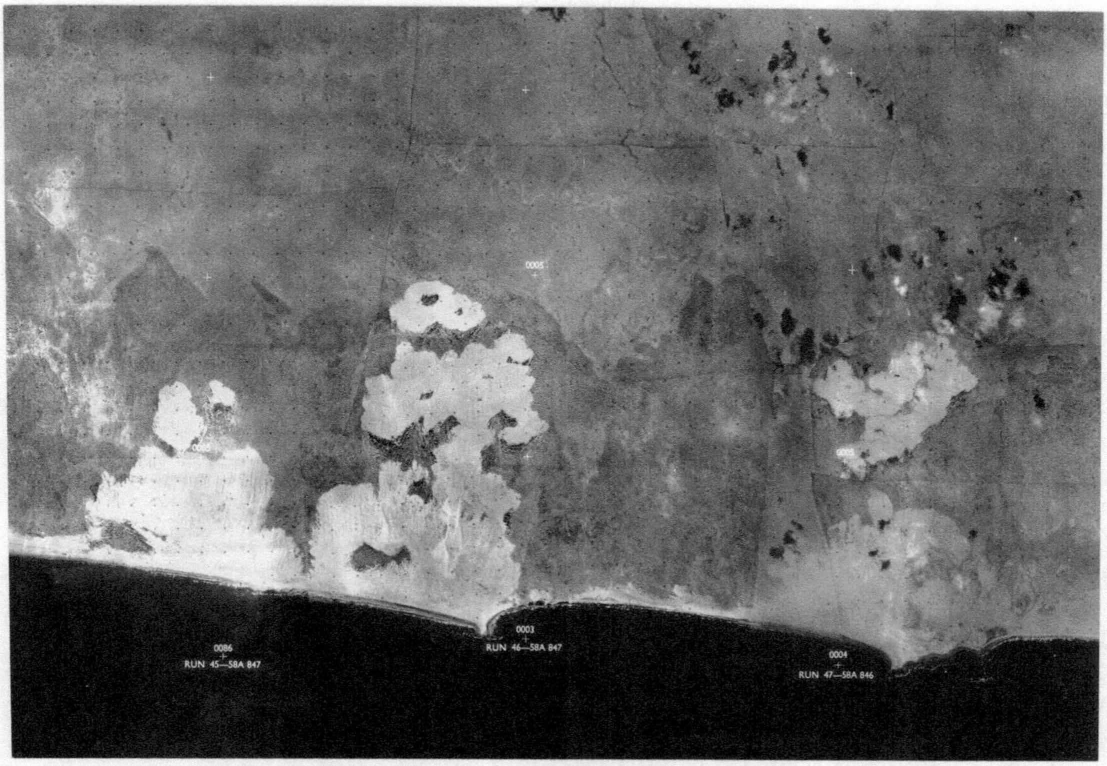

Fig. 14.32 **a** Dunes along southern Somalia. The mosaic (Block 1, Sheet 9 at 1:200,000) was made by HSL in 1970 using 1958 RAF photography. *Source* RN Munro AP archives. During 2020, the original negative will be donated to NCAP, Edinburgh, UK. **b** Google Earth image of same area. Images 2011–2017. Minor modification of the principal parabolic dune blowouts. Interior edge of older parabolic dunes suggests an abrupt phase of stabilisation and rubefaction. Google Earth ©

gullies are clear on the 1958 AP, but are not indicated on an Italian military map of 1910: but even if they existed, then the small scale of 1:200,000 would likely make delineation difficult. It is possible there exist more detailed Italian maps that could be utilised in a future exercise. The formation of the gullies must have been related to intense rainfall cutting rills to expose deep soil of the stabilised dunes. These then were deepened and expanded with headward erosion. On the top of the sand ridge today near Brava, there is considerable encroachment on the bushlands with change to rainfed farmland (top right of Fig. 14.35 b). This encroachment is on the flatter parts of the sand ridge, but nearer to Mogadishu these sand ridges are extensively, and for long, cultivated. At Brava, where the dunes rise up to 110 m above sea level and may catch more moisture, it is a more recent development as the city has expanded from a tiny port in 1985. There are red and white dunes: the red dunes are formed from iron-coated sands eroded out of the ancient dune ridges; the white sands come from the sea: these active sands ride up and over the ancient dune ridge. There will be mixing of both in some places. The stabilisation of the gullies happened after 1985, as RNM visited them at that time (Fig. 14.37), when *Opuntia* (Prickly pear) was seen growing as a stabiliser, reducing erosive rainfall, and likely too as a discouragement to livestock to enter the thickets.

Along another part of the dune belt closer to Mogadishu in 1977, the movement of sands on the landward side of the dune ridge was giving some concern to those planning the proposed development of the Genale-Bulo Marerta irrigation scheme (MMP 1978a. Annex 1: Soils). The Genale-Bulo Marerta Project covered 67,400 ha of lands located on the floodplains of the lower Wabi Shabelle some 100 km south-west of Mogadishu. It was bounded on the south side by the coastal aeolian sand dunes ridges and on north by old channel landforms. The Genale (Janaale) area was first developed in 1926 by Italy, with canals established by 1927 over some 30,000 ha. MPP noted that no other records from this Italian period could be found in Mogadishu. A concern in 1978 was that the coastal dunes inland of Merca would move onto the farmlands of the Genale—Bulo Marerta

b

Fig. 14.32 (continued)

scheme and also bury the air strip south-west of Shalambood (Fig. 14.38a). But this did not happen partly due to conservation efforts on the sands and also because the air strip was built parallel to the prevailing wind directions and sand movement. The main drift of sand is to the south-west and the edge of the dunes have been vegetated. The air strip remains free of sand (Fig. 14.38b). Additional comparisons in Shalambood area a little to the north (Fig. 14.39a, b) show that between 1968 when active dunes were threatening to move across the road and 2020, a large effort was made to stabilise the dunes on the landward side. We have not found the details of how this was achieved but was clearly a great success. In general, and in southern Somalia, the juxta-positioning of winds and the presence of shrubby vegetation cover on the sands appear to have limited any major expansion of these dune groups for the last 60 years. These time-lapsed pairs show there has been noticeable afforestation of active dunes, all very promising.

Far to the north-east along the Somali coastline, in a much more arid climate, lies the peninsula of Ras Xaafuun (Hafuun) with a shallow lagoon to the east. Up to the 1940s, this was a very productive area for salt evaporation by the Italians and local communities, but destroyed by Allied forces in 1940. The remains of the pans occur around the lagoon near Hordio in the 1958 mosaic (Fig. 14.40a) and the modern Google Image (Fig. 14.40b). The southern margin of the lagoon is a sand bar that was covered by the 26 December 2004 tsunami: since then, a new town of Xaafuun has been built on the higher ground of Ras Xaafuun, actually quite close to the old Italian settlement of Hafuun that they called Dante. The modern village is secure from any future events, and efforts have been made to assist the communities and hopefully revitalise the salt industry.

The mosaic print shown in Fig. 14.41 is a part of Sheet G-13, Block 4, in the D-6294 mosaic series of Northern Somalia. Made from 1958 RAF photography, it lies just

a

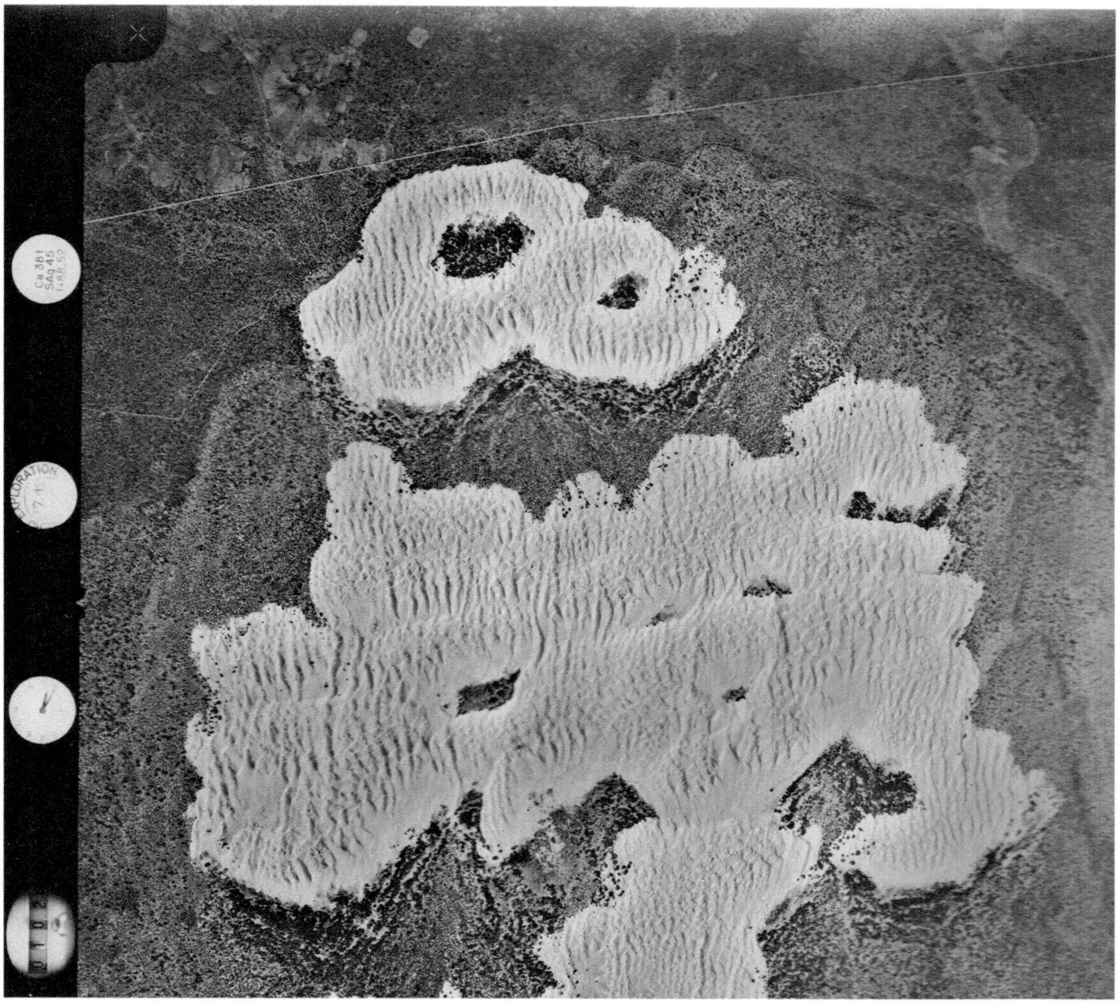

Fig. 14.33 **a** Aerial photograph from 17 December 1968. Enlargement of coastal dunes seen in Photo 32 and 33. *Source* RNMunro AP archives donated to NCAP. **b** Google Earth © image of the same area appears very similar with slight modification after 46 years. December 2014. Google Earth ©

north of the Ras Xaafuun (Hafuun) lagoon. The coastline at extreme right is at approximately 10.955323° N, 51.117796° E. Immediately north of the lagoon ephemeral streams enter the coastal plain. These drain the interior plains underlain by Eocene to Miocene limestone, evaporites and sandstones. The latter contribute some sands that will deflate, as source-bordering dunes, but it is clear that most of the barchan dunes are derived from sands emerging from the sea and drift northwards. Enlargements of the 1958 mosaic (Fig. 14.42a) and recent Google Earth imagery (Fig. 14.42b) show this process as sands emerge from the sea and drift over the sand bar. Along this part of the coast, there is a north-easterly movement of barchan dunes along the coastal plain that then re-enter the sea again—at the right-hand edge of this mosaic. Comparing the mosaic with the situation now, the accuracy of this uncontrolled print lay down mosaic is good; the direction and numbers of barchan dunes appear similar, and in monochrome rather clearer than in the colour imagery of Google Earth; and the density of vegetation in wadi lines and on sandy plains appears much the same.

14.8.2 Dust Hot Spots

A recent global dust assessment (UNEP, WMO, UNCCD 2016) identified two Somali dust hot spots, regarded as of likely anthropogenic origin: the Nogal Valley (that they also call the Nogal Lake though there is no permanent water feature there) and the Hobyo coastline between Mogadishu and Eil (Eyl). It was not stated if any fieldwork was made to support these claims.

For the coastal zone, the Hobyo comprises grasslands and shrublands and is a desert and xeric scrubland ecoregion: an anthropogenic origin seems doubtful as it is a very dry,

Fig. 14.33 (continued)

mostly sparsely populated and narrow hot coastal desert region. Google Earth imagery shows that sands and finer materials are drifting onshore, in places along a narrow strip backed by steep terrain that precludes movement of sand inland. Elsewhere, sand sheets drift northwards from here up to 60 km inland before entering the sea again close to the most easterly part of the Nogal (Tug Der) Valley.

Inland, the Nogal Valley broadens out. It lies in a tectonic half-graben up to 200 km long (Ali and Lee 2019), cut into Eocene and Pliocene sediments; elevations rise to ca 500 m asl. It is a very dry pastoral area, devoid of vegetation arcs, but with scattered bushes set amidst very pale bedrock and Quaternary fills. The Tug Der is incised into the valley floor and closer to the ocean occupies a narrow canyon. An anthropogenic reason for dust generation in the Nogal basin seems not proven.

14.8.3 Northern Escarpment

In the Horn of Africa, the most north-easterly part of the region, the landscape is arid with little scope for irrigation or water harvesting except very locally. We examine the Ferio Range (10°18'N, 45°30'E). One repeat scene shows the relatively unchanging nature of a desert landscape: Fig. 14.43a is a USAAF Trimetrogon K-17 right oblique

a

Fig. 14.34 a Part of 1:50,000 scale sheet F° 1 (Muriole) from *Carta della Somalian Italiana*, surveyed in 1911 by the Instituto geografico militare. Publication date not given. At right the mobile dune ridge of Bacari with its defined crest, now lost, and vegetated dunes to its west, south and at far right. An old forested course of the Shabelle is at upper left. **b**. The same area is shown in Fig. 14.14. The active dune fronts shown in 1911 later advanced inland as parabolic dunes from west during severe storm events between 1911 and 1958 and also since then, probably as intense and short-lived storms: historical records from cyclones may indicate just when. Traces of older parabolic dune fronts with rubified soils and some cultivation, and more recent paler sands covered by scrub, are clear. Google Earth © image dated 7 February 2011

photo from 1945 of the Ferio Range looking south. Figure 14.43b is from approximately the same viewpoint (27 May 2013) and shows the Ferio Range with the Somaliland plateau in background. These oblique views are made from a point some 40 km east of Berbera. In general, the two images 68 years apart seem initially to be much the same, as might be expected in an arid zone. However, the arboreal vegetation in the streams at lower left appears denser than in 1945, and stream course changes can be observed on the alluvial fan complex at right centre. An additional feature is that red aeolian sands at the base of the Cretaceous Yesomma Sandstone cliffs at lower left move northwards as sand sheets and barchans amidst tussock grassland. The active grass tussock dunes move northwards, from one of the few inland sand source areas on the escarpment, while other dunes move southwards off the sea: these meet in a zone of star dunes up to 50 m high. Off-screen from Fig. 14.43b, these sands merge with a series of southwards-moving onshore dunes from the coast forming chains of east–west aligned "mini" star dune complexes. These areas received torrential storms from Tropical Cyclone Pawan in December 2019 (IFRC 2019).

b

Fig. 14.34 (continued)

Further east, Brook et al. (1990a, 1990b, 1998) examined massive and inactive stalactites and stalagmites in two caves on the Golis Range, and $^{230}Th/^{234}U$ Uranium series dating of speleothems and tufa formations showed these were formed during several phases (at 116,000–113,000; 87,000–75,000; 12,000–4000 yr B.P., though with large uncertainties) during interglacial periods, and that isotope stages 1 and 5 were here represented by increased moisture and a strong south-west monsoonal air flow. Analysis of pollen data, however, suggested that the Golis area did not support rain forest or montane forest vegetation during the time periods of the dated speleothems. They considered that the few occurrences of *Juniperus* pollen probably relate to patches of dry woodland with *Juniperus*: similar forest occurs today on the Golis escarpment edge north of Erigavo (Pichi-Sermolli 1955).

14.8.4 Conclusions

The huge blowout features of the coastal dune ridge along the coastline show minor change since the 1950s. The prevailing wind directions today mean that movement further inland of the dunes onto the cultivated alluvial soils is minimal, despite the menacing appearance of the blowouts. All that could change in a world of global change that could include changes to wind directions and strengths. This has happened in the past when mobile sands covered the margins of the Shabelle alluvial lands. While sand is coming onshore at various locations, such as at Brava, attempts to stabilise active sands have been quite successful: such work needs to be maintained. Natural stabilisation of gully systems is noted between Brava and Kismayu in the higher rainfall zone, but

a

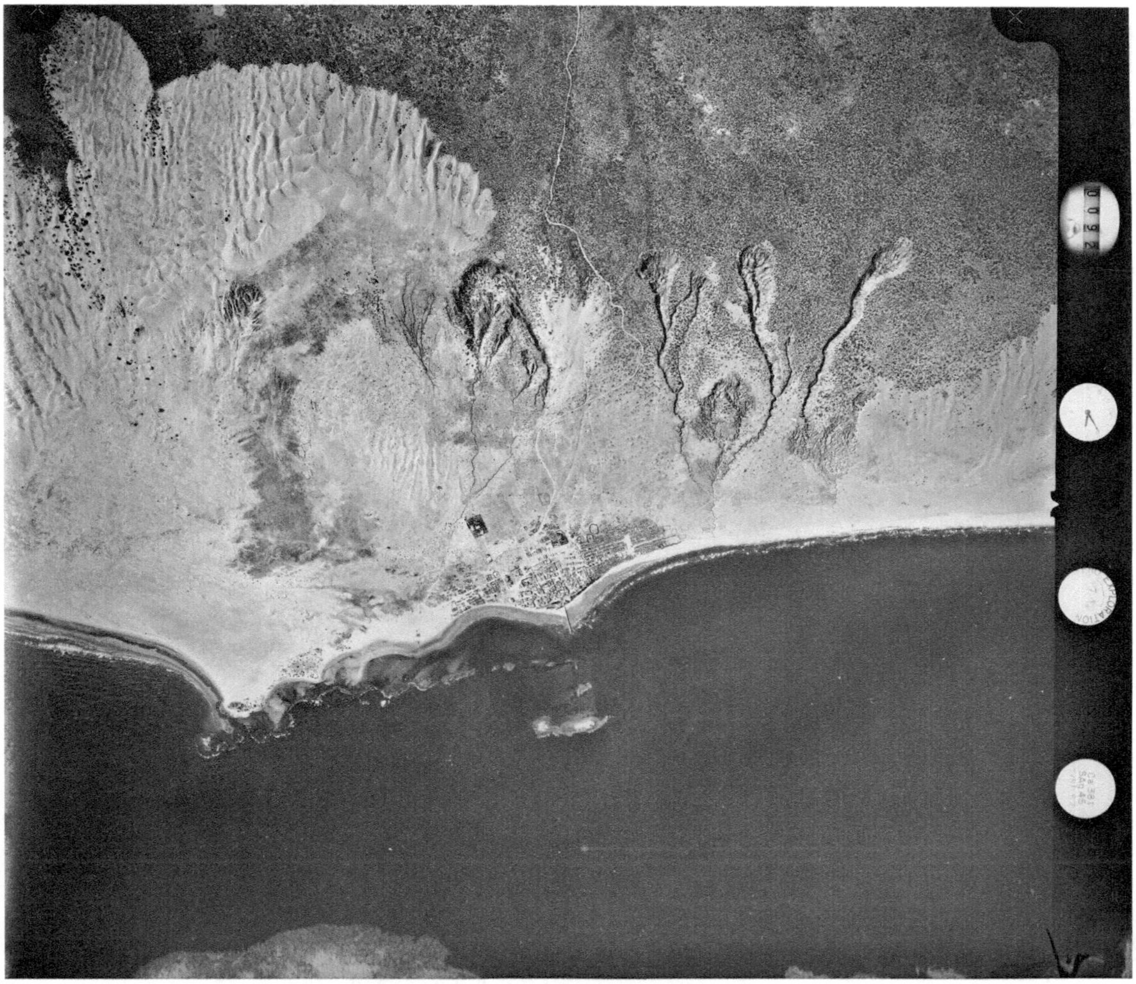

Fig. 14.35 a Gullies in the ancient dunes at Brava. December 1968 aerial photography. The gullies appear to have formed, in part, along an old track over the dunes. *Source* RNMunro AP archives donated to NCAP. Photo 38 was taken at head of one of gullies at right. (Poesen Gully Index: P1). **b** Google Earth © imagery dated 30 August 2012. Gullies have not enlarged and are healing through vegetation, but parabolic dune at right has advanced inland before being converted to transverse dune ridges. Google Earth ©

there is severe rill and gully erosion from Merca to Mogadishu, a zone where the vegetation is highly degraded.

A deeper appreciation is needed for this remarkable feature and its importance to rangeland and pastoral systems. The use of modern analytical techniques, such as Optical Stimulated Luminescence Dating (OSL), will assist in establishing the chronological framework necessary to determine its environmental history.

The coastal sand belt has a rich and varied vegetation cover with plant communities on mobile and fixed dunes (Pichi-Sermolli 1955), and a serious debate should be made concerning any future ideas for extracting minerals from the sands that would likely disturb the biodiversity of this important pastoral region.

14.9 Case Study: Soil and Water Conservation and Vegetation Arcs in Somalia

14.9.1 Background

The authors have made use of archival materials, recent imagery and reports of contemporary travellers in the north to examine aspects of two distinct land cover types—enhancement of indigenous rainfed water-harvesting lands in the Arabsiyo catchment area and vegetation arcs that still cover large areas of Northern Somalia and the Ogaden.

b

Fig. 14.35 (continued)

14.9.2 Arabsiyo Valley Soil and Water Conservation, 1963–64

The USAID implemented a project in Northern Somalia, now Somaliland, in the early 1960s. The Arabsiyo Valley was chosen as a pilot project. On the adjacent plains, the project constructed some 2800 earth bunds, whereas contour earth dykes, in total amounting to 380 km, were made by bulldozer to utilise the meagre rainfall (then 450–500 mm) within fields and control field erosion and reduce headward erosion of gullies.

In the late 1940s, the Somaliland Agricultural Department had started soil and water conservation measures and devised a system that would hold water but allow large storms to pass with minimal damage. The WB (1981) reported that "earth bunds, up to 120 m long and 1 m high on a 2 m base, were built along contour lines. Each bund was turned uphill at its tail end for a vertical distance of 30 cm above the base contour allowing 30 cm of water to be trapped. Each bund was set 60 cm higher than the bund below and field slope varied from 1 to 6%. Lands with a slope of 2% received bunds 30 m apart; and about 300 m of bunds would be needed per ha. In 1954 a bunding hire service was set up with pairs of oxen pulling scraper boards. The scheme offered bunding services free of charge provided they were matched by a similar number of teams operated by the farmers". The scheme was very popular, and the USAID project of new bund construction started in 1963, with tractors replacing oxen.

On the USAID project, a number of reports were written by Calvin Wixom (Wixom 1963, 1964). In the 1963 annual report, an oblique image was shown of the Arabsiyo Valley Soil and Water Conservation Project, reproduced in

Fig. 14.36 Ground photo of rills and pinnacle erosion of ancient sands on edge of gully above Brava. $CaCO_3$ soil concretions occur in the sandy clay loams–sandy loams of the relict aeolian dune soils below 3 m. *Photo* Munro ©, 1985

Fig. 14.44a. Using Google Earth, we are able to repeat this view in an image dated 23 November 2015 (Fig. 14.44b). The view taken by Wixom was towards the east and from about 9.561600°N, 43.69700° (some 14.4 km SSW of Arabsiyo and 40 km due west of Hargeisa). The landscape is a weakly dissected undulating plateau on the Cretaceous Yessoma Sandstone. On the plateau, there is a pale-coloured surface pan over a deep, dark brown soil. There is a marked change into paler bedrock of the Yessoma, shown as pink in the image. The main streams have become wooded, but a badlands zone, called *kerrib* in Sudan, and actually on weathered bedrock, remains largely unvegetated. Zooming in with Google Earth, one can observe that the edge of the soil cover on the plateau is, not surprisingly, still being gullied in parts (Fig. 14.45) but in general the steep sloping lands appear more vegetated. On the undulating plateau lands, the fields are aligned along the contours and show that the USAID project structures remain extant and appear to be maintained. In places, some fields appear to have been afforested.

The project, that ran from 1963 to 66, was reviewed by McCarthy et al. (1985) 17 years after it ceased, and this assessment was also incorporated in a general review of USAID projects by Erdmann (1993). The purpose of the Arabsiyo bunds was to improve soil moisture availability, increase infiltration of rainfall, decrease sheet and gully erosion; increase natural regeneration along barriers and increase soil organic carbon (SOM). Erdmann (1993) noted that impacts of the bunds on yields were evident after the first season, with sorghum yields doubled in the early years of the project: from 700 kg/ha to 1400 kg/ha (McCarthy et al. 1985). This doubling of the yield gave a typical farmer some 2100 kg surplus sorghum. The following is from McCarthy et al. (1985): they noted that at the time of the evaluation, farmers reported that production in treated areas was still significantly superior to that of non-treated areas, and indeed, yields were still 40–60% higher in the fields with the bunds. Another effect of the bunds was to increase sorghum stover production and, consequently, fodder for the livestock; after the project departed, a lucrative market for

Fig. 14.37 Use of *Opuntia* helps stabilise the Brava dunes by reducing impact of erosive rainfall on the sands. Above Brava. *Photo* Munro ©, 1985. Ground repeat photography cannot be made at present

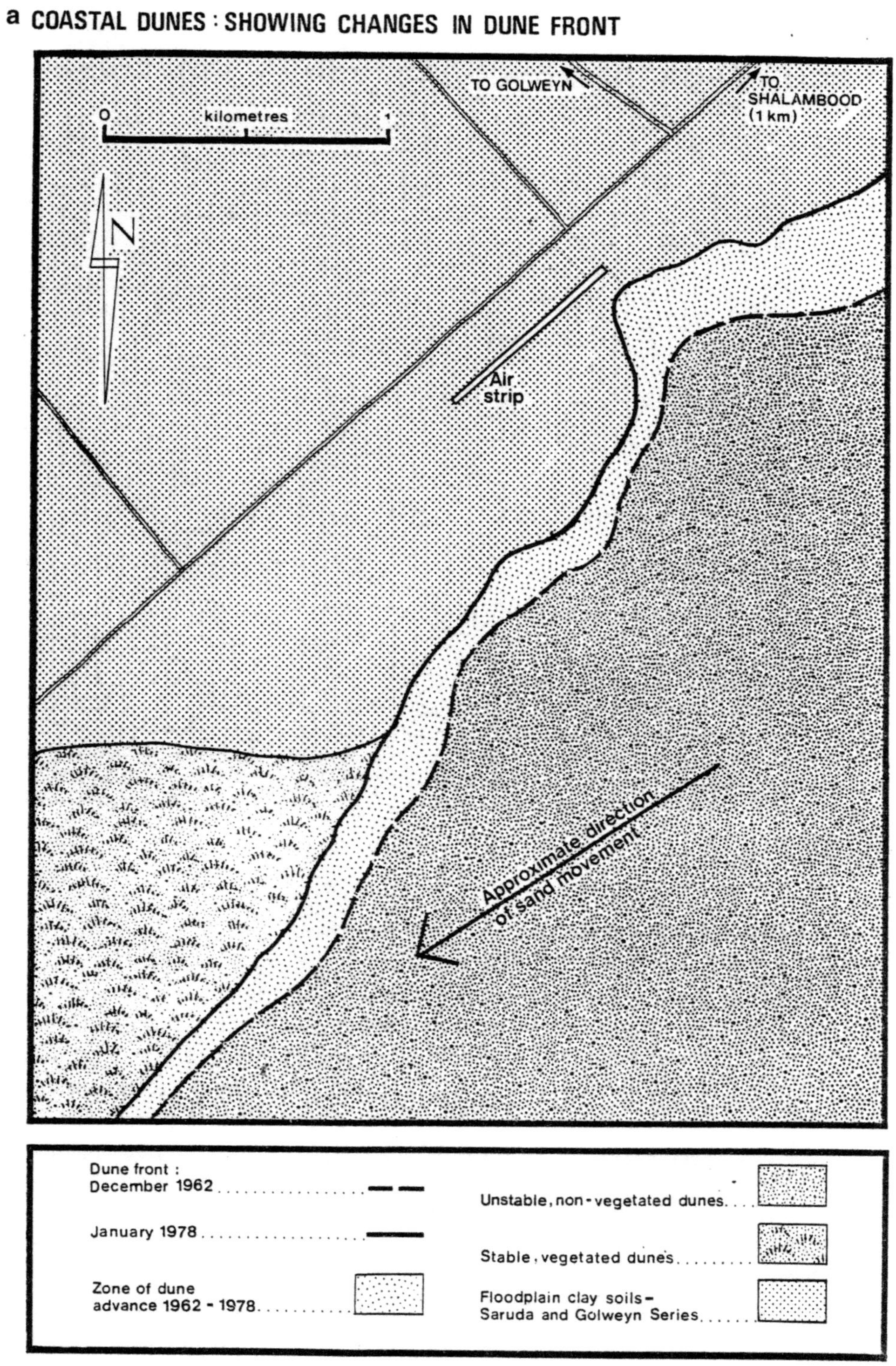

Fig. 14.38 **a** Dune fronts in 1978. Observations made by KJ Virgo, Fig F.1, Annex I, Soil (MMP, 1978a, b). Report archived at WOSSAC. **b** Google Earth image of same situation today. Google Earth ©

b

Fig. 14.38 (continued)

chat (*Catha edulis*) was initiated, and many farmers subsequently converted their food crop fields into *chat* plantations; 17 years after the end of the project, maintenance of the bunds was described as spotty and only the largest breaches in the structures appeared to have been repaired; maintenance of these earthen structures was described as demanding.

The WB project appraisal (1976) of the NWADEP agreed that bund construction by tractors was appropriate, but complementary use of oxen could be supported by favourable credit terms to farmers who opted for it. It suggested that a general reduction in bund height from 1 m height could be made in most parts. Further in the WB's Economic Memorandum (WB 1977), the expansion of the bunding project with WB funds was regarded as a priority area. The USAID-funded initiative was discussed by Reij et al. (1988) who said that the NWADEP was also using bulldozers to create the 1-m-high bunds (Abdi 1986; Critchley 1987).

Since then, this form of water harvesting has been extended over a very area of country to the west and south and extends into Ethiopia also as far as Jigjiga. South of Gunbarah these cultivated strips merge with linear patterned terrain, variously NNW-SSE and N-S aligned, that are natural vegetation arcs. Some of these contain farmsteads with water harvesting in the valleys, and on lower slopes of hills, water tanks have been constructed to harvest run-off.

On the rangelands of the Somaliland plateau, the future of livestock and range management was investigated by Hartmann et al. (2009) and most recently by Pfeifer et al. (2018). The latter noted that climate change is reducing the availability of biomass for livestock, while human behaviour is further reducing biomass production. The loss of animals in rangelands has resulted in pastoralists enclosing parts of the range for rainfed cropping and blocking of transhumance routes to water points. Vegetation arcs will suffer also. Land degradation is widespread in the country and involves soil erosion on unprotected lands, loss of vegetation by grazing, deforestation mainly for charcoal. Land use planning policies that aim to manage the region are formulated (Somalia Ministry of National Resources 2013) but need updating: in a region likely to suffer adversely from ongoing climate change, a more pragmatic participatory rangeland management approach is required. This can include changes such as restricted access to rangelands in wet season and catchment-based SWC interventions. This later was the approach of USAID in 1960s at Arabsiyo. Pfeifer et al.

a

Fig. 14.39 **a** Shalambood area. 17 December 1962. 1:30,000. *Source* RNMunro AP archives donated to NCAP. **b** The SW edge of the Genale scheme. Dunes at bottom left have been completely stabilised and road alignment unchanged since 1962. Google Earth ©

(2018) warned also against using high-resolution imagery without fully understanding the methods of the informal and closed information systems in pastoral communities. We can add that the late Dr Murray Watson, a pioneer in Africa of using oblique photography from low-flying light aircraft to identify households, livestock and wildlife (Watson et al. 1979; Watson 1982, Watson and Nimmo 1985), always backed up his flying with simultaneous ground surveys to verify the sampling strategy, the ecological strata and the statistical methods that produced the data sets from these strata. Modern high-resolution imagery is able to duplicate, at lower costs and more frequent passes, the photographic/imaging part of the RMR methodology. Future practitioners, using drones no doubt, should adopt similar rigour in identifying ecological strata, but making the ground verification is currently difficult.

14.9.3 Vegetation Arc and Water Harvesting in Somalia

The vegetation patterns or arcs of Somalia have long fascinated ecologists and biogeographers. The vegetation associations were first reported by Gillett (1941) who did not

b

Fig. 14.39 (continued)

have the advantage of aerial photographic cover or flights, and their extensive distribution was first reported by Geologist W. A. Macfadyen (1950) who utilised USAF Trimetrogon oblique photography and then made ground checks to describe the process. The DOS 1:125,000 maps of Somaliland used RAF AP flown 1946–1955 and record the presence of vegetation arcs.

The vegetation arcs of Somalia comprise shrubs and bushes and are different in composition but likely of similar origin to the grass patterns in Sudan described by Worrall (1959) in the Butana area, and by Wickens and Collier (1971) in Kordofan; in Somalia other papers followed (Boaler and Hodge 1962, 1964; Hemming 1965). A global study of patterns, including North America, Mauretania, Niger, Mali, Saudi Arabia and Jordan by White (1970, 1971), summarised the mechanisms by which they formed: the arcs occur on gentle slopes in the 125–300 mm rainfall zone—thus a much more drier zone than the water-harvesting field patterns shown in Fig. 14.44 and 14.45—and are not formed by addition of windblown materials as the vegetation under the arcs grown on soils is higher in clay and permits greater moisture penetration than the bare aprons between each arc. The bare area is a run-off zone for water and eroded clay + silt material, and the arc then appears to migrate up-slope. With woody vegetation, this will be slow. Boaler and Hodge (1964) considered that the Somali arcs represented the degraded residues of more extensive cover in a former wetter climate, but presented no data to support. The regular symmetry of all these arcs, not just in Somalia, with orientation at right angles to direction of water flow and ground observations that they are active with species zoning, would suggest they are a natural feature of the Sahelian region. The remaining Somali arcs lie outside the rainfed agriculture zone and so are not at risk of being ploughed up. They are important to pastoralist communities who appear to utilise them carefully in Somalia and the neighbouring terrains of the Ethiopian Ogaden. This is an encouraging indicator of land being protected and preserving carbon in a fragile landscape.

Macfadyen (1950) used USAAF Trimetrogon ("TMG") photography flown in March 1945 by the 19th Photo Charting Squadron (19CS) that flew B-24 Liberators converted to photo-reconnaissance at 23,000 ft with simultaneous exposures of one vertical and obliques to left and right. Figure 14.46a, from Macfadyen (1950), shows TMG right oblique #61 of roll 283. The TMG films are at the US National

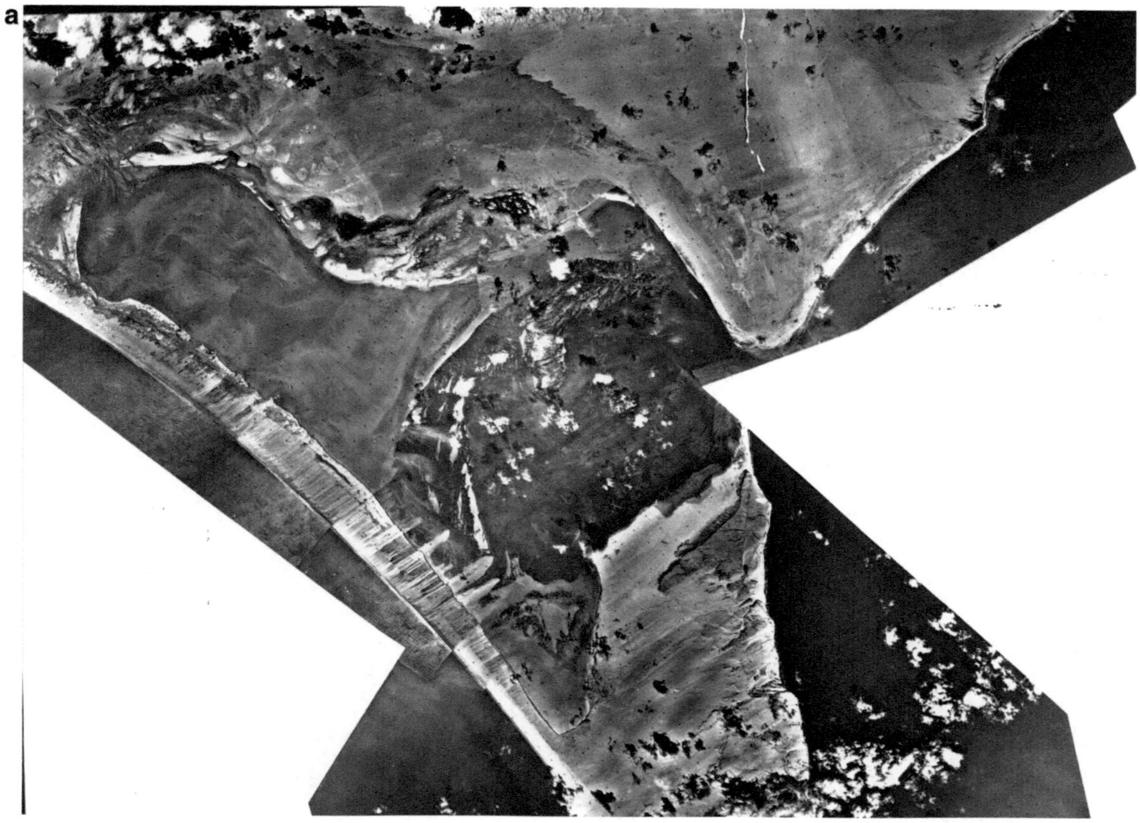

Fig. 14.40 a Xaafuun lagoon and Ras Xaafuun headlands in 1958 (RAF photography). Scans made of one large HSL mosaic print. Ref: Block 4, Sheet F13, 1:60,000, RAF 1958 photography. *Source* RNMunro AP archives. During 2020, the original negative will be donated to NCAP, Edinburgh, UK. **b** Ras Xaafuun and Hordio lagoon. Google Earth imagery 2012–2019. Slightly different orientation than Fig. 58. This entire coastline was struck by the 26 December 2004 tsunami. Sands stream onshore from the south (left), from coastal source-bordering barchan dunes and sand sheets that then move north for 65 km, along a "sand corridor" before re-entering the ocean to the north. The westerly edge of drifting sand backed up against old coastal cliffs on Miocene Limestone is quite clear. Google Earth ©

Archives. The view today on Google Earth shows a view that appears essentially the same (Fig. 14.46b) except for several prominent vehicle routes of today. Of course, a detailed examination is needed on the ground or vertical aerial photography to assess vegetation change more precisely.

In the Hiraan area of the Central Rangelands of Somalia, west of the Shabelle, are entirely indigenous traditional small-scale water-harvesting systems known as *Caag* and *Gawan.* These are practised by agropastoralists and on slopes over 0.5%. (Reij et al. 1988; Reij 1991; Critchley et al. 1991, 1992a, 1992b). Rainfall in Hiraan amounts to 150–300 mm and the system has developed over several decades to grow food crops (mostly sorghum but with cow peas) in response to drought. The *Caag* involves diverting run-off into basins with guiding arms. The *Gawan* is more of an *in situ* system with rainfall concentrated in small basins on flatter slopes with less run-off fed. West of the Shabelle natural vegetation arcs are found on the plateau lands, and a typical view is shown in Fig. 14.47a, copied from a 1:60,000 mosaic made from 1958 RAF aerial photography by HSL in 1960. Examining Google Earth imagery in an area some 38 to 45 km NW of Bulobarde, there are numerous enclosed fields set within extensive vegetation arc belts. At 4.121490 N, 45.26270 E and 188 m asl, for example, what appears to be a *Gawan* system has been developed on very gently sloping lands of natural vegetation arcs (Fig. 14.47 b), and this may be how the system has developed in general, as discussed above. Examples of *Caag* with longer run-on zones are common in valley lines.

In another area, the situation in 1958 is shown in Fig. 14.48 a, from a 1:250,000 scale mosaic. The same area has been examined on Google Earth (Fig. 14.48b); an enlargement (Fig. 14.49) shows woody vegetation on the arc that indicates a permanency and long life of many of these features, while on other adjacent arcs the woody vegetation has disappeared. The settlement is some 56 km NE of Qardho and 980 km NNE of Mogadishu in the Bur Aden hills (Google Earth © 2004 imagery). The impression from

b

Fig. 14.40 (continued)

studying these images is that they appear much the same and cover vast areas of Somalia and the Ogaden in Ethiopia. In making closer look by zooming in with Google Earth, it is apparent that shrubby and arboreal vegetation, often 10 m in diameter, are common. Locally, there is degradation and vegetation has disappeared. That much remains is encouraging for the maintenance of biodiversity in these drylands. It is interesting to find views of Earth, from the Somali deserts, that appear to look unchanged, but closer examination does show there have been modifications to the band widths of arcs and general degradation where human activity has increased. These deserve more study on the ground.

The mechanism of arc formation has been addressed in a number of mathematical modelling studies made over the past twenty years (e.g. Valentin et al. 1999; d'Herbès et al. 2001; Sheratt 2005; Borogogno et al. 2009; Deblauwi et al. 2012; Foti and Ramirez 2013; Chandrasekaran 2017; Gowda et al. 2018). Substantial change was attributed by Gowda et al. (2018) to the increases in human activity, and they conclude that mathematical modelling can account for the creation of arcs by an interactive process of self-organisation between vegetation and rainfall run-off. Further, they noted that bandwidth is shown to increase in some areas, and thus the vegetation cover is greater. In areas where there has been minimal human interference, the bands appear much the same. The change is thought due to species composition changes due to human pressures. The review by Valentin et al. (1999) is particularly useful, and the authors note that arcs form in rainfall zones of 50–750 mm; slopes range from 0.2 to 2.0%; dust is likely to add fertility to the soil and thus leads to the increases in biomass; they act as natural bench structures or terraces to control soil erosion; and the whole ecosystem including bare areas and the vegetation bands is highly resilient to climate change and is thus, crucially, self-sustainable. Some workers consider that these were part of a once more extensive vegetation cover that has degraded as the climate became drier. Evidence for this is not yet proved, but attempts to afforest the bare inter-band areas

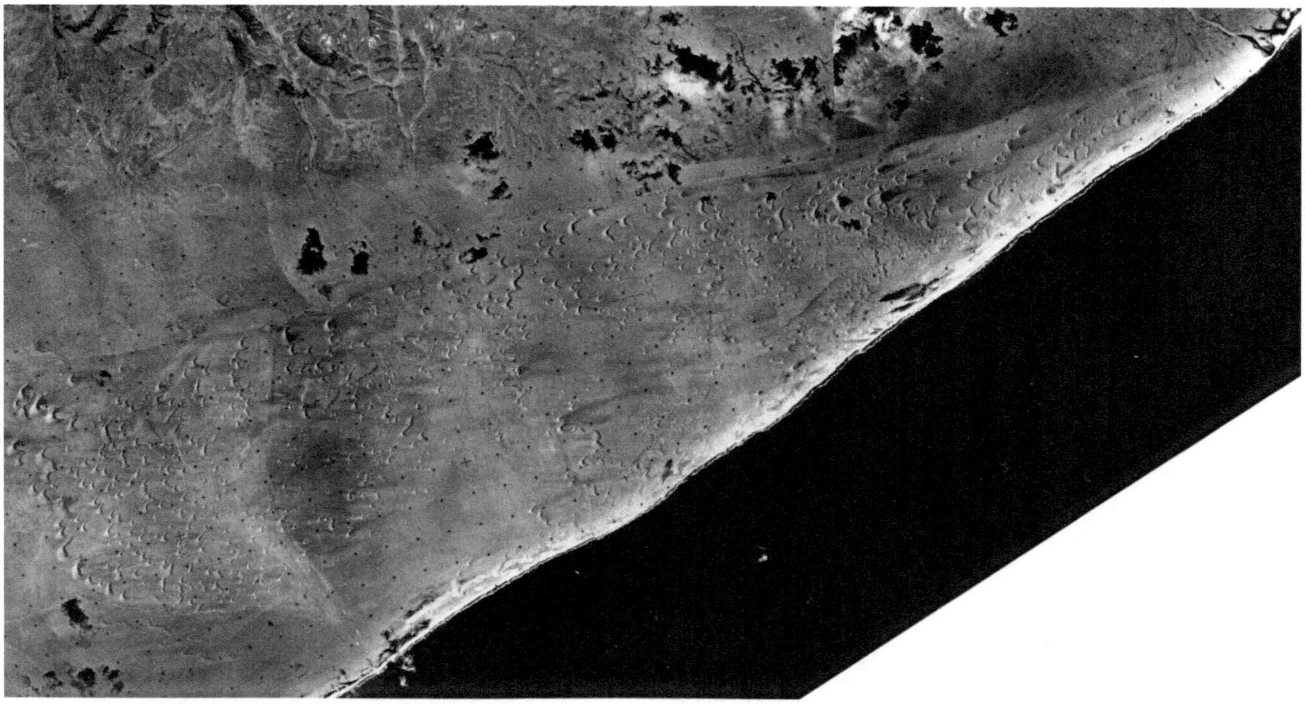

Fig. 14.41 HSL mosaic of Northern Somalia, Block 4, sheet G-13, 1:60,000 scale. The interior westerly edge of the "sand corridor" is clearly visible against cliffs of Miocene Limestone. The barchan dunes march across the coastal plain only to re-enter the sea at extreme right on this image. *Source* Munro AP archives. During 2020 the original negative will be donated to NCAP, Edinburgh, UK

a

Fig. 14.42 **a** Close-up—RAF 1958 photography mosaiced by HSL in 1970s. *Source* RNMunro AP archives. During 2020, the original negative will be donated to NCAP, Edinburgh, UK. **b** Close-up 2015. Onshore movement of sand shows same wind direction. This low-lying strand was devastated by the 2004 tsunami. Google Earth ©

b

Fig. 14.42 (continued)

often fail and lead to an increase in degradation of the land under both the planted trees and arcs, as moisture is less available.

Since 2004, Munro has made observations in the Butana of Central Sudan where the bands are of grass and has used Trimetrogon photography to examine the change (Munro et al. in preparation for DAI, Berlin). The grass patterns of the Butana and Kordofan (Worrall 1959; Ruxton and Berry 1960; White 1971; Wickens and Collier 1971) remain in a few places but since the 1970s much has been ploughed up and the grasslands converted to rainfed sorghum cropping. The erratic nature of the rainfall, however, is such that cropping can often fail, and a valuable grazing resource has been degraded: a double failure of the land use planning regulatory process.

In other countries though—e.g. Australia and Niger—decision-makers and land managers are aware of the importance of these natural systems for soil and water conservation, water harvesting and maintenance of nutrients in the soil (Kusserow and Haenisch 1999). In Northern Somalia (Somaliland), the water-harvesting terraces (shown above) and indeed very similar features, the *teras* (van Dijk 1993), on the Inter-Riverine zone of the Atbara-Gash rivers near Kassala in Sudan, are likely to have been developed by early farmers in many marginal semi-arid areas to copy-cat the ecosystem of the vegetation arc. In the lands west of the Arabsiyo study area in Somaliland, and extending into the Ethiopian Ogaden towards Jigjiga, there is a gradual change from water-harvesting structures to vegetation arcs and linear features of various shapes. The inference is that in this region, the natural vegetation was gradually displaced by crops—notably sorghum and millet, while in the lower rainfall and less populated areas of the Ethiopian Ogaden and Somali Haud shrubby bushland has survived in the arcs.

a

Fig. 14.43 **a** USAF Trimetrogon K-17 right oblique aerial photograph from 1945. Shows the Ferio Range. View looking south onto the Haud. Cretaceous Yesomma Sandstones pass up in Eocene Auradu Limestone capped by Taleh Evaporites and Dolomites. *Photo Source* Govt of Somaliland, 1954. **b** Approximately same view of Ferio Range with Somaliland plateau in background. Oblique from a point some 40 km East of Berbera. In such very arid areas, floods are rare and the landscape appears timeless. Google Earth © mosaiced images dated 27 May 2013 and later

14.10 Conclusions

14.10.1 Water Harvesting

In NW Somaliland, a suggestion was made at the 1986 workshop on water harvesting (Critchley 1987) to establish fodder banks and trees in this region using water-harvesting technology on the bunded strips. Google Earth imagery now shows the presence of numerous wooded strips, and the arable landscape has changed for the better.

The Somalia contribution to the Global Forest Resource assessment (FAO 2014) noted that Mesquite, *P. juliflora* (and the very similar *P. chilensis)* is dominating large areas of sands along the coast. *Prosopis* seedlings for sand dune fixation had been introduced by the National Range Agency prior to the outbreak of the civil conflict (Zollner 1986; Fagotto 1987; Bowen 1988; Pye and Tsoar 2009). One should note though it was one of various other species, mostly local, that were being used. Due to lack of control in planted areas, and the attraction of the Mesquite's highly nutritious seed pod to goats and other livestock, it spread uncontrolled over huge areas. Sand stabilisation was a common reason for the introduction of *Prosopis* in many countries, but foresters forgot the recommendations made in the Sudan by FAO-UNDP (Bosshard 1966; Wunder 1966)

b

Fig. 14.43 (continued)

that it could spread easily as a serious weed unless carefully managed. Unmanaged this voracious invasive plant has spread and out-manoeuvred native flora.

Meroni et al. (2017) reported on the invasion of the exotic tree mesquite *Prosopis juliflora* in NW Somalia and its possible threat for farmland, but that it is confined mainly to the non-arable steep lands, and it has not spread onto the bunded farmlands by livestock who have eaten the seeds: it is controlled by farmers here who know that if a seedling is extracted within a week or two after emergence it will not survive and re-emerge elsewhere. The long-term answer is to manage *Prosopis* by (medicinal) charcoal manufacture, closed access and milling of seed pods as a food for livestock and communities. In some areas of Sudan, for example, these practices were developed in the 1940s and are being reintroduced after a long gap.

a

Fig. 14.44 **a** Oblique of the Arabsiyo Valley and gully edge (from Wixom 1963). **b** Arabsiyo Valley repeat, using Google Earth ©. Image dated 23 November 2018

b

Fig. 14.44 (continued)

Fig. 14.45 Enlargement of part of the Arabsiyo. The contour bunds introduced by USAID in 1960s remain extant and much more widespread as work has been replicated. Headward erosion onto the plateau is reduced. 23 November 2018 Google Earth ©

a

Fig. 14.46 a USAF Trimetrogon 1945. Roll 283 R61. This is an oblique right view, facing south, of the Dufea'au area. © Royal Geographical Society, 1950. Reprinted from Macfadyen, 1950 with permission of the RGS. Original Trimetrogon images are available from US National Archives. **b** Google Earth imagery and Vegetation arcs—Dufea'au area. The view is remarkably similar. Groundwork could determine if vegetation is same. Google Earth ©

14.10.2 Vegetation Arcs of Somalia and the Ogaden

In the drier areas of Northern Somalia, in the Ethiopian Ogaden and to the Juba at Luuq, vegetation arcs and patterns appear to be surviving over extensive areas. This is encouraging. We notice a gradual change or evolution into water-harvesting field systems, where rainfall becomes higher which suggests that water-harvesting systems for agricultural use originated when early farmers seeking to grow crops in these dry areas understood that the natural process of vegetation arc formation could be adapted for arable use and thus mimicked nature. Archaeological investigations could help understand their origins. In general, though, much needs to be done at all levels of management to understand and preserve these extraordinary ecosystems in a changing world. While assessments of land cover change, such as made by Pricope et al. (2013), provide

b

Fig. 14.46 (continued)

a broad picture over a vast area of the Horn of Africa, the sample areas did not cover any arc area in Somalia, and this needs to be addressed with supporting studies in the field.

14.10.3 Northern Escarpment

Due to the closure of the AP archives during COVID-19, we were unable to examine any vertical or oblique aerial photographs of the remarkable mist forests along some 300 km of the Golis Range on the north-facing Somaliland escarpment, but this area lends itself to a future repeat photography study. Trimetrogon photography in particular provides the means to do this. The escarpment lies between the longitudes of Berbera and Bosaso, at elevations of 1000–2400 m above sea level and benefits from moisture-laden clouds brought inland by monsoon winds off the Indian Ocean. It is very similar to the mountains of Dhofar in Oman where in places a dense arboreal vegetation exists. The Golis Range forests comprise some of the important remaining centres of biological diversity and species endemism in Somalia, but civil war and lack of control have led to some degradation of these forests (see Box 2 in Min. of Fisheries and Marine Resources 2014).

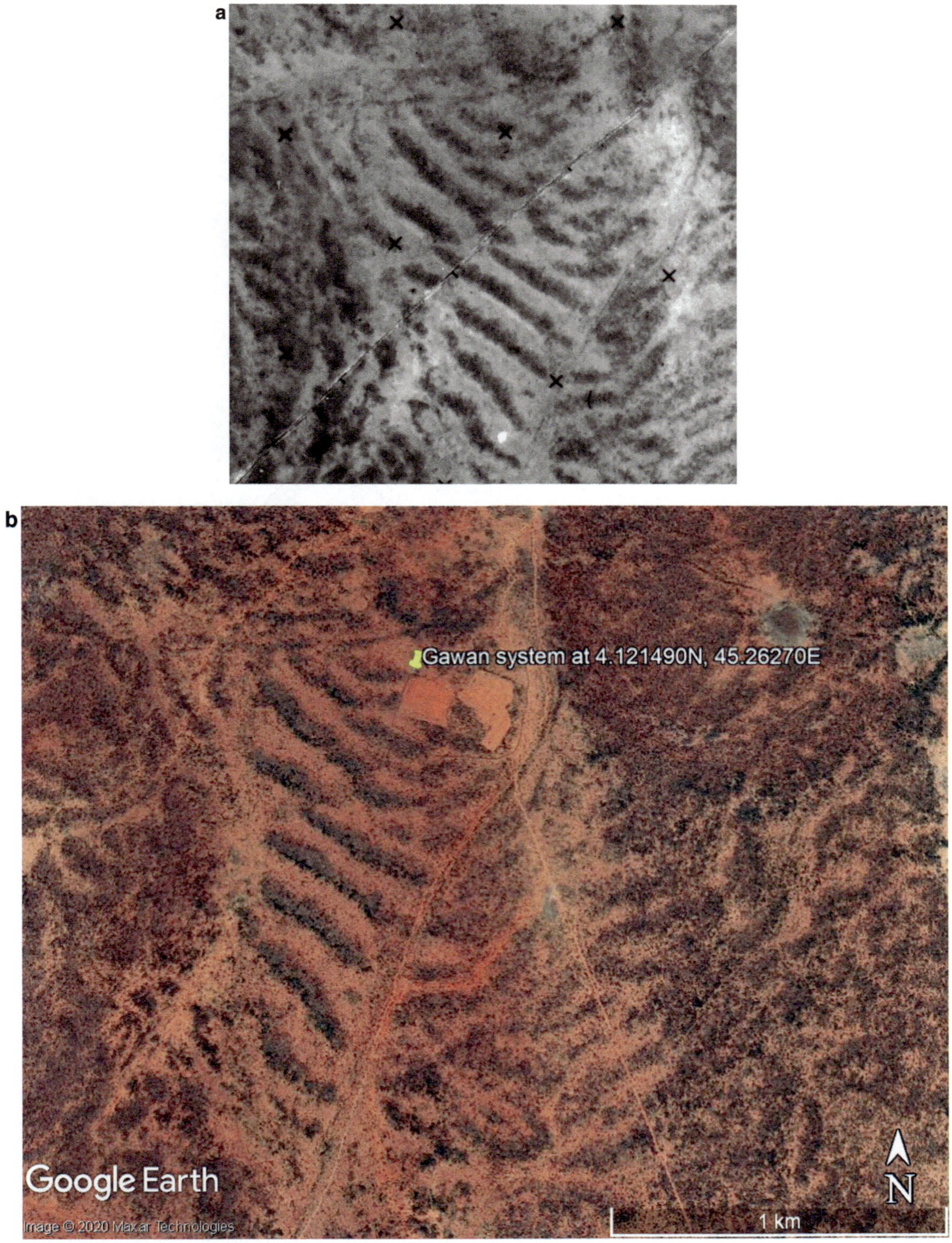

Fig. 14.47 **a** Vegetation arcs in Central Somalia 45 km NW of Bulo Barde. 1958 RAF photography. The line from top right to bottom left is the join between adjacent APs. Copied from negative mosaic at 1:60,000 scale made by Hunting Surveys for the FAO land and Water Surveys (Mosaic Series D-5890, Sheet 17). Same area is shown in Photo 55. *Source* Munro AP archives. During 2020, the original negative will be donated to NCAP, Edinburgh, UK. **b** Water harvesting in Central Somalia 45 km NW of Bulo Barde. Shows conversion of vegetation arcs to *Gawan* basin system of cropping. Vegetation arcs remain largely the same as 1958

a

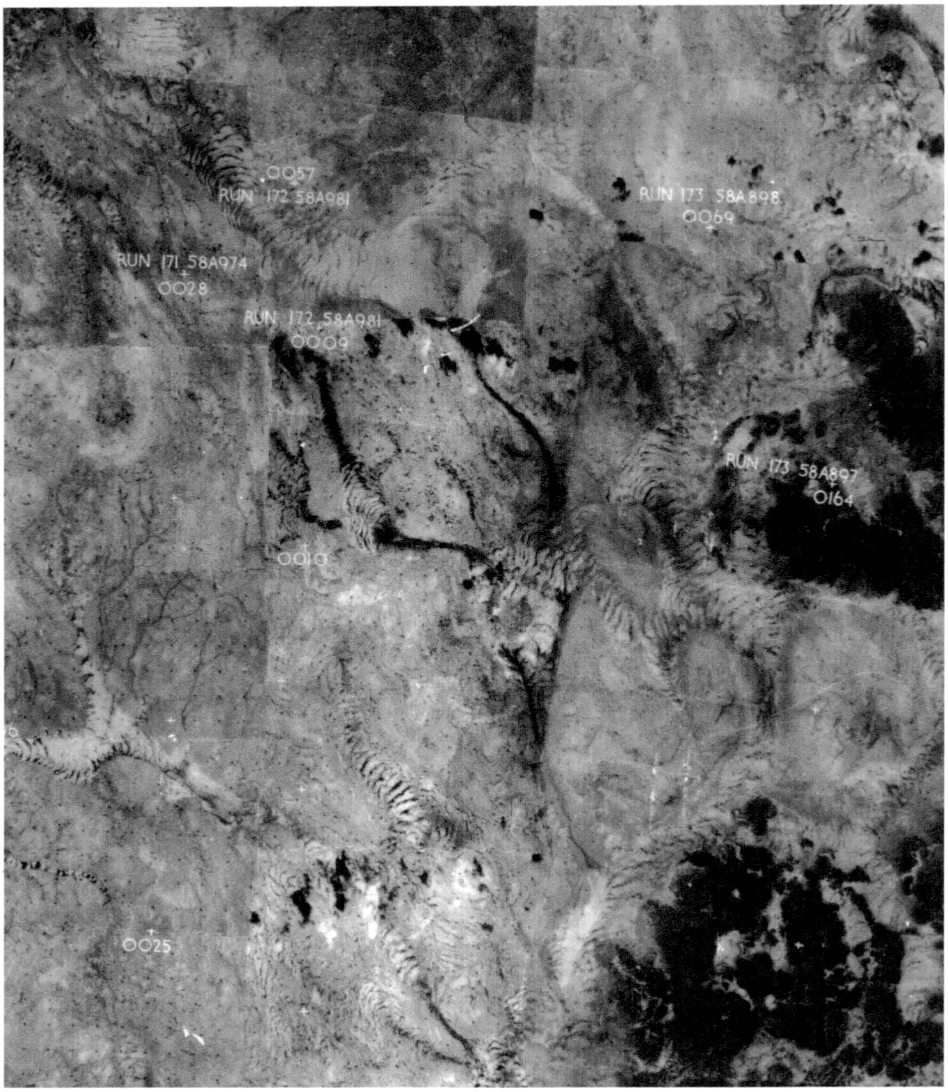

Fig. 14.48 **a** From 1:250,000 scale HSL Mosaic made in 1973 using RAF 1958 vertical photography. *Source* Munro AP archives. During 2020 the original negative will be donated to NCAP, Edinburgh, UK. **b** Same area Google Earth 7 March 2004. Google Earth ©

14.10.4 Use of Archival Materials

A pertinent and interesting point made by Erdmann (1993) was that for a thorough case study review, examination of project and (in that case) USAID Mission literature is essential, but as much of this information is grey matter literature, it will be "difficult to make a reference collection": the archives at WOSSAC and SWALIM have done much to redress that sort of information gap.

This assessment has used archive materials, many carefully kept for 40 years. Although the civil war led to loss of most of the archives in Somalia, a large number of documents have long been stored in the WOSSAC archive, and FAO-SWALIM has sought out materials in other countries.

SWALIM also has georeferenced archival soil studies and with limited fieldwork in the south analysed new soil profiles during 2007 in the Juba and Shabelle. Vargas et al. (2007) noted that while remote sensing can be very useful in

b

Fig. 14.48 (continued)

examining land cover and land use over the soil, it has limitation if one wishes to examine, for example, soil depth: this is of course quite true! Soil and vegetation surveys on the ground are essential; groundwater is best assessed by drilling water wells and modelling seasonal changes. Vargas et al. (2007) also said that some of the soil survey data from 1968 through 1985 that could locate soil sites had been lost, but in fact we believe most of the key studies are available in the WOSSAC archives: that is all for the future to make more detailed reviews.

The online archive at WOSSAC with only 17 items digitised out of a Somalia collection of 514 items represents an untapped resource to be digitised and made available. While some of these are available in SWALIM online, the originals should all be digitally captured at WOSSAC—funds are needed to digitise all materials that sit in the archives waiting to be scanned!

The present paper has introduced the use of a huge archive of Somalia mosaics, held now at the UK's National Collection of Aerial Photography. These need future funding for digitising, dissemination and a programme to assess change. The Royal Air Force AP films from the late 1940s through 1958 are available at NCAP, and an assessment of change over all of Somalia is possible.

There is a scope in this region for future assessments to utilise the archival AP and grey literature archives—the latter all too frequently ignored—to assess ecological change. There are various ecological baselines to re-examine, but obviously groundwork will be essential in any quantitative assessment of faunal and floral change. The assessment of the formation and history of the ancient dunes and sand formations requires a modern approach. Repeat photography using additional photographic archives including 1945 USAF Trimetrogon can be made. Soil sites

Fig. 14.49 Enlargement of a set of arcs from Fig. 14.48b. The settlement (at 9°38′46.18‶N, 49° 34′15.52′E) is near the Bur Aden hills, some 56 km NE of Qardho. A pastoral area too dry for rainfed farming, tanks have been dug in the bedrock to harvest run-off. Woody vegetation occupies some of the arcs showing the stable nature of the arc, but some have lost their vegetation. Google Earth © 2004 imagery

have been georeferenced by SWALIM from the various maps made on the studies noted, above but additional sites on the Lockwood-FAO map can be georeferenced and re-examined to assess change in morphology and chemistry including soil organic carbon (SOC) in the rainfed agricultural lands; variations in soil salinity/sodicity; and variations to the quality and depth to subsurface water tables. The original Lockwood aerial photos with locations have been lost (one presumes), but careful use of the maps can relate the approximate location of sites for future analysis.

After the civil war in Somalia ended, considerable effort was made by donor agencies and NGOs to assist in the rehabilitation of the country, and a very considerable amount of grey literature is available online. While ground access has been possible over much of the north Somaliland and Puntland, the southern part, the Inter-Riverine area remains largely inaccessible for security reasons. The appraisal by SWALIM of land cover of the older development schemes in the south has largely based on remote sensing; the land cover maps shown in Monaci et al. (2007) did not appear to have used any of the land cover studies in the Inter-Riverine area (e.g. the studies by Madgwick 1988, 1989; MMP 1985), and also are of low resolution making it difficult to be used in a comparison with archival aerial photography.

An option in the near future will be to use drones for repeat views. At low altitude, these are often used at archaeological sites, but oblique repeats at a higher altitude (such as 20,000 ft, the height of the Trimetrogon) are feasible and may even be permitted. A repeat position can be located with Google Earth or equivalent and the oblique view coordinates fed into the photo-drone's flight plan. But the risk of anything flying being shot by foe or friend down remains high in Somalia. In this work, we have no more than just skimmed the surface of a remarkable resource of RAF aerial photography, a timeline of the state of the landscape of Somalia in 1958, and provided examples that show how the landscape has changed since then. Much more can and should be done to utilise the thousands of aerial photographs from earlier periods.

Acknowledgements Munro dedicates this work to the efforts of colleagues at Hunting Technical Services (HTS) and Sir M MacDonald and Partners (MMP) who from mid-1960s to mid-80s, produced many of the key and fine agricultural development planning studies on Somalia: from HTS, Peter Humphreys†, Robin James, Rodney Tucker, Robin Goodyear, Harry Piper†, Fred Collier and Aviator/Ecologist Murray Watson†; from MMP: engineers, Mostyn Morgan, Peter Bray†, Berry Kenny, Mark Brett, Mike Pollitzer, Peter Drury†, Colin Fielder†; and all local staff (1980–85). Thanks also to Peter Whiteman in Tigrai 1975; Jon E. Kalb† in Addis Ababa, 1972–75; ornithologists John and Jonquil Ash†, Mogadishu 1985; FAO hydrologists Harry Underhill† in Rome and Brian Gemmell in Mogadishu, 1981; rainfed farmers Andrew Macpherson and Chris Gardner in Somalia, 1985; LRDC tsetse project ecologist Eric Trump†, Mogadishu, 1985; Hilary and Maggie Currey†, Juba Sugar, 1985. In 2019–20 thanks are due to: George Brook (Center for Archaeological Sciences, University of Georgia, Athens) and Keith Braithwaite (Glasgow) for information; Kevin MacLaren (NCAP, Edinburgh) for Somalia AP cover; Brian W. Kerr and Ian C. Baillie (WOSSAC, Cranfield Univ.) for grey literature assistance; Martin Williams (Adelaide Univ.) on Juba memories; Antje Kuepper (GAF AG, Munich) for status of the Copernicus Emergency Management Service; Lucy McCann (Archivist, Bodleian Library, Oxford) for information on their AP holdings: the Royal Geographical Society for permission to reprint the 1945 USAF Trimetrogon photograph, shown in *Photo 5*, from Geographical Journal (1950); Keith J. Virgo†, Fred W Collier and Gordon Page for Somali memories; Hanne Kusserow, Seppe Deckers, Jean Poesen and Paolo Paron gave support; Dr Bill Adams (Geography, Cambridge) is thanked for a critical reading of a draft; sincere thanks to Gordon Petrie† (Glasgow) who provided details of the Italian mapping, his own work in Somaliland in 1950s and commented on a draft shortly before his death in April 2020.

References

Abbate E, Sagri M, Sassi, FP (1994) Geological map of Somalia. 1:500,000. Somali National University and the Ministry of Mineral and Water Resources, Mogadishu. Italian Ministry for University and Scientifica and Technological Research: National Research Council, Italy. S.E.L.C.A, Florence

Abdi AM (1986) Water harvesting systems in the North-western region of Somalia. Paper presented at: World Bank workshop on water harvesting in Sub-Saharan Africa. Kenya, Baringo, 13–20 Oct 1986

Abel NOJ, Kille ME (1976) Conservation and exploitation of the wildlife of south-eastern Somalia. Field document 5. (FO: DP SOM/72/012) Strengthening of forestry and wildlife management project. FAO/UNDP, Mogadishu

Agrisystems Overseas Ltd. (1984) Furjano agricultural settlement project. Appraisal report to UNHCR Somalia, Mogadishu

Ali Kasim M, Carmignani L, Conti P, Fantozzi PL (2002) Geology of the Mesozoic-Tertiary sedimentary basins in southwestern Somalia. J Afr Earth Sci 34:3–20

Angelucci A, Abdulkadir MM, Robba E (1983) A preliminary report on the quaternary sequence in the coastal area of Benadir (central Somalia). Quaderni Di Geologia Della Somalia 7:69–74

Angelucci A, de Gennaro M, de Magistris MA, Di Girolamo P (1994) Economic aspects of red sands from the southern coast of Somalia. Int Geol Rev 36:884–889

Angelucci A, de Gennaro M, de Magistris MA, Di Girolamo P (1995) Mineralogical, geochemical and sedimentological analysis on Recent and Quaternary sands of the Littora region between Mogadishu and Merka (southern Somalia) and their economic implication. Geol Romana 31:249–263

Artini E (1915) Intorno alla composizione mineralogica di alcune sabbie ed arenarie raccolte dalla missione scientifica Stefanini-Paoli nella Somalia Italiana. Pavia

Artini E (1926) Sulla composizione mineralogica di quattro campioni di sabbia provienti dalle dune dei dintorni di Chisimaio nell'Oltre Giuba. Agricoltura Coloniale, 20

Barkhadle AMI (1993) The Somali Traditional "Deegaan" ecological classification system. Revista di Agricoltura Subtropicale e Tropicale 87(1). Instituto Agronomico per l'Oltremare, Firenze, Italy

Barkhadle AMI, Ongaro L, Pignatti S (1994) Pastoralism and plant cover in the lower Shabelle region, Southern Somalia. Landscape Ecol 9(2):79–88

Bingle J. PTY Ltd. (1981–84) Technical reports of semi-mechanised rainfed agriculture farms at Sablaale and Kurtun Waarey. Settlement Development Agency, Mogadishu

Boaler SB, Hodge CAH (1962) Vegetation stripes in Somaliland. J Ecol 50:465–474

Boaler SB, Hodge CAH (1964) Observation on vegetation arcs in Northern Region, Somali Republic. J Ecol 52:511–544

Booker McConnell Ltd. (1976) Planning and design study for the Juba Sugar project. Available at www.wossac.com Ref ID # 45137

Bosshard WC (1966) Tree species for the Arid Zone of the Sudan. FREP Pamphlet No. 33. Forestry Research and Education Project Pamphlet (FREP). Forests Department and UNDP, Sudan. Forest Research Institute, Soba–Khartoum

Bourn DM (1977) Report of a tsetse reconnaissance mission to Somalia. Centre for Overseas Pest Research, College House, London

Bowen MR (1988) A survey of tree planting in Somalia 1925–1985. Oxford Forestry Institute Occasional papers no. 36. 40pp. Oxford Forestry Institute

Braithwaite CJR (1984) Depositional history of the late Pleistocene limestones of the Kenya coast. J Geol Soc London 141:685–699

Brandt SA (2000) A tale of two world bank-financed dam projects in the horn of Africa. In: Brandt SA, Hassan F (eds) Dams and cultural heritage management, pp 33–36. Final report of working paper submitted to World Commission of Dams

Brandt SA, Brook GA (1984) Archaeological and Paleoenvironmental research in Northern Somalia. Curr Anthropol 25(1):119–121

Brandt SA, Gresham TH (1989) An archaeological survey of the proposed Baardheere Reservoir, Upper Jubba River, Southern Somalia., Nyama Akuma 32:28–30 (1989)

Brook GA (2002) Yardangs in ancient, reddened Aeolian sands south of Mogadishu, Somalia. Arab World Geogr 5(3):141–155

Brook GA, Burney DA, Cowart JB, Ellwood BB (1990a) Present and former deserts: evidence of environmental change from cave sediments in east Africa and the American Southwest. Publication of IGCP 252. Studia Carsologica 2:19–28

Brook SA, Burney DA, Cowart JB (1990b) Desert paleoenvironmental data from cave speleothems with examples from the Chihuahuan, Somali-Chalbi, and Kalahari deserts. Palaeogeogr Palaeoclimatol Palaeoecol 76(3–4):311–329

Brook GA, Cowart JB, Ford DC (1996) Raised marine terraces along the Gulf of Aden coast of Somalia. Phys Geogr 17(4):297–312

Brook GA, Cowart JB, Brandt SA (1998) Comparison of quaternary environmental change in eastern and southern Africa using cave speleothem, tufa and rock shelter sediment data. In: Alsharan AS, Glennie KW, Whittle GL, Kendall CG StC (eds) Quaternary deserts and climatic change. Balkema, Rotterdam, pp 239–249

Brook GA, Cowart JB, Brandt SA, Scott L (1997) Quaternary climate change in southern and eastern Africa during the last 300 ka: the evidence from caves in Somalia and the Transvaal region of South Africa. Zeitschrift Fur Geomorphologie Neue Folge, Supplementband 108:15–48

Burgess N, Clarke GP (eds) (2000) Coastal forests of Eastern Africa. IUCN. Available at: http://coastalforests.tfcg.org/

Carbone F, Carush MC (1981) Lithofacies map of the quaternary deposits of Benadir coast from El Adde to Gesira (Somalia). Somali State Printing Agency, Mogadiscio

Carbone F, Matteucci R, Arush MA (1984) Schema geologico Della costa del Benadir tra Gesira ed El Adde (Somalia centro-meridionale). Boll Soc Geol It 103:439–445

Carbone F, Accordi G (2000) The Indian Ocean Coast of Somalia. Mar Pollut Bull 41(1–6):141–159

Chandrasekaran L (2017) Modelling vegetation patterns in vulnerable ecosystems. SIAM News. Society of industrial and applied mathematics. Downloaded October 2019 from: https://sinews.siam.org/Details-Page/modeling-vegetation-patterns-in-vulnerable-ecosystems

Clark JD (1972) The prehistoric cultures of the Horn of Africa. Octagon Books, New York

Critchley WRS (ed) (1987) Some lessons from water harvesting in Sub-Saharan Africa. Report from a workshop held in Baringo, Kenya. 13–17 October 1986. World Bank, Washington

Critchley WRS, Reij CP, Turner SD (1992a) Soil and Water Conservation in Sub-Saharan Africa—Towards sustainable production by the rural poor. p 110. Report by the Centre for Development Cooperation Services, Free University, Amsterdam for IFAD, Rome

Critchley WRS, Reij CP, Seznec A (1992b) Water harvesting for plant production. Volume II. Case studies and conclusions for Sub-Saharan Africa. WB technical paper number 157

Citaco (1976) Soblaale and Kurtun Waarey Agricultural Resettlement and Irrigation Scheme. Citaco, Rome

d'Herbès J-M, Valentin C, Tongway DJ, Leprun J-C (2001) Banded vegetation patterns and structures and related structures. Ecol Stud 149:1–19

De Mûelenaere S, Frankl A, Mitiku H, Poesen J, Deckers J, Munro RN, Veraverbeke S, Nyssen J (2014) Historical landscape photographs for calibration of Landsat land use/cover in the northern Ethiopian highlands. Land Degrad Dev 25:319–335

Deshmukh I (1987) Terrestrial Baseline Studies (TEBS) of Jubba Environmental and Socio-economic Studies (JESS). Associates in Rural Development, Vermont

Douthwaite RJ (1987) Lowland forest resources and their conservation in Southern Somalia. Environ Conserv 14(1):29–35

Erdmann TK (1993) An analysis of 10 African natural resource management practices. Division of food, agriculture and resources analysis, office of analysis, research and technical support Bureau for Africa, *USAID*. Tech paper 8. 84 pp. Oct 1993. AMEX int./USAID

FAO (1980) Northern Rangelands Project, Somalia. Interim report. Ag/SOM/0221/(SOM). 27p. FAO, Rome

FAO (2014) Global Forest Resources Assessment 2015. Country report Somalia. Authors: Hashi Osama Mohamed, Abdiwali Omar Yusuf, Ibrahim Abdi-nur Yakub. FAO, Rome. http://www.fao.org/3/a-az337e.pdf

Fagotto F (1987) Sand-dune fixation in Somalia. Environ Conserv 14 (2):157–163

Fantoli A (1965) Contributo alia climatologia Della Somalia, riassunto dei risultati e tabelle meteorologiche e pluviometriche. Minist. Affari Esteri, Cooperaz. Scient. e Tecnica. CXII +478 pp. Roma

Gemmell BAP (1982) Hydrological data collection and upgrading of the national hydrometric network on the Juba and Shebelli Rivers. FAO, Mogadishu and Rome

Gillett J (1941) The plant formations of western British Somaliland and the Harar province of Abyssinia. Kew Bull 2:37–75

Gowda K, Iams S, Silber M (2018) Signatures of human impact on self-organized vegetation in the Horn of Africa. Nat Sci Rep 8 (3622):1–8

GTZ (1982) Evaluation of agricultural settlement projects at Kurtun Waarey and Sablaale and Proposals for Future development. GTZ, Mogadishu

GTZ (1984) Reconnaissance survey of potential settlement areas for refugee settlement in the lower Shabelle Region of Somalia. Authors: W Hannover and D. Waffenschmidt. GTZ, Mogadishu

Hadden RL (2007) The geology of Somalia: a selected bibliography of Somalian geology, geography and earth science. Topographic Engineering Center, US Army Corps of Engineers

Hartmann I, Sugulle AJ, Awale AI (eds) (2009) The impact of climate change on pastoral societies of Somaliland. Candlelight for Health, Education and Environment. Heinrich Boll Foundation

Hemming CF (1966) The vegetation of the Northern region of the Somali Republic. Proc Linn Soc London 177(2):173–250

Hemming CF (1971) Recommendations on enclosures. Survey of the Northern Rangelands, Somali Democratic Republic. UNDP/SF/73/005, FAO, Rome

Hendy CRC (1985) National tsetse and trypanosomiasis control project land use survey. LRDC, Surbiton, UK. Ministry of Livestock, Forestry and Range. Wossac Ref ID # 41939

HMSO (1942) The Abyssinian Campaigns. London, HMSO. Available at www.wossac.com

Holmes DA, Western S (1969) Soil texture patterns in the alluvium of the lower Indus plain. J Soil Sci 20(1):23–37

HTS (1969) Project for the water control and management of the Shabelle River. Several vols. Hunting Technical Services Ltd, Borehamwood; Sir M. MacDonald and Partners. London. (Soils author: KJ Virgo). Available at www.wossac.com Ref ID # 3

HTS (1977a) Inter riverine agricultural study. Hunting Technical Services Ltd. Land Resources author: Robin James. Settlement Development Agency, Mogadishu. HTS, Borehamwood, UK. Available at www.wossac.com Ref ID # 7131, 7132, Maps: 27578, 27579

HTS (1977b) Soils of the Dujuma Settlement Area. Hunting Technical Services Ltd. (Soils author: RB Tucker). Settlement Development Agency, Mogadishu. HTS, Borehamwood, UK. Available at www.wossac.com Ref ID # 2008

HTS (1978) Fanoole Settlement project phase 1—reconnaissance report. Hunting Technical Services Ltd. (Soils author: RB Tucker). Settlement Development Agency, Mogadishu. HTS, Borehamwood, UK. Available at www.wossac.com Ref ID # 85, 42382

HTS and GRM (1978) Somalia Livestock Sector Review and Project Identification, vol 4. Kuwait Fund for Arab Economic Development. HTS Ltd, Gunn Rural Management (PTY) Ltd. (GRM), Sudanese Investments and Consultants. Wossac IDs: 7134, 7135, 7136, 7137. Available at www.wossac.com 7134, 7135, 7136, 7137

HTS (1979) Homboy irrigated settlement project. Hunting Technical Services Ltd. Settlement Development Agency, Mogadishu. HTS, Borehamwood, UK. Available at www.wossac.com Ref ID # 3865–3869

HTS (1982) Bay region agriculture development project. BRADP, Ministry of Agriculture, Mogadishu. HTS, Borehamwood, UK. (Soils author: RB Tucker). Available at www.wossac.com Ref ID # 6208, 6209, 6210

HTS (1983) Bay region agriculture development project. Project mid-term review. Ministry of Agriculture, Mogadishu. HTS, Borehamwood, UK. Available at www.wossac.com Ref ID # 3831, 3832, 3833, 3834

HTS (1988) National tsetse and trypanosomiasis control project—Somalia. ODA/NTTCP Transition Phase Operation 1987–1989. Interim Report. Hunting Technical Services Ltd./Agricair (PVT) Ltd. Author: W.R. Woof. Wossac 11333

HTS (1999) Aeolian sand formations of the Tihama: geomorphology and assessment of sand stabilisation programmes. (103 pp.; author: RN Munro) Tihama Environment Protection Programme, Tihama Development Authority, Hodeidah, Republic of Yemen. Hunting Technical Services, Hemel Hempstead, UK

Hunt JA (1952) A general survey of the Somaliland Protectorate 1944–1950. Colonial Development and Welfare Scheme Report D.484. Hargeisa

Iannelli P (1984) The principles of pasture improvement and range management and their application in Somalia. FAO Pasture and Fodder Crop Studies No. 9. 213p. www.wossac.com Ref ID # 25124

IBRD (1957) The economy of the trust territory of Somaliland. Report of a Mission at request of Govt. of Italy. 138p. International Bank for Reconstruction and Development, Washington. Jan 1957

IFRC (2019) Emergency plan of action: Somalia Tropical Cyclone Pawan. DREF Operation: MDRSO008 Glide n°. TC-2019-000165-SOM Date of issue: 20 December 2019. Int. Federation of Red Cross and Red Crescent Societies. www.ifrc.org

IGM (1939) L'Istituto Geografico Militare in Africa Orientale 1885–1937. 261 pp. Instituto Geografico Militare, Florence

IUCN (1991) Somalia, pp 255–263; In: Protected Areas of the World. A review of national systems. Volume 3: Afrotropical. Somalia, pp 255–263. International Union for the Conservation of Nature (IUCN), Gland and Cambridge

IUCN (1997) Natural resource and socio-economic assessment of Somalia's Central Indian Ocean coastal plain: for identification of potential integrated multi-sectoral natural resources management interventions. IUCN Eastern Africa Programme, Nairobi

IUCN (2000) An ecological assessment of the coastal plains of North Western Somalia (Somaliland). By Malte Sommerlatte and Abdi Umar. IUCN, Eastern African Somali Natural Resources Management, Nairobi

IUCN (2006) Country environmental profile for Somalia. Draft Report made for the European Union, Nairobi, Kenya

IUCN (2007) Environmental and natural resources strategy. International Union for the Conservation of Nature (IUCN), Gland, Switzerland

Jones A, Breunig-Madsen H, Brossard M, Dampha A, Deckers J, Dewitte O, Gallali T, Hallett S, Jones R, Kilasara M, Le Roux P, Micheli E, Montanarella L, Spaargaren O, Thiombiano L, Van Ranst E, Yemefack M, Zougmoré R (eds) (2013) Soil Atlas of Africa. Publications Office of the European Union, Luxembourg, European Commission, p 176

Kalb JE et al (2000) Bibliography of the earth sciences for the horn of Africa: Ethiopia, Eritrea, Somalia, and Djibouti 1620–1993. American Geological Institute, Alexandria, Virginia. 149p

Kusserow H, Haenisch H (1999) Monitoring the dynamics of "tiger bush" (Brousse tigrée) in the West African Sahel (Nigel) by a combination of Landsat MSS and TM, SPOT, aerial and kite photographs. Photogrammetrie Fernerkundung Geoinformation 2 (1999):S77-94

Le Houérou HN (1972) Report on Improvement of the rangelands and Related Problems in Northern Somalia. FAO, Rome

Lee CW (1969) Aerial applications of insecticides for tsetse fly control in East Africa. Bull World Health Org 41:261–268

Livingstone RG (1964) A history of military mapping Camera Development. Photogramm Eng 32(1964):97–110

Lockwood/FAO (1968) Somalia: agricultural and water surveys. Lockwood survey corporation/FAO, Rome. (With Landform and Great Soil Group map at 1:500,000 scale. Available at www.wossac.com and SWALIM

LRDC (1983) Ecological conditions and tsetse distribution in Southern Somalia. LRDC Project Record 82

McCarthy JW, Clapp-Wincek C, Londer S, Thomas A (1985) A soil and water conservation project in two sites in Somalia: Seventeen Years Later. A.I.D. USAID Project Evaluation Impact report No. 62. Washington, D.C.: USAID. Available at www.wossac.com Ref ID # 2666

Sir MacDonald M, Partners (MMP) (1970) Flood investigations report. Lower Juba Aerial Reconnaissance, 28 June 1970. Sir M. MacDonald and Partners, London

MMP (1978a) Genale-Bulo Marerta Project. Annex 1: Soils (author: KJ Virgo); Annex II: Water Resources. Sir M. MacDonald and Partners. Associate: Hunting Technical Services Ltd. Ministry of Agriculture, Mogadishu. Available at www.wossac.com ref ID # 2, 7215

MMP (1978b) Jowhar Sugar Estate. Drainage and Reclamation Study. 2 vols. Sir M. MacDonald and Partners. Associate: Hunting Technical Services Ltd. (Soils author: KJ Virgo). Ministry of Industry, Mogadishu. Available at www.wossac.com. Ref # 1

MMP (1979) Mogambo Irrigation Project. Supplementary Feasibility Study. Sir M MacDonald and Partners Limited, Cambridge UK. (Soils author: KJ Virgo). Available from SWALIM

MMP and HTS (1980) Reconnaissance and Pre-Feasibility Studies in Jalalaqsi and Jowhar Districts and Gedo Region. 3 vols. Sir M. MacDonald and Partners. Associate: Hunting Technical Services Ltd. (Soils author: RN Munro). National Refugee Commission, Mogadishu. Funded by UNHCR. Available SWALIM

MMP and HTS (1981) Flood Damage Assessment Report. 2 vols and album of 12 maps and river profiles. Sir M. MacDonald and Partners, Cambridge UK and Associate, Hunting Technical Services Ltd. (Agriculture and geomorphology author: RN Munro). Ministry of Agriculture, Mogadishu and Land and Water development Division, FAO, Rome. Available at www.wossac.com Ref IOD # 8191, 8192, 8193

MMP (1983) Lower Shabelle Swamps Reconnaissance Survey Report. National Tsetse and Trypanosomiasis Central Project. LRDC Project Record SOMAL-01-8/Rec-84/83. (Authors: Mostyn Morgan and David Thorn). MMP, Cambridge, UK

MMP and HTS (1985) Farjano Settlement Project Land Evaluation. (Soils, land suitability and land use maps at 1:25,000 scale). Sir M. MacDonald and Partners, Cambridge, in association with Hunting Technical Services, Borehamwood. National Refugee Commission, Mogadishu, and UNHCR, Geneva. (Available at www.wossac.com Ref ID 3801; and SWALIM)

MMP and IoH (1986) Hydrometry Project Somalia—Mission Report, Stage 2, Feb-June 1986. Sir M. MacDonald and Partners, Cambridge and Institute of Hydrology, UK

MMP (1987) Mogambo irrigation project. Groundwater and salinity study, October–November 1987. Sir M MacDonald and Partners Limited, Cambridge UK and John Bingle PTY Ltd. (Available at SWALIM)

Macfadyen WA (1950) Vegetation patterns in the semi-desert plains of British Somaliland. Geograph J 116(4–6):199–211

Macpherson A (1983) A semi mechanised system for rainfed agriculture in semi-Arid areas of East Africa. Settlement Development Agency, Mogadishu

Madgwick J (1984) Somalia research project: expedition publicity leaflet (2 pp) Ecology and Conservation Unit, Dept. of Botany, University College London

Madgwick J, Maunder M, Varty N, Wood B (1988) An ecological study of the remaining areas of riverine forest in the Jubba Valley, Southern Somalia. Final report Madgwick J (ed). Somalia research project, University College London. Available at www.wossac.com Ref ID # 41934, 41953

Madgwick J (1988) A Riverine forest in the Jubba Valley: a vegetation analysis and comments on forest conservation. Biogeographia 14:67–88

Madgwick J (1989) Somalia's threatened forests. Oryx 23(2):94–101

Madgwick J (2000) Focus on Somalia—status and vegetation. In: Burgess ND, Clarke GP (eds) Coastal forests of Eastern Africa. IUCN (2000). Available at: http://coastalforests.tfcg.org/

Mainguet M (1984) Inventory of sand movement in Somalia. UNSSO/National Range Agency. UNSO/DES/SOM/80/001. Available from WOSSAC

Matthiessen P, Douthwaite B (1985) The impact of tsetse fly control campaigns on African wildlife. Oryx 19(4):202–209

Mbara CJ, Gadain HM, Muthusi FM (2007) Status of medium to large irrigation schemes in Southern Somalia, FAO_SWALIM Technical Report No W-05. FAO-SWALIM, Nairobi, Kenya. Available at SWALIM

Merla G, Abbate E, Azzaroli A, Bruni P, Canuti P, Fazzuoli M, Sagri M, Tacconi P (1979) A geological map of Ethiopia and Somalia (1973) 1:2,000,000 and comment with a map of major landforms. 95p. Dept. Geology and Paleontology, University of Florence, Italy. CNR, Italy

Meroni M, Ng W-T, Rembold F et al (2017) Mapping prosopis Juliflora in West Somaliland with landsat 8 satellite imagery and ground information. Land Degrad Dev 28:494–506

Ministry of Planning and Co-Ordination (1969) Conclusions and recommendations on development of water resources and irrigated land along the Shebelli flood plain region. Mogadishu, 49p

Ministry of Planning, Investment and Economic Development (MoPIED) (2018) Somalia Drought Impact and Needs Assessment (DINA), vol 1. Synthesis report. MoPIED and Min Humanitarian affairs and disaster management. Supported by World Bank, European Union, and Global Facility for Disaster Reduction and Recovery (United Nations)

Ministry of Fisheries and Marine Resources (2014) 5th National report on the implementation of the convention of biological diversity of Somalia. With contributions from numerous Government Ministries and NGOs. Mogadishu

Morcos SA, Varley A (eds) (1990) A bibliography on oceanographic and marine environmental research, pp 1775–1984. Red Sea, Gulf of Aden and Suez Canal. CANALALECSO-PERSGA-UNESCO, Paris

Mott MacDonald (1996) Middle Shabelle flood control study. Mott MacDonald Group, Cambridge, England

Munro RN, Deckers J, Haile M, Grove AT, Poesen J, Nyssen J (2008) Soil landscapes, land cover change and erosion features of the Central Plateau Region of Tigrai, Ethiopia: Photo-monitoring with an interval of 30 years. CATENA 75:55–64

Munro RN, Mohammed AMI, Hussien A, Ahmed MA, Kreutzer H, Rout R, Jaafer A (2018) In preparation. Sand, wind and soil at the pyramids of Begrawiya—a preliminary assessment. In a monograph on the Meroe Gezira expected to be published in 2021 by Deutsches Archeological Institute (DAI), Berlin

Munro RN, Mohammed AM, Ibrahim HA, Babiker E-H (2012) Aeolian sand landforms in parts of the Sudan and Nubia. Origins and impacts on past and present land use. Sudan Nubia 16:140–154. Sudan Archaeological Research Society, c/o British Museum

Munro RN, Woldegerima T, Hailu B, Zenebe A, Gebremedhin Z, Hailemichael A, Nyssen J (2019) A history of soil and water conservation in Tigray. In: Nyssen J, Jacob M, Frankl A (eds) Geo-Trekking in Ethiopia's tropical mountains, the Dogu'a Tembien District. Springer GeoGuide

Munro RN, Walkington H, Franks S, Wilkinson TJ, Sanderson DC (2013) Aspects of late Cainozoic Aeolian landscapes in Arabia: implications or early man. In:Al-Ansary A, Al-Muaikel KI, Alsharekh AM, co-editor Swaneh M (eds) Man and environments in the Arab World in light of archaeological discoveries. Proceedings of 2nd Adumatu Forum, Sakaka, Jawf, Saudi Arabia, 4–6 May 2010. Sakaka: Abdul Rahman Al-Sudairy Foundation, pp 7–46

Munro RN, Wilkinson Tony J (2007) Environment, landscapes and archaeology of the Yemeni Tihamah. In: Starkey J, Starkey P, Wilkinson TJ (eds) Natural resources and cultural connections of the Red Sea, pp 13–33. Proceedings of Red Sea Project III, British Museum, October 2006. British Archaeological Reports (Int. Series), 1661, 13–33. Society for Arabian Studies Monographs No. 5. Oxford: Archaeopress

Munro RN, el Tom O (2009) Six dams reconnaissance soils studies on Main Nile between Sabaloka and Egypt. Unpublished Report made for Lahmeyer International and Dams Implementation Unit, Sudan

Mussi M (1975) État des connaissances sue le Quaternaire de la Somalie. Quaternaria 18:161–183

Newtech-HTSPE (2008) Study of the sustainable development of semi-mechanized rain-fed farming, Sudan. Ministry of Agriculture, Khartoum. Newtech Engineering Group and HTSPE, UK

Nyssen J, Munro RN, Haile M, Poesen J, Descheemaeker K, Nigussie H, Moeyersons J, Govers G, Deckers J (2007) Understanding the environmental changes in Tigray: a photographic record over 30 years. Tigray Livelihood papers no 3, VLIR—Mekelle University IUC Programme and Zala-Daget Project, 82 p. ISBN 978-90-8826-016-2

Nyssen J, Poesen J, Descheemaeker K, Nigussie H, Haile M, Moeyersons J, Frankl A, Govers G, Munro RN, Deckers J (2008) Effects of region-wide soil and water conservation in semi-arid areas: the case of northern Ethiopia. Zeitschrift für Geomorphologie 52:291–315

Nyssen J, Frankl A, Haile M, Munro RN, Posen J, Grove AT, Deckers J (2009a) North Ethiopian landscapes re-photographed. Liber Amicorum 15(48):217–234

Nyssen J, Haile M, Naudts J, Munro N, Posen J, Moeyersons J, Frankl A, Deckers J, Pankhurst R (2009b) Desertification? Northern Ethiopia Re-photographed after 140 Years. Sci Total Environ 407:2749–2755

Nyssen J, Frankl A, Munro RN, Billi P, Haile M (2010a) Digital photographic archives for environmental and historical studies: an example from Ethiopia. Scottish Geograp J 126(3):185–207

Nyssen J, Poesen J, Haile M, Munro RN, Grove AT, Deckers J (2010b) Repeat photography challenges received wisdom on land degradation in the northern Ethiopian Highlands. In: Webb RH, Boyer DE, Turner RM (eds) Repeat photography—methods and applications in natural sciences. Island Press, Washington, pp 186–196. ISBN: 9781597267120

Nyssen J, Frankl A, Haile M, Hurni H, Descheemaeker K, Crummey D, Ritler A, Portner B, Nievergelt B, Moeyersons J, Munro RN, Deckers J, Billi P, Poesen J (2014) Environmental conditions and human drivers for changes to north Ethiopian mountain landscapes over 145 years. Sci Total Environ 485–486(2014):164–179

Paolo P, Munro RN (2011) Analysis of environmental change in Somalia using historical aerial photography archives. In: Poster presented by Paron P, Munro RN (eds) AIG regional conference, Addis Ababa, 18–22 February, 2011. Abstracts Volume (ed: Asrat A, Dramis F, Nyssen J, Umer M), p 114

Paron P, Vargas R (2007) Landform of selected study areas in Somaliland and Southern Somalia. Integrated landform mapping approach at semi-detailed scale using remote sensing and GIS techniques. In: FAO-SWALIM. Project report L-02. Nairobi, Kenya. Available at SWALIM

Parry DE (1974) A natural resource evaluation of ERTS-1 imagery of the Central Afar region in Ethiopia. Photogram Rec 8(43):65–80

Parry DE, Wickens GE (1981) The Qozes of Southern Darfur, Sudan Republic. Geogr J 147:307–320

PERGSA/GEF (2002) A bibliography of oceanographic and marine environmental research 1985–1998. Red Sea and Gulf of Aden Region. PERSGA technical series no 2. PERSGA, Jeddah, Saudi Arabia. 230pp

Pfeifer C, Crane, TA, Mugunieri L, Farah AA, Dubad AB, Mohamed A, Isman AI, Ahmed MA, Ibrahim SJ (2018) The dynamics of natural resources in Somaliland—implications for livestock production. ILRI Discussion paper 35. Nairobi, Kenya: ILRI

Pichi-Sermolli REG (1955) Tropical East Africa (Ethiopia, Somaliland, Kenya, Tanganyika). In: Plant ecology reviews of research, vol 6. UNESCO, Paris, pp 302–360

Piccoli G, Boccaletti M, Angelucci A, Robba E, Arush MA, Cabdulqadir MM (1986) Geological history of central and southern Somalia since the Triassic. Mem Soc Geol Ital 31:415–425

Pignatti S, Warfa AM (1983) The *boscaglia* vegetation complex in Southern Somalia. Tuxenia 3:159–168

Pignatti S, Moggi G, Raimondo FM (1993) Dry coastal ecosystems of Somalia. In: van E. van Der Maarel (ed) Ecosystems of the world, vol 2B, dry coastal deserts. Elsevier, pp 31–36

Pratt DJ (1972) Survey of the Northern Rangelands: a plan for national range service for Somalia. AGP.SF/SOM/012. FAO, Rome

Pricope NG, Husak G, Lopez-Carr D, Funk C, Michaelson J (2013) The climate-population nexus in the East African Horn: Emerging degradation trends in rangeland and pastoral livelihood zones. Glob Environ Change 23:1525–1541

Pye K, Tsoar H (2009) Aeolian sand and sand Dunes. Reprint edition. Springer

Rees DJ, Omar AM, Rodol O (1991) Implications of the rainfall climate of southern Somalia for semi-mechanised rain-fed crop production. Agric Meteorol 56:21–33

Reij C (1991) Indigenous soil and water conservation in Africa. IIED, Gatekeeper series no 27, London, IIED

Reij C, Mulder M, Begemann L (1988, second printing 1990) Water harvesting for plant production, vol 1: a comprehensive review of the literature. World Bank technical paper: No. 91. Washington

Relief Web (2020) Somalia floods update (26 May 2020). FAO and SWALIM. (3 pp). Available at: https://reliefweb.int/sites/reliefweb.int/files/resources/Somalia_Flood_Update-26052020.pdf

Ruxton BP, Berry L (1960) The Butana grass patterns. J Soil Sci 11:61–62

Selchozpromexport (1965) The Giuba River Scheme. Moscow. [www.wossac.com ID: 41725]

Siegert K (1986) Water harvesting in the Bay and Lower Shabeele Region of Somalia. Paper presented at: World Bank workshop on water harvesting in Sub-Saharan Africa. Kenya, Baringo, 13–20 Oct 1986

Sinclair ARE, Fryxell JM (1985) The Sahel of Africa: ecology of a disaster. Can J Zool 63:987–994

Somalia Ministry of National Resources (2013) National adaptation program of action on climate change for Somalia. Mogadishu

Somaliland Oil Exploration Company (1954) A Geological reconnaissance of the sedimentary deposits of the protectorate of British Somaliland. Government of Somaliland Protectorate

SWALIM (2016) Disaster Management Plan for the Juba and Shabelle Basin in Somalia. Report no. W-27. 39p. Somalia Water and Land Information Management (SWALIM), FAO, Nairobi

TAMS/FTS (1977) Mogambo irrigation project. TAMS, New York and Financial Technical Services, Cairo

Technital SpA (1975) Juba River Valley Development Study, vol III Geology, Hydrology and Hydraulic Engineering. Technital, Rome

Traversi C (1964) L'Italia In Africa. Storia Della Cartografia Coloniale Italiana. Serie Scientifico-Culurale, Comitato Per La Documentazione Del'Opera Dell'Italia In Africa, Ministero Degli Affari Esteri. Instituto Poligrafico Dello Stato, Roma, 294 pp

Ullah S, Gadain H (2016) National biodiversity strategy and action plan for Somalia. FAO Somalia, Nairobi Kenya

UNDP (2016) Somalia National action programme for the United Nations convention to combat desertification. UNDP Somalia. Written by Harum Mukhayer and Hassan Abdirizak (UNDP Somalia)

UNEP (2005) The state of the environment in Somalia: a desk study. UN Environment, Geneva

UNEP, WMO, UNCCD (2016) Global assessment of sand and dust storms. United Nations Environment Programme, Nairobi

United Nations (1989) Country paper: Somalia. In: Ground water in Eastern, Central and Southern Africa, pp 237–250. Natural Resources/Water Series No. 10. ST/TDC/9. Dept. Technical Co-operation for development, UN, New York

Valentin C, d'Herbès JM, Poesen J (1999) Soil and water components of banded vegetation patterns. CATENA 37(1999):1–24

Van Dijk JA (1993) Opportunities for expanding water harvesting in Sub-Saharan Africa: the case of the *Teras* of Kassala. IIED Gatekeeper series no SA40, IIED, London

Vannini M, Chelazzi G, Chelazzi L, Ercolini A, Ferrara F, Messana G, Messeri P, Pardi L (1977) Researches on the coast of Somalia: the shore and the dune of Sar Uanle. Monitore Zoologico Italiano. Supplemento 9(1):249–271. https://doi.org/10.1080/03749444.1977.10736851

Vargas RR, Omuto C, Alim MS (2007) Soil survey of the Juba and Shabelle riverine areas in Southern Somalia. FA0-SWALIM. Project report L-08. Nairobi, Kenya. Available at SWALIM

Varty N (1988) Mammals and problems of their conservation in the riverine forests of the Jubba Valley, southern Somalia. Trop Zool 1 (2):179–192

Varty N, Hill JE (1988) Notes on a collection of bats from the riverine forests of the Jubba Valley, southern Somalia. Mammalia 52 (4):533–540

Varty N (1990) Ecology of the small mammals in the riverine forests of the Jubba Valley, Southern Somalia. J Trop Ecol 6(2):179–189

Watson RM, Tippett CI, Beckett JJ, Scholes V (1979) Central Rangeland survey. Somali Democratic Republic. Resource Management and Research, London

Watson, R. M. (1982) Northern Rangelands Survey. Somali Democratic Republic. 4 VOLS. London: Resource Management and Research

Watson RM, Nimmo JM (1985) Southern Rangelands Survey. Somali Democratic Republic. Resource Management and Research, London

Webb RH, Boyer DE, Turner RM (eds) (2010) Repeat photography – methods and applications in natural sciences. Island Press, Washington

White LP (1970) Brousse tigrée patterns in southern Niger. J Ecol 58:549–553

White LP (1971) Vegetation stripes on sheet wash surfaces. J Ecol 59:615–622

Whiteman PTS (1975) Moisture conservation by fallowing in Botswana. Exp Agric 11:305–314

Whiteman PTS (1984) Importance of residual moisture reserves for increasing the reliability of cereal yields in East Africa. East Africa Agric. for. J 44:p57

Wickens GE, Collier FW (1971) Some vegetation patterns in the Republic of Sudan. Geoderma 6(1):43–59

Wickens GE (1976) The Flora of jebel Marra (Sudan Republic) and its geographical affinities. Royal Botanic Gardens, Kew. Kew Bulletin Additional Series V. London, HMSO

Wieland RG, Werger MJA (1985) and types and vegetation in the Luuq District of South-Western Somalia. J Trop Ecol 1:65–87

Wieland R, (1983) Soil survey, land evaluation and vegetation correlation for land use planning in the Luuq District Refugee Area. Wossac Ref ID # 41940, p 81

Williams MAJ (1995) Late quaternary palaeohydrology of the Jubba River, Somalia. In: Abstracts, XIV congress, international union for quaternary research. Berlin, 1995, p 298

Williams MAJ (2014a) Climate change in deserts. Cambridge University Press, Past, Present and Future
Williams MAJ (2014b) Potential impacts of damming the Juba valley, western Somalia: insights from geomorphology and alluvial history. Geophysical Research Abstracts vol 16, EGU2014b-1302
Williams MAJ (2016) Nile waters, Saharan sands. Adventures of a Geomorphologist at Large. Springer Biographies
Wixom C (1963) Arabsiyo soil and water conservation project (Northern Region): first annual report calendar year 1963. USAID
Wixom C (1964) Arabsiyo soil and water conservation project (Northern Region): 2nd annual report calendar year 1964. USAID
World Bank (1977) Somalia: country economic memorandum. Report no 1421a-SO. Washington
World Bank (1979) Somalia. Central Rangelands Development Project. Staff Appraisal Report and Implementation Volume. Washington, p 52
World Bank (1980) Somalia. Agricultural sector review, vol 3. Washington
World Bank (1984) Somalia. Mechanical rainfed agriculture. MFC-SOM1/31
World Bank Group and FAO (2018) Rebuilding resilient and sustainable agriculture in Somalia. Somalia Country Economic Memorandum, vol 1. IBRD/WB, FAO Rome, Washington
Worrall GA (1959) The Butana grass patterns. J Soil Sci 10:34–53
Wunder WG (1966) Prosopis juliflora in the Arid Zones of Sudan. FREP Pamphlet No. 36. Forestry Research and Education Project Pamphlet (FREP). Forests Department and UNDP, Sudan. Forest Research Institute, Soba–Khartoum
Zollner D (1986) Sand dune stabilization in central Somalia. For Ecol Manage 16:223–232

Index
Page numbers followed by *f* and *t* indicate figures and tables, respectively

P. Billi (ed.), *Landscapes and Landforms of the Horn of Africa*, World Geomorphological Landscapes,
https://doi.org/10.1007/978-3-031-05487-7

I

J

K

GPSR Compliance

The European Union's (EU) General Product Safety Regulation (GPSR) is a set of rules that requires consumer products to be safe and our obligations to ensure this.

If you have any concerns about our products, you can contact us on ProductSafety@springernature.com

In case Publisher is established outside the EU, the EU authorized representative is:

Springer Nature Customer Service Center GmbH
Europaplatz 3
69115 Heidelberg, Germany

Batch number: 10731667

Printed by Printforce, the Netherlands